Novikov/Conformal Gearing

Gear industry
Gear science
NO, thank you !!
We are too busy !!

Stephen P. Radzevich

Novikov/Conformal Gearing

Scientific Theory and Practice

Springer

Stephen P. Radzevich
Southfield Innovation Center
EATON Corp.
Sterling Heights, MI, USA

ISBN 978-3-031-10018-5 ISBN 978-3-031-10019-2 (eBook)
https://doi.org/10.1007/978-3-031-10019-2

This Springer imprint is published by the registered company Springer Nature Switzerland AG
The registered company address is: Gewerbestrasse 11, 6330 Cham, Switzerland

In memoriam of Professor Mikhail L. Novikov,
the inventor of the ***Novikov Gear System****,*
whose 110th anniversary is celebrated in 2025

Introduction

Knowledge is power (*Scientia potestas est*) –Francis Bacon (1561–1626)

Historical Background

Gears and gear drives have a long-lasting history of evolution. Cog wheels evolved to a high precision involute gear capable of transmitting a huge amount of power at high rotations.

Present-day gear designs are based on the latest accomplishments in the theory of gearing. Ultimately, these yield a very high predictability of the designed gear-sets.

Novikov/conformal and *high-conformal gearing* represent a special and relatively new kind of gearing. Gearing of this kind features a high-power-density gearing – this is a present-day trend of in the gear design and production.

Invented in the mead of 1950s, gears of this design features "convex-to-concave" contact between the tooth flanks of a gear and a mating pinion. Favorable tooth contact of this kind enables favorable stress distribution at contact points of the tooth flanks.

Despite of enormous amount of the research in the field has been carried out in the past, the gear science and gear practice have been further evolved in the recent decades.

Uniqueness of This Publication

This book is an invaluable source of information on the *Novikov/conformal* and *high-conformal gearing*. Involute gears of conventional design are also considered here. Gear experts all around the world have contributed their latest accomplishments in the field of gearing.

The main features of *Novikov/conformal* and *high-conformal gearing* are covered in this volume. This includes but is not limited to the specific design parameters of gears of this particular design: Novikov boundary circle, one (or two) straight pseudo-lines of action, pseudo-path of contact, instantaneous lines of action, and so forth.

Intended Audience

High-power density (or, in other words, *high power-to-mass ratio*) is a key feature of *Novikov/conformal gearing*, especially of *high-conformal gearing*. Therefore, gear experts those involved in the design, production, inspection, and in application of advanced design of gear transmissions are among those whom this book is addressed to. Most of gear engineers and gear researchers from the industry, as well as graduate students, will be benefited by the book.

Organization of This Book

This book is composed of eleven chapters. The list of the chapters, along with a brief description of each chapter, can be found out immediately below. Also, several appendices are appended at the end of the book.

Chapter 1: Titled as "Novikov Gearing Is a Kind of Involute Gearing," this chapter is contributed by Prof. *Stephen P. Radzevich* (USA).

This chapter of the book deals with *Novikov gearing*. Nowadays, gearing of this kind is commonly referred to as *Novikov/conformal* gearing. If properly designed, manufactured, and applied, *Novikov gearing* features the higher power density compared to that in conventional involute gearing of a similar size. Therefore, the usage of *Novikov gearing* allows for transmitting more power by means of gearboxes of a smaller size. An increase of the power density is one of the strategic trends of evolution in the gear industry.

A brief historical overview of *Novikov gearing* is presented in this chapter of the book. It is shown that *Novikov gearing* is a reduced case of involute gearing. The concept of the *boundary Novikov circle* (introduced by the author, and referred to as the *boundary N-circle*, for simplicity) enables one clear analysis of the geometry and kinematics of *Novikov gearing*. The necessary and sufficient conditions for perfect operation of the conformal gearing (i.e., *Novikov gearing*) and *high-conformal gearing* are outlined. Use of the concept of *reversibly enveloping surfaces* (or just R_e*–surfaces*, for simplicity) makes it possible to state that neither *Novikov gearing* (more generally) nor *high-conformal gearing* can be finish-machined in continuous-indexing machining process; that is, the gears cannot be finish-cut by hobs, by shapers and rack-cutters, by worm grinding wheel, and so forth.

Chapter 2: Titled as "Meshing Theory for Abnormal Novikov Helical Gears," this chapter is contributed by Prof. *Yaping Zhao* and by *Siyu Liu* (both of China).

The *Novikov gear* in the traditional sense is enveloped by a rack-cutter with an arc tooth profile, and the tooth surface of the gear is very complex. This chapter presents a new type of circular arc gear drive with circular tooth profile in the normal section. The gear is designed as a new gear whose normal section is arc-shaped, and the tooth surface equation of the arc-shaped gear is established by the spherical vector function. The tooth profile is relatively simple.

Based on the transmission ratio condition, the augmented contact equations are written to determine the datum point where the transmission ratio error is zero. The instantaneous contact points are determined by solving nonlinear equations established in the light of common point and common normal conditions, also known as improved TCA method. The tooth width condition is used to determine the endpoints of the contact paths of two tooth surfaces, and the internal contact points are obtained. The calculation formula of relative principal curvature is also derived.

The numerical cases show that the gears proposed are instantaneously in point contact, tooth surface meshing belongs to point contact, and the contact trace is roughly along the direction of tooth length. The gear pair can accomplish constant ratio transmission, and its velocity ratio can completely be equal to the nominal transmitting ratio.

Chapter 3: Titled as "Helical Bevel Novikov Gears," this chapter is contributed by Dr. *Michał Batsch* (Poland).

This chapter presents a mathematical model of convexo-concave helical bevel Novikov gear meshing. A topological modification of this kind of gearing was introduced. Results of tooth contact analyses aiming at the determination of the effect of gear parameters on contact pattern and transmission error were presented. In addition, a comparison between the *Novikov gear* and a conventional spiral bevel gear generated with the aid of the duplex helical method was provided. It was found that the *Novikov gear* displayed larger instantaneous contact pattern than the conventional one.

Chapter 4: Titled as "Hyperboloidal-Type Normal Circular-Arc Gearing," this chapter is contributed by Dr. *Houjun Chen*, *Zhilan Ju*, *Xiaoping Zhang*, and *Chang Qu* (all of China). Discussed in this chapter hyperboloidal-type normal circular-arc gearing (HNCGing) is a latest development of *Novikov gearing* to transmit the motion and power between two orthogonally crossed axes. The basic principle of molding-surface conjugation, which represents the kinematics and the geometry of conjugation by a pair of directrices of conjugate surfaces rather than the conjugate surfaces themselves, was presented, and both the conditional equation of conjugation and the structural condition of conjugation were established to provide a theoretical foundation for constructing novel types of gearings. The generation of the pinion and the mating gear in HNCGing was analyzed, mathematical models of conjugate tooth surfaces were established, and the induced curvature characteristics of conjugate surfaces were explored. This helps people get ready to acquire the relationship of design parameters, calculate the Hertzian contact between the two mating teeth, and carry out the interfere-free tool-path programming for NC machining of tooth surface. In order to meet the needs of digital manufacturing, an integrated manufacturing software system of HNCGing was developed, which connects computer-aided design, computer-aided manufacturing, and computer-aided engineering in a single application platform. This software system supports three-dimensional modeling, adaptive tool-path programming for five-axis forming milling of tooth surface, and simulation of meshing and contact for the tooth contact analysis.

Chapter 5: Titled as "Modern Methods of Estimating and Increasing of the Load-Bearing Capacity of Novikov Gearing," this chapter is contributed by Dr. *Viktor I. Korotkin* (Russia). The use of *Novikov gearing* with teeth of different hardness for various purposes is considered.

For gearboxes of general machine-building application, variants of transmissions with hardened teeth are offered, for which a special basic profile has been developed, which is part of the state standard.

A fundamental approach to solving the problem of determining the main parameters for calculating the strength of these gears, consisting of two stages, is presented:

(a) Determination of both the main bending, contact stresses and stiffness of the teeth, and the phase, taking into account the coefficients of influence of the ends of the gear rim
(b) Modeling of the process of real multi-pair engagement with finding criterial stress in dangerous areas of the teeth that determine the load-bearing capacity and service life of the gearbox

Implementing this approach, generalized results of solving the spatial problem of the stress-strain state of the teeth at any position of the contact pad along their length are obtained, and the way of obtaining partial forces at the contact sites and criterial stress are indicated. An engineering method for calculating the strength of *Novikov gearing* is proposed, on the basis of which gearboxes of general mechanical engineering with high-hard *Novikov* gears of relatively low degrees of accuracy are created. This made it possible to replace high-precision expensive involute gearing on the strained output stages of the gearboxes with less time-consuming in the manufacture hardened *Novikov gearing* of coarser degrees of accuracy and reduce the specific gravity of the gearbox by 1.5 times compared to the involute analogues. Preliminary calculations of the created promising hardened high-precision *Novikov gearing* are given, and their significant advantage in comparison with involute analogues is shown.

Known designs of modified *Novikov gears* with longitudinal modification of the working surfaces of thermally improved teeth of drive gearboxes of widely used for are oil pumping units described.

On the example of one of the models of the gearbox, it is shown that the use of the proposed longitudinal modification of the working surfaces of the teeth in combination with an increase in the gear module provides a significant impact on reducing contact and bending stresses as well as increasing the load-bearing capacity and service life of the gearbox compared to the serial version produced by the domestic industry. The achieved effect at different stages of the gearbox is to reduce the stress of the teeth – from 1.75 to 2.54 times, increase the load-bearing capacity of the gearbox – from 2.81 to 4.1 times, and increase the service life of the gearbox – by an order of magnitude.

Chapter 6: Titled as "Some Features of the Contact Strength of *Novikov Gearing*," this chapter is contributed by Dr. *Viktor I. Korotkin* and by *Nikolay P. Onishkov* (both of Russia).

It is known that *Novikov gears* were created as having high contact strength, which is mainly confirmed by the practice of their testing and operation. This allows in many cases to replace involute analogues, which are often destroyed due to progressive pitting of the working surfaces of the teeth.

Studies in recent years have shown that in some cases, with some geometric parameters of the basic profile and gearing, breakdowns and chips of the teeth occur, the type of which does not correspond to the classical picture of fatigue fractures. Moreover, such destruction is observed also in the area of the tooth heads, where the bending moments are insignificant. In addition, with the increase in tension at the border of the contact pad, especially in materials with reduced plastic properties, there are also radial microcracks that do not close when the load is removed and grow with repeated, even insignificant, stresses.

The flat nature of the destruction assumed the possibility of inadmissible shear stresses at the boundaries of local contact areas.

This assumption was tested with the involvement of the well-established *Lebedev-Pisarenko* criterion about the limit state of the material. It has been established that stress at the border of the contact pad increases with the decrease in the plastic properties of the material. Thus, in the case of circular contact, the excess of equivalent stresses according to the *Lebedev-Pisarenko* criterion at the spot boundary compared to the center for plastic steels – normalized and thermally improved – is about 12%, for hardened – up to 20%. In the case of quasi-elliptical contact, this excess depends on the ellipticity coefficient.

Thus, there is reason to believe that the stresses at the border of the contact pad may be limiting the working capacity of the transmission, and this requires more in-depth additional research.

The study of the influence of the state of the surface layers of the material of teeth on the contact strength allowed to recommend hardness material of teeth not more 58 by *Rockwell*, preventing the increasing embrittlement of the material, and to provide for shot blasting for especially stressed gears.

In order to reduce the danger of chips of the teeth occur, it is desirable to porvide the longitudinal modification of the working surfaces of the teeth, earlier proposed by us the longitudinal modification of the working surfaces of the teeth.

Chapter 7: Titled as "The Tooth Relieving of Worm Hobs for Cutting Novikov Gears with Double Lines of Action," this chapter is contributed by Dr. *Aleksander Sandler* (Russia).

Worm hobs for cutting *Novikov gears* have the profile of a generating worm, which varies not only in a radius but also in a sign of curvature. A technique for relieving tooth flanks of these hobs by disk grinding wheel is developed. It allows for bringing together grinding and ground profiles in an axial section of the generating worm as close as possible. A method for calculating setting-ups and profile coordinates of the grinding wheel is given for real production conditions.

Chapter 8: Titled as "Mikhail L. Novikov: The Inventor of *Novikov Gear System*," this chapter is contributed by Prof. *Stephen P. Radzevich* (USA). The chapter presents the author's point of view on the role and contribution of Dr. *M. L. Novikov* in the development of a new type of gearing, later called *Novikov Gear System*. Brief biographical information about Dr. *M.L. Novikov* and Dr. *E. Wildhaber*, who is often and undeservedly called the co-developer of the *Novikov gearing*, is provided. The geometry and kinematics of *Novikov gearing*, both, are analyzed, and its fundamental difference from the gearing with a circular tooth profile previously proposed by Dr. *E. Wildhaber* is shown. When writing this text, the aim was to draw the attention of an interested reader to the issue of the priority of Dr. *M.L. Novikov* in the development of a new type of gearing. To do this, we briefly analyze the situation around the *Novikov Gear System* over 60 years after it was invented. Possible ways to further improve the *Novikov gearing* are briefly outlined.

Discussion of the issue of the priority by Dr. *M.L. Novikov* in the development of a new type of gearing, namely, of *Novikov Gear System*, is currently acquiring special meaning in connection with the upcoming 110th anniversary (in 2025) from the birth of the inventor and scientist.

Chapter 9: Titled as "Poor Understanding of the Scientific Theory of Gearing by the Majority of Gear Scientists and Engineers," this chapter is contributed by Prof. *Stephen P. Radzevich* (USA).

A brief discussion on the *Scientific Theory of Gearing* is represented in this chapter of the book. "There is nothing more practical than a good theory," J.-C. Maxwell said. It is illustrated the advantage of what kind can be taken from a good gear theory. It is also shown what troubles gear scientists and engineers are faced with if ignoring the results obtained in the scientific theory of gearing. A recently published article (May 2019) in *Gear Solutions* magazine is taken for this purpose solely as an illustrative example.

Chapter 10: Titled as "Gear Manufacturing Accuracy Prediction, Control and Management," this chapter is contributed by Prof. *Mykola E. Terniuk*, Dr. *Anatolii V. Kryvosheia*, Dr. *Oleksandr M. Krasnoshtan*, Dr. *Pavlo M. Tkach*, and Dr. *Serhii V. Lutsky* (all of Ukraine).

The modelling of the main conformities of real teeth geometry generation process is carried out. A method and algorithms for analytical prediction of teeth errors are developed. A systemic model of real teeth geometry generation process is proposed. General equations of the real teeth profiles of the machined gears are derived. The correlation between the increments of coordinates and the rate gear accuracy indicators is established. Direct measurement methods, diagnostics and factorial methods, as well as computational-probabilistic and computational-adaptive methods are proposed for the relative primary error functions values calculation.

Control methods and systems for real teeth geometry generation process are described. A full set of possible ways of the gear accuracy control is established as well as the classification of possible ways is made. The error changes analysis, and the defective part missing riskiness estimation is carried out for various control systems in relation to element error control, factor methods and for combined systems.

Technological methods for the accuracy control of gears machining process are considered as well as general description of the accuracy control methods of gears machining process is given. The systematization of levels and types of accuracy regulation is carried out. The possibilities of accuracy regulation at the technological process level, including the possibility of accuracy control of two – and three-tool processing methods within gears base surfaces position changes, are revealed. The process initial parameters influence on the final gear accuracy established. The analysis of accuracy control at the level of operations and transitions is carried out. While analyzing, the factors allowing variation were established, and compensatory methods of adaptive control were considered.

The principles of assurance of gears optimal accuracy according to the given criteria are formulated.

Chapter 11: Titled as "Elliptical Gear Drives," this chapter is contributed by Dr. *Boris M. Klebanov* (Israel, USA). This chapter is devoted to gears with an elliptical shape of centrode. The issues of geometry, kinematics, and dynamics of gear drives with elliptical gears are considered. The method of designing elliptical gears is proposed, including the dimensioning of centrodes and determination of module and number of teeth depending on the load, rotational speed, and strength of the material the gear rim is made of. The text is accompanied by design formulas and diagrams that make the design process relatively easy and quick. Examples of design and calculations contribute to better understanding of proposed method.

The appendices are contributed by Prof. *Stephen P. Radzevich*.

It is likely this book is not free from omissions or mistakes, or that it is as clear and ambiguous as it should be. If you have any constructive suggestions, please communicate them to the editors via e-mails: radzevich@usa.com.

Sterling Heights, MI, USA
June 1, 2022

Stephen P. Radzevich

Contents

Editor and Contributors

About the Editor

Stephen P. Radzevich (USA) is a professor of Mechanical Engineering and a professor of Manufacturing Engineering. He received M.Sc. (1976), Ph.D. (1982), and Dr.(Eng)Sc. (1991) – all in mechanical engineering. He has extensive industrial experience in gear design and manufacture. He has developed numerous software packages dealing with *CAD* and *CAM* of precise gear finishing for a variety of industrial sponsors. His main research interest is *Kinematic Geometry of Surface Generation*, particularly with the focus on (a) precision gear design, (b) high-power density gear trains, (c) torque share in multi-flow gear trains, (d) design of special purpose gear cutting/finishing tools, (e) design and machining (finishing) of precision gears for low-noise/noiseless transmissions of cars, light trucks, etc. He has spent about 45 years developing software, hardware, and other processes for gear design an optimization. Besides his work for industry, he trains engineering students at universities and gear engineers in companies. He authored and coauthored over 40 monographs in the field, handbooks, and textbooks. The monographs entitled *Generation of Surfaces* (2001), *Kinematic Geometry of Surface Machining* (*CRC* Press, 2008), *CAD/CAM of Sculptured Surfaces on Multi-Axis NC Machine: The DG/K-Based Approach* (M&C Publishers, 2008), *Gear Cutting Tools: Funda-*

mentals of Design and Computation (*CRC* Press, 2010, 2nd edition 2017), *Precision Gear Shaving* (*Nova Science Publishers*, 2010), *Dudley's Handbook of Practical Gear Design and Manufacture* (2nd edition, *CRC* Press, 2012, 3rd edition 2016, 4th edition 2021), *Theory of Gearing: Kinematics, Geometry, and Synthesis* (*CRC* Press, 2012, 2nd edition 2018), *Geometry of Surfaces: A Practical Guide for Mechanical Engineers* (Willey, 2013, 2nd edition Springer, 2019), *Advances in Gear Design and Manufacture* (*CRC* Press, 2019), *High-Conformal Gearing: Kinematics and Geometry* (*CRC* Press, 2015, 2nd edition Elsevier, 2020), and *Advances in Gear Design and Manufacture* (CRC Press, 2019), *Recent Advances in Gearing: Scientific Theory and Applications* (Springer, 2021) are among the recently published volumes. He also authored and coauthored about 400 scientific papers and holds over 260 patents on invention in the field, both USA patents and patents of other countries.

Contributors

Michał Batsch (Poland), B.Eng., M.Sc., Ph.D., has gained the title of Engineer at the Faculty of Mechanical Engineering and Aeronautics at the Rzeszów University of Technology (Poland) in 2012 (field of automatic control and robotics). In 2013, he graduated in the field of mechanics and mechanical engineering (specialty mechanical drives), gaining the title of Master of Science. In 2015, he received the title of Doctor of Engineering for the thesis titled *Analysis of the Contact Region Geometry of the Cylindrical Convexo-Concave Novikov Gear* within the discipline of design and exploitation of machines. Since then, he has been employed as Assistant Professor in Department of Mechanical Engineering at the Rzeszów University of Technology (Poland) and teaches classes in gear geometry and kinematics, machine elements, and computer-aided design. His scientific interests cover the design and manufacturing technology of gears with unconventional tooth profiles. He is also employed in the Gear Research Laboratory where he is involved in research of spiral bevel gear grinding technology. He is an author of over twenty scientific publications in the field of gear drives. In addition, he has participated as a manager or member of the research team in many research projects, in which he was responsible for developing the original constructional solutions implemented in the industry.

Houjun Chen of China is an assistant professor of Mechanical Engineering at Nantong University. He received B.Eng. (2000), M.Eng. (2003), and Ph.D. (2009) degrees – all in mechanical engineering. He has spent about 20 years developing the theory and the technology of gearing and is the first one who proposed to discuss the kinematics and the geometry of conjugation by a pair of directrices of conjugate surfaces rather than the conjugate surfaces themselves. He invented many novel types of gear drives and developed an integrated *CAD/CAM/CAE* software system for hyperboloidal-type normal circular-arc gears. His main research interest is *Theory of Gearing and Its Applications*, particularly with the focus on (a) geometric theory of conjugate surfaces, (b) design of novel gearings, (c) dynamics of gearing, (d) digital manufacturing of gears, and (e) noise and vibration control of gearing. He has successively presided over or participated in the completion of more than 10 projects, such as project supported by the National Natural Science Foundation of China, project supported by the Natural Science Foundation of Jiangsu Province, China, etc. He authored and coauthored over 10 scientific papers in international journals, such as *Mechanism and Machine Theory*, *International Journal of Advanced Manufacturing Technology*, *Frontiers of Mechanical Engineering*, *Journal of Industrial and Production Engineering*, etc. He also trains graduate students at Nantong University of China in the field of mechanical engineering.

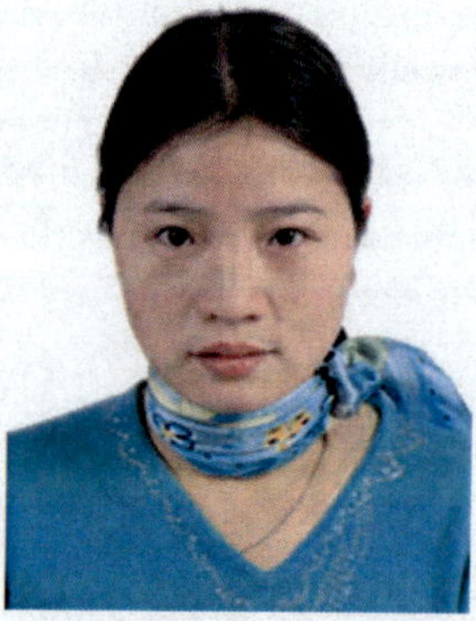

Zhilan Ju (China) is an assistant professor of Mechanical Engineering at Nantong University. She received B. Eng. (2000) and M.Eng. (2003) degrees – all in mechanical engineering. Her main research interest is in *Ultraprecision Machining*, particularly with the focus on (a) CNC machining, (b) *CAD/CAM/CAE*, and (c) manufacture of novel gearings. She has presided over or participated in the completion of more than 10 projects, such as project supported by the National Natural Science Foundation of China, Natural Science Fund for Colleges and Universities in Jiangsu Province, China, etc. She also authored and coauthored about 10 scientific papers in the field.

Boris M. Klebanov (Israel/USA) is a professional designer of machines and mechanisms with 60 years of experience in heavy and aviation industries. He received MSc (1959) and PhD (1969) degrees in mechanical engineering. He is coauthor of four monographs: "Gears in Mechanical Engineering," "Mashinostroenie," Leningrad, 1978; "Calculation and Design of Reduction Gears," "Politecnica," Sanct-Peterburg, 1993; "Machine Elements: Life and Design," CRC Press, Boca Raton – London – New York, 2008; "Power Mechanisms of Rotational and Cyclic Motions," CRC Press, Boca Raton – London – New York, 2016.

He has authored dozens of scientific articles on design and calculation of machine elements for strength and rigidity and participated in collective collections of articles "Advances in Gear Design and Manufacture," CRC Press, Boca Raton – London – New York, 2019, and "Dudley's Handbook of Practical Gear Design and Manufacture," Fourth Edition, CRC Press, 2022. He also has many inventions in the field of machinery.

Viktor I. Korotkin (Russia), scientific activity is focused on the theoretical and experimental studies of gears, including Novikov gears with different hardness of tooth surfaces and types of heat treatment. Theoretical studies include the solution of spatial problems of the stress-strain state of bodies of complex shape and finite sizes, developments in the field of spatial theory of gearing, and the creation of computational programs. The obtained results of the research were used in the development of optimal parameters of the basic profiles of Novikov gears, modern calculations of their geometry, bending, surface, and deep contact strength, which made it possible to create methods for radically increasing the load-bearing capacity of these gears used in various industries.

Dr. Korotkin is the organizer and active participant of long-term bench and operational comparative tests of involute and Novikov gears with nitrocarburized and heat-improved teeth, which made it possible to introduce the latter into production.

Main achievements: Inventor of the USSR, winner of the Gold Medal at the International Exhibition of Machine Tools in Brno (Czechoslovakia) for the development of the Gear Cutting Machine 5A250 for straight-toothed conical wheels with barrel-shaped teeth, author of five monographs (one of which is published in the USA), about 200 publications, granted with 18 patents on inventions, participant of numerous international and domestic exhibitions, where the general and special purpose gearboxes developed by the author with improved technical and economic indicators, the main coauthor of the Interstate Standard for the basic profile for Novikov gears with high-hard tooth surfaces and methodological recommendations for their strength calculation were presented.

Alexander M. Krasnoshtan (Ukraine) is graduated (2004) from the Vinnytsya National Technical University with a degree of Mechanical Engineer in "Automobiles and transport management."

He defended (2007) his Ph.D. thesis "Modeling and determination of the main characteristics of a car transmission based on a gear-lever variator" at the Kharkiv National Automobile and Highway University. Starting 2008, he occupies a position of associated professor at National Transport University in Kyiv.

He proposed to use the general algorithm for solving the problem of complex structural-parametric optimization of multi-operational technological systems in sphere of transport systems. Independently or with coauthors, he developed theory of using the gear-lever variator in construction of vehicles.

He has authored over 30 scientific and scientific journalistic works, including 5 patents on inventions.

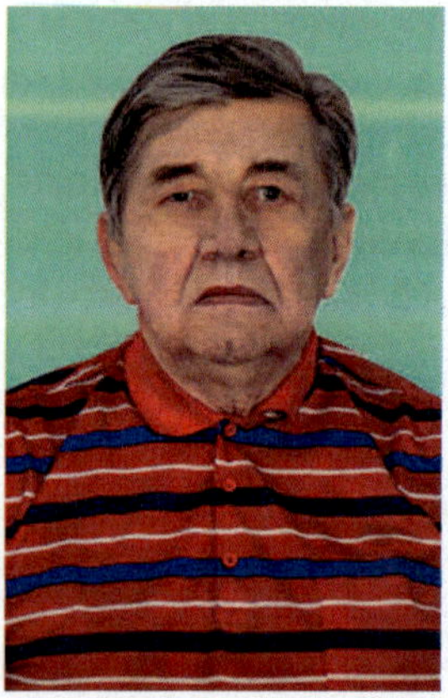

Anatoly V. Krivosheya (Ukraine) is a Doctor of Mechanical Engineering Sciences. He received both his master's degree in 1971 and Ph.D. in 1981 in mechanical engineering. He works at the Institute for Superhard Materials of the National Academy of Sciences of Ukraine and deals with the design of tools made of superhard materials, along with other modern tool materials for machining complex-shaped products, including gears of different classes and types. He has extensive experience in designing gear cutting tools using shaping theory, affine space mapping theory, set theory, and gear manufacturing. He has developed many software packages related to computer-aided design (CAD) of gear cutting tools and precision automated machining (CAM) of gears for various industrial machines and mechanisms. His main research interests are related to the kinematics and geometry of gear shaping, in particular with an emphasis on precision machining of gears by a group of sequentially working tools, taking into account their regrinding, as well as designing and machining of precision gears for low-noise transmissions of cars, trucks, tractors, etc. He has spent nearly 50 years developing software, hardware, and other processes to design and optimize gear manufacturing processes. In addition to working in the industry, he teaches undergraduate and graduate students in engineering at universities and specialists in various companies.

He is a coauthor of the monograph "3-D Modeling of Tools, Shaping, and Stock Removal During Cutting" (Kharkiv: NSU "KhPI," 2001, 304 p.). He has also authored and coauthored about 150 scientific papers, including 22 patents on inventions.

Siyu Liu (China) is a postgraduate student at Northeastern University China under the supervision of Prof. Yaping Zhao (see below).

Sergey V. Lutskii (Ukraine) is a professor of Mechanical Engineering. He received a Master of Science degree (1981) and Ph.D. (1992) in the area of automation of technological processes. He is a full member (academician) of the Ukrainian Academy of Economic Cybernetics (2015). He has extensive experience in the machine-building industry, the engineering path has gone from a designer of wheeled-tracked vehicles in machine-building production, and the head of the formation and maintenance of databases of automated control systems for design documentation of design bureaus for the production of wheeled-tracked vehicles to the director of the scientific and technical center of a research institute for mechanical engineering.

His main research interest is the implementation of new developed system information principles and approaches for solving production problems.

In addition to working for industry, he trains students in engineering specialties in wheeled-tracked vehicles and metrology at universities.

He has authored and coauthored over 120 scientific papers and is granted with 6 patents on inventions, both patents of Ukraine and international patents.

Nikolay P. Onishkov (Russia) is a known expert in the field of gearing. His research is focused on the geometry and strength of gears, including *Novikov gears*. Most of his studies are devoted to superficially (thermal and chemical-thermal) hardened gears, for which a method of calculation for the prevention of destruction due to insufficient deep contact strength has been developed.

The *Lebedev-Pisarenko theory* is applied and developed based on the use of limit state criteria for structurally heterogeneous material, taking into account changes in its plastic properties. The obtained results of the research are of great applied importance in assessing the stress state and load-bearing capacity of the on-the-top zones of cemented and nitro-cemented gears widely used in the drives of machines for various purposes. A calculation technique was created, which made it possible to significantly reduce for *Novikov gears*. The standard regulatory coefficients recommended for involute analogues.

On the basis of extensive experimental data, a calculation model was developed for the first time ever, with the help of which it was possible to significantly clarify the assessment of hardened gears with local contact, in particular, *Novikov gears*, according to the strength of the working surfaces of the teeth, taking into account a significant increase in the criterional stresses at the border of the instant contact while reducing the permissible stresses. This model allows to predict the potential destruction of teeth as a result of their chips and other types of breakdowns, including in the areas of action of relatively minor bending moments.

Dr. *Onishkov* has authored and coauthored over 50 scientific works, including 2 monographs; also, he is granted with 5 patents of the Russian Federation.

Chang Qu (China) is a professor of Mechanical Engineering at Nantong University. She received B.Eng. (1989) and M.Eng. (2001) degrees – all in mechanical engineering. Prof. Qu has extensive industrial experience in the design and manufacture of auto parts and components. Her main research interest is *CAD and its applications*, particularly with the focus on (a) computer graphics, (b) mechanical CAD, and (c) kinect application. Prof. Qu has presided over or participated in the completion of more than 30 projects. She trains graduate students at Nantong University in the field of mechanical engineering. She also authored and coauthored about 40 scientific papers and holds over 15 patents on invention in the field.

Stephen P. Radzevich (USA) is a professor of Mechanical Engineering and a professor of Manufacturing Engineering. He received his M.Sc. in 1976, Ph.D. in 1982, and Dr.(Eng)Sc. in 1991, all in mechanical engineering. Dr. Radzevich has extensive industrial experience in gear design and gear manufacture. He has developed numerous software packages dealing with computer-aided design (CAD) and computer-aided machining (CAM) of precise gear finishing for a variety of industrial sponsors. His main research interest is the field of *Kinematics and Geometry of Gearing*, particularly with a focus on precision gear design, high-power-density gear trains, torque share in *split-power-transmission systems* (*SPTS*), design of special purpose gear cutting/finishing tools, and design and machine (finish) of precision gears for low-noise and noiseless transmissions of cars, light trucks, and so forth.

Dr. Radzevich has spent about 45 years developing software, hardware, and other processes for gear design and optimization. Besides his work for industry, he trains engineering students at universities and gear engineers in companies.

He has authored and coauthored over 40 monographs in the field, handbooks, and textbooks and about 400 scientific papers, and holds over 260 patents on invention in the field: both USA patents and patents of other countries.

Aleksandr Sandler date of birth October 17, 1947 (Moscow, Russia). In 1970 he graduated from the Moscow Machine-Tool Institute (STANKIN) with a degree in Mechanical Engineering and PhD in 1980. Thesis topic: Research on the issues of profiling grinding wheels for processing hob cutters with a large angle of elevation of the producing surface (in relation to the manufacture of wheels of multi-thread coarse-modular worm gears).

In 1972, he has started research in the area of abrasive processing, grinding tools, and grinding processes, as a set of factors for choosing the characteristics of an abrasive tool and processing modes. Since 1974 he began to engage in precision machining of hob cutters for worm gears – sharpening of the face helical surfaces and relief of the rear flank surfaces of the teeth. In 1978, the subject of scientific work expanded to the technology of production of multi-thread large-module worm cutters for precision worm wheels.

During his postgraduate studies, STANKIN was engaged in the technological advancement of worm gears with a hydrostatic lubrication effect based on engagement with closed contact lines (ZLK). In 1978, the direction of the dissertation work was changed to continue the work begun earlier on sharpening and relief of precision hob cutters.

Since 2016, he has taken up the issues of grinding helical and relief surfaces with a profile of significant and variable curvature in order to increase the accuracy of the profile of the surface to be ground and maintain the profile of the cutting surface of the cutter during regrinding. In most detail, these issues have been resolved in relation to milling cutters for cutting gears of gears with *Novikov engagement*.

Has authored and coauthored over than 40 publications in total.

Mykola E. Terniuk (Ukraine) is graduated (1972) from the Kharkiv National Technical University "KhPI" with a degree of Mechanical Engineer in the field of "Automation and complex mechanization of mechanical engineering."

In 1976 at the Kharkiv National Technical University "KhPI," he defended his Ph.D. thesis titled "Control of the accuracy in production of cylindrical gears."

In 1983 at the Tula Polytechnic Institute, he defended his doctoral dissertation on the topic "Fundamentals of integrated optimization of technological systems for the production of gear wheels."

Professor Mykola Ternyuk designed and introduced a general algorithm for solving the problem of complex structural-parametric optimization of multi-operational technological systems. Independently (or with coauthors), he has developed a number of new types of gears and technological equipment for manufacturing them, as well as the foundations of the theory of structures and design solutions for multi-parameter gears and gear variators. The revealed regularity of the structure of the Universe in the form of a System of periodic systems of structures of objects of the visible material world has been published. The type and place of the Periodic Table of Technical Elements in its structure has been shown. Based on the above mentioned, he proposed an algorithm for a formalized directed synthesis of technical innovations.

He has authored over 700 scientific works, including 250 patents on inventions.

Pavlo M. Tkach (Ukraine) is a Ukrainian scientist whose professional research is focused on the reduction spur and helical gears for variety of application. The topic of his Ph.D. thesis (defended in 2004) is the lifetime extension of reduction spur gears through tooth crowning, as well as tooth profile modification. He has authored over 100 scientific publications, including four monographs: *Highly Loaded Cylindrical Gears with Biconvex and Biconcave Teeth* (V. Dahl EUNU Publishing, 2005), *Intro to Machines Theory* (V. Dahl EUNU Publishing, 2009), *Short Course on Machine Elements* (V. Dahl EUNU Publishing, 2010), and

Machine Elements. Competitive Tasks (*V. Dahl* EUNU Publishing, 2014). Also, he is granted with 15 patents on inventions. In the meantime, his research is focused mainly on the wear resistance increase, and the tooth strength improvement of spur gears utilizing for this purpose non-standard tooth profiles.

From 2002 to 2014, Dr. Tkach occupied a position of assistant professor and, later on, a position of associate professor at *Volodymyr Dahl* East Ukraine National University (city of Luhansk, Ukraine). In the meantime, he is the Head of the Science Management Department at E.O. Paton Electric Welding Institute of the National Academy of Sciences of Ukraine (city of Kyiv, Ukraine).

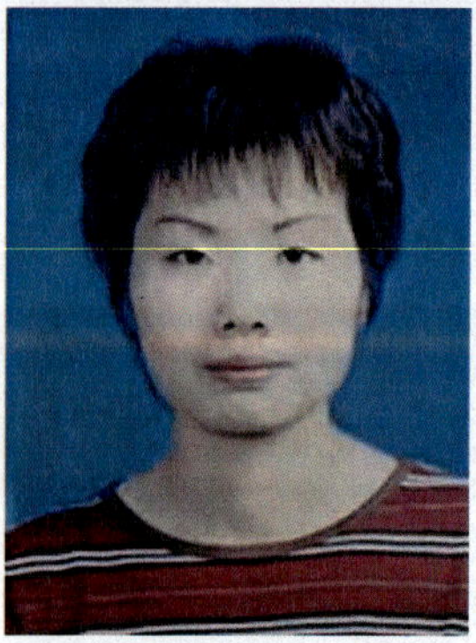

Xiaoping Zhang (China) is a professor of Mechanical Engineering at Nantong University. She received B. Eng. (1995) and M.Eng. (2002) degrees – all in mechanical engineering. Her main research interest is in *energy-saving equipment use and development*, particularly with the focus on (a) *CAD*, (b) sea-water desalinating unit, and (c) dynamics of gearing. Prof. *Zhang* has presided over or participated in the completion of more than 20 projects. She trains graduate students at Nantong University in the field of mechanical engineering. She also authored and coauthored about 20 scientific papers and holds over 15 patents on invention in the field.

Yaping Zhao (China) is a professor of Mechanical Engineering in Northeastern University, China. His research interests lie in meshing theory for mechanical transmissions, research and development of novel types of gear drives, engineering differential geometry, and so forth.

His academic titles include New Century Excellent Talent awarded by Chinese Ministry of Education, Excellent Talent in Institutions of Higher Learning in Liaoning Province, Jianlong specially appointed professor of NEU, Leading Talent of Shenyang, Promising Young Person of Science and Technology of University, and others.

He was a research fellow and a visiting scholar in Nottingham Trent University, England, and Northwestern University, USA, for 1.5 years in total.

As the corresponding or the first author, he has published more than 30 papers in the international SCI journals, for example, in *Mechanism and Machine Theory*, *ASME journal of Mechanical Design*, *Computer-Aided Design*, *Applied Mathematical Modelling*, *International Journal of Mechanical Sciences*, *Journal of Vibration and Control*, *Proc. IMechE C JMES*, Forschung im Ingenieurwesen, *Tribology Transactions*, and *Science China Technological Sciences*. For several times, he was invited to contribute special chapter for the monograph in the book series of mechanisms and machine science published by Springer. He was repeatedly rewarded for best paper by the academic community and government agency.

He won China Production-Study-Research Cooperation Innovation Prize. He holds four China patents on the worm drive invention. He made oral presentation on many international academic conferences on gearing, for instance PTG in the USA, MPT of JMSE in Japan, IGC in France, ICG in Munich of Germany, EuCoMeS in European, MTM and Robotics in Romania, and IFToMM World Congress. Moreover, he often served as session chair and contributing reviewer for these conferences.

Dr. Zhao presided over a lot of scientific research projects on gearing, whose representative supporter is NSFC.

Dr. Zhao is a senior member of CMES and a member of ASME. He serves as the deputy chairperson of the TC of gearing and transmissions in IFToMM and the research grant proposal assessor in the field of gear drive for many institutes who manage the scientific research fund. He also serves as the reviewer for a number of international SCI journals, and the editorial board member of *Mechanical Engineering Technology* and *Advances in Computational Design*.

Chapter 1
Novikov Gearing Is a Kind of Involute Gearing

1.1 Introduction

Invented at the beginning of 1950, Novikov gearing has been extensively investigated by both Soviet and Western scientists and engineers. A large body of published scientific work on Novikov gearing is available in the public domain. The gearing is successfully used in helicopter transmissions[1] (see Fig. 1.1) [1, 2] as well as in other applications.

Despite the several efforts taken to investigate Novikov gearing, numerous key issues in the gear kinematics and gear geometry are still not clear to the majority of gear experts all around the world. A lack of understanding of the kinematics and geometry of Novikov gearing has resulted in confusions and even absurdity in this relatively narrow area of mechanical engineering. Many scientists and researchers in the field of gearing still loosely refer to Novikov gearing as Wildhaber–Novikov gearing, or just W-N gearing, which is completely incorrect and indicates insufficient training in the theory of gearing, especially the absence of an understanding of the kinematics and geometry of both Novikov gearing and Wildhaber gearing. In this chapter, the differences between the two gear systems are clearly outlined. It is shown that Novikov gearing and Wildhaber gearing must only be considered separately and that combining the two gear systems in a common system (that is, W-N gearing) is a huge mistake. Potential improvements to Novikov gearing are disclosed.

[1] These gears in Westland Helicopters, Ltd. were finish cut by a "disk-type" grinding wheel.

S. P. Radzevich, *Novikov/Conformal Gearing*,
https://doi.org/10.1007/978-3-031-10019-2_1

Fig. 1.1 A Novikov gear pair for helicopter transmission. (Westland Helicopters, Ltd.)

1.2 Historical Overview

The Novikov gear system was invented in the mid-1950s by Colonel Mikhail L. Novikov. At that time, M.L. Novikov was a professor at the Zhukovsky Military Aviation Engineering Academy (MAEA), Moscow, USSR. Dr. Novikov, M.L. was carrying out extensive research and development of high-power-density gear transmission systems.

Transmission and transformation of rotary motion are the two main purposes of gearing of all kinds. The axes of the rotation of the driving and driven gears can be arbitrarily oriented in relation to one another. From this perspective, three kinds of gearings are commonly distinguished:

- Parallel-axes gearing (or P_a– gearing, for simplicity)
- Intersected-axes gearing (or I_a– gearing)
- Crossed-axes gearing (or C_a– gearing)

For simplicity, but without loss of generality, the consideration below is limited to only P_a– gearing.

In the past, pin gearing and other primitive kinds of gearings were in use. None of them is capable of smoothly transmitting an input rotary motion. Born in Switzerland, the mathematician and mechanician Leonhard Euler[2] proposed involute gearing – the only kind of parallel-axes gearing that is capable of smoothly

[2]The proposed involute gearing by L. Euler (at around 1760) deserves to be called Euler gearing, or just *Eu*–gearing, for simplicity.

transmitting a uniform rotation from a driving shaft to a driven shaft. This is because the base pitches of mating gears are equal to one another. Gearing of no other kind is capable of smoothly transmitting a uniform input rotation. The equality of the base pitches of a gear and of its mating pinion is observed only in involute gearing.

In external involute parallel-axes gearing, a convex involute tooth profile of the driving member interacts with a convex involute tooth profile of the driven member. In other words, external involute P_a– gearing features "convex-to-convex contact" of the mating tooth flanks. As the contacting tooth profiles are convex, this imposes a strong constraint on the bearing capacity of the involute gearing because of high contact stress. It is highly desirable to replace two convex contacting tooth profiles of the gear teeth with their convex-to-concave contact. In conventional involute gearing, this is not permissible, as it inevitably entails violation of the three fundamental laws of gearing [3].

As early as 1956,[3] a breakthrough invention in the realm of gearing was made by Dr. M.L. Novikov, who proposed a novel kind of gearing [4]. The concept of the proposed gear system is illustrated in Fig. 1.2 [4]. Later on, Novikov gearing was investigated in his doctoral thesis [5] and then summarized in a monograph [6].

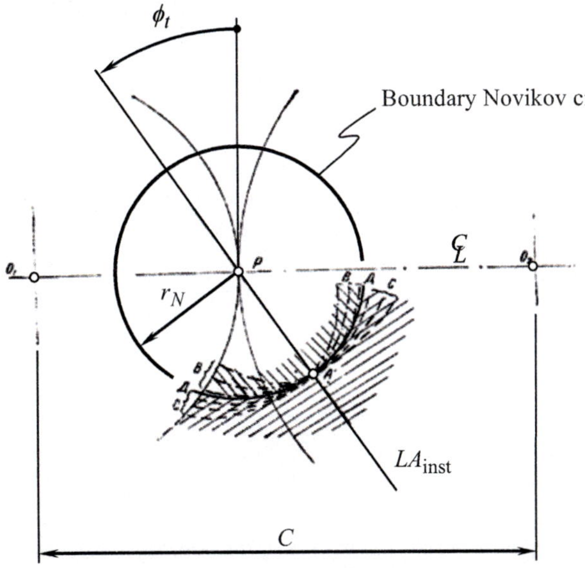

Fig. 1.2 On the concept of Novikov gearing. (After Dr. M.L. Novikov; USSR Pat. No. 109,113, 1956), the boundary Novikov circle of a radius r_N was introduced later (~2008) by Prof. S.P. Radzevich

[3]It should be mentioned here that the first pair of Novikov gears made out of an aluminum alloy (a pre-prototype) had been cut on April 25, 1954 by means of a disk-type mill cutter. In all, 15 gear pairs for testing purposes had been machined in the summer of 1954 by means of a disk-type mill cutter.

Below, we do not follow the obsolete approach used by Dr. M.L. Novikov to design a gear pair of novel design. Instead, the concept of Novikov gearing is derived on the basis of conventional external parallel-axes involute gearing (*Eu*–gearing). Euler gearing is chosen for the derivation, as it has been proven [3, 4] that the "Novikov gear system" is a reduced case of the "involute gear system."

1.3 Principal Design Features of Novikov Gearing

For designing a pair of Novikov gears, let us assume that the location and orientation of the axes of rotation of the driving and driven members of the gear pair to be designed have been specified and the gear ratio provided. The desirable value of the transverse pressure angle is also known. With that said, a pair of Novikov gears can be designed following the routine briefly outlined below.

1.3.1 Gear Vector Diagram of a Novikov Gear Pair

Designing a pair of Novikov gears begins with the construction of a gear vector diagram of the gear pair to be designed. The rotation vector[4] of the gear $\boldsymbol{\omega}_g$ is pointed along the axis of rotation O_g of the gear. The magnitude ω_g of the rotation vector $\boldsymbol{\omega}_g$ equals to $\omega_g = |\boldsymbol{\omega}_g|$. The rotation vector of the mating pinion $\boldsymbol{\omega}_p$ is pointed along the axis of rotation O_p of the pinion. The magnitude ω_p of the rotation vector $\boldsymbol{\omega}_p$ is $\omega_p = |\boldsymbol{\omega}_p|$. The magnitudes ω_g and ω_p relate to one another as: $u = \omega_p/\omega_g$.

The axes of rotation O_g and O_p are at a certain center distance C. The rotation vectors $\boldsymbol{\omega}_g$ and $\boldsymbol{\omega}_p$ form a crossed-axes angle Σ, that is: $\Sigma = \angle(\boldsymbol{\omega}_g; \ \boldsymbol{\omega}_p)$. In the case of external P_a– gearing, the angle Σ always equals $\Sigma = 180°$.

The principle of inversion of rotations can be implemented in the gear pair to be designed.

Let us assume that both the axes of the rotations O_g and O_p are rotated together with the rotation vector $-\boldsymbol{\omega}_g$. Because the identity $\boldsymbol{\omega}_g + (-\boldsymbol{\omega}_g) \equiv 0$ is observed, the gear becomes stationary under the additional rotation $-\boldsymbol{\omega}_g$. The pinion is rotated with the rotation: $\boldsymbol{\omega}_{pl} = (\boldsymbol{\omega}_p - \boldsymbol{\omega}_g)$. The vector of instant rotation $\boldsymbol{\omega}_{pl}$ of the pinion in relation to the gear is pointed along the axis of instantaneous rotation, i.e., P_{ln}. The gear vector diagrams are discussed in more detail in the study by Radzevich [3].

[4] It is instructive to remember here that rotations, by nature, are not vectors at all. However, if special care is taken, then rotations can be treated as vectors.

1.3.2 Plane of Action in Parallel-Axes Novikov Gearing

The plane of action PA in parallel-axes Novikov gearing is a plane through the axis of instantaneous rotation P_{ln}. This plane forms a transverse pressure angle ϕ_{t}, which is perpendicular to the plane through the axes of rotation O_{g} and O_{p} of the gear and pinion, respectively. The base diameter of the gear $d_{\text{b.g}}$ and that of the pinion $d_{\text{b.p}}$ can be expressed in terms of the pitch radii r_{g} and r_{p}, respectively, of the pitch cylinders and the transverse pressure angle ϕ_t:

$$d_{\text{b.g}} = 2\,r_{\text{g}}\cos\phi_{\text{t}} \tag{1.1}$$

$$d_{\text{b.p}} = 2\,r_{\text{p}}\cos\phi_{\text{t}} \tag{1.2}$$

Once the base cylinders are determined, the transmission of a rotation from the driving member to the driven member of the gear pair can be interpreted with the help of the so-called "equivalent pulley-and-belt transmission" shown in Fig. 1.3.

Either Eqs. (1.1) or (1.2) can be used for the derivation of an expression for the calculation of the base pitch p_{b} in a transverse section of the gear pair:

$$p_{\text{b}} = \frac{\pi\,d_{\text{b.g}}}{N_{\text{g}}} = \frac{\pi\,d_{\text{b.p}}}{N_{\text{p}}} \tag{1.3}$$

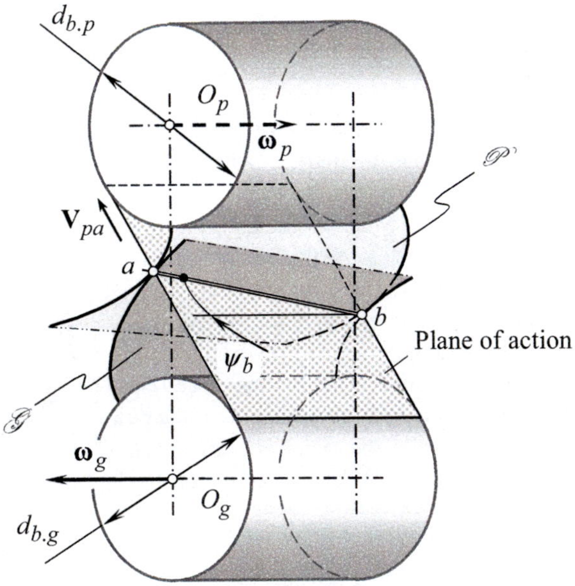

Fig. 1.3 The key elements of the principal kinematics of parallel-axes gearing

Equation (1.3) is valid for parallel-axes gearing that is capable of smoothly transmitting a rotation.

1.3.3 A Desirable Line of Contact in Parallel-Axes Gearing

The tooth flank of the gear $\mathcal{G}$ and that of the pinion $\mathcal{P}$ make contact along a desirable line of contact, i.e., LC_{des} (see Fig. 1.3), or just a line of contact, i.e., LC. The line of contact LC is a planar curve of a favorable geometry that is entirely located within the plane of action PA. In Fig. 1.3, the straight-line segment ab is the line of contact. The tooth flanks $\mathcal{G}$ and $\mathcal{P}$ of the gear and the mating pinion, respectively, interact only within the active portion of the plane of action, as shown in Fig. 1.4.

As referred to in Fig. 1.4a, the straight-line segment N_gN_p is the total length of the plane of action. In reality, the active portion of the plane of action PA is of a shorter length Z_{pa} (see Fig. 1.4b). In the case of Novikov gearing, the equality $Z_{pa} = 0$ is observed.

In involute helical parallel-axes gearing, the desirable line of contact LC between the tooth flanks of the gear $\mathcal{G}$ and the pinion $\mathcal{P}$ (one should remember that the tooth flanks $\mathcal{G}$ and $\mathcal{P}$ are not yet constructed) is a straight-line segment that forms a base helix angle ψ_b with the axis of instant rotation P_{ln}.

The total contact ratio m_t in the gear pair can be expressed in terms of the transverse contact ratio m_p and the face contact ratio m_F:

$$m_t = m_p + m_F \tag{1.4}$$

where $m_p = Z_{pa}/p_b$ and $m_F = F_{pa} \tan \psi_b/p_b$.

The inequality $m_t \geq 0$ must be observed for any and all parallel-axes gear pairs.

When the base cylinders of diameters $d_{b.g}$ and $d_{b.p}$ rotate, the desirable line of contact LC travels (together with the plane of action PA) in relation to the reference systems, one of which is associated with the gear and the other with the pinion. In such a motion, the tooth flank of the gear $\mathcal{G}$ (as well as the tooth flank of the pinion $\mathcal{P}$) can be interpreted as a family of consecutive positions of the desirable line of contact in the corresponding reference system.

In the example (see Fig. 1.4b), the active portion ab of the gear tooth profile[5] is shaped in the form of an involute of a circle. The profile ab is specified by the radii of the outer cylinders of the gear and of the pinion, i.e., $r_{o.g}$ and $r_{o.p}$, respectively. Point *a* corresponds to the "start-of-active-profile" point (SAP– point), whereas point *b* corresponds to the "end-of-active-profile" point (EAP– point).

[5]The tooth involute profile is called "involute" because the active portion of the tooth profile is shaped in the form of an involute of a circle, regardless of whether the actual geometry of the rest of the gear tooth profile (fillets, bottom lands) is involute.

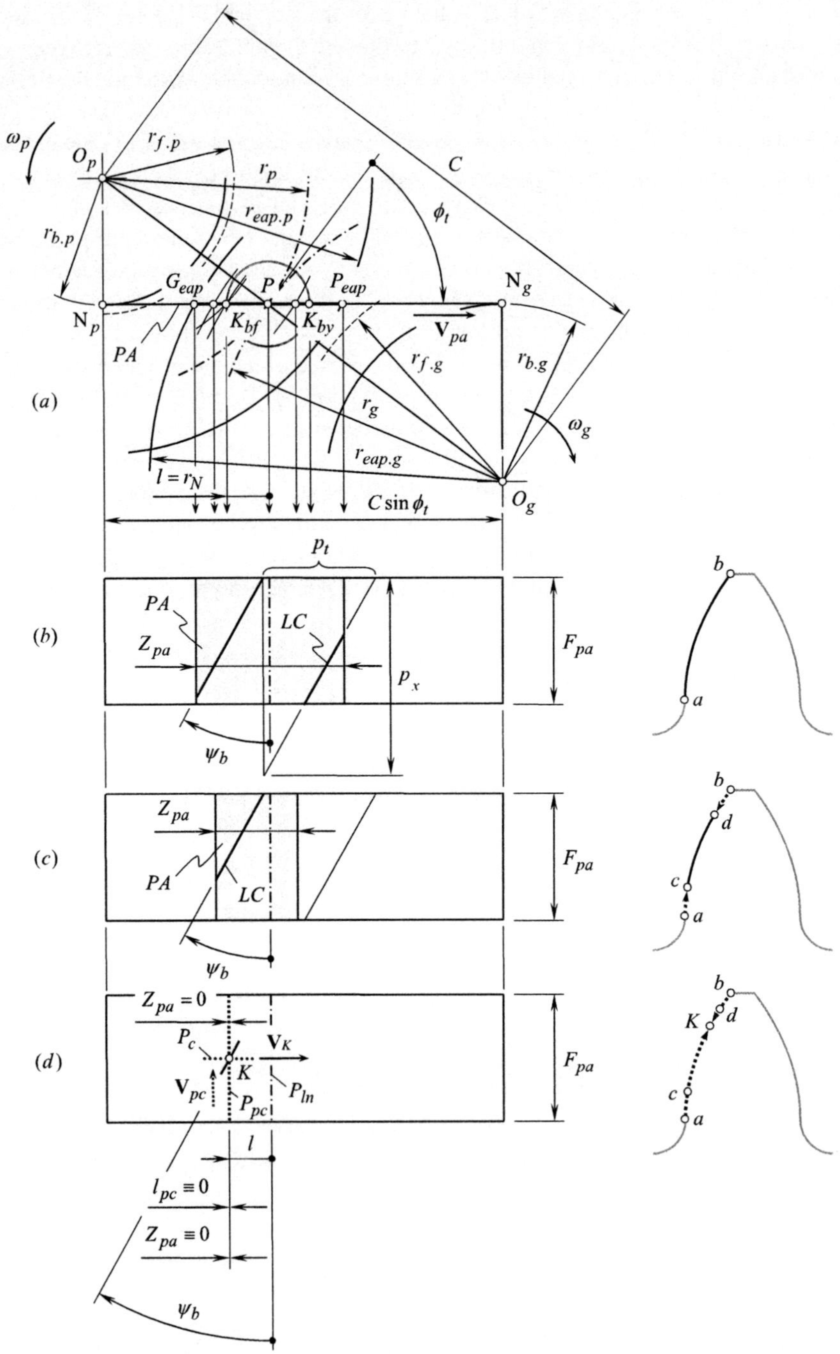

Fig. 1.4 The elements of a parallel-axes gear pair that features a zero transverse contact ratio ($m_p = 0$)

For both members of a gear pair – that is, for the gear and the pinion – the radius r_{eap} of the EAP– circle can be smaller than the outer radius of the gear $r_{\mathrm{o.g}}$ (or smaller than the outer radius of the pinion $r_{\mathrm{o.p}}$, for example) due to chamfering. In such a scenario (see Fig. 1.4c), the active portion of the plane of action gets narrower. The SAP– point c and the EAP– point d become closer to one another: the active portion cd of the involute tooth profile is shorter than that of ab, as illustrated in Fig. 1.4b. This gives a certain freedom to the gear designer when selecting the geometry of the nonactive portions ac and bd of the tooth profile. As these portions of the tooth profile do not interact with one another, the geometry of the segments ac and bd is not restricted by conditions of meshing of the tooth profiles (which is a must for the active portion cd).

In extreme cases, the EAP– circles of the gear and of the pinion can pass through a certain point K within the straight-line segment $P_{\mathrm{g}}P_{\mathrm{p}}$. Because of this, the length Z_{pa} of the active portion of the plane of action becomes zero ($Z_{\mathrm{pa}} = 0$) and the active portion of the involute tooth profile shrinks to point K. This point is referred to as the "involute tooth point." The nonactive portions aK and bK of the tooth profile meet each other at point K. These portions of the tooth profile are not subjected to conditions of meshing, and, thus, this gives additional freedom to the gear designer when selecting the geometry of the nonactive portions aK and bK of the tooth profile (see Fig. 1.4d).

As the width of the active portion of the plane of action is zero ($Z_{\mathrm{pa}} = 0$) and the involute tooth profile is shrunk to a point, the transverse contact ratio m_{p} results in a zero value. In order to meet the inequality $m_{\mathrm{t}} \geq 0$, the following inequality must be met:

$$m_{\mathrm{t}} = m_{\mathrm{p}} + m_{\mathrm{F}} = 0 + m_{\mathrm{F}} = m_{\mathrm{F}} > 0 \tag{1.5}$$

The point system of parallel-axes gearing (see Fig. 1.4d) gives much freedom when designing the nonactive portions of the tooth profiles of the gear and the pinion, as the geometry of these portions is free of constraints imposed by conditions of the meshing of two conjugate tooth profiles.

1.3.4 Design Features of Novikov Gearing

The concept of Novikov gearing is based on the schematic depicted in Fig. 1.4d. For this case, Dr. M.L. Novikov proposed replacing the convex-to-convex contact of the tooth profiles with their convex-to-concave contact. This replacement is only possible in cases in which the active portion of the involute tooth profile is shrunk to a point (and it is infeasible in cases in which the active portion of the involute tooth profile is of a certain length [3]).

Shown in Fig. 1.2, the first (in time) schematic that illustrates the concept of Novikov gearing [4] is far from being the best and most consistent.

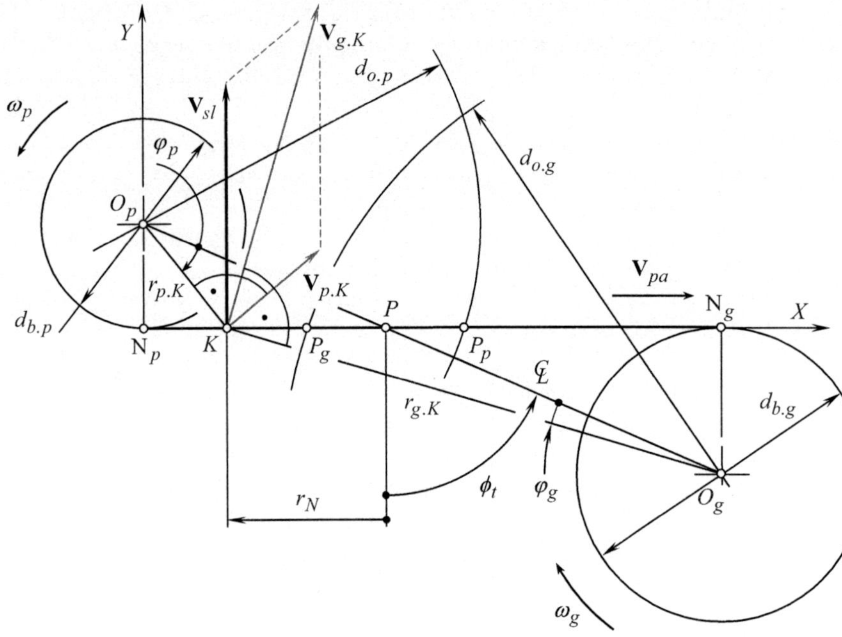

Fig. 1.5 Tooth profile sliding in parallel-axes Novikov gearing

The point of contact K of the tooth flanks $\mathcal{G}$ and $\mathcal{P}$ is located within the straight line of action LA. The larger the distance of the contact point K from the pitch point P, the more freedom there is for the gear designer in selecting the radii of curvature of the interacting tooth profiles. At the same time, the larger the distance of the contact point K from the pitch point P, the higher the losses on friction between the tooth flanks $\mathcal{G}$ and $\mathcal{P}$ and the higher the tooth flanks are worn down (see Fig. 1.5). Ultimately, the actual location of the contact point K is a tradeoff between the two factors just mentioned.

Furthermore, let us assume that the pinion is stationary and that the gear performs an instant rotation in relation to the pinion. The axis P_{ln} of the instantaneous rotation $\boldsymbol{\omega}_{\mathbf{pl}}$ is a straight line through the pitch point P. When the pinion is viewed motionless, the contact point K traces a “boundary Novikov circle”[6] of a radius r_N, which is centered at the pitch point P as illustrated in Fig. 1.2.

The pinion tooth profile $\mathcal{P}$ can either align with a circular arc of the boundary Novikov circle of a radius r_N or relieve into the bodily side of the pinion tooth. The pitch point is included in the interval, whereas the contact point K is not. On the other hand, the location of the center of curvature of the concave gear tooth profile $\mathcal{G}$

[6]It is instructive to mention here that Dr. M.L. Novikov did not use the concept of the “boundary circle.” The concept of the “boundary N−circle” was introduced by Prof. S.P. Radzevich [3] in the early 2000s. Prof. S.P. Radzevich also proposed referring to this circle as the “boundary Novikov circle” or just as the N−circle of a limiting radius $r_{\mathbf{N}}$.

within the straight line LA is limited to an open interval $P \to \infty$. Theoretically, the pitch point P can be included in this interval for K. The radius of curvature r_p of the convex pinion tooth profile $\mathcal{P}$ is smaller than r_g, which is the radius of curvature of the concave gear tooth profile $\mathcal{G}$. Thus, the inequality $r_p < r_g$ is observed.[7]

In Novikov gearing, both the pinion and the gear are helical. The helices are of the opposite hand. No spur Novikov gearing is feasible at all. Due to this, when the gears rotate, the contact point K "pseudo-travels" axially parallel to the axis of instantaneous rotation P_{ln}. In this straight "pseudo-motion," the contact point traces a straight line P_{pc}, which is called the "pseudo-path of contact" P_{pc} (as the equality $Z_{\mathrm{pa}} = 0$ is observed, the length of the path of contact P_c at every transverse section is always zero ($P_c \equiv 0$)).

The radial position of the contact point K on both the gear and pinion tooth flanks, i.e., $\mathcal{G}$ and $\mathcal{P}$, respectively, remains the same. It is therefore fundamental to the operation of gears that contact occurs nominally at a point and that the point of contact "pseudo-travels" axially across the full face width of the gears during rotation. It should be stated as a condition of the operation of Novikov gearing that, for a given profile, the tooth surfaces should not interact before or after culmination when rotated at angular speeds that are in the gear ratio.

The transverses contact ratio m_p in Novikov gearing is zero ($m_p \equiv 0$). Because of this, geometrically, the meshing of the gear teeth in the transverse section is instantaneous. The face contact ratio m_F of the gear pair is always greater than one ($m_F = m_t > 1$).

1.3.5 *Principal Design Parameters of Novikov Gearing*

From a historical perspective, it is interesting to consider the calculation of the principal design parameters of a Novikov gear pair following the approach proposed by Dr. M.L. Novikov [6]. The calculation of the design parameters of a Novikov gear pair with a circular-arc tooth profile is considered below as an example.

For the calculation of the design parameters of a Novikov gear pair, the center distance C and the gear ratio $u = \omega_p/\omega_g$ of the gear pair are required.

The radius of the pitch circles of a gear R_g and that of a mating pinion R_p can be expressed in terms of the center distance C and the tooth ratio u, as follows:

$$R_g = C \cdot \frac{u}{1+u} \tag{1.6}$$

[7]It should be pointed out here that there are no constraints to designing a gear pair with a convex gear tooth profile and a concave pinion tooth profile.

$$R_{\text{p}} = C \cdot \frac{1}{1+u} \tag{1.7}$$

The distance l at which the pseudo-path of contact P_{pc} is remote from the pitch point P as well as the transverse pressure angle ϕ_t in the gear mesh must be specified.

The displacement[8] l is the principal design parameter of a Novikov gear pair. Many of the design parameters of a Novikov gear pair can be expressed in terms of the displacement, $l = KP$.

For the calculation of the radii of curvature of the tooth profile of the gear r_{g} and that of the pinion r_{p}, the following empirical formulas are used:

$$r_{\text{g}} = l \cdot (1 + k_{\text{rg}}) \tag{1.8}$$

$$r_{\text{p}} = l \cdot (1 + k_{\text{rp}}) \tag{1.9}$$

The actual value of the factor k_{rp} should fulfill the inequality $k_{\text{rp}} \geq 0$. However, it is practical to set the factor k_{rp} equal to zero; then, the equality $r_{\text{p}} = l$ is observed. The factor k_{rg} is within the range $k_{\text{rg}} = 0.03\ldots0.10$.

The radius of the outside circle of the pinion $R_{\text{o.p}}$ is calculated from the following formula:

$$R_{\text{o.p}} = R_{\text{p}} + (1 - k_{\text{po}}) \cdot l \tag{1.10}$$

The addendum factor k_{po} of the pinion depends on the following:

(a) The actual value of the pressure angle, ϕ_{t}
(b) The absolute dimensions of the gear pair
(c) The accuracy of machining
(d) Conditions of lubrication

It is common practice to set the pinion addendum factor k_{po} in the range:

$$k_{\text{po}} = 0.1 \div 0.2 \tag{1.11}$$

The radius of the root circle of the pinion $R_{\text{f.p}}$ is calculated from the equation:

$$R_{\text{f.p}} = R_{\text{p}} - a_{\text{g}} - \delta \tag{1.12}$$

In Eq. (1.12), the following are designated:

a_{g}	as the dedendum of the mating gear ($a_{\text{g}} = (0.1\ldots0.2) \cdot l$)
δ	as the radial clearance in the gear pair ($\delta = l \cdot k_{\text{po}}$)

[8] It is instructive to note here that the equality $l = r_{\text{N}}$ is observed.

It is proven to be practical to set the fillet radius ρ_p in the range $\rho_p = 0.3 \cdot l$. The radius of the root circle of the gear $R_{f.g}$ is given as follows:

$$R_{f.g} = C - R_{o.p} \tag{1.13}$$

The radius of the outer circle of the gear $R_{o.g}$ is calculated from the expression:

$$R_{o.g} = R_g + a_g \tag{1.14}$$

The corner of the gear tooth addendum needs to be rounded with a radius ρ_g, the actual value of which is less than the fillet radius ρ_p of the pinion ($\rho_g < \rho_p$).

The following relations among the design parameters of a Novikov gear pair are recommended by Dr. M.L. Novikov in his study [6]: $r_p = l$, $r_g \leq 1.10 \cdot r_p$, $\rho_p = 0.3 \cdot l$, $m_t/l = 0.8$, $t_p/t_g = 1.5$, $\phi_t = 30°$, $\lambda = 60\ldots80°$ ($\psi = 10\ldots30°$), and circular pitch of the teeth $p = t_g + t_p + B$, where backlash $B = 0.2\ldots0.4$ mm.

The effective face width F_{pa} in the gear pair is given by:

$$F_{pa} = (1.1 \div 1.2) \cdot p \cdot \tan\lambda \tag{1.15}$$

For a preliminary analysis of a Novikov gear pair, the following empirical expression returns a practical value for the displacement l:

$$l = (0.05 \div 0.20) \cdot R_p \tag{1.16}$$

The performance of a parallel-axes Novikov gear pair strongly depends on the following three design parameters:

- The displacement l
- The transverse pressure angle ϕ_t
- The lead angle λ

It should be stressed here one more time that a smooth rotation of the driven shaft under a uniform rotation of the driving shaft is possible only if the transverse contact ratio of the Novikov gear pair is equal to zero ($m_p \equiv 0$) and the face contact ratio is greater than one ($m_t = m_F > 1$).

The application of a Novikov gear pair (namely, that of Novikov gearing of N_{by}–mesh, in particular), featuring geometries of the tooth profiles known so far, makes it possible to increase the contact strength of the gear teeth by up to 2.0 ÷ 2.1 times and the bending strength by up to 1.3 ÷ 1.5 times compared to involute helical gearing. The friction losses are up to 2.0 ÷ 2.5 less, and the tooth wear is 3 ÷ 4 times less in Novikov gearing [7]. All these application data are obtained for Novikov gearing with the hardness of the tooth surfaces in the range up to *HB* 350. During the years that Novikov gearing was actively being investigated, Novikov gearing with harder tooth flanks was not profoundly investigated.

Fig. 1.6 Gearbox manufactured by Westland Helicopters, Ltd. (After Astridge, D. G., et al., "Tribology of High Conformity Gears," *Institution of Mechanical Engineers*, 1987, pp. 819–825)

The application of Novikov gearing makes the weight reduction of gearboxes possible by up to 1.3 times (on average).

A uniform rotation of the shafts in Novikov gearing is only attained due to the face overlap of the gear teeth. Geometrically, meshing of the gear teeth in a transverse cross section is instantaneous.

Geometrically, the functional portions of the tooth flanks of the gear and the mating pinion in Novikov gearing are represented by two conjugate helices, that is, by two spatial curves. Under the applied load, these portions spread over the helical strips along the helices.

The nonfunctional portions of the tooth flanks are not conjugate to one another. Moreover, they are not envelopes to one another. This gives more freedom to the gear designer to make the gear teeth stronger.

An example of a Novikov gear pair is illustrated in Fig. 1.1. This is a gear pair manufactured by Westland Helicopters, Ltd. and is implemented in the design of gear transmission, as shown in Fig. 1.6 [8].

In current terminology and designations, the calculation of the principal design parameters of a Novikov gear pair is considered in the study by Radzevich [3].

The approach disclosed in this chapter can be used to enhanc Novikov/conformal gear pairs that feature other geometries of the tooth profile in the transverse section of the gear pair.

1.4 High-Conformal Gearing

The condition under which a convex-to-concave contact between the tooth flanks of a gear and a mating pinion becomes feasible is the fundamental achievement of Dr. M.L. Novikov. Once the active tooth profiles in a parallel-axes involute gearing shrinks to a point (namely, to the involute tooth point), then a favorable contact between the tooth flanks $\mathcal{G}$ and $\mathcal{P}$ can be attained. A capability to accommodate for the manufacturing errors and for the displacements under an operating load is the only consideration when determining the geometry of the interacting tooth profiles $\mathcal{G}$ and $\mathcal{P}$ by Dr. M.L. Novikov.

1.4.1 Critical Degree of Conformity in Novikov Gearing

It was assumed by Dr. M.L. Novikov from the very beginning that a convex-to-concave contact between the tooth flanks of a gear and a mating pinion is "sufficient" for a significant increase of the bearing capacity of the contact area between the tooth flanks $\mathcal{G}$ and $\mathcal{P}$. A later analysis revealed that in the case of Novikov gearing, a convex-to-concave contact is "necessary" but "not sufficient" for a significant increase of the bearing capacity of parallel-axes gearing. A certain critical degree of conformity at the point of contact K of the tooth flanks $\mathcal{G}$ and $\mathcal{P}$ must be attained in order to make the convex-to-concave contact beneficial. An increase in the degree of conformity below its critical value – that is, from $\delta^{\mathrm{a}}_{\mathrm{cnf}}$ to $\delta^{\mathrm{b}}_{\mathrm{cnf}}$ – makes a limited increase of the bearing capacity of the gear pair possible. A low increase of the bearing capacity is because both the values of the degree of conformity (that is, $\delta^{\mathrm{a}}_{\mathrm{cnf}}$ and $\delta^{\mathrm{b}}_{\mathrm{cnf}}$) are smaller compared to its critical value $[\delta_{\mathbf{cnf}}]$, and, thus, the inequalities $\delta^{\mathrm{a}}_{\mathrm{cnf}} < [\delta_{\mathrm{cnf}}]$ and $\delta^{\mathrm{b}}_{\mathrm{cnf}} < [\delta_{\mathrm{cnf}}]$ are observed.

However, when the actual degree of conformity $\delta^{\mathrm{c}}_{\mathrm{cnf}}$ becomes larger than the critical value ($\delta^{\mathrm{c}}_{\mathrm{cnf}} > [\delta_{\mathrm{cnf}}]$), then even a small increase in the degree of conformity at the point of contact of the tooth flanks $\mathcal{G}$ and $\mathcal{P}$ causes a significant increase in the bearing capacity of the tooth flanks in Novikov gearing. Therefore, the substitution of a convex-to-convex contact between the tooth flanks in parallel-axes gearing with their convex-to-concave contact (as in Novikov gearing) is necessary but not sufficient for a significant increase in the power capacity of a parallel-axes gear pair. In addition to that, a certain critical degree of conformity, $[\delta_{\mathbf{cnf}}]$, at the point of contact between the tooth flank of the gear $\mathcal{G}$ and that of the mating pinion $\mathcal{P}$ must be exceeded.

Gearing for the degree of conformity $\delta_{\mathbf{cnf}}$ at the point of contact of the tooth flanks $\mathcal{G}$ and $\mathcal{P}$ that exceeds its critical value $[\delta_{\mathbf{cnf}}]$ – that is, the gearing for which the inequality $\delta_{\mathbf{cnf}} > [\delta_{\mathbf{cnf}}]$ is valid – is referred to as "high-conformal gearing."

The intuitively understood qualitative term "degree of conformity" can be quantified. For this purpose, a characteristic curve called the "indicatrix of conformity," $\mathrm{Cnf}_{\mathrm{R}}(\mathcal{G}/\mathcal{P})$, at the point of contact of the tooth flank of a gear $\mathcal{G}$ and that of its mating pinion $\mathcal{P}$ is commonly used [3, 9]. The indicatrix of conformity $\mathrm{Cnf}_{\mathrm{R}}(\mathcal{G}/\mathcal{P})$ is a planar, centrally symmetrical curve of the fourth order. The position vector of the point r_{cnf} of the indicatrix of conformity corresponds to the degree of conformity of the tooth flanks $\mathcal{G}$ and $\mathcal{P}$ in a corresponding direction through the contact point K [3, 9]. The smaller the radius $r_{\mathbf{cnf}}$, the larger is the degree of conformity at the point of contact of the interacting surfaces, and vice versa.

1.4.2 Minimum Required Degree of Conformity at the Point of Contact of Two Interacting Tooth Flanks

Favorable conditions of the point of contact of the tooth flanks of a gear and a pinion are the main anticipated advantage of a high-conformal gear pair. The higher the degree of conformity, the higher is the load-carrying capacity of the contacting tooth flanks. Therefore, a minimum possible mismatch in the curvature of the teeth of the gear and pinion is desired.

In reality, the tooth flanks of a gear and its mating pinion in a high-conformal gear pair are displaced from their desired position. The undesired displacements are mostly because of the following:

(a) Manufacturing errors
(b) The elastic deflections of the gear teeth, of the gear shafts, of the housing (that occur under the applied load), thermal expansions of the components, and so forth.

High-conformal gearing is sensitive toward tooth flank displacements.

To accommodate for the inevitable displacements, a certain degree of mismatch in the curvature of the teeth of the gear and pinion is required. An insufficiently small mismatch can be incapable of accommodating the displacements. However, as the mismatch increases, the contact stresses increase as well. High contact stress may lead to various forms of surface failures such as heavy wear, pitting, or scuffing damage. Therefore, a minimum required degree of mismatch in the curvature of the teeth of the gear and pinion is necessary to be determined. Otherwise, one of two scenarios could be observed.

First, the gear pair is capable of absorbing the inevitable displacements of the tooth flanks, but the degree of conforming of the contacting tooth flanks is not sufficient for the high load-carrying capacity of the gear pair.

Second, the gear pair features a sufficient degree of conformity of the tooth flanks but is not capable of accommodating the tooth flank displacements.

In both cases, the gear pair has no chance of being successfully used in practice.

Shown in Fig. 1.7 is a three-dimensional (3D) plot of the function $\delta_{\mathrm{cnf}} = \delta_{\mathrm{cnf}}\left(k, \overline{K}\right)$. Fig. 1.7 relates to the cases of the convex-to-concave contact of the tooth flanks of the gear $\mathcal{G}$ and the pinion $\mathcal{P}$.

The performed analysis of the 3D plot allows for drawing the following conclusions.

The sections of the surface $\delta_{\mathrm{cnf}} = \delta_{\mathrm{cnf}}\left(k, \overline{K}\right)$ by planes $k_i = \mathrm{Const}$ (see Fig. 1.7) are represented by curves that have asymptotes. For a particular curve $k_i = \mathrm{Const}$, shown in Fig. 1.7 in bold line, the axis $\boldsymbol{\delta}_{\mathbf{cnf}}$ and the straight line $\boldsymbol{\delta}_{\mathbf{cnf}} = 1$ are the asymptotes.

The greatest possible degree of mismatch in the curvature of the teeth of the gear and pinion corresponds to the parameter $\overline{K} \to -\infty$. The interval of alteration to the parameter $\overline{K}$ starting from $-\infty$ and going up to approximately $\overline{K} = -2$ is convenient to accommodate any desired displacement of the tooth flanks $\mathcal{G}$ and $\mathcal{P}$ from their

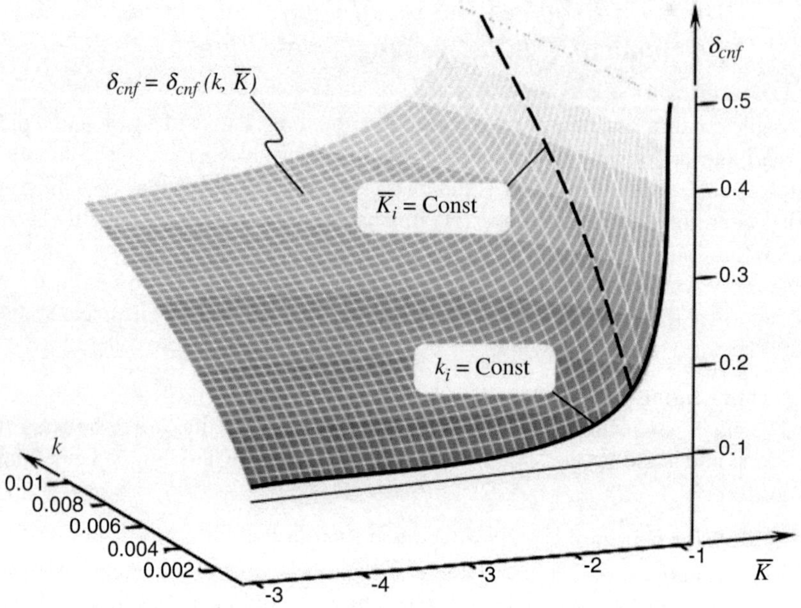

Fig. 1.7 A three-dimensional plot of the function $\delta_{\mathrm{cnf}} = \delta_{\mathrm{cnf}}\left(k, \overline{K}\right)$ constructed for the convex-to-concave kind of contact between the tooth flanks of a gear $\mathcal{G}$ and its mating pinion $\mathcal{P}$ in a conformal gear pair

correct configuration. However, within the interval $(-\infty < \overline{K} < -2)$ of alteration in the $\overline{K}-$ parameter, an increase of the degree of conforming of the tooth profiles $\mathcal{G}$ and $\mathcal{P}$ is negligibly small. Within this interval of the $\overline{K}-$ parameter, the load-carrying capacity of a conformal gear pair approximately remains in the same range. Therefore, just the convex-to-concave contact between the tooth flanks of a gear and its mating pinion provides a limited improvement in the load-carrying capacity of a gear pair. Being convex-to-concave, an additional requirement needs to be fulfilled in order to get not just conformal gearing but also high-conformal gearing.

On the other hand, even a small change to the actual value of the $\overline{K}-$ parameter within the interval $-2 < \overline{K} < -1$ results in a significant increase of the degree of conformity of the teeth flanks $\mathcal{G}$ and $\mathcal{P}$. This immediately entails a corresponding increase in the load-carrying capacity of the gear pair.

In the example considered above, the value of the $\overline{K}-$ parameter (that is, the value of $\overline{K} \approx -2$) can be referred to as its critical value $\overline{K}_{\mathrm{cr}}$ (or "threshold," in other words). This allows for distinguishing between just conformal gearing (for which $-\infty < \overline{K} < \overline{K}_{\mathrm{cr}}$) from high-conformal gearing (for which $\overline{K}_{\mathrm{cr}} \leq \overline{K} < -1$).

Without going into the details of the analysis, it is evident that gears for high-conformal gearing require tighter tolerances for any possible displacements of the tooth flanks $\mathcal{G}$ and $\mathcal{P}$ from their desirable location and orientation. Otherwise, there could be no future for the applications of high-conformal gear systems.

Based on the results of the performed analysis, the following statement is drawn up: conformal gearing and high-conformal gearing meet all the three fundamental laws of gearing [3, 9, 10]. Gearing of both the gear systems is capable of smoothly transmitting an input uniform rotation. As a consequence, both conformal gearing and high-conformal gearing have the following features:

- The transverse contact ratio is identical to zero ($m_p \equiv 0$).
- The total contact ratio m_t is equal to the face contact ratio m_F and is greater than one ($m_t = m_F > 1$).
- The tooth profile of one member of the gear pair is convex, whereas that of the mating gear is concave.
- The convex tooth profile of one member of the gear pair is entirely located within the interior of the boundary $N-$ circle, whereas the concave tooth profile of another member of the gear pair is entirely located within the exterior of the boundary $N-$ circle.
- The difference between the magnitudes of the radius of curvature of the concave tooth profile and that of the convex tooth profile in the gear pair is equal to or smaller than the given threshold beyond which a higher conformity of the interacting tooth profiles contributes much to the bearing capacity of the gear pair.

The principal difference between Novikov gearing and Wildhaber gearing is clearly illustrated in Fig. 1.8.

In Novikov gearing, the common unit perpendicular $\mathbf{n}_g$ is always aligned with the instantaneous line of action LA_{inst}, as illustrated in Fig. 1.8a. The gear tooth flank $\mathcal{G}$ is properly configured in relation to the boundary $N-$ circle of a radius r_N – and, thus, the fundamental laws of gearing are fulfilled.

In Wildhaber gearing, the common unit perpendicular $\mathbf{n}_g$ is not aligned with the instantaneous line of action LA_{inst}, as illustrated in Fig. 1.8b. The gear tooth flank $\mathcal{G}^*$ is improperly configured in relation to the boundary $N-$ circle in a gear pair – and, thus, the fundamental laws of gearing are violated.

Helical gearing with a circular-arc tooth profile proposed by Dr. E. Wildhaber [11] does not meet the three fundamental laws of gearing. An insufficient understanding of gearing of this particular kind clearly follows from the paper by T. Allen [12]. The comparison between Novikov gearing [4] and Wildhaber gearing [11] that started decades ago by Chironis [13] is incomplete. To the best possible extent, the comparison was later accomplished in my own work as summarized in the studies by Radzevich [3, 9, 10, 14, 15].

Two points need to be mentioned here.

First, neither gears for Novikov gearing nor gears for high-conformal gearing can be finish cut by the continuous indexing method (generating method) of gear machining. That is, the gears cannot be hobbed, shaped, or ground by a worm grinding wheel.

This is because the fundamental laws of gearing are not fulfilled in the gear machining mesh [9]. Only R_e- surfaces can be accurately generated in the continuous indexing method of gear machining. Cutting tools for machining gears for both

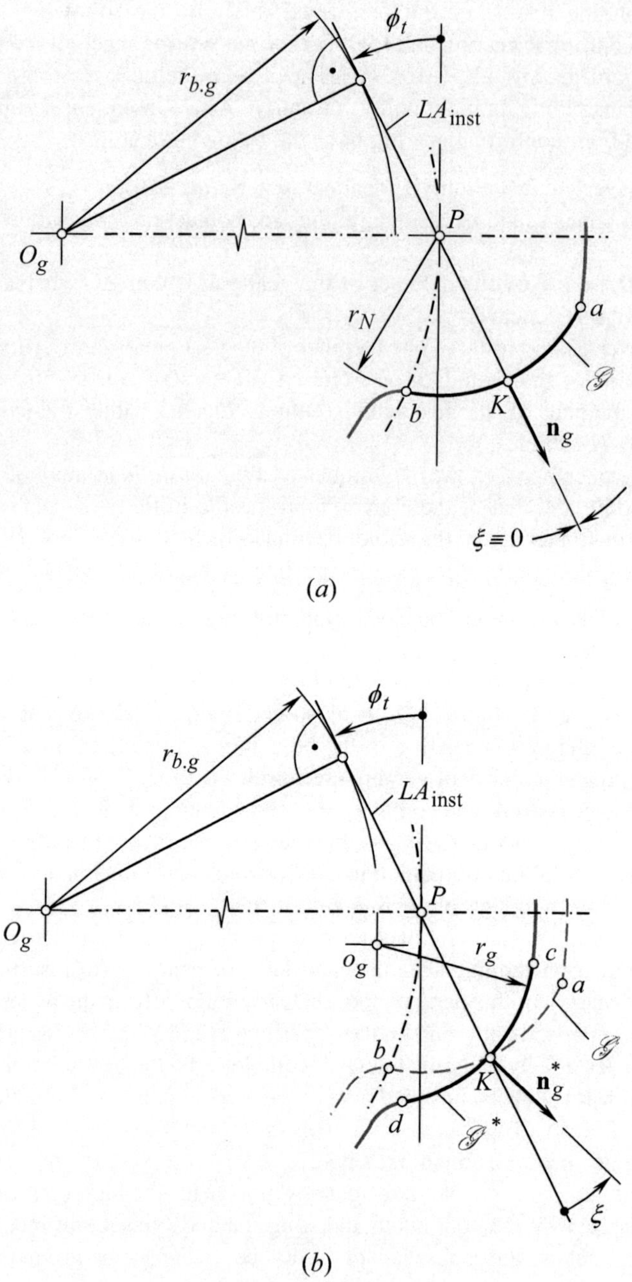

Fig. 1.8 Correct $\mathcal{G}$ (**a**) and incorrect $\mathcal{G}^*$ (**b**) configuration of the circular-arc tooth profile of a gear in Novikov gearing (the common perpendicular $\mathbf{n}_g$ is aligned with the instantaneous line of action LA_{inst}) and in Wildhaber gearing (the common perpendicular $\mathbf{n}_g$ does not align with the instantaneous line of action LA_{inst})

Novikov gearing and high-conformal gearing are considered in the studies by Radzevich [15] and [10].

Second, neither tooth profile modification nor longitudinal modification of the tooth flanks of a gear and a mating pinion is permissible to Novikov gearing as well as to high-conformal gearing in a more general case.

1.5 Conformal Gearing with Intermediate Balls

The theoretical analysis of the concept of Novikov/conformal gearing performed in the previous sections of this chapter provides the readers with systemized knowledge of the field and is helpful to gear designers. Of course, not all potential applications of the obtained results are discussed here. The main goal of this chapter is to illustrate a few more examples of implementation of the theory, which are slightly beyond a straight interpretation of it. Certain examples can illustrate a possible way of evolution of the concept of Novikov/conformal gearing.

A brief look at the schematic of a parallel-axes Novikov/conformal/high-conformal gear pair with the boundary $N-$ circle that overlaps the gears inspired the author[9] to investigate the possibility of replacement of sliding between the gear tooth flanks with rolling of the interacting elements. Potentially, such a replacement could be useful as the friction losses in rolling are commonly smaller compared to that those in sliding.

As illustrated in Fig. 1.9, in a section by a transverse plane, both the driving pinion and the driven gear feature a concave tooth profile (see Fig. 1.9a). For simplicity, but without loss of generality, the radius of curvature ρ_g of the gear tooth flank $\mathcal{G}$ and the radius of curvature ρ_p of the pinion tooth flank $\mathcal{P}$ are set equal to the radius r_N of the boundary $N-$ circle. The line of action LA forms a transverse pressure angle $\boldsymbol{\phi}_t$ perpendicular to the centerline $\mathbf{Ȼ}$ in the gear pair.

The balls of a radius $r_{ball} = r_N$ interact with the gear and the pinion tooth flanks $\mathcal{G}$ and $\mathcal{P}$ at two points that are labeled as K_g and K_p, respectively. The parallel-axes conformal gear pair with intermediate balls features two pseudo-paths of contact, $P_{pc.g}$ and $P_{pc.p}$. In the case under consideration, the pseudo-paths of contact are two straight lines through the contact points K_g and K_p, which are perpendicular to the plane of drawing in Fig. 1.9a.

When the gears rotate, the balls travel in the axial direction of the gear pair parallel to the axis of instantaneous rotation P_{ln} (the axis of instantaneous rotation P_{ln} is a straight line through the pitch point P and is perpendicular to the plane of drawing in Fig. 1.9a). At this time, when the balls are situated within the face width F_{pa} of the gear pair, they are solely supported by the tooth flanks $\mathcal{G}$ and $\mathcal{P}$ of the gear and the mating pinion, respectively. After a ball is passed up through a channel formed by two helical surfaces $\mathcal{G}$ and $\mathcal{P}$, it is returned to its initial position by the

[9] On March 2, 2018.

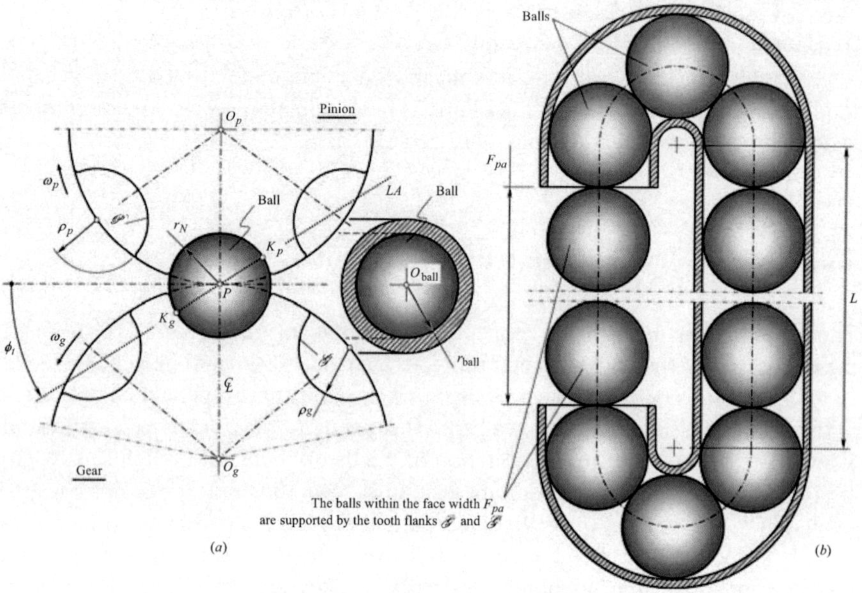

Fig. 1.9 Conformal gearing with intermediate balls: (**a**) schematic and (**b**) a device for returning the balls

device shown in Fig. 1.9b. The device features a gap between the ends, the width of which is equal to or slightly exceeds the face width F_{pa}. The length L in Fig. 1.9b can be expressed in terms of the ball diameter $2r_N$, the face width F_{pa}, and the pitch helix ψ. Other means to return the balls can be designed as well.

The forces that act over a ball when it is supported by the tooth flanks $\mathcal{G}$ and $\mathcal{P}$ are analyzed in Fig. 1.10. Here, in Fig. 1.10, the driving pinion tooth flank $\mathcal{P}$ pushes the ball with a certain resultant force $\mathbf{F}_\Sigma$. This force is resolved onto a normal component $\mathbf{F}_n$ and a tangential component $\mathbf{F}_t$. The components $\mathbf{F}_n$ and $\mathbf{F}_t$ of the resultant force $\mathbf{F}_\Sigma$ are considered (normal/tangential) in relation to the ball.

In order to ensure straight motion of a ball parallel to the axis of instantaneous rotation P_{ln}, the design parameters of the tooth flanks $\mathcal{G}$ and $\mathcal{P}$ of the gear and the mating pinion, respectively, have to be designed so as to form the so-called "pocket" for a ball that is currently interacting with the tooth flanks $\mathcal{G}$ and $\mathcal{P}$. This can be done considering the local geometry of the interacting surfaces $\mathcal{G}$ and $\mathcal{P}$ with the ball.

Every point of a ball is a convex umbilic point, for which the Dupin indicatrix Dup (ball) can be graphically interpreted as a circle of a certain radius. The circle centers at the contact point with the gear (from one side) and with the pinion (from the opposite side). Every point of the helical tooth flanks $\mathcal{G}$ and $\mathcal{P}$ is a saddle-like point (hyperbolic point) that features a section by a normal plane of both kinds, that is, convex and concave plane sections are observed (depending on the actual configuration of the section plane).

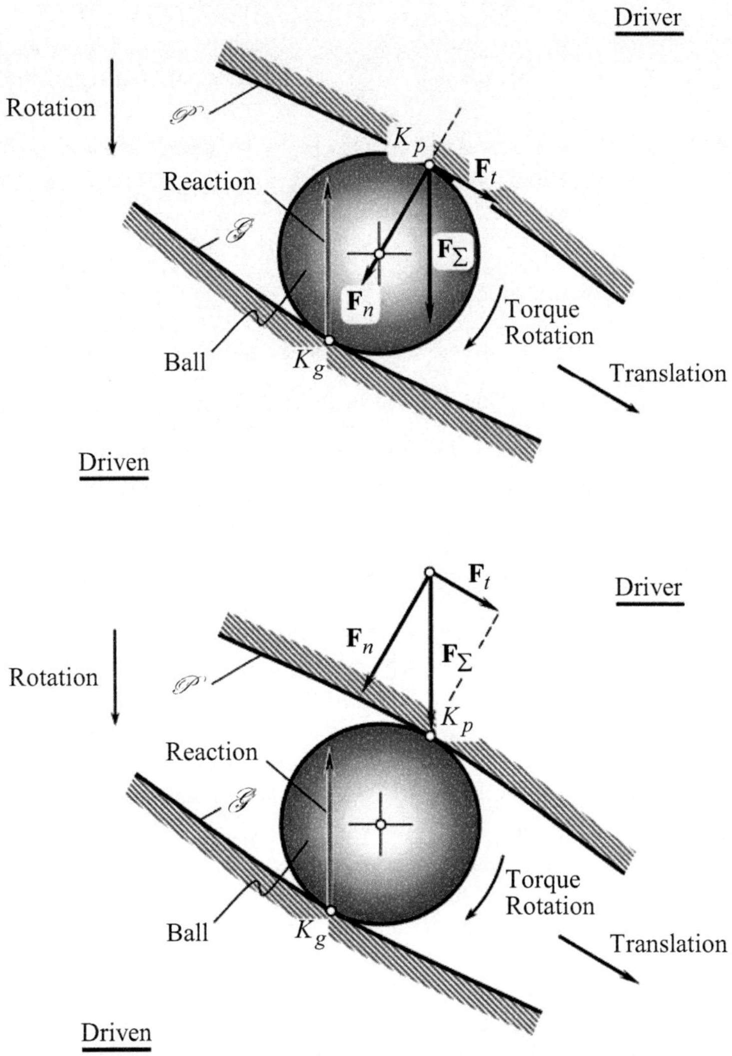

Fig. 1.10 Forces that act in the gear mesh in conformal gearing with intermediate balls

The gear and the mating pinion tooth flanks $\mathcal{G}$ and $\mathcal{P}$, respectively, have to be designed so as to ensure a concave normal plane section of both surfaces $\mathcal{G}$ and $\mathcal{P}$ in the direction that the ball travels. Under such a scenario, the ball is locked in the pocket and travels exactly parallel to the axis of instantaneous rotation P_{ln}. The said is illustrated in Fig. 1.11.

The Dupin indicatrix Dup($\mathcal{G}$) is constructed at point K_{g} of the gear tooth flank $\mathcal{G}$ (see Fig. 1.11a). The point K_{g} is the point of contact of the ball with the gear tooth flank $\mathcal{G}$. The sector with positive ($\rho_{\text{g}} > 0$) and negative ($\rho_{\text{g}} < 0$) values of the radius

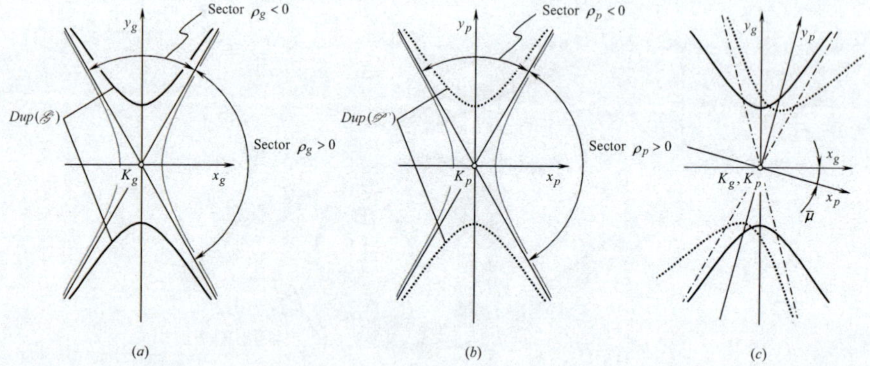

Fig. 1.11 Determination of the permissible direction of the pseudo-paths of contact $P_{pc.\,g}$ and $P_{pc.\,p}$ in conformal gearing with intermediate balls: (**a**) Dupin indicatrix Dup$(\mathcal{G})$ at point K_g of the gear tooth flank $\mathcal{G}$, (**b**) Dupin indicatrix Dup$(\mathcal{P})$ at point K_p of the pinion tooth flank $\mathcal{P}$, and (**c**) the Dupin indicatrices Dup$(\mathcal{G})$ and Dup$(\mathcal{P})$ overlap one another

of curvature ρ_g of the sections of the tooth flank $\mathcal{G}$ by normal planes are determined by means of the Dupin indicatrix Dup$(\mathcal{G})$.

Similarly, a Dupin indicatrix Dup$(\mathcal{P})$ is constructed at point K_p of the pinion tooth flank $\mathcal{P}$ (see Fig. 1.11b). The point K_p is the point of contact of the ball with the pinion tooth flank $\mathcal{P}$. The sector with positive ($\rho_p > 0$) and negative ($\rho_p < 0$) values of the radius of curvature ρ_p of the sections of the tooth flank $\mathcal{P}$ by normal planes are determined by means of the Dupin indicatrix Dup$(\mathcal{P})$.

Ultimately, the permissible directions of the pseudo-paths of contact $P_{pc.g}$ and $P_{pc.p}$ in Novikov/conformal and high-conformal gearing with intermediate balls are determined as illustrated in Fig. 1.11c. Viewing along the straight line K_gK_p, the tooth flanks $\mathcal{G}$ and $\mathcal{P}$ are turned in relation to one another through an angle $\overline{\mu}$. The angle $\overline{\mu}$ is similar to the angle of the local relative orientation μ of two smooth regular surfaces at the point of their contact[10] [3].

It is clear from the analysis in Fig. 1.11c (together with Fig. 1.10) that when the balls are located within the face width F_{pa} of the gear pair, they are spaced from one another at a distance that equals to the axial pitch p_x of the tooth helix.

Determination of the instantaneous kinematics of a ball inside the "pocket" is illustrated in Fig. 1.12.

The Dupin indicatrices Dup$(\mathcal{G})$ and Dup$(\mathcal{P})$ at points K_g and K_p of the tooth flanks $\mathcal{G}$ and $\mathcal{P}$, respectively, are employed for the derivation of the instantaneous

[10]The difference between the angles μ and $\overline{\mu}$ is solely because of a plane within which the angles are measured. The angle of the local relative orientation, μ, of two smooth regular surfaces $\mathcal{G}$ and $\mathcal{P}$at the point of their contact is measured between the principal directions within a common tangent plane to the surfaces. The angle $\overline{\mu}$ is measured between the projections of the principal directions onto a plane that is perpendicular to the straight line K_gK_p. This plane is parallel to a common tangent plane in the contact "ball-to-$\mathcal{G}$." It is also parallel to a common tangent plane in the contact "ball-to-$\mathcal{P}$."

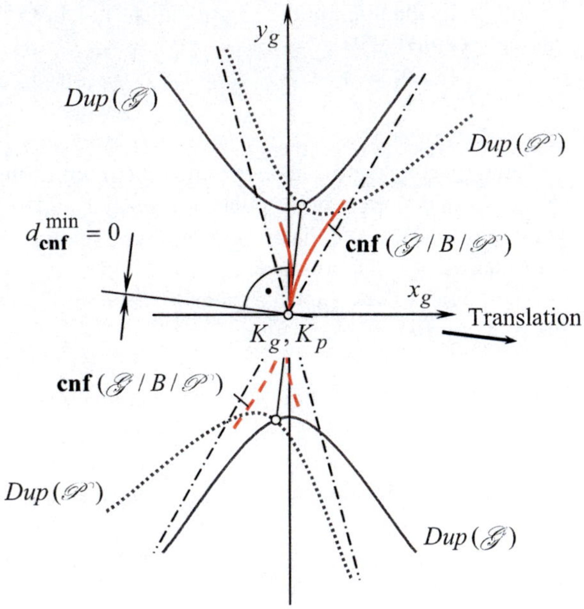

Fig. 1.12 Instantaneous kinematics of a ball inside the "pocket"

kinematics of a ball inside the "pocket." As that same ball makes contact with each of the tooth flanks $\mathcal{G}$ and $\mathcal{P}$, the Dupin indicatrix of the ball Dup(B) can be eliminated from further analysis.

Consider a plane π_g through the contact point K_g, which is perpendicular to the common perpendicular $\mathbf{n}_g$. The Dupin indicatrix Dup($\mathcal{G}$) is constructed within the plane π_g. Another plane π_p is a plane through the contact point K_p, which is perpendicular to the common perpendicular $\mathbf{n}_p$. The Dupin indicatrix Dup($\mathcal{P}$) is constructed within the plane π_p. The unit normal vectors $\mathbf{n}_g$ and $\mathbf{n}_p$ are along a line through the center of the ball. Therefore, the planes π_g and π_p are parallel to one another.

The third plane π_b is an arbitrary plane that is parallel to the planes π_g and π_p. In particular, the plane π_b can be a plane through the center of the ball. Furthermore, both Dupin indicatrices Dup($\mathcal{G}$) and Dup($\mathcal{P}$) are projected onto the plane π_b. The "composite indicatrix of conformity, **cnf** $(\mathcal{G}/B/\mathcal{P})$" is constructed on the premise of the projections of the Dupin indicatrices Dup($\mathcal{G}$) and Dup($\mathcal{P}$) onto the plane π_b.

The direction of translation of the ball inside the "pocket" can be specified in terms of the minimum diameter of the composite indicatrix of conformity **cnf** $(\mathcal{G}/B/\mathcal{P})$, constructed at points K_g and K_p of the tooth flanks $\mathcal{G}$ and $\mathcal{P}$.

In polar coordinates, the point of the Dupin indicatrix Dup($\mathcal{G}$) at a point of the tooth flank $\mathcal{G}$ can be specified as:

$$r_{dup.g} = r_{dup.g}(\varphi) \quad (1.17)$$

Similarly, the point of the Dupin indicatrix $\mathrm{Dup}(\mathcal{P})$ at a point of the tooth flank $\mathcal{P}$ can be specified as:

$$r_{\mathrm{dup.p}} = r_{\mathrm{dup.p}}(\varphi, \overline{\mu}) \tag{1.18}$$

Here, $r_{\mathrm{dup.g}}$ is the distance at which the point of the Dupin indicatrix $\mathrm{Dup}(\mathcal{G})$ is remote from the origin of the polar coordinated, and $r_{\mathrm{dup.p}}$ is the distance at which the point of the Dupin indicatrix $\mathrm{Dup}(\mathcal{P})$ is remote from the origin of the polar coordinated; φ is the polar angle.

The point of the composite indicatrix of conformity $\mathbf{cnf}\,(\mathcal{G}/\mathcal{B}/\mathcal{P})$ constructed at points K_{g} and K_{p} is specified as:

$$r_{\mathbf{cnf}}(\varphi, \overline{\mu}) = r_{\mathrm{dup.g}}(\varphi) - r_{\mathrm{dup.p}}(\varphi, \overline{\mu}) \tag{1.19}$$

The direction of translation of the ball is perpendicular to a straight line along which the minimum diameter $d_{\mathbf{cnf}}^{\min}$ is measured. In the particular case under consideration, this straight line is a straight line through the points of intersection of the Dupin indicatrices $\mathrm{Dup}(\mathcal{G})$ and $\mathrm{Dup}(\mathcal{P})$. The latter yields certain simplifications.

The main advantages of the discussed concept of Novikov/conformal and high-conformal gearing with intermediate balls are summarized immediately below:

- The tooth addendum can be extremely short (reasonably, the shortest possible), and the teeth can be made short and thick; the bending strength problem can be reduced or even eliminated due to this.
- The ball diameter can be reduced, that is, the radius r_{N} of the boundary $N-$ circle can be a smaller value compared to that in conventional Novikov/conformal and high-conformal gearing: the smaller the radius r_{N}, the less is the friction, and vice versa.
- Besides there are two pseudo-paths of contact, $P_{\mathrm{pc.g}}$ and $P_{\mathrm{pc.p}}$, a gearing of the proposed design that is neither a $N_{\mathrm{bf}}-$ type of the conformal gear pair nor a $N_{\mathrm{by}}-$ type of the conformal gear pair; this is a novel type of conformal/high-conformal gearing; the pseudo-paths of contact $P_{\mathrm{pc.g}}$ and $P_{\mathrm{pc.p}}$ are not identical to the pseudo-paths of contact $P_{\mathrm{pc.bf}}$ and $P_{\mathrm{pc.bf}}$ in conventional conformal/high-conformal gearing with two pseudo-paths of contact.
- The applied load is shared between contact points K_{g} and K_{p}; therefore, the contact strength can be doubled.
- As the balls are not rigidly connected either to a gear nor to a mating pinion, the accuracy requirements can be drastically reduced.
- The smaller the ball diameter, the smaller is the axial dimension required for the gear drive.
- The gear drive is geometrically accurate, that is, both the base pitches of a gear and of a mating pinion are equal to the operating base pitch of the gear pair (like in involute gearing).
- This gearing is designed to transmit a high torque at low and average rotations.

- When necessary, the friction between the balls and the tooth flanks of a gear and of a mating pinion can also be incorporated into the analysis.
- The tooth flanks can be hardened and precisely ground after the hardening (hobbing can be used for roughing as a preliminary machining; finish grinding is performed either by disk-type or by end-type grinding wheels).
- The tooth flanks of a gear and a mating pinion form a "pocket" for the ball; when the gears rotate, the "pocket travels axially"; the ball travels together with the pocket.
- When torque is transmitted by a ball, the ball is locked in the pocket that features negative curvatures of the interacting surfaces at both contact points.
- The geometry of the pipe (of the "channel") can be optimized: centrifugal forces, acting over the balls; wear of the channel walls; semi-cubic parabola, and so forth.
- After the gears are worn down, the balls can be replaced with balls of a larger diameter.
- Production of precision balls is well-established in the industry: in the production of ball bearings, the balls can be machined to a high level of accuracy.
- The gears can be machined to a high level of accuracy based on the experience gained in production screws for "ball screw-and-nut" pairs.
- No separator is required for the balls.
- One or two balls simultaneously are engaged in the transmission of power/rotation/torque. If two are used, then the axial distance between the two balls is equal to the axial pitch in the gear pair. The rest of the balls are located close to one another.
- The forces act only within the plane of action PA. No forces act perpendicular to the plane of action PA as there is no motion of the balls in a direction perpendicular to the plane of action. (The direction of the acting forces can be altered if the friction forces are taken into consideration.)

A more in-detail analysis can reveal the additional advantages of conformal/high-conformal gearing with intermediate balls.

1.6 Conclusion

This chapter deals with Novikov gearing, or, more generally, with Novikov/conformal, and high-conformal gearing. The key features of Novikov gearing are discussed, and the kinematics and geometry of gearing of this kind are considered.

The differences between Novikov gearing and Wildhaber gearing are described in this chapter, making a case for why the two systems should "not" be combined as Wildhaber–Novikov gearing or W-N gearing.

It is stressed here that a poor understanding of the kinematics and geometry of the novel kind of gearing proposed by Dr. M.L. Novikov is the root cause of loosely combining Novikov gearing with Wildhaber gearing. These two types of gearings must be only considered separately from one another. The terms "Wildhaber–

Novikov gearing" and "W-N gearing" are incorrect by nature and thus must be eliminated from scientific communications on gearing.

It is a mistake to refer to the Novikov gear system as gears with a circular-arc tooth profile. Novikov gearing is not a kind of gearing with a circular-arc tooth profile like gears in the Wildhaber gear system. Novikov gearing is a reduced kind of involute gearing. In Novikov gearing, the functional involute tooth profile is shrunk to a point. This point is referred to as the "involute tooth point." The rest of the tooth profile is inactive and thus can be shaped with no constraints imposed by the fundamental laws of gearing.

The acting standards on Novikov gearing (in Russia, China, and elsewhere) and on gear cutting tools for cutting gears in Novikov gearing are generally incorrect. Gears in Novikov gearing cannot be cut (finish cut) by hobs, shape cutters, shavers, worm grinding wheels, and others. No generating finish machining of gear tooth flanks are permissible. Only disk-type mill cutters, disk-type grinding wheels, and so forth, can be used for this purpose.

High-conformal gearing has a huge potential for application in high-power-density gear transmission systems and in low-noise gear transmissions.

References

1. Shotter, B.A.: Experiences with conformal/W-N gearing, Proceedings of World Congress on Gearing, Paris, France, June 22–June 24, 1977, pp. 527–540 (1977)
2. Shotter, B.A.: The Lynx Transmission and Conformal Gearing, SAE Technical Paper 781041 (1978)
3. Radzevich, S.P.: Theory of Gearing: Kinematics, Geometry, and Synthesis, 2, revised and expanded edn. CRC Press, Boca Raton, FL (2018) 934 pages. [1st Edition: Radzevich, S.P., Theory of Gearing: Kinematics, Geometry, and Synthesis, CRC Press, Boca Raton, Florida, 2012, 760 pages]
4. Pat. No. 109,113, (USSR), Gearing and Cam Mechanisms Having Point System of Meshing, M.L. Novikov, National Classification 47h, 6; Filed: April 19, 1956, published in Bull. of Inventions No.10 (1957)
5. Novikov, M.L.: The principles of the geometric theory of point meshing of gearing for the purpose of transmitting of high power, Doctoral Thesis, Moscow, Zhukovskii Military Aviation Engineering Academy (MAEA) (1955)
6. Novikov, M.L.: Gearing that Features Novel Kind of Meshing. Zhukovskii Military Aviation Engineering Academy (MAEA), Moscow (1958) 186 pages
7. Cheng, D.X.: Chinese Standard, pp. 158–187. China Machine Press (1991)
8. Astridge, D.G., et al.: Tribology of high conformity gears, pp. 819–825. Institution of Mechanical Engineers (1987)
9. Radzevich, S.P.: Geometry of Surfaces: a Practical Guide for Mechanical Engineers, 2nd edn. Springer International Publishing (2019) 466 pages. [1st edition: Radzevich, S.P., Geometry of Surfaces: A Practical Guide for Mechanical Engineers, Wiley, 2013, 264 pages]
10. Radzevich, S.P.: High-Conformal Gearing: Kinematics and Geometry, 2nd edn. Elsevier, Amsterdam (2020) 506 pages. [1st edition: Radzevich, S.P., High-Conformal Gearing: Kinematics and Geometry, 1nd edition, CRC Press, Boca Raton, Florida, 2015, 332 pages]
11. Pat. No. 1,601,750, (USA). Helical Gearing./E. Wildhaber, Filed: November 2, 1923, published October 5, 1926

12. Allan, T.: Some aspects of the design and performance of Wildhaber-Novikov gearing. Proc. Inst. Mech. Engrs, Part I. **179**(30), 931–954 (1964/1965)
13. Chironis, N.: New tooth shape taking over? Design of Novikov Gears. In: Chironis, N. (ed.) Gear Design and Application, pp. 124–135. McGraw-Hill Book Company, New York (1967) 374p
14. Radzevich, S.P.: Dudley's Handbook of Practical Gear Design and Manufacture, 4th edn. CRC Press, Boca Raton (2021) 1144 pages. [Radzevich, S.P., Dudley's Handbook of Practical Gear Design and Manufacture, 3rd edition, CRC Press, Boca Raton Florida, 2017, 606 pages. Radzevich, S.P., Dudley's Handbook of Practical Gear Design and Manufacture, 2nd edition, CRC Press, Boca Raton Florida, 2012, 880 pages]
15. Radzevich, S.P.: Gear Cutting Tools: Science and Engineering, 2nd edn. CRC Press, Boca Raton (2017) 606 pages. [1st edition: Radzevich, S.P., Gear Cutting Tools: Fundamentals of Design and Computation, CRC Press, Boca Raton Florida, 2010, 786 pages

Bibliography

Coy, J.J., Townsend, D.P., Zaretsky, E.V.: Gearing, 69 pages. NASA Reference Publication 1152, AVSCOM, Technical Report 84-C-15 (1985)

Dyson, A., Evans, H.P., Snidle, R.W.: Wildhaber-Novikov circular arc gears: some properties of relevance to their design. Proc. Royal Soc. A Math. Phys Eng. Sci. **425**, 341–363 (1989)

Evans, H.P., Snidle, R.W.: Wildhaber-Novikov circular arc gears: Elastohydrodynamics. J. Tribol. **115**(3), 487–492 (1993)

Chapter 2
Meshing Theory for Abnormal Novikov Helical Gears

Yaping Zhao and Siyu Liu

2.1 Introduction

Since Euler put forward an involute tooth profile in 1765, involute gear transmission has been widely used. With the development of production, there are higher requirements for gears under a high load state. In 1922, Vickers, Bostock, and Bramley studied a gear with concave and convex tooth profile meshing and called it VBB. The gear had a large radius of curvature, and the contact strength was much larger than that of an involute gear. However, due to the small thickness of the concave tooth addendum, the gear teeth broke when used. In 1926, Wildhaber proposed the arc tooth profile, which also caused an accident.

In 1956, Novikov put forward a new circular-arc gear, which was cut by a reference rack tool. Under the same parameters, the radius of curvature of the circular-arc gear was 10 to more than 200 times larger than that of an involute helical gear. The contact stress of the tooth surface was greatly reduced, and the bearing capacity was greatly improved. The bearing capacity of a single-arc gear is higher than that of an involute gear, but the bending strength of the tooth root is lower than that of an involute gear. In order to improve the root strength of the circular-arc gear, the former Soviet Union put forward a common tangent double-circular-arc gear in the 1960s. This tooth profile was prone to pitting near the pitch line, and, in order to avoid pitting and improve the root strength, a graded double-circular-arc profile was proposed. Its bearing capacity was greatly improved than that of a single-arc gear. The circular-arc gear used in industry is processed by a hob, the normal tooth profile of the oblique rack cutter is an arc, and the processed gear does not have an arc tooth profile. Due to the widespread use of gear grinding machines, the strength of the hard involute tooth surface is higher than that of an arc gear, which cannot be ground, and, so, arc gears are not widely used.

Y. Zhao · S. Liu
Northeastern University, Shenyang, China

S. P. Radzevich, *Novikov/Conformal Gearing*,
https://doi.org/10.1007/978-3-031-10019-2_2

This chapter presents a new type of circular-arc gear with a circular tooth shape in its normal section. In the traditional sense, a circular-arc gear is enveloped by a rack cutter, the tooth profile of which is circular. The tooth surface of a gear is extremely complicated. Based on this, a mathematical model of the helical surface of a normal circular-arc gear is established.

The meshing principle of normal circular-arc gear transmission is analyzed. First, the coordinate system of the gear pair's relative motion is established, and the meshing function of the gear pair is deduced by coordinate transformation. The instantaneous contact point is determined by solving nonlinear equations based on common points and common normal conditions.

In this chapter, the calculation principle of a normal circular-arc gear is studied and numerical simulation of the conjugate region is carried out through concrete examples. The calculation methods of contact reference points, points on the tooth width boundary, and other contact points inside the boundary are determined. The specific meshing parameters of each point are obtained through numerical examples.

2.2 Equation and Normal Vector for a Normal Circular-Arc Helical Surface

The main aim of this chapter is to establish a mathematical model of the helical surface of a normal double-circular-arc cylindrical gear. First, the formation principle of the helical surface of the normal double-circular cylindrical gear is introduced. The helical surface equation of the concave and convex tooth profile of the gear is written by the spherical vector and circular vector coordinates. Finally, according to the knowledge of differential geometry, the corresponding unit normal vectors of the helical surfaces of two gears are obtained. Since each side of the gear tooth has concave and convex teeth, two points are in contact when they mesh. Because the research method of meshing of the two parts is the same, here, we choose to study the part that participates in meshing.

2.2.1 A Mathematical Model of the Left Convex Tooth Surface of Normal Circular-Arc Gear 1

The design is carried out based on the basic tooth profile parameters of a GB/T12759-1991 double-circular-arc cylindrical gear. The relative coordinate system for calculating the tooth surface equation of a double-circular-arc gear is established. Let the double-circular gear be located in the right-hand Cartesian coordinate system $\sigma_1\left\{\mathrm{O}_1;\ \vec{i}_1,\ \vec{j}_1,\ \vec{k}_1\right\}$. In this coordinate system, the axis direction of the double-circular-arc gear is the axis $\vec{k}_1$ and the center point on the axis of the gear is O_1. The spiral angle is set as β, where β is a positive value, and gear 1 is set as a right-handed

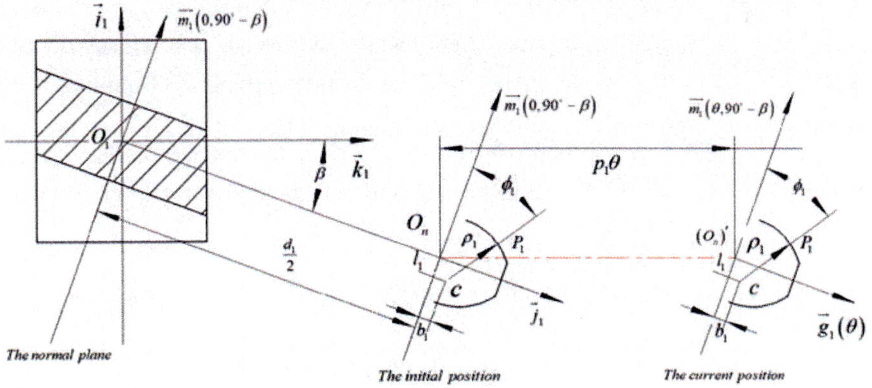

Fig. 2.1 A coordinate system diagram formed by a left convex tooth surface

gear. The working tooth profile on each side of a tooth of a double-arc gear is divided into two sections: a convex arc (outside the pitch circle) and a concave arc (inside the pitch circle), which are connected by a transition arc. The connecting arc is tangent to the convex tooth arc and intersects the concave tooth arc. For details, please refer to the study by Lu and Shang [1]. We can calculate its radius and center position according to geometric relations:

$$r_j = \frac{t^2 + n^2}{2(t\cos\delta_1 - n\sin\delta_1)}$$

where

$$t = 0.5\pi m + l_1 + l_2 - \rho_1\cos\delta_1 - \rho_2\cos\delta_2,\ n = \rho_1\sin\delta_1 + b_1 + \rho_2\sin\delta_2 - b_2$$

Figure 2.1 A right-hand gear. The tooth surface equation is deduced based on the left convex arc tooth profile.

The tooth surface is a spiral tooth surface formed by the spiral motion of the working arc tooth profile. In the coordinate system, the intersection point of the convex tooth symmetric axis and indexing (node) cylinder is O_n, and O_n is taken as the circle point. Using the knowledge of spherical vector function in differential geometry, the normal coordinate system $\sigma_n\left\{O_n; \vec{j}_1, \vec{m}_1, \vec{k}_1\right\}$ can be obtained. At the initial position, its basis vectors are $\vec{j}_1$ and $\vec{m}_1(0, 90° - \beta)$, respectively. The circular tooth profile is located on the surface to make a spiral motion to form the tooth surface. When the angle parameter of the gear rotating along the axis is θ, point O_n moves to point O_n' along the axis direction, and the axial distance between the two points is $\left|\overrightarrow{O_nO_n'}\right| = p_1\theta_1$, where p_1 is the spiral parameter of the normal double-circular-arc gear, $p_1 = \frac{d_1}{2}\cot\beta$. After finishing the spiral motion, according to the knowledge of sphere vector function and circle vector function, it can be concluded that the vector $\vec{j}_1$ of the unit basis becomes $\vec{g}_1(\theta_1)$ and the vector $\vec{m}_1(0, 90° - \beta)$

of the unit basis becomes $\vec{m}_1(\theta, 90° - \beta)$. According to the geometric relationship shown in the figure, in the right-handed Cartesian coordinate system $\sigma_1\left\{O_1; \vec{i}_1, \vec{j}_1, \vec{k}_1\right\}$, point P_1 is any point on the working tooth profile of the convex teeth of the normal double-arc gear and the directed angle between the path vector $\overrightarrow{cP_1}$ and the $\vec{m}_1(\theta_1, 90° - \beta)$ axis is set as ϕ_1. The radial diameter of the center O_1 to any point P_1 of the working arc is the equation of the tooth surface of the working arc (left convex tooth):

$$\left(\vec{r}_1\right)_1 = \overrightarrow{O_1P_1} = \overrightarrow{O_1O_n} + \overrightarrow{O_nO_n'} + \overrightarrow{O_n'P_1} = (\rho_1 \cos\phi_1 - l_1)\cos\beta\,\vec{e}_1(\theta_1)$$
$$+\left(\rho_1 \sin\phi_1 + b_1 + \frac{d_1}{2}\right)\vec{g}_1(\theta_1) + [(\rho_1 \cos\phi_1 - l_1)\sin\beta$$
$$+p_1\theta_1]\vec{k}_1 = x_1\vec{i}_1 + y_1\vec{j}_1 + z_1\vec{k}_1 \tag{2.1}$$

where

$$x_1 = (\rho_1 \cos\phi_1 - l_1)\cos\beta\cos\theta_1 - \left(\rho_1 \sin\phi_1 + b_1 + \frac{d_1}{2}\right)\sin\theta_1$$
$$y_1 = (\rho_1 \cos\phi_1 - l_1)\cos\beta\sin\theta_1 + \left(\rho_1 \sin\phi_1 + b_1 + \frac{d_1}{2}\right)\cos\theta_1, z_1$$
$$= (\rho_1 \cos\phi_1 - l_1)\sin\beta + p_1\theta_1$$
$$\phi_1 \in \left(0, \frac{\pi}{2}\right)$$

It can be seen from the Eq. (2.1) of the helical surface of the normal double-circular-arc gear that all parameters p_1, ρ_1, l_1, b_1, and β in the equation are constant values, whereas parameters θ_1 and ϕ_1 are variables and serve as the binary coordinates of the tooth surface. Therefore, the helical surface equation of a gear is a binary equation about the angle parameter θ_1 and the tooth profile parameter ϕ_1.

According to the above-calculated tooth surface eqution (2.1), it can be seen that the equation is about two variable parameters and continuous differentiability, so the partial derivatives of the equation can be obtained with respect to these two parameters. The partial derivative of the left convex tooth of a normal double-circular-arc gear is solved as follows:

$$\frac{\partial\left(\vec{r}_1\right)_1}{\partial\theta_1} = -\left(\rho_1 \sin\phi_1 + b_1 + \frac{d_1}{2}\right)\vec{e}_1(\theta_1) + (\rho_1 \cos\phi_1 - l_1)\cos\beta\,\vec{g}_1(\theta_1) + p_1\vec{k}_1 \tag{2.2}$$

$$\frac{\partial\left(\vec{r}_1\right)_1}{\partial\phi_1} = -\rho_1 \sin\phi_1 \cos\beta\,\vec{e}_1(\theta_1) + \rho_1\cos\phi_1\,\vec{g}_1(\theta_1) - \rho_1\sin\phi_1\sin\beta\,\vec{k}_1 \tag{2.3}$$

According to the definition of a unit normal vector in gear meshing theory, the normal unit vector of any point on the working tooth surface of this section can be obtained as:

$$\left(\vec{n}_1\right)_1 = \frac{\frac{\partial(\vec{r}_1)_1}{\partial\theta_1} \times \frac{\partial(\vec{r}_1)_1}{\partial\phi_1}}{\rho_1 D_1}$$

$$= \frac{\rho_1 \begin{vmatrix} \vec{e}_1(\theta_1) & \vec{g}_1(\theta_1) & \vec{k}_1 \\ -\left(\rho_1 \sin\phi_1 + b_1 + \frac{d_1}{2}\right) & (\rho_1 \cos\phi_1 - l_1)\cos\beta & p_1 \\ -\sin\phi_1 \cos\beta & \cos\phi_1 & -\sin\phi_1 \sin\beta \end{vmatrix}}{\rho_1 D_1}$$

$$= \frac{n_{1e}}{D_1}\vec{e}_1(\theta_1) + \frac{n_{1g}}{D_1}\vec{g}_1(\theta_1) + \frac{n_{1k}}{D_1}\vec{k}_1 \tag{2.4}$$

where we denote:

$$n_{1e} = -(\rho_1 \cos\phi_1 - l_1)\sin\beta\cos\beta\sin\phi_1 - p_1\cos\phi_1$$

$$n_{1g} = -\left(\rho_1 \sin\phi_1 + b_1 + \frac{d_1}{2}\right)\sin\beta\sin\phi_1 - p_1\cos\beta\sin\phi_1$$

$$n_{1k} = -\left(\rho_1 \sin\phi_1 + b_1 + \frac{d_1}{2}\right)\cos\phi_1 + (\rho_1\cos\phi_1 - l_1)\cos^2\beta\sin\phi_1$$

$$D_1{}^2 = C_{10}\sin^2\phi_1 + 2\rho_1 C_{11}\sin\phi_1 + 2b_1 l_1\cos^2\beta\sin\phi_1\cos\phi_1 + \left(b_1 + \frac{d_1}{2}\right)^2 + p_1^2 \tag{2.5}$$

where

$$C_{10} = \rho_1{}^2\sin^2\beta + \left(l_1{}^2 - b_1^2\right)\cos^2\beta, C_{11} = \frac{d_1}{2} + b_1\sin^2\beta$$

The above is the calculation of the unit normal vector of the working tooth profile curve of the left convex tooth engaged in meshing. It can be seen from Eqs. (2.4) and (2.5) that the unit normal vector $\left(\vec{n}_1\right)_1$ contains only variables θ_1 and ϕ_1 and that its three components n_{1e},n_{1g}, and n_{1k} contain only one variable ϕ_1.

2.2.2 *Mathematical Model of the Left Concave Tooth Surface of Normal Circular-Arc Gear 2*

The analysis shows that the left rotary gear's left concave tooth surface should mesh with the left convex tooth surface of the gear, which should satisfy the same helical angle and opposite direction. A similar method is used to solve the tooth surface equation of the left concave tooth of gear 2, and the vector coordinate system is established, as shown in Fig. 2.2.

In the coordinate system $\sigma_2\left\{O_2; \vec{i}_2, \vec{j}_2, \vec{k}_2\right\}$, the coordinate origin O_2 is located at the midpoint of the gear axis and the unit base vector $\vec{k}_2$ is along the direction of the gear axis. Gear 2 is a left normal double-circular-arc tooth surface, and its helical angle is also β. Similar to gear 1, at the initial position, as shown in Fig. 2.2, the base vector of the normal section of the curved normal double-circular-arc helical plane is composed of unit vectors $\vec{j}_2$ and $\vec{n}_2(0, \beta)$. The radius of the concave tooth is ρ_2, and the position parameters of the center of the circle are similar to those of gear 1. Any position can be regarded as obtained by moving $\left|\overrightarrow{O_nO_n'}\right| = p_2\theta_2$ at the initial position, p_2 is the spiral parameter of the normal double-circular helical surface, and θ_2 is the angle of rotation around the axis $\vec{k}_2$. At any position, its normal section is spanned by the unit vectors $\vec{g}_2(-\theta_2)$ and $\vec{n}_2(-\theta_2, \beta)$. According to the geometric relationship shown in Fig. 2.2, in the right-handed Cartesian coordinate system $\sigma_2\left\{O_2; \vec{i}_2, \vec{j}_2, \vec{k}_2\right\}$, point P_2 is any point on the left working tooth profile of the concave tooth of the normal double-arc gear and the directed angle between the path vector $\overrightarrow{cP_2}$ and the negative direction of the

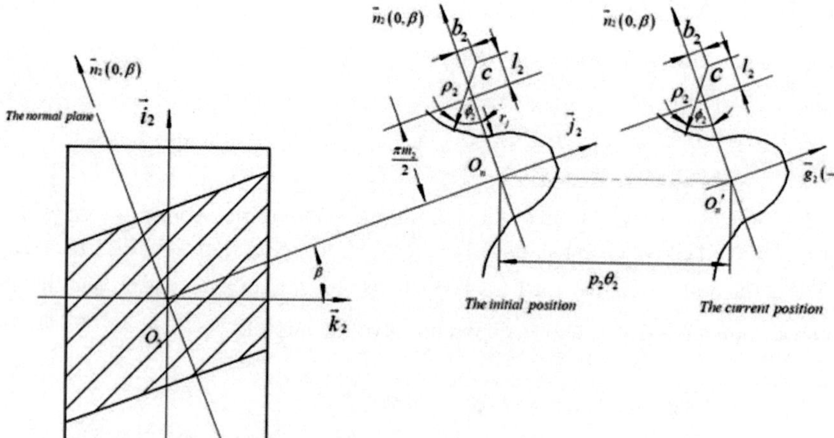

Fig. 2.2 A coordinate system diagram formed by a left concave tooth surface

$\vec{n}_2(-\theta_2, \beta)$ axis is ϕ_2. Then, the radial diameter of the origin O_2 to any point P_2 of the working arc is the equation of the tooth surface of the working arc:

$$\begin{aligned}\left(\vec{r}_2\right)_2 &= \overrightarrow{O_2P_2} = \overrightarrow{O_2O_n} + \overrightarrow{O_nO_n'} + \overrightarrow{O_n'P_2} \\ &= \left(\frac{\pi m_2}{2} + l_2 - \rho_2 \cos\phi_2\right)\cos\beta\,\vec{e}_2(-\theta_2) \\ &\quad - \left(\rho_2 \sin\phi_2 - b_2 - \frac{d_2}{2}\right)\vec{g}_2(-\theta_2) \\ &\quad + \left[p_2\theta_2 - \left(\frac{\pi m_2}{2} + l_2 - \rho_2\cos\phi_2\right)\sin\beta\right]\vec{k}_2 = x_2\vec{i}_2 + y_2\vec{j}_2 + z_2\vec{k}_2\end{aligned} \tag{2.6}$$

where

$$x_2 = \left(\frac{\pi m_2}{2} + l_2 - \rho_2\cos\phi_2\right)\cos\beta\cos\theta_2 - \left(\rho_2\sin\phi_2 - b_2 - \frac{d_2}{2}\right)\sin\theta_2$$

$$y_2 = -\left(\frac{\pi m_2}{2} + l_2 - \rho_2\cos\phi_2\right)\cos\beta\sin\theta_2 - \left(\rho_2\sin\phi_2 - b_2 - \frac{d_2}{2}\right)\cos\theta_2$$

$$z_2 = p_2\theta_2 - \left(\frac{\pi m_2}{2} + l_2 - \rho_2\cos\phi_2\right)\sin\beta$$

$$\phi_2 \in \left(0, \frac{\pi}{2}\right)$$

It can be seen from the Eq. (2.6) of the helical surface of the normal double-circular-arc gear that all parameters p_2, ρ_2, l_2, b_2, and β in the equation are constant values, whereas parameters θ_2 and ϕ_2 are variables and serve as the binary coordinates of the tooth surface. Therefore, the helical surface equation of the gear is also a binary equation about the angle parameter θ_2 and the tooth profile parameter ϕ_2.

According to the above-calculated tooth surface equation (2.6), it can be seen that the equation is about two variable parameters and continuous differentiability. So, the partial derivatives of the equation can be obtained with respect to these two parameters. The partial derivative of the left concave tooth of a normal double-circular-arc gear is solved as follows:

$$\begin{aligned}\frac{\partial\left(\vec{r}_2\right)_2}{\partial\theta_2} &= -\left(\rho_2\sin\phi_2 - b_2 - \frac{d_2}{2}\right)\vec{e}_2(-\theta_2) \\ &\quad - \left(\frac{\pi m_2}{2} + l_2 - \rho_2\cos\phi_2\right)\cos\beta\,\vec{g}_2(-\theta_2) + p_2\vec{k}_2\end{aligned} \tag{2.7}$$

$$\frac{\partial\left(\vec{r}_2\right)_2}{\phi_2} = \rho_2 \sin\phi_2 \cos\beta\, \vec{e}_2(-\theta_2) - \rho_2 \cos\phi_2\, \vec{g}_2(-\theta_2) - \rho_2 \sin\phi_2 \sin\beta\, \vec{k}_2 \tag{2.8}$$

According to the definition of a unit normal vector in gear meshing theory, the normal unit vector of any point on the working tooth surface of this section can be obtained as:

$$\left(\vec{n}_2\right)_2 = \frac{\frac{\partial\left(\vec{r}_2\right)_2}{\partial\theta_2} \times \frac{\partial\left(\vec{r}_2\right)_2}{\partial\phi_2}}{\rho_2 D_2} = \frac{\begin{vmatrix} \vec{e}_2(-\theta_2) & \vec{g}_2(-\theta_2) & \vec{k}_2 \\ \rho_2 - \left(\rho_2 \sin\phi_2 - b_2 - \frac{d_2}{2}\right) & -\left(\frac{\pi m_2}{2} + l_2 - \rho_2\cos\phi_2\right)\cos\beta & p_2 \\ \sin\phi_2\cos\beta & -\cos\phi_2 & -\sin\phi_2\sin\beta \end{vmatrix}}{\rho_2 D_2}$$

$$= \frac{n_{2e}\,\vec{e}_2(-\theta_2) + n_{2g}\,\vec{g}_2(-\theta_2) + n_{2k}\,\vec{k}_2}{D_2} \tag{2.9}$$

where

$$n_{2e} = \left(\frac{\pi m_2}{2} + l_2 - \rho_2\cos\phi_2\right)\sin\beta\cos\beta\sin\phi_2 + p_2\cos\phi_2$$

$$n_{2g} = -\left(\rho_2\sin\phi_2 - b_2 - \frac{d_2}{2}\right)\sin\beta\sin\phi_2 + p_2\cos\beta\sin\phi_2$$

$$n_{2k} = \left(\rho_2\sin\phi_2 - b_2 - \frac{d_2}{2}\right)\cos\phi_2 + \left(\frac{\pi m_2}{2} + l_2 - \rho_2\cos\phi_2\right)\cos^2\beta\sin\phi_2$$

$$\mathrm{D}_2{}^2 = C_{20}\sin^2\phi_2 - C_{21}\sin\phi_2 - C_{22}\sin\phi_2\cos\phi_2 + (b_2 + r_2)^2 + p_2{}^2 \tag{2.10}$$

where

$$C_{20} = \left[\left(\frac{\pi m_2}{2} + l_2\right)^2 - b_2{}^2\right]\cos^2\beta + \rho_2{}^2\sin^2\beta, C_{21} = 2\rho_2\left(b_2\sin^2\beta + \frac{d_2}{2}\right),$$

$$C_{22} = 2b_2\left(\frac{\pi m_2}{2} + l_2\right)\cos^2\beta$$

The above is the calculation of the unit normal vector of the working tooth profile curve of the left concave tooth engaged in meshing. It can be seen from Eqs. (2.9) and (2.10) that the unit normal vector $\left(\vec{n}_2\right)_2$ contains only variables θ_2 and ϕ_2 and that its three components n_{2e},n_{2g}, and n_{2k} contain only one variable ϕ_2.

Above, we establish the relative coordinate system of the normal double-circular spiral surface, deduce the corresponding equations of the concave and convex tooth profile spiral surface, and obtain the unit normal vector of the surface. The theoretical derivation and calculation in this chapter have established the foundation for studying the meshing principle of a normal double-circular-arc gear pair and help calculate the formula of the meshing performance parameters of gear pairs.

2.3 System of Nonlinear Equations to Determine Instantaneous Contact Point

In the previous chapter, the helical surface equation of a normal double-circular-arc gear was established. The equation is a binary equation composed of two variables, and the basic parameters of the concave and convex surface on the left side are provided. In this chapter, the meshing theory of the tooth surface equation is used to further study the meshing of a normal double-circular-arc gear. A normal double-circular-arc gear has point contact form, so it is necessary to establish the relative motion coordinate system between them, calculate the relative motion parameters of the two gears, then deduce the tooth surface meshing equation, and calculate the nonlinear equations by the elimination method, which lays the foundation for solving the contact point in the following.

2.3.1 *Equation and Normal Vector of Surface Families in a Static Coordinate System*

In order to study normal double-circular-arc gear transmission, all coordinate systems related to the two meshing gears are shown in Fig. 2.3 during the meshing analysis of a normal double-circular-arc gear pair.

According to the relative position relationship of meshing, two static coordinate systems, namely, $\sigma_{o1}\left\{O_{o1};\ \vec{i}_{o1},\ \vec{j}_{o1},\ \vec{k}_{o1}\right\}$ and $\sigma_{o2}\left\{O_{o2};\ \vec{i}_{o2},\ \vec{j}_{o2},\ \vec{k}_{o2}\right\}$, are established, and the initial position before meshing is represented by the corresponding parameters related to the two tooth surfaces. The dynamic coordinate systems $\sigma_1\left\{O_1;\ \vec{i}_1,\ \vec{j}_1,\ \vec{k}_1\right\}$ and $\sigma_2\left\{O_2;\ \vec{i}_2,\ \vec{j}_2,\ \vec{k}_2\right\}$, connected with the gear phase, are established to represent the current position of the meshing of the two gears, respectively. The centers of the axes of the two gears are located at point O_1 and point O_2, respectively. The center distance of the gear pair is the straight-line distance of two points: $\left|\overrightarrow{O_1O_2}\right| = a$. If gear 1 is a dextral gear, then gear 2 is left-handed. When the two gears move relative to each other, they rotate along their axes $\vec{k}_1$ and $\vec{k}_2$, respectively, and $\vec{i}_{o1} = \vec{i}_{o2}$.

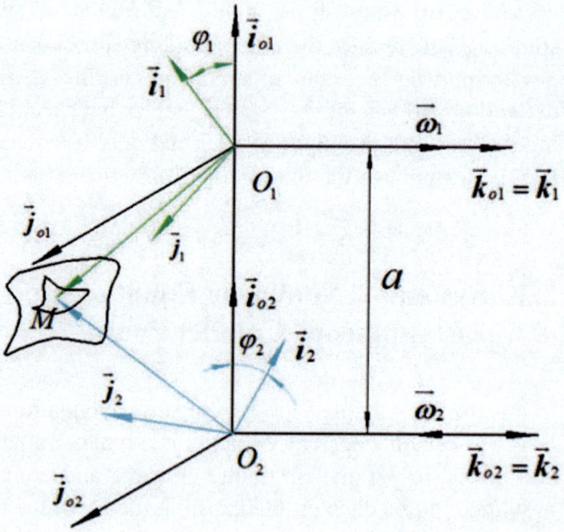

Fig. 2.3 The relative motion coordinate system of a gear pair

In the precious chapter, the coordinate system in which the tooth surface equation is established is located in the moving coordinate system, and the position represented is the current position of two meshing gears. When two normal double-circular-arc gears move relative to each other, they rotate around their respective axes. Moreover, at any moment when the two gears are meshing, they rotate from their initial position to their current position, and, at the current position, the rotation angle relative to its initial position is expressed by angles φ_1 and φ_2, respectively. At the same time, the two tooth surfaces contact each other at the unique contact point M.

The moving coordinate system of the normal double-circular-arc gear 1 is the equation after gear 1 rotates around its axis $\vec{k}_1$ by an angle of φ_1 at the initial moment. Therefore, using the basic knowledge of coordinate transformation in differential geometry, the left convex tooth surface equation of the normal double-circular-arc gear 1 is converted from the driven coordinate system $\sigma_1\left\{O_1;\ \vec{i}_1,\ \vec{j}_1,\ \vec{k}_1\right\}$ to the static coordinate system $\sigma_{o1}\left\{O_{o1};\ \vec{i}_{o1},\ \vec{j}_{o1},\ \vec{k}_{o1}\right\}$. Combined with the equation of the left convex tooth surface of the normal double-circular-arc gear, we can get:

$$\begin{aligned}\left(\vec{r}_1\right)_{o1} &= R\left[\vec{k}_{o1}, \varphi_1\right]\left(\vec{r}_1\right)_1 = \begin{bmatrix} \cos\varphi_1 & -\sin\varphi_1 & 0 \\ \sin\varphi_1 & \cos\varphi_1 & 0 \\ 0 & 0 & 1 \end{bmatrix}\left(\vec{r}_1\right)_1 \\ &= x_{o1}\vec{i}_{o1} + y_{o1}\vec{j}_{o1} + z_{o1}\vec{k}_{o1}\end{aligned} \tag{2.11}$$

where

$$x_{o1} = (\rho_1 \cos\phi_1 - l_1)\cos\beta\cos(\theta_1 + \varphi_1) - \left(\rho_1 \sin\phi_1 + b_1 + \frac{d_1}{2}\right)\sin(\theta_1 + \varphi_1)$$

$$y_{o1} = (\rho_1 \cos\phi_1 - l_1)\cos\beta\sin(\theta_1 + \varphi_1) + \left(\rho_1 \sin\phi_1 + b_1 + \frac{d_1}{2}\right)\cos(\theta_1 + \varphi_1)$$

$$z_{o1} = z_1 = (\rho_1 \cos\phi_1 - l_1)\sin\beta + p_1\theta_1$$

Matrix $R\left[\vec{k}_{o1}, \varphi_1\right]$ is the rotation matrix of the rotation angle φ_1 of the vector $\left(\vec{r}_1\right)_{o1}$ around the axis $\vec{k}_1$.

After the coordinate transformation of the tooth surface equation, it is necessary to use the same method to convert the unit normal vector $\left(\vec{n}_1\right)_1$ of the left convex tooth surface of the gear from the driven coordinate system $\sigma_1\left\{O_1; \vec{i}_1, \vec{j}_1, \vec{k}_1\right\}$ to the static coordinate system $\sigma_{o1}\left\{O_{o1}; \vec{i}_{o1}, \vec{j}_{o1}, \vec{k}_{o1}\right\}$. The unit normal vector $\left(\vec{n}_1\right)_1$ is a unit vector with bases $\vec{e}_1(\theta_1)$, $\vec{g}_1(\theta_1)$, and $\vec{k}_1$. After the coordinate transformation, the coordinate expression of the unit normal vector in the static system is obtained:

$$\left(\vec{n}_1\right)_{o1} = \mathrm{R}\left[\vec{k}_{o1}, \varphi_1\right]\vec{n}_1 = \frac{n_{xo1}}{D_1}\vec{i}_{o1} + \frac{n_{yo1}}{D_1}\vec{j}_{o1} + \frac{n_{k1}}{D_1}\vec{k}_1 \tag{2.12}$$

where

$$n_{xo1} = n_{1e}\cos(\theta_1 + \varphi_1) - n_{1g}\sin(\theta_1 + \varphi_1), n_{yo1} = n_{1e}\sin(\theta_1 + \varphi_1) + n_{1g}\cos(\theta_1 + \varphi_1)$$

$$n_{1k} = -\left(\rho_1 \sin\phi_1 + b_1 + \frac{d_1}{2}\right)\cos\phi_1 + (\rho_1\cos\phi_1 - l_1)\cos^2\beta\sin\phi_1$$

We can see from the above-transformed formula that the base vector of the unit normal vector of gear 1 has changed after the coordinate transformation. It can be seen that the coordinate system in which the unit normal vector is located is rotated by an angle of $(\theta_1 + \varphi_1)$, and the unit normal vector will be represented by θ_1, ϕ_1, and φ_1, three unknown variables.

Then, using the same method, the equation of the tooth surface of the left concave tooth of the normal double circular-arc gear 2 is transformed from the driven coordinate system $\sigma_2\left\{O_2; \vec{i}_2, \vec{j}_2, \vec{k}_2\right\}$ to the static coordinate system $\sigma_{o2}\left\{O_{o2}; \vec{i}_{o2}, \vec{j}_{o2}, \vec{k}_{o2}\right\}$, which can be obtained as:

$$
\begin{aligned}
\left(\vec{r}_2\right)_{o2} &= R\left[\vec{k}_{o2}, -\varphi_2\right]\left(\vec{r}_2\right)_2 = \begin{bmatrix} \cos\varphi_2 & \sin\varphi_2 & 0 \\ -\sin\varphi_2 & \cos\varphi_2 & 0 \\ 0 & 0 & 1 \end{bmatrix}\left(\vec{r}_2\right)_2 \\
&= x_{o2}\vec{i}_{o2} + y_{o2}\vec{j}_{o2} + z_{o2}\vec{k}_{o2}
\end{aligned} \tag{2.13}
$$

where

$$
\begin{aligned}
x_{o2} &= \left(\frac{\pi m_2}{2} + l_2 - \rho_2\cos\phi_2\right)\cos\beta\cos(\theta_2+\varphi_2) \\
&\quad - \left(\rho_2\sin\phi_2 - b_2 - \frac{d_2}{2}\right)\sin(\theta_2+\varphi_2) \\
y_{o2} &= -\left(\frac{\pi m_2}{2} + l_2 - \rho_2\cos\phi_2\right)\cos\beta\sin(\theta_2+\varphi_2) \\
&\quad - \left(\rho_2\sin\phi_2 - b_2 - \frac{d_2}{2}\right)\cos(\theta_2+\varphi_2) \\
z_{o2} &= z_2 = p_2\theta_2 - \left(\frac{\pi m_2}{2} + l_2 - \rho_2\cos\phi_2\right)\sin\beta
\end{aligned}
$$

Matrix $R\left[\vec{k}_{o2}, -\varphi_2\right]$ is the rotation matrix of the rotation angle φ_2 of the vector $\left(\vec{r}_2\right)_{o2}$ around the axis $\vec{k}_2$.

Then, the coordinate expression of the normal vector in the static system is obtained after coordinate transformation:

$$
\left(\vec{n}_2\right)_{o2} = R\left[\vec{k}_{o2}, -\varphi_2\right]\left(\vec{n}_2\right)_2 = \frac{n_{xo2}}{D_2}\vec{i}_{o2} + \frac{n_{yo2}}{D_2}\vec{j}_{o2} + \frac{n_{2k}}{D_2}\vec{k}_2 \tag{2.14}
$$

where

$$
\begin{aligned}
n_{xo2} &= n_{2e}\cos(\theta_2+\varphi_2) + n_{2g}\sin(\theta_2+\varphi_2), n_{yo2} = -n_{2e}\sin(\theta_2+\varphi_2) \\
&\quad + n_{2g}\cos(\theta_2+\varphi_2) \\
n_{2k} &= \left(\rho_2\sin\phi_2 - b_2 - \frac{d_2}{2}\right)\cos\phi_2 + \left(\frac{\pi m_2}{2} + l_2 - \rho_2\cos\phi_2\right)\cos^2\beta\sin\phi_2
\end{aligned}
$$

We can see from the above-transformed formula that the base vector of the unit normal vector of gear 2 has changed after the coordinate transformation. It can be seen that the coordinate system in which the unit normal vector is located is rotated by an angle of $(\theta_2 + \varphi_2)$, and the unit normal vector will be represented by θ_2, ϕ_2, and φ_2, three unknown variables.

The relative motion of the two gears is analyzed, and the corresponding tooth surface equations and unit normal vectors are transformed, which provides an important basis for calculating the contact point.

2.3.2 Derivation and Elimination of Tooth Surface Contact Equations

As shown in Fig. 2.3, there is a contact point M between tooth surface 1 and tooth surface 2 in the meshing process, and the position of the contact point M will naturally change along the contact path within one meshing cycle of the gear pair. In the meshing process of point contact gear, it is necessary to ensure the continuous tangent of the two tooth surfaces. The condition of continuous tangent is that the two tooth surfaces have a common normal at the contact point M at any time. Accordingly, nonlinear contact equations for determining the contact point M can be established according to a common point and common normal condition:

$$\begin{cases} \left(\vec{r}_1\right)_{o2} - \left(\vec{r}_2\right)_{o2} + \left(\overrightarrow{O_2O_1}\right)_{o2} = 0 \\ \left(\vec{n}_1\right)_{o2} - \left(\vec{n}_2\right)_{o2} = 0 \end{cases} \tag{2.15}$$

For the second equation in the system (2.15), we should judge the sign on the left side of the equal sign. We should consider the positive sign if the two-unit normal vectors point to the same point at the contact point. The negative sign is considered if the direction is opposite at the contact point. Here, when judging the pointing of the unit normal vector, we can observe the direction of the two first-order partial derivatives and use the right-hand rule to judge the pointing of the normal vector.

We can see that the normal vector points to the outside of gear 1 from the internal entity. Similarly, the normal vector can be seen pointing from the external space of gear 2 to the internal entity. Therefore, we can conclude from the above inference that the sign in the second equation in the system (2.15) is minus.

There are six unknowns in the system (2.15), which are θ_1, θ_2, ϕ_1, ϕ_2, φ_1, and φ_2.

When solving a nonlinear equation group iteratively, it is well-known that the less amount of its unknowns, the easier is the implementation of its iterative solving process. However, there are six unknowns in system (2.15), which are difficult to solve directly. For the purpose of simplifying the solving process, we should transform system (2.15) into its simplest equivalent form with the least unknowns by means of the elimination artifice. The concrete eliminating steps are as follows:

Substituting the expressions of the components of the normal vectors $\left(\vec{n}_1\right)_{o2}$ and $\left(\vec{n}_2\right)_{o2}$ in the second equation of system (2.15), i.e., the alleged common normal equation, yields three equations as follows:

$$\begin{cases} \dfrac{n_{1e}}{D_1}\cos(\theta_1+\varphi_1) - \dfrac{n_{1g}}{D_1}\sin(\theta_1+\varphi_1) = \dfrac{n_{2e}}{D_2}\cos(\theta_2+\varphi_2) + \dfrac{n_{2g}}{D_2}\sin(\theta_2+\varphi_2) \\ \dfrac{n_{1e}}{D_1}\sin(\theta_1+\varphi_1) + \dfrac{n_{1g}}{D_1}\cos(\theta_1+\varphi_1) = -\dfrac{n_{2e}}{D_2}\sin(\theta_2+\varphi_2) + \dfrac{n_{2g}}{D_2}\cos(\theta_2+\varphi_2) \\ \dfrac{n_{1k}}{D_1} = \dfrac{n_{2k}}{D_2} \end{cases} \tag{2.16}$$

The first two equations in (2.16) can be regarded as those containing unknown trigonometric function variables, $\sin(\theta_1 + \varphi_1)$ and $\cos(\theta_1 + \varphi_1)$, and can be solved and expressed as follows:

$$\sin(\theta_1+\varphi_1)=\frac{D_1\left[\left(n_{1e}n_{2g}-n_{2e}n_{1g}\right)\cos(\theta_2+\varphi_2)-\left(n_{1e}n_{2e}+n_{1g}n_{2g}\right)\sin(\theta_2+\varphi_2)\right]}{D_2\left({n_{1e}}^2+{n_{1g}}^2\right)} \tag{2.17}$$

$$\cos(\theta_1+\varphi_1)=\frac{D_1\left[\left(n_{1e}n_{2g}-n_{2e}n_{1g}\right)\sin(\theta_2+\varphi_2)+\left(n_{1e}n_{2e}+n_{1g}n_{2g}\right)\cos(\theta_2+\varphi_2)\right]}{D_2\left({n_{1e}}^2+{n_{1g}}^2\right)} \tag{2.18}$$

Using the common point equation, the components of the tooth surface equation are put into the first equation of the system (2.15), and three equations can be obtained:

$$\begin{cases}\left(\frac{\pi m_2}{2}+l_2-\rho_2\cos\phi_2\right)\cos\beta\cos(\theta_2+\varphi_2)-\left(\rho_2\sin\phi_2-b_2-\frac{d_2}{2}\right)\sin(\theta_2+\varphi_2)\\ =(\rho_1\cos\phi_1-l_1)\cos\beta\cos(\theta_1+\varphi_1)\\ \quad-\left(\rho_1\sin\phi_1+b_1+\frac{d_1}{2}\right)\sin(\theta_1+\varphi_1)+a-\left(\frac{\pi m_2}{2}+l_2-\rho_2\cos\phi_2\right)\cos\beta\sin(\theta_2+\varphi_2)\\ \quad-\left(\rho_2\sin\phi_2-b_2-\frac{d_2}{2}\right)\cos(\theta_2+\varphi_2)=(\rho_1\cos\phi_1-l_1)\cos\beta\sin(\theta_1+\varphi_1)\\ \quad+\left(\rho_1\sin\phi_1+b_1+\frac{d_1}{2}\right)\cos(\theta_1+\varphi_1)p_2\theta_2-\left(\frac{\pi m_2}{2}+l_2-\rho_2\cos\phi_2\right)\sin\beta\\ =(\rho_1\cos\phi_1-l_1)\sin\beta+p_1\theta_1\end{cases} \tag{2.19}$$

By substituting the previously solved expressions (2.17) and (2.18) in the first two equations of (2.19) respectively, the following two equations can be obtained:

$$\begin{cases}\left[\left(\frac{\pi m_2}{2}+l_2-\rho_2\cos\phi_2\right)\cos\beta-A\right]\cos(\theta_2+\varphi_2)\\ \quad-\left[B+\left(\rho_2\sin\phi_2-b_2-\frac{d_2}{2}\right)\right]\sin(\theta_2+\varphi_2)-a\quad=0\\ \left[-\left(\frac{\pi m_2}{2}+l_2-\rho_2\cos\phi_2\right)\cos\beta-A\right]\sin(\theta_2+\varphi_2)\\ \quad-\left[B+\left(\rho_2\sin\phi_2-b_2-\frac{d_2}{2}\right)\right]\cos(\theta_2+\varphi_2)\quad=0\end{cases} \tag{2.20}$$

where

$$A = \frac{D_1\left[(\rho_1 \cos\phi_1 - l_1)\cos\beta\left(n_{1e}n_{2e} + n_{1g}n_{2g}\right) - \left(\rho_1 \sin\phi_1 + b_1 + \frac{d_1}{2}\right)\left(n_{1e}n_{2g} - n_{2e}n_{1g}\right)\right]}{D_2\left({n_{1e}}^2 + {n_{1g}}^2\right)}$$

$$B = \frac{D_1\left[(\rho_1 \cos\phi_1 - l_1)\cos\beta\left(n_{1e}n_{2g} - n_{2e}n_{1g}\right) + \left(\rho_1 \sin\phi_1 + b_1 + \frac{d_1}{2}\right)\left(n_{1e}n_{2e} + n_{1g}n_{2g}\right)\right]}{D_2\left({n_{1e}}^2 + {n_{1g}}^2\right)}$$

Through the above steps, we eliminate the trigonometric function variables sin $(\theta_1 + \varphi_1)$ and $\cos(\theta_1 + \varphi_1)$ in (2.19) so that the first two equations of (2.19) contain only $(\theta_2 + \varphi_2)$. Then, it can be solved through Eq. (2.20):

$$\sin(\theta_2 + \varphi_2) = -a\frac{\left[B + \left(\rho_2 \sin\phi_2 - b_2 - \frac{d_2}{2}\right)\right]}{\left[B + \left(\rho_2 \sin\phi_2 - b_2 - \frac{d_2}{2}\right)\right]^2 + \left[\left(\frac{\pi m_2}{2} + l_2 - \rho_2 \cos\phi_2\right)\cos\beta - A\right]^2} \tag{2.21}$$

$$\cos(\theta_2 + \varphi_2) = a\frac{\left[\left(\frac{\pi m_2}{2} + l_2 - \rho_2 \cos\phi_2\right)\cos\beta - A\right]}{\left[\left(\frac{\pi m_2}{2} + l_2 - \rho_2 \cos\phi_2\right)\cos\beta - A\right]^2 + \left[B + \left(\rho_2 \sin\phi_2 - b_2 - \frac{d_2}{2}\right)\right]^2} \tag{2.22}$$

Here, using the trigonometric function that the sum of squares is one, that is, $\sin^2(\theta_2 + \varphi_2) + \cos^2(\theta_2 + \varphi_2) = 1$, we can obtain:

$$\left[B + \left(\rho_2 \sin\phi_2 - b_2 - \frac{d_2}{2}\right)\right]^2 + \left[\left(\frac{\pi m_2}{2} + l_2 - \rho_2 \cos\phi_2\right)\cos\beta - A\right]^2 - a^2 = 0 \tag{2.23}$$

So far, we eliminated $(\theta_1 + \varphi_1)$ and $(\theta_2 + \varphi_2)$, leaving only two unknowns ϕ_1 and ϕ_2 in the equivalent equation (2.23) after elimination. Since one equation cannot be solved with two unknowns, we observe that the third equation in (2.16) also contains only two unknown parameters ϕ_1 and ϕ_2. Therefore, by combining the third term in Eq. (2.16) with Eq. (2.23), we can obtain a binary nonlinear system of equations about ϕ_1 and ϕ_2:

$$\begin{cases} f_1(\phi_1, \phi_2) = \left[B + \left(\rho_2 \sin\phi_2 - b_2 - \frac{d_2}{2}\right)\right]^2 + \left[\left(\frac{\pi m_2}{2} + l_2 - \rho_2 \cos\phi_2\right)\cos\beta - A\right]^2 - a^2 = 0 \\ f_2(\phi_1, \phi_2) = D_2\left[(\rho_1 \cos\phi_1 - l_1)\cos^2\beta \sin\phi_1 - \left(\rho_1 \sin\phi_1 + b_1 + \frac{d_1}{2}\right)\cos\phi_1\right] \\ -D_1\left[\left(\rho_2 \sin\phi_2 - b_2 - \frac{d_2}{2}\right)\cos\phi_2 + \left(\frac{\pi m_2}{2} + l_2 - \rho_2 \cos\phi_2\right)\cos^2\beta \sin\phi_2\right] = 0 \end{cases} \tag{2.24}$$

So far, there are still three unknown parameters and two angle sum equations in the contact equation, and the solution of the new system is the same as that of the original system. Therefore, to make the system well-conditioned, we need to add an equation or assign a specific value to one of the three variables, θ_1,φ_1, and φ_2, to obtain the other parameters to determine each contact point.

2.4 Instantaneous Transmission Ratio of a Gear Pair

In gear transmission, the instantaneous transmission ratio i_{12}^* should be equal to the nominal transmission i_{12}. The instantaneous transmission ratio we can obtain from the movement condition of gear transmission is expressed as follows:

$$\vec{V}_{12}^{(M)} \cdot \vec{n}_1 = 0 \tag{2.25}$$

Since the axial positions of the two gears are relatively unchanged, $\overrightarrow{O_1O_2}$ is a constant vector and the relative velocity at the contact point M is:

$$\vec{V}_{12}^{(M)} = \left(\vec{\omega}_1\right)_{o1} \times \left(\vec{r}_1\right)_{o1} - \left(\vec{\omega}_2\right)_{o2} \times \left(\vec{r}_2\right)_{o2} \tag{2.26}$$

According to the relationship shown in Fig. 2.3, it can be seen that:

$$\left(\vec{\omega}_1\right)_{o1} = \left|\vec{\omega}_1\right| \vec{k}_{o1}, \left(\vec{\omega}_2\right)_{o2} = -\left|\vec{\omega}_2\right| \vec{k}_{o2} \tag{2.27}$$

In combination with (2.44), (2.45), and (2.46), the instantaneous transmission ratio i_{12}^* can be expressed as:

$$i_{12}^* = \frac{\left|\vec{\omega}_1\right|}{\left|\vec{\omega}_2\right|} = -\frac{\left[\vec{k}_{o2}, \left(\vec{r}_2\right)_{o2}, \left(\vec{n}_2\right)_{o2}\right]}{\left[\vec{k}_{o1}, \left(\vec{r}_1\right)_{o1}, \left(\vec{n}_1\right)_{o1}\right]} \tag{2.28}$$

By bringing in the various components of $\left(\vec{r}_1\right)_{o1}, \left(\vec{r}_2\right)_{o2}, \left(\vec{n}_1\right)_{o1}, \left(\vec{n}_2\right)_{o2}, n_{1e}, n_{1g}, n_{2e}, n_{2g}, p_1, p_2$ we can get:

$$i_{12}^* = -\frac{\left[\vec{k}_{o2} \times \left(x_{o2}\vec{i}_{o2} + y_{o2}\vec{j}_{o2} + z_{o2}\vec{k}_{o2}\right)\right] \cdot \left(\frac{n_{xo2}}{D_2}\vec{i}_{o2} + \frac{n_{yo2}}{D_2}\vec{j}_{o2} + \frac{n_{2k}}{D_2}\vec{k}_2\right)}{\left[\vec{k}_{o1} \times \left(x_{o1}\vec{i}_{o1} + y_{o1}\vec{j}_{o1} + z_{o1}\vec{k}_{o1}\right)\right] \cdot \left(\frac{n_{xo1}}{D_1}\vec{i}_{o1} + \frac{n_{yo1}}{D_1}\vec{j}_{o1} + \frac{n_{1k}}{D_1}\vec{k}_1\right)}$$

$$= -\frac{\left(y_{o2}\vec{i}_{o2} - x_{o2}\vec{j}_{o2}\right)\cdot\left(\frac{n_{xo2}}{D_2}\vec{i}_{o2} + \frac{n_{yo2}}{D_2}\vec{j}_{o2} + \frac{n_{2k}}{D_2}\vec{k}_2\right)}{\left(y_{o1}\vec{i}_{o1} - x_{o1}\vec{j}_{o1}\right)\cdot\left(\frac{n_{xo1}}{D_1}\vec{i}_{o1} + \frac{n_{yo1}}{D_1}\vec{j}_{o1} + \frac{n_{1k}}{D_1}\vec{k}_1\right)}$$

$$= -\frac{y_{o2}\frac{n_{xo2}}{D_2} - x_{o2}\frac{n_{yo2}}{D_2}}{y_{o1}\frac{n_{xo1}}{D_1} - x_{o1}\frac{n_{yo1}}{D_1}} = -\frac{D_1}{D_2}\frac{y_{o2}n_{xo2} - x_{o2}n_{yo2}}{y_{o1}n_{xo1} - x_{o1}n_{yo1}} \tag{2.29}$$

where

$$\begin{aligned} y_{o2}n_{xo2} - x_{o2}n_{yo2} = & \left[-\left(\frac{\pi m_2}{2} + l_2 - \rho_2\cos\phi_2\right)\cos\beta\sin(\theta_2+\varphi_2)\right. \\ & \left. - \left(\rho_2\sin\phi_2 - b_2 - \frac{d_2}{2}\right)\cos(\theta_2+\varphi_2)\right]\left[n_{2e}\cos(\theta_2+\varphi_2) + n_{2g}\sin(\theta_2+\varphi_2)\right] \\ & - \left[\left(\frac{\pi m_2}{2} + l_2 - \rho_2\cos\phi_2\right)\cos\beta\cos(\theta_2+\varphi_2)\right. \\ & \left. - \left(\rho_2\sin\phi_2 - b_2 - \frac{d_2}{2}\right)\sin(\theta_2+\varphi_2)\right]\left[-n_{2e}\sin(\theta_2+\varphi_2) + n_{2g}\cos(\theta_2+\varphi_2)\right], \end{aligned}$$

$$\begin{aligned} y_{o1}n_{xo1} - x_{o1}n_{yo1} = & \left[(\rho_1\cos\phi_1 - l_1)\cos\beta\sin(\theta_1+\varphi_1)\right. \\ & \left. + \left(\rho_1\sin\phi_1 + b_1 + \frac{d_1}{2}\right)\cos(\theta_1+\varphi_1)\right]\left[n_{1e}\cos(\theta_1+\varphi_1) + n_{2g}\sin(\theta_2+\varphi_2)\right] \\ & - \left[(\rho_1\cos\phi_1 - l_1)\cos\beta\cos(\theta_1+\varphi_1)\right. \\ & \left. - \left(\rho_1\sin\phi_1 + b_1 + \frac{d_1}{2}\right)\sin(\theta_1+\varphi_1)\right]\left[n_{1e}\sin(\theta_1+\varphi_1) + n_{1g}\cos(\theta_1+\varphi_1)\right] \end{aligned}$$

So, (2.29) can be written as

$$\begin{aligned} i_{12}^{*} &= -\frac{D_1}{D_2}\frac{-\left(\frac{\pi m_2}{2} + l_2 - \rho_2\cos\phi_2\right)\cos\beta p_2\cos\beta\sin\phi_2 - \left(\rho_2\sin\phi_2 - b_2 - \frac{d_2}{2}\right)p_2\cos\phi_2}{(\rho_1\cos\phi_1 - l_1)\cos\beta p_1\cos\beta\sin\phi_1 - \left(\rho_1\sin\phi_1 + b_1 + \frac{d_1}{2}\right)p_1\cos\phi_1} \\ &= -\frac{D_1}{D_2}\frac{d_2}{d_1}\frac{-\left(\frac{\pi m_2}{2} + l_2 - \rho_2\cos\phi_2\right)\cos^2\beta\sin\phi_2 - \left(\rho_2\sin\phi_2 - b_2 - \frac{d_2}{2}\right)\cos\phi_2}{(\rho_1\cos\phi_1 - l_1)\cos^2\beta\sin\phi_1 - \left(\rho_1\sin\phi_1 + b_1 + \frac{d_1}{2}\right)\cos\phi_1} \end{aligned} \tag{2.30}$$

In combination with the third equation in (2.16), it can be concluded that:

$$i_{12}^{*} = -\frac{D_1}{D_2}\frac{\mathrm{d}_2}{d_1}\frac{-\left(\frac{\pi m_2}{2} + l_2 - \rho_2 \cos\phi_2\right)\cos^2\beta \sin\phi_2 - \left(\rho_2 \sin\phi_2 - b_2 - \frac{d_2}{2}\right)\cos\phi_2}{\left(\rho_1 \cos\phi_1 - l_1\right)\cos^2\beta \sin\phi_1 - \left(\rho_1 \sin\phi_1 + b_1 + \frac{d_1}{2}\right)\cos\phi_1}$$

$$= -\frac{D_1}{D_2}\frac{\mathrm{d}_2}{d_1}\frac{-n_{2k}}{n_{1k}} = \frac{\mathrm{d}_2}{d_1} \tag{2.31}$$

According to the Eq. (2.31), we can see that the transmission type of a double-circular-arc gear pair is continuous ratio transmission. Moreover, the instantaneous transmission ratio is equal to the nominal transmission ratio.

2.5 Fundamental Quantities of Two Tooth Surfaces

According to Eqs. (2.2) and (2.3), the first and second fundamental quantities of the helical surface of the left convex tooth of the normal double-circular-arc gear can be obtained as follows:

$$E_1 = \frac{\partial\left(\vec{\boldsymbol{r}}_1\right)_1}{\partial\theta_1}\cdot\frac{\partial\left(\vec{\boldsymbol{r}}_1\right)_1}{\partial\theta_1} = \left(\rho_1 \sin\phi_1 + b_1 + \frac{d_1}{2}\right)^2 + \left(\rho_1 \cos\phi_1 - l_1\right)^2 \cos^2\beta_1 + p_1{}^2 \tag{2.32}$$

$$F_1 = \frac{\partial\left(\vec{\boldsymbol{r}}_1\right)_1}{\partial\theta_1}\cdot\frac{\partial\left(\vec{\boldsymbol{r}}_1\right)_1}{\partial\phi_1} = \rho_1 \cos\beta\left(b_1 \sin\phi_1 - l_1 \cos\phi_1 + \rho_1\right) \tag{2.33}$$

$$G_1 = \frac{\partial\left(\vec{\boldsymbol{r}}_1\right)_1}{\partial\phi_1}\cdot\frac{\partial\left(\vec{\boldsymbol{r}}_1\right)_1}{\partial\phi_1} = \rho_1{}^2 \tag{2.34}$$

The second partial derivatives of θ_1 and φ_1 can be obtained through the tooth surface equations:

$$\frac{\partial^2\left(\vec{\boldsymbol{r}}_1\right)_1}{\partial\theta_1{}^2} = -\left(\rho_1 \cos\phi_1 - l_1\right)\cos\beta\,\vec{\boldsymbol{e}}_1(\theta_1) - \left(\rho_1 \sin\phi_1 + b_1 + \frac{d_1}{2}\right)\vec{\boldsymbol{g}}_1(\theta_1) \tag{2.35}$$

$$\frac{\partial^2\left(\vec{\boldsymbol{r}}_1\right)_1}{\partial\theta_1\partial\phi_1} = -\rho_1 \cos\phi_1\,\vec{\boldsymbol{e}}_1(\theta_1) - \rho_1 \sin\phi_1 \cos\beta\,\vec{\boldsymbol{g}}_1(\theta_1) \tag{2.36}$$

$$\frac{\partial^2\left(\vec{\boldsymbol{r}}_1\right)_1}{\partial\phi_1{}^2} = -\rho_1 \cos\phi_1 \cos\beta\,\vec{\boldsymbol{e}}_1(\theta_1) - \rho_1 \sin\phi_1\,\vec{\boldsymbol{g}}_1(\theta_1) - \rho_1 \cos\phi_1 \sin\beta\,\vec{\boldsymbol{k}}_1 \tag{2.37}$$

According to the relevant theories of differential geometry and according to (2.35), (2.36), and (2.37), the second fundamental quantity of the convex tooth helical surface can be obtained by the dot product:

$$\begin{aligned} L_1 &= \frac{\rho_1 E_1 \sin\beta \sin\phi_1 + p_1 F_1}{D_1 \rho_1}, \\ M_1 &= \frac{F_1 \sin\beta \sin\phi_1 + \rho_1 p_1}{D_1}, \\ N_1 &= \frac{-\rho_1 n_{1g}}{D_1 \sin\phi_1} \end{aligned} \tag{2.38}$$

According to (2.7) and (2.8), the first fundamental quantity of the helical surface of the left concave tooth of the normal double-circular-arc gear can be obtained as follows:

$$\begin{aligned} E_2 = \frac{\partial\left(\vec{r}_2\right)_2}{\partial\theta_2} \cdot \frac{\partial\left(\vec{r}_2\right)_2}{\partial\theta_2} &= \left(\rho_2 \sin\phi_2 - b_2 - \frac{d_2}{2}\right)^2 \\ &+ \left(\tfrac{\pi m_2}{2} + l_2 - \rho_2 \cos\phi_2\right)^2 \cos^2\beta + {p_2}^2 \end{aligned} \tag{2.39}$$

$$\begin{aligned} F_2 &= \frac{\partial\left(\vec{r}_2\right)_2}{\partial\theta_2} \cdot \frac{\partial\left(\vec{r}_2\right)_2}{\partial\phi_2} \\ &= -\rho_2 \cos\beta \left[\rho_2 - b_2 \sin\phi_2 - \left(\frac{\pi m_2}{2} + l_2\right) \cos\phi_2\right] \end{aligned} \tag{2.40}$$

$$G_2 = \frac{\partial\left(\vec{r}_2\right)_2}{\partial\phi_2} \cdot \frac{\partial\left(\vec{r}_2\right)_2}{\partial\phi_2} = {\rho_2}^2 \tag{2.41}$$

Similarly, the second partial derivative of the concave tooth surface can be obtained according to the above steps.

$$\frac{\partial^2\left(\vec{r}_2\right)_2}{\partial{\theta_2}^2} = -\left(\frac{\pi m_2}{2} + l_2 - \rho_2 \cos\phi_2\right) \cos\beta\, \vec{e}_2(-\theta_2) + \left(\rho_2 \sin\phi_2 - b_2 - \frac{d_2}{2}\right) \vec{g}_2(-\theta_2) \tag{2.42}$$

$$\frac{\partial^2\left(\vec{r}_2\right)_2}{\partial\theta_2 \partial\phi_2} = -\rho_2 \cos\phi_2\, \vec{e}_2(-\theta_2) - \rho_2 \sin\phi_2 \cos\beta\, \vec{g}_2(-\theta_2) \tag{2.43}$$

$$\frac{\partial^2\left(\vec{r}_2\right)_2}{\partial{\phi_2}^2} = \rho_2 \cos\phi_2 \cos\beta\, \vec{e}_2(-\theta_2) + \rho_2 \sin\phi_2\, \vec{g}_2(-\theta_2) - \rho_2 \cos\phi_2 \sin\beta\, \vec{k}_2 \tag{2.44}$$

The second fundamental quantity of the concave tooth helical surface can be obtained as:

$$L_2 = \frac{-\rho_2 E_2 \sin\beta \sin\phi_2 + p_2 F_2}{\rho_2 D_2}, M_2 = \frac{-F_2 \sin\beta \sin\phi_2 - p_2 \rho_2}{D_2}, N_2 = \frac{\rho_2 n_{2g}}{D_2 \sin\phi_2} \tag{2.45}$$

2.6 Moving Frames on Two Tooth Surfaces

At any contact point M, we can use the relative positions of the two coordinate systems to calculate the relative principal curvature of the gear pair. They are $\sigma_M^{(1)}\left\{M; \left(\vec{\boldsymbol{\alpha}}_\xi^{(1)}\right)_1, \left(\vec{\boldsymbol{\alpha}}_\eta^{(1)}\right)_1, \left(\vec{\boldsymbol{n}}_1\right)_1\right\}$ and $\sigma_M^{(2)}\left\{M; \left(\vec{\boldsymbol{\alpha}}_\xi^{(2)}\right)_2, \left(\vec{\boldsymbol{\alpha}}_\eta^{(2)}\right)_2, \left(\vec{\boldsymbol{n}}_2\right)_2\right\}$.

The two main directions on the spiral plane are solved as follows. In the frame $\sigma_1\left\{O_1; \vec{\boldsymbol{i}}_1, \vec{\boldsymbol{j}}_1, \vec{\boldsymbol{k}}_1\right\}$, in order to obtain the normal curvature and short-range torsion on the helical surface of the normal double-circular-arc gear more easily, it is necessary to determine two mutually perpendicular principal directions, i.e., $\left(\vec{\boldsymbol{\alpha}}_\xi^{(1)}\right)_1$ and $\left(\vec{\boldsymbol{\alpha}}_\eta^{(1)}\right)_1$, on the helical surface. Since the helical surface is a binary equation about θ_1 and ϕ_1, the unit tangent vector $\left(\vec{\boldsymbol{\alpha}}_\xi^{(1)}\right)_1$ of the fixed surface is taken as the unit tangent vector of its line ϕ_1; then, at any point M on the helical surface, the principal direction $\left(\vec{\boldsymbol{\alpha}}_\xi^{(1)}\right)_1$ can be obtained by the following formula:

$$\left(\vec{\boldsymbol{\alpha}}_\xi^{(1)}\right)_1 = \frac{\frac{\partial(\vec{\boldsymbol{r}}_1)_1}{\partial\phi_1}}{\left|\frac{\partial(\vec{\boldsymbol{r}}_1)_1}{\partial\phi_1}\right|} = -\sin\phi_1\cos\beta\,\vec{\boldsymbol{e}}_1(\theta_1) + \cos\phi_1\,\vec{\boldsymbol{g}}_1(\theta_1) - \sin\phi_1\sin\beta\,\vec{\boldsymbol{k}}_1 \tag{2.46}$$

Then, the other base vector $\left(\vec{\boldsymbol{\alpha}}_\eta^{(1)}\right)_1$ of the movable mark frame $\left\{M; \left(\vec{\boldsymbol{\alpha}}_\xi^{(1)}\right)_1, \left(\vec{\boldsymbol{\alpha}}_\eta^{(1)}\right)_1, \vec{\boldsymbol{n}}_1\right\}$ on the helical surface of the gear, that is, the other unit tangent vector of the helical surface of the gear, can be determined according to the right-hand helical rule:

$$\left(\vec{\boldsymbol{\alpha}}_\eta^{(1)}\right)_1 = \vec{\boldsymbol{n}}_1 \times \left(\vec{\boldsymbol{\alpha}}_\xi^{(1)}\right)_1 = \left(\alpha_{\eta x}^{(1)}\right)_1 \vec{\boldsymbol{e}}_1(\theta_1) + \left(\alpha_{\eta y}^{(1)}\right)_1 \vec{\boldsymbol{g}}_1(\theta_1) + \left(\alpha_{\eta z}^{(1)}\right)_1 \vec{\boldsymbol{k}}_1 \tag{2.47}$$

where

$$\left(\alpha_{\eta x}^{(1)}\right)_1 = -\left(\cos\phi_1 \frac{n_{1k}}{D_1} + \sin\phi_1 \sin\beta \frac{n_{1g}}{D_1}\right), \left(\alpha_{\eta y}^{(1)}\right)_1$$
$$= \left(\sin\phi_1 \sin\beta \frac{n_{1e}}{D_1} - \sin\phi_1 \cos\beta \frac{n_{1k}}{D_1}\right), \left(\alpha_{\eta z}^{(1)}\right)_1$$
$$= \left(\sin\phi_1 \cos\beta \frac{n_{1g}}{D_1} + \cos\phi_1 \frac{n_{1e}}{D_1}\right)$$

The two main directions on the spiral plane are solved as follows. In the frame $\sigma_2\left\{O_2;\ \vec{i}_2,\ \vec{j}_2,\ \vec{k}_2\right\}$, in order to obtain the normal curvature and geodesic torsion on the helical surface of the normal double-circular-arc gear more easily, it is necessary to determine two mutually perpendicular principal directions, i.e., $\left(\vec{\boldsymbol{\alpha}}_{\xi}^{(2)}\right)_2$ and $\left(\vec{\boldsymbol{\alpha}}_{\eta}^{(2)}\right)_2$, on the helical surface. Since the helical surface is a binary equation about θ_2 and ϕ_2, the unit tangent vector $\left(\vec{\boldsymbol{\alpha}}_{\xi}^{(2)}\right)_2$ of the fixed surface is taken as the unit tangent vector of its line ϕ_2; then, at any point M on the helical surface, the principal direction $\left(\vec{\boldsymbol{\alpha}}_{\xi}^{(2)}\right)_2$ can be obtained by the following formula:

$$\left(\vec{\boldsymbol{\alpha}}_{\xi}^{(2)}\right)_2 = \frac{\frac{\partial(\vec{r}_2)_2}{\partial\phi_2}}{\left|\frac{\partial(\vec{r}_2)_2}{\partial\phi_2}\right|} = \sin\phi_2 \cos\beta\, \vec{\boldsymbol{e}}_2(-\theta_2) - \cos\phi_2\, \vec{\boldsymbol{g}}_2(-\theta_2) - \sin\phi_2 \sin\beta\, \vec{\boldsymbol{k}}_2 \tag{2.48}$$

Then, the other base vector $\left(\vec{\boldsymbol{\alpha}}_{\eta}^{(2)}\right)_2$ of the movable mark frame $\left\{M;\ \left(\vec{\boldsymbol{\alpha}}_{\xi}^{(2)}\right)_2,\ \left(\vec{\boldsymbol{\alpha}}_{\eta}^{(2)}\right)_2,\ \vec{\boldsymbol{n}}_2\right\}$ on the helical surface of the gear, that is, the other unit tangent vector of the helical surface of the gear, can be determined according to the right-hand helical rule:

$$\left(\vec{\boldsymbol{\alpha}}_{\eta}^{(2)}\right)_2 = \vec{\boldsymbol{n}}_2 \times \left(\vec{\boldsymbol{\alpha}}_{\xi}^{(2)}\right)_2 = \left(\alpha_{\eta x}^{(2)}\right)_2 \vec{\boldsymbol{e}}_2(-\theta_2) + \left(\alpha_{\eta y}^{(2)}\right)_2 \vec{\boldsymbol{g}}_2(-\theta_2) + \left(\alpha_{\eta z}^{(2)}\right)_2 \vec{\boldsymbol{k}}_2 \tag{2.49}$$

where

$$\left(\alpha_{\eta x}^{(2)}\right)_2 = \cos\phi_2 \frac{n_{2k}}{D_2} - \sin\phi_2 \sin\beta \frac{n_{2g}}{D_2},$$
$$\left(\alpha_{\eta y}^{(2)}\right)_2 = \sin\phi_2 \cos\beta \frac{n_{2k}}{D_2} + \sin\phi_2 \sin\beta \frac{n_{2e}}{D_2},$$
$$\left(\alpha_{\eta z}^{(2)}\right)_2 = -\left(\sin\phi_2 \cos\beta \frac{n_{2g}}{D_2} + \cos\phi_2 \frac{n_{2e}}{D_2}\right)$$

Given that $\left(\vec{\boldsymbol{n}}_1\right)_{o2} = \left(\vec{\boldsymbol{n}}_2\right)_{o2}$ at the contact point M, the four vectors $\left(\vec{\boldsymbol{\alpha}}_\xi^{(1)}\right)_{o2}$, $\left(\vec{\boldsymbol{\alpha}}_\eta^{(1)}\right)_{o2}$, $\left(\vec{\boldsymbol{\alpha}}_\xi^{(2)}\right)_{o2}$, and $\left(\vec{\boldsymbol{\alpha}}_\eta^{(2)}\right)_{o2}$ lie on the common tangent plane of the two meshing tooth surfaces. Therefore, the relative positions of the two frames can be determined by the directed angles from $\left(\vec{\boldsymbol{\alpha}}_\xi^{(1)}\right)_{o2}$ to $\left(\vec{\boldsymbol{\alpha}}_\xi^{(2)}\right)_{o2}$.

The principal directions are the expressions after a coordinate transformation. The transformation is as follows:

$$\left(\vec{\boldsymbol{\alpha}}_\xi^{(1)}\right)_{o1} = R\left[\vec{\boldsymbol{k}}_{o1}, \varphi_1\right]\left(\vec{\boldsymbol{\alpha}}_\xi^{(1)}\right)_1 = \alpha_{\xi x}^{(1)}\vec{\boldsymbol{i}}_{o1} + \alpha_{\xi y}^{(1)}\vec{\boldsymbol{j}}_{o1} + \alpha_{\xi z}^{(1)}\vec{\boldsymbol{k}}_{o1} \tag{2.50}$$

Because the coordinate system is parallel,

$$\left(\vec{\boldsymbol{\alpha}}_\xi^{(1)}\right)_{o2} = \left(\vec{\boldsymbol{\alpha}}_\xi^{(1)}\right)_{o1} = \alpha_{\xi x}^{(1)}\vec{\boldsymbol{i}}_{o2} + \alpha_{\xi y}^{(1)}\vec{\boldsymbol{j}}_{o2} + \alpha_{\xi z}^{(1)}\vec{\boldsymbol{k}}_{o2} \tag{2.51}$$

where

$$\begin{aligned}
\alpha_{\xi x}^{(1)} &= -\sin\phi_1\cos\beta\cos(\theta_1+\varphi_1) - \cos\phi_1\sin(\theta_1+\varphi_1),\\
\alpha_{\xi y}^{(1)} &= -\sin\phi_1\cos\beta\sin(\theta_1+\varphi_1) + \cos\phi_1\cos(\theta_1+\varphi_1),\\
\alpha_{\xi z}^{(1)} &= -\sin\phi_1\sin\beta
\end{aligned}$$

$$\left(\vec{\boldsymbol{\alpha}}_\xi^{(2)}\right)_{o2} = R\left[\vec{\boldsymbol{k}}_{o2}, -\varphi_2\right]\left(\vec{\boldsymbol{\alpha}}_\xi^{(2)}\right)_2 = \alpha_{\xi x}^{(2)}\vec{\boldsymbol{i}}_{o2} + \alpha_{\xi y}^{(2)}\vec{\boldsymbol{j}}_{o2} + \alpha_{\xi z}^{(2)}\vec{\boldsymbol{k}}_{o2} \tag{2.52}$$

where

$$\begin{aligned}
\alpha_{\xi x}^{(2)} &= \sin\phi_2\cos\beta\cos(\theta_2+\varphi_2) - \cos\phi_2\sin(\theta_2+\varphi_2),\\
\alpha_{\xi y}^{(2)} &= -\sin\phi_2\cos\beta\sin(\theta_2+\varphi_2) - \cos\phi_2\cos(\theta_2+\varphi_2),\\
\alpha_{\xi z}^{(2)} &= -\sin\phi_2\sin\beta
\end{aligned}$$

$$\left(\vec{\boldsymbol{\alpha}}_\eta^{(2)}\right)_{o2} = R\left[\vec{\boldsymbol{k}}_{o2}, -\varphi_2\right]\left(\vec{\boldsymbol{\alpha}}_\eta^{(2)}\right)_2 = \alpha_{\eta x}^{(2)}\vec{\boldsymbol{i}}_{o2} + \alpha_{\eta y}^{(2)}\vec{\boldsymbol{j}}_{o2} + \alpha_{\eta z}^{(2)}\vec{\boldsymbol{k}}_{o2} \tag{2.53}$$

where

$$\begin{aligned}
\alpha_{\eta x}^{(2)} &= \left(\alpha_{\eta x}^{(2)}\right)_2\cos(\theta_2+\varphi_2) + \left(\alpha_{\eta y}^{(2)}\right)_2\sin(\theta_2+\varphi_2),\\
\alpha_{\eta y}^{(2)} &= -\left(\alpha_{\eta x}^{(2)}\right)_2\sin(\theta_2+\varphi_2) + \left(\alpha_{\eta y}^{(2)}\right)_2\cos(\theta_2+\varphi_2),\\
\alpha_{\eta z}^{(2)} &= \left(\alpha_{\eta z}^{(2)}\right)_2
\end{aligned}$$

2.7 Curvature Parameters of Two Tooth Surfaces

According to the solution method of the curvature parameters in gear meshing theory, the normal curvature and mean curvature of any point on the surface can be obtained by two fundamental quantities of this point:

$$k_n = \frac{Ld\theta^2 + 2Md\theta d\phi + Nd\phi^2}{Ed\theta^2 + 2Fd\theta d\phi + Gd\phi^2} \tag{2.54}$$

$$H = \frac{LG - 2MF + NE}{2(EG - F^2)} \tag{2.55}$$

In (2.54), let $d\theta = 0$; then, the principal curvature along the principal direction $\left(\vec{\boldsymbol{\alpha}}_{\xi}^{(1)}\right)_1$ can be obtained as:

$$k_{\xi}^{(1)} = \frac{N_1}{G_1} = \frac{-n_{1g}}{D_1 \sin \phi_1 \rho_1} \tag{2.56}$$

According to the relationship between the normal curvature and average curvature, the principal curvature along the principal direction $\left(\vec{\boldsymbol{\alpha}}_{\eta}^{(1)}\right)_1$ can be expressed as:

$$k_{\eta}^{(1)} = 2H_1 - k_{\xi}^{(1)} \tag{2.57}$$

Given that $d\theta = 0$, the expression of the main direction $\left(\vec{\boldsymbol{\alpha}}_{\xi}^{(1)}\right)_1$ can be rewritten as:

$$\left(\vec{\boldsymbol{\alpha}}_{\xi}^{(1)}\right)_1 = \frac{d\left(\vec{\boldsymbol{r}}_1\right)_1}{ds} = \frac{\partial\left(\vec{\boldsymbol{r}}_1\right)_1}{\partial\theta}\frac{d\theta}{ds} + \frac{\partial\left(\vec{\boldsymbol{r}}_1\right)_1}{\partial\phi}\frac{d\phi}{ds} = \frac{\partial\left(\vec{\boldsymbol{r}}_1\right)_1}{\partial\phi}\frac{d\phi}{ds} \tag{2.58}$$

Dotting both sides with $\frac{d\phi}{ds}$:

$$\frac{d\phi_1}{ds} = \frac{\left(\vec{\boldsymbol{\alpha}}_{\xi}^{(1)}\right)_1 \cdot \frac{\partial\left(\vec{\boldsymbol{r}}_1\right)_1}{\partial\phi_1}}{\left|\frac{\partial\left(\vec{\boldsymbol{r}}_1\right)_1}{\partial\phi_1}\right|^2} = \frac{\frac{\partial\left(\vec{\boldsymbol{r}}_1\right)_1}{\partial\phi_1} \cdot \frac{\partial\left(\vec{\boldsymbol{r}}_1\right)_1}{\partial\phi_1}}{\rho_1\left|\frac{\partial\left(\vec{\boldsymbol{r}}_1\right)_1}{\partial\phi_1}\right|^2} = \frac{1}{\rho_1} \tag{2.59}$$

Therefore, the geodesic torsion on the helicoid of the convex teeth along the main direction $\left(\vec{\boldsymbol{\alpha}}_{\xi}^{(1)}\right)_1$ can be obtained:

$$\begin{aligned}\tau_{\xi}^{(1)} &= \frac{1}{\rho_1 D_1}\begin{vmatrix}\left(\frac{d\phi_1}{ds}\right)^2 & 0 & 0\\ E_1 & F_1 & G_1\\ L_1 & M_1 & N_1\end{vmatrix} = \left(\frac{d\phi_1}{ds}\right)^2 \frac{F_1 N_1 - {\rho_1}^2 M_1}{\rho_1 D_1}\\ &= \frac{-1}{{\rho_1}^2 D_1^2}\left[\frac{F_1\left(n_{1g} + \rho_1 \sin\beta \sin^2\phi_1\right)}{\sin\phi_1} + {\rho_1}^2 p_1\right]\end{aligned} \tag{2.60}$$

Similarly, the curvature parameters of the concave tooth surface can be obtained according to the above steps.

In (2.54), let $d\theta = 0$; then, the principal curvature along the principal direction $\left(\vec{\boldsymbol{\alpha}}_{\xi}^{(2)}\right)_2$ can be obtained:

$$k_{\xi}^{(2)} = \frac{N_2}{G_2} = \frac{-n_{1g}}{D_1 \sin\phi_1 \rho_1} \tag{2.61}$$

The principal curvature along the principal direction $\left(\vec{\boldsymbol{\alpha}}_{\eta}^{(2)}\right)_2$ can be expressed as:

$$k_{\eta}^{(2)} = 2H_2 - k_{\xi}^{(2)} \tag{2.62}$$

Therefore, we can calculate the geodesic torsion of the concave tooth along the main direction $\left(\vec{\boldsymbol{\alpha}}_{\xi}^{(2)}\right)_2$:

$$\tau_{\xi}^{(2)} = \left(\frac{d\phi_2}{ds}\right)^2 \frac{F_2 N_2 - {\rho_2}^2 M_2}{\rho_2 D_2} = \frac{1}{{\rho_2}^2 D_2^2}\left[\frac{F_2\left(n_{2g} - \rho_2 \sin\beta \sin^2\phi_2\right)}{\sin\phi_2} - p_2 {\rho_2}^2\right] \tag{2.63}$$

2.8 Relative Curvature Parameters of the Gear Drive

At any contact point M, we can use the relative positions of the two coordinate systems to calculate the relative principal curvature of the gear pair. They are $\sigma_M^{(1)}\left\{M; \left(\vec{\boldsymbol{a}}_{\xi}^{(1)}\right)_1, \left(\vec{\boldsymbol{a}}_{\eta}^{(1)}\right)_1, \left(\vec{\boldsymbol{n}}_1\right)_1\right\}$ and $\sigma_M^{(2)}\left\{M; \left(\vec{\boldsymbol{a}}_{\xi}^{(2)}\right)_2, \left(\vec{\boldsymbol{a}}_{\eta}^{(2)}\right)_2, \left(\vec{\boldsymbol{n}}_2\right)_2\right\}$, the relative position of which is shown in Fig. 2.4.

The directed angle γ from $\vec{\boldsymbol{\alpha}}_{\xi}^{(1)}$ to $\vec{\boldsymbol{\alpha}}_{\xi}^{(2)}$, measured along the common normal $\vec{\boldsymbol{n}}_1$, can be used to determine the relative location between the two frames $\sigma_M^{(1)}$ and $\sigma_M^{(2)}$.

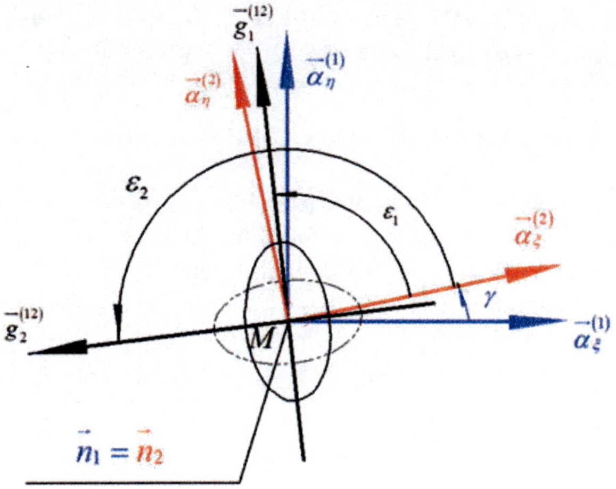

Fig. 2.4 The relative position between two tooth surfaces

Obviously, the value range of the directed angle is $[0, 2\pi)$, and its sine and cosine values can be respectively obtained by the following two formulas:

$$\sin\gamma = -\left(\vec{a}_\xi^{(1)}\right)_{o2} \cdot \left(\vec{a}_\eta^{(2)}\right)_{o2} = -\left(a_{\xi x}^{(1)} a_{\eta x}^{(2)} + a_{\xi y}^{(1)} a_{\eta y}^{(2)} + a_{\xi z}^{(1)} a_{\eta z}^{(2)}\right) \tag{2.64}$$

$$\cos\gamma = \left(\vec{a}_\xi^{(1)}\right)_{o2} \cdot \left(\vec{a}_\xi^{(2)}\right)_{o2} = a_{\xi x}^{(1)} a_{\xi x}^{(2)} + a_{\xi y}^{(1)} a_{\xi y}^{(2)} + a_{\xi z}^{(1)} a_{\xi z}^{(2)} \tag{2.65}$$

Through the generalized Euler formula and Bertrand formula, we can calculate the relative curvature along $\vec{\boldsymbol{\alpha}}_\xi^{(2)}$ and $\vec{\boldsymbol{\alpha}}_\xi^{(1)}$ at the contact point M of the gear pair:

$$\widehat{k}_\xi^{(1)} = k_\xi^{(1)} \cos^2\gamma + k_\eta^{(1)} \sin^2\gamma + 2\tau_\xi^{(1)} \sin\gamma\cos\gamma \tag{2.66}$$

$$\widehat{k}_\eta^{(1)} = k_\xi^{(1)} \sin^2\gamma + k_\eta^{(1)} \cos^2\gamma - 2\tau_\xi^{(1)} \sin\gamma\cos\gamma \tag{2.67}$$

$$\widehat{\tau}_\xi^{(1)} = \left(k_\eta^{(1)} - k_\xi^{(1)}\right) \sin\gamma\cos\gamma + \tau_\xi^{(1)}\left(\cos^2\gamma - \sin^2\gamma\right) \tag{2.68}$$

The relative curvature parameters of the gear pair along with $\vec{\boldsymbol{\alpha}}_\xi^{(2)}$ and $\vec{\boldsymbol{\alpha}}_\eta^{(2)}$ can be solved as follows:

$$k_\xi^{(12)} = \widehat{k}_\xi^{(1)} - k_\xi^{(2)}, k_\eta^{(12)} = \widehat{k}_\eta^{(1)} - k_\eta^{(2)}, \tau_\xi^{(12)} = \widehat{\tau}_\xi^{(1)} - \tau_\xi^{(2)} \tag{2.69}$$

2.9 Relative Principal Curvature and the Relative Principal Direction of the Gear Drive

Calculating the relative principal curvature of the contact surface can provide a theoretical basis for judging whether there is curvature interference in point contact gear pairs and can help determine the contact ellipse at the contact point. In the previous chapter, we defined the angular position relation of the two moving coordinate systems at the contact point. In this section, we propose a more reasonable and effective method to calculate the relative principal curvature based on the traditional local synthesis method.

Based on the generalized Euler formula, two relative principal curvatures $k_1^{(12)}$ and $k_2^{(12)}$ of gear pairs along their principal directions $\vec{\boldsymbol{g}}_1^{(12)}$ and $\vec{\boldsymbol{g}}_2^{(12)}$, respectively, can be obtained. The results are as follows:

$$k_1^{(12)} = k_\xi^{(12)} \cos^2 \varepsilon_1 + k_\eta^{(12)} \sin^2 \varepsilon_1 + \tau_\xi^{(12)} \sin 2\varepsilon_1 \tag{2.70}$$

$$k_2^{(12)} = k_\xi^{(12)} \cos^2 \varepsilon_2 + k_\eta^{(12)} \sin^2 \varepsilon_2 + \tau_\xi^{(12)} \sin 2\varepsilon_2 \tag{2.71}$$

in which the angles ε_1 and ε_2 are the directed angles from the vector $\vec{\boldsymbol{\alpha}}_\xi^{(2)}$ to the two relative principal directions $\vec{\boldsymbol{g}}_1^{(12)}$ and $\vec{\boldsymbol{g}}_2^{(12)}$ of the gear pair measured along the common normal, respectively. The values of ε_1 and ε_2 are determined according to their value ranges as:

$$\varepsilon_1 = \begin{cases} \dfrac{1}{2} \arctan 2\left|\dfrac{\tau_\xi^{(12)}}{k_\eta^{(12)} - k_\xi^{(12)}}\right|, & \dfrac{\tau_\xi^{(12)}}{k_\eta^{(12)} - k_\xi^{(12)}} \le 0 \\ 90^\circ - \dfrac{1}{2} \arctan \dfrac{2\tau_\xi^{(12)}}{k_\eta^{(12)} - k_\xi^{(12)}}, & \dfrac{\tau_\xi^{(12)}}{k_\eta^{(12)} - k_\xi^{(12)}} \ge 0 \end{cases}, \quad \varepsilon_2 = \varepsilon_1 + 90^\circ \tag{2.72}$$

Moreover, the two relative principal directions $\vec{\boldsymbol{g}}_1^{(12)}$ and $\vec{\boldsymbol{g}}_2^{(12)}$ can be obtained by means of the two unit vectors $\vec{\boldsymbol{\alpha}}_\xi^{(2)}$, $\vec{\boldsymbol{\alpha}}_\eta^{(2)}$ and the two directed angles ε_1 and ε_2, respectively, as below:

$$\left(\vec{\boldsymbol{g}}_1^{(12)}\right)_{o2} = \cos \varepsilon_1 \left(\vec{\boldsymbol{\alpha}}_\xi^{(2)}\right)_{o2} + \sin \varepsilon_1 \left(\vec{\boldsymbol{\alpha}}_\eta^{(2)}\right)_{o2} \tag{2.73}$$

$$\left(\vec{\boldsymbol{g}}_2^{(12)}\right)_{o2} = \cos \varepsilon_2 \left(\vec{\boldsymbol{\alpha}}_\xi^{(2)}\right)_{o2} + \sin \varepsilon_2 \left(\vec{\boldsymbol{\alpha}}_\eta^{(2)}\right)_{o2} \tag{2.74}$$

Since the direction of $\vec{\boldsymbol{n}}_1$ is from the inside to its outside, the two relative principal curvatures $k_1^{(12)}$ and $k_2^{(12)}$ must satisfy $k_1^{(12)} \le 0$ and $k_2^{(12)} \le 0$ to avoid the curvature interference around the contact point M.

Theoretically, the momentary contact of the gear drive at a contact point will extend to an elliptical region due to the elasticity of its tooth surfaces. From Hertz theory, the size (semi-major axis a_e and semi-minor axis b_e) of a contact ellipse can be determined as follows:

$$a_e = \sqrt{\frac{2\Delta\zeta}{\min\left(\left|k_1^{(12)}\right|, \left|k_2^{(12)}\right|\right)}} = \sqrt{\frac{2\Delta\zeta}{\left|k_1^{(12)}\right|}} \tag{2.75}$$

$$b_e = \sqrt{\frac{2\Delta\zeta}{\max\left(\left|k_1^{(12)}\right|, \left|k_2^{(12)}\right|\right)}} = \sqrt{\frac{2\Delta\zeta}{\left|k_2^{(12)}\right|}} \tag{2.76}$$

in which the symbol Δ_ζ is a constant, the value of which is usually taken to be $\Delta_\zeta = 0.00635\ mm$ for the gear drive. Moreover, the contact ellipse with the solid line (the major axis coincides with $\vec{g}_1^{(12)}$) corresponds to the case of $k_1^{(12)} < k_2^{(12)}$ as shown in Fig. 2.4, whereas the contact ellipse with the double dots line (the major axis coincides with $\vec{g}_2^{(12)}$) corresponds to the case of $k_1^{(12)} > k_2^{(12)}$.

So, the corresponding radius vectors of the two end points A and B of the long axis of the instantaneous contact ellipse are:

$$\left(\vec{r}_{1A}\right)_{o2} = \left(\vec{r}_1\right)_{o2} + a_e\left(\vec{g}_1\right)_{o2}, \left(\vec{r}_{1B}\right)_{o2} = \left(\vec{r}_1\right)_{o2} - a_e\left(\vec{g}_1\right)_{o2} \tag{2.77}$$

$$\left(\vec{r}_{2A}\right)_{o2} = \left(\vec{r}_2\right)_{o2} + a_e\left(\vec{g}_1\right)_{o2}, \left(\vec{r}_{2B}\right)_{o2} = \left(\vec{r}_2\right)_{o2} - a_e\left(\vec{g}_1\right)_{o2} \tag{2.78}$$

2.10 Numerical Example Study

This chapter determines the helical surface equation of a normal double-circular-arc gear, and the corresponding theoretical formula of tooth surface meshing of a gear pair is established. This chapter will introduce how to calculate the contact path boundary of a gear pair, determine the meshing reference point, analyze the method of obtaining each contact point, and put forward the corresponding calculation principle.

2.10.1 Main Technical Parameters

Based on the results of tooth contact analysis, a numerical example of the meshing characteristics of normal double-circular-arc gear transmission is provided. The main parameters of the normal double-circular-arc tooth profile are designed by referring to the basic tooth profile parameters of the double-circular-arc gear GB/T12759-1991 in the gear manual.

Table 2.1 The basic parameters of a normal double-arc gear pair

Parameter	Code and formula	The numerical
Center distance	a (mm)	181
Normal module	m (mm)	5
The number of teeth in gear 1	Z_1	24
The number of teeth in gear 2	Z_2	46
Drive ratio	i_{12}	23/12
Helical angle	$\beta = \arccos\left(\frac{m(z_1+z_2)}{2a}\right)(°7.)$	14.7937
Diameter of indexing circle of gear 1	$d_1 = \frac{mz_1}{\cos\beta}$ (mm)	124.1143
Diameter of indexing circle of gear 2	$d_2 = \frac{mz_2}{\cos\beta}$ (mm)	237.8857
Helical parameters of gear 1	$p_1 = d_1 \cot\beta/2$	234.9791
Helical parameters of gear 2	$p_2 = d_2 \cot\beta/2$	450.3766
Coefficient of tooth width	ϕ_a	0.8
Tooth width	$b = \phi_a a$ (mm)	144.8
Arc radius of convex tooth surface	$\rho_1 = 1.3m$ (mm)	6.5
Arc radius of concave tooth surface	$\rho_1 = 1.68m$ (mm)	8.4
Center offset of convex profile	$l_1 = 0.5298m$ (mm)	2.6445
Center offset of concave surface	$l_2 = 0.6994m$ (mm)	3.4970
Displacement of center of convex surface	$b_1 = 0.02m$ (mm)	0.12
Displacement of center of concave surface	$b_2 = 0.03m$ (mm)	0.15

In this chapter, a set of circular-arc gear transmission parameters are selected to carry out a numerical example study of transmission. The basic parameters of the gear pair in the calculation example are shown in Table 2.1.

2.10.2 *Calculating Method for Instantaneous Contact Point*

According to the binary nonlinear equation (2.24) obtained above, two unknowns ϕ_1 and ϕ_2 can be solved. By substituting the results in (2.21) and (2.22), the values of $\sin(\theta_2 + \varphi_2)$ and $\cos(\theta_2 + \varphi_2)$, respectively, can be obtained. Thus, $(\theta_2 + \varphi_2)$ can be obtained using the inverse trigonometric function. Then, by substituting the values of $\sin(\theta_2 + \varphi_2)$ and $\cos(\theta_2 + \varphi_2)$ in Eqs. (2.17) and (2.18), the values of $\sin(\theta_1 + \varphi_1)$ and $\cos(\theta_1 + \varphi_1)$, respectively, and the value of $(\theta_1 + \varphi_1)$ can be obtained.

Then, through the last equation in the common point equation (2.19), the relation between θ_1 and θ_2 can be obtained as follows:

$$\theta_2 = f(\theta_1) = \frac{\left(\rho_1 \cos\phi_1 - \rho_2 \cos\phi_2 - l_1 + l_2 + \frac{\pi m_2}{2}\right)\sin\beta + p_1\theta_1}{p_2}$$
$$= \frac{p_1}{p_2}\theta_1 + \frac{1}{p_2}(\rho_1 \cos\phi_1 - \rho_2 \cos\phi_2) + \frac{1}{p_2}\left(\frac{\pi m_2}{2} - l_1 + l_2\right) \quad (2.79)$$

So far, there are still three unknown parameters and two angle sum equations in the contact equation, and the solution of the new system is the same as that of the original system. Therefore, to make the system well-conditioned, we need to add an equation or assign a specific value to one of the three variables, θ_1,φ_1, and φ_2, to obtain the other parameters to determine each contact point.

Before the numerical simulation, we should determine the tooth surface boundary equation. First, we should establish the plane Cartesian coordinate system $O_1 - x_{R1}y_{R1}$, where point O_{o1} is located at the center of the gear, the x_{R1} axis is the axis direction of the gear, and y_{R1} is the radial direction of the gear. r_{a1}, r_1, and r_{f1} represent the radius of the gear apex circle, indexing circle, and root circle, respectively. h_a and h_f represent the gear's apex height and root height, respectively.

The coordinates of any point on the helical surface of the gear satisfy the following mapping:

$$x_{R1} = z_{o1} = (\rho_1 \cos\phi_1 - l_1)\sin\beta + p_1\theta_1 \tag{2.80}$$

$$y_{R1} = \sqrt{x_{o1}^2 + y_{o1}^2} = \sqrt{(\rho_1 \cos\phi_1 - l_1)^2 \cos^2\beta + \left(\rho_1 \sin\phi_1 + b_1 + \frac{d_1}{2}\right)^2} \tag{2.81}$$

Two major boundary equations of the tooth surface can be obtained:
Left-hand side boundary equation:

$$x_{R1} = -L_b/2 \tag{2.82}$$

Right-hand side boundary equation:

$$x_{R1} = L_b/2 \tag{2.83}$$

A single variable ϕ_1 determines the values of the addendum and root equations. In the previous chapter, we solved the binary nonlinear equation and obtained the value of ϕ_1 to determine the position of the contact point in the direction of the tooth height.

The contact end of the contact path is along the tooth length direction. Therefore, we can determine the value range of variable θ_1 using the tooth width equation as follows:

$$\theta_1^{(\max)} = \frac{L_b/2 - (\rho_1 \cos\phi_1 - l_1)\sin\beta}{p_1} \tag{2.84}$$

$$\theta_1^{(\min)} = \frac{-L_b/2 - (\rho_1 \cos\phi_1 - l_1)\sin\beta}{p_1} \tag{2.85}$$

Equations (2.84) and (2.85) are the equations for determining the two end points of the contact path, which we should guarantee to be within the range of the desired contact points during numerical calculation.

After determining the two contact end points of the contact path, we can assign values within the value range of variable θ_1 and determine the internal contact points by solving contact equations. Then, we can get the values of θ_2,φ_1, and φ_2.

2.10.3 Numerical Results

When solving nonlinear contact equations, there are generally two critical problems to be solved: one is to determine the existence of the solution and the other is how to provide an appropriate initial value for an iterative solution. If the technical parameters are set improperly, then their answers will not exist in the given solution domain. This means that the two tooth surfaces are separated and that the scheme is unreasonable. After the process parameters are reset, the nonlinear contact equations need to be solved again. Even if the system has no solution result, it may have obvious physical significance and the judgment of the existence of the solution can provide a theoretical basis for the selection and setting of parameters.

First, the nonlinear equation (2.24) are solved. By putting each parameter value into the formula and using MATLAB to plot the intersection of two nonlinear equations, the initial iteration value can be determined. Then, the F solve function is used for iteration to obtain the exact value of the solution. So, we can find the nominal pressure angles ϕ_1 and ϕ_2. As shown in Fig. 2.5 below:

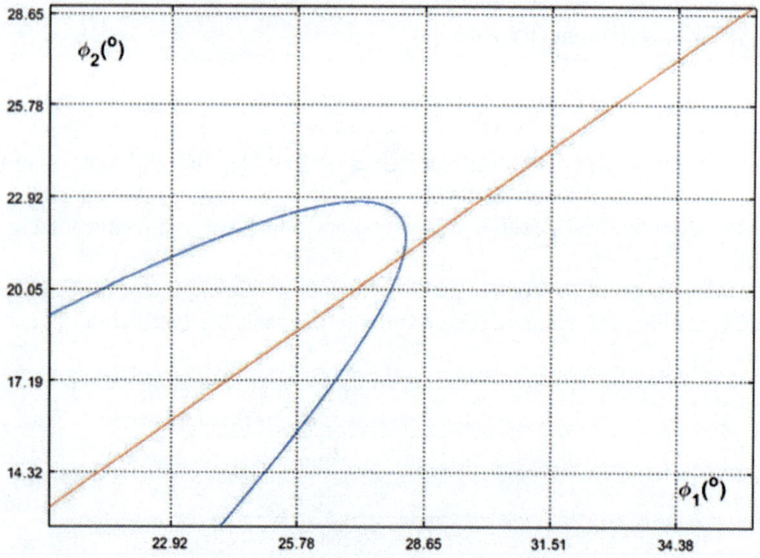

Fig. 2.5 A graph of the intersection of two nonlinear equations

The data of ϕ_1 and ϕ_2 are shown in Table 2.2:

The values of $\sin(\theta_2 + \varphi_2)$ and $\cos(\theta_2 + \varphi_2)$ can be obtained by substituting the results in Table 2.2 in (2.21) and (2.22), respectively. Then, combined with (2.17) and (2.18), we can get the values of $\sin(\theta_1 + \varphi_1)$ and $\cos(\theta_1 + \varphi_1)$, which are listed in Table 2.3:

After obtaining the result of θ_1 using (2.84) and (2.85), the numerical result of θ_2 can be obtained by combining Eq. (2.79), which is listed in Table 2.4:

The meshing interval is determined according to the maximum and minimum values. We can obtain multiple instantaneous contact points by bisecting the area with the reference points as the center. Therefore, the contact points are numbered, and the numerical solution is solved iteratively using MATLAB. So, we can get the values of φ_1 and φ_2 at particular points. The calculation results of each point are displayed in Table 2.5.

Each point is plotted on the coordinate diagram of the gear, and the contact trace is plotted by the interpolation method. The results are displayed in Figs. 2.6 and 2.7.

Besides, the sizes of the instantaneous contact ellipses are determined from (2.77) and (2.78) based on the relative principal curvatures obtained above. The major axes of contact ellipses are also drawn in Figs. 2.6 and 2.7. Not only that, the values of the directed angle γ at all the contacts are obtained from (2.64) and (2.65). Here, $\sin\gamma = 0.0277$ and $\cos\gamma = 0.9994$. The value of γ is constant at every contact point, $\gamma = 1.5876°$.

We have expounded that, in the proximity of a contact point, the condition of avoiding curvature interference is that both of the relative principal curvatures $k_1^{(12)}$ and $k_2^{(12)}$ be less than 0. Consequently, the values of the relative principal curvatures in Table 2.5 indicate that the bearing contact patterns of the gear pair do not have curvature interference. After observing Figs. 2.6 and 2.7, it is easy to discover that the contact paths of the tooth surface couples can cover the whole thread length of the gear. This further implies that this kind of a gear pair has a larger contact ratio.

Table 2.2 The numerical results of ϕ_1 and ϕ_2

$\phi_1(°)$	28.1494
$\phi_2(°)$	21.0791

Table 2.3 Sine and cosine values of two tooth surfaces

Tooth face 1	$\sin(\theta_1 + \varphi_1)$	$\cos(\theta_1 + \varphi_1)$
	0.9966	0.0829
Tooth face 2	$\sin(\theta_2 + \varphi_2)$	$\cos(\theta_2 + \varphi_2)$
	0.9986	0.0504

Table 2.4 The values of θ_1 and θ_2 at particular points

	The left endpoint	The right endpoint
$\theta_1(°)$	−17.8476	17.4638
$\theta_2(°)$	−9.0986	9.3278

Table 2.5 The parameter calculation results of contact points on two tooth surfaces

Contact point	①	②	③	④	⑤	⑥
$\phi_1(°)$	28.1494	28.1494	28.1494	28.1494	28.1494	28.1494
$\phi_2(°)$	21.0791	21.0791	21.0791	21.0791	21.0791	21.0791
$\theta_1(°)$	−17.8476	−14.3171	−10.7865	−7.2559	−3.7254	−0.1948
$\theta_2(°)$	−9.0980	−7.2559	−5.4139	−3.5719	−1.7298	0.1122
$\varphi_1(°)$	104.9601	101.4296	97.8990	94.3684	90.8379	87.3073
$\varphi_2(°)$	94.3426	92.5006	90.6586	88.8165	86.9745	85.1325
$\vec{k}_1^{(12)}$ (μm)	−0.6085	−0.6085	−0.6085	−0.6085	−0.6085	−0.6085
$\vec{k}_2^{(12)}$ (μm)	−35	−35	−35	−35	−35	−35
z_{o1}(mm)	−72.4	−57.9272	−43.4478	−28.9683	−14.4889	0
$\sqrt{x_{o1}^2+y_{o1}^2}$(mm)	65.3029	65.3029	65.3029	65.3029	65.3029	65.3029
z_{o2}(mm)	−72.4	−57.9272	−43.4478	−28.9683	−14.4889	0
$\sqrt{x_{o2}^2+y_{o2}^2}$(mm)	116.1718	116.1718	116.1718	116.1718	116.1718	116.1718
Contact point	⑦	⑧	⑨	⑩	⑪	⑫
$\phi_1(°)$	28.1494	28.1494	28.1494	28.1494	28.1494	28.1494
$\phi_2(°)$	21.0791	21.0791	21.0791	21.0791	21.0791	21.0791
$\theta_1(°)$	2.7483	5.6914	8.6345	11.5776	14.5207	17.4638
$\theta_2(°)$	1.6477	3.1833	4.7188	6.2543	7.7898	9.3254
$\varphi_1(°)$	84.3642	81.4211	78.4780	75.5349	72.5918	69.6487
$\varphi_2(°)$	83.5969	82.0614	80.5259	78.9904	77.4548	75.9193
$\vec{k}_1^{(12)}$ (μm)	−0.6085	−0.6085	−0.6085	−0.6085	−0.6085	−0.6085
$\vec{k}_2^{(12)}$ (μm)	−35	−35	−35	−35	−35	−35
z_{o1}(mm)	12.0606	24.1307	36.2008	48.2708	60.3409	72.4
$\sqrt{x_{o1}^2+y_{o1}^2}$(mm)	65.3029	65.3029	65.3029	65.3029	65.3029	65.3029
z_{o2}(mm)	12.0606	24.1307	36.2008	48.2708	60.3409	72.4110
$\sqrt{x_{o2}^2+y_{o2}^2}$(mm)	116.1718	116.1718	116.1718	116.1718	116.1718	116.1718

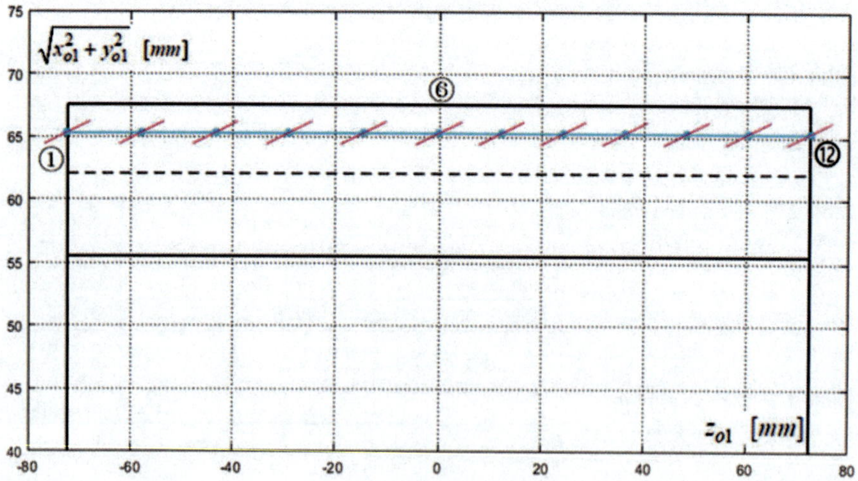

Fig. 2.6 A projection drawing of the contact path on gear 1

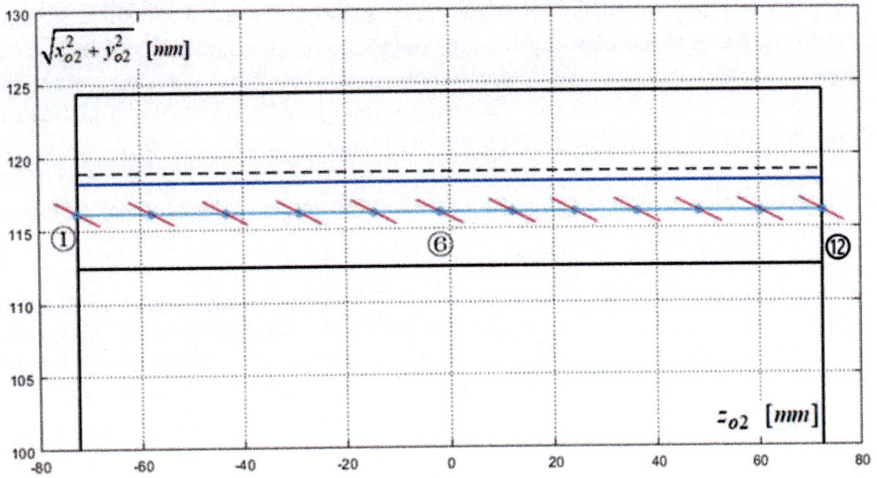

Fig. 2.7 A projection drawing of the contact path on gear 2

The lengths of the major axes of the contact ellipses and their orientations in Figs. 2.6 and 2.7 are determined from (2.82) and (2.83), respectively. Figures 2.6 and 2.7 depict that the orientations of the major axes of the contact ellipses are roughly along the contact path. Moreover, the length difference between the contact paths of the two tooth surface couples is almost 0, which indicates that the meshing of these two tooth surface couples is almost symmetric.

2.11 Conclusion

This chapter presents a new type of a circular-arc gear drive with a circular tooth profile in its normal section. The gear is designed as a new gear, the normal section of which is arc-shaped, and the tooth surface equation of the arc-shaped gear is established by the spherical vector function. The tooth profile is relatively simple, and its point-contact meshing theory is well-established and investigated. The tooth surface equations, the normal vector equations, the meshing equations, and so on, are all acquired for this type of a circular-arc gear drive. In accordance with the tooth contact analysis principle, the contact equation groups are established to ascertain its momentary contact points. To expediently solve the preceding equations with six unknowns, the elimination artifice is adopted to transform it into its equivalent form with three unknowns, thus making the angular position relationship between the two moving frames located at the contact point clear, and, based on this, the relative principal curvature of the gear pair is correctly determined.

The numerical cases show that the gears proposed are instantaneously in point contact, tooth surface meshing belongs to point contact, and the contact trace is roughly along the direction of the tooth length. The gear pair can accomplish constant ratio transmission, and its velocity ratio can be completely equal to the nominal transmitting ratio by rigorous derivation. The calculation results of the relative principal curvature show that there is no curvature interference in tooth surface meshing. Its meshing behavior is good in terms of its geometry using its bearing contact pattern.

Acknowledgments This work was supported by the National Natural Science Foundation of China (52075083 and 52172401) and the Open Fund of the Key Laboratory for Metallurgical Equipment and Control of Education Ministry in Wuhan University of Science and Technology (MECOF2021B02 and MECOF2020B03).

References

1. Lu, X., Shang, J.: Meshing Principle of Arc Gear [M]. China Machine Press (1990)

Bibliography

Chen, W.: Gear Meshing Theory [M]. China Coal Industry Press, Beijing (1986)
Chen, Z. (ed.): Circular Arc Tooth Cylindrical Gear Transmission [M]. Higher Education Press, Beijing (1995)
Cheng, D.: Mechanical Design Manual [M], vol. 3, 4th edn. Chemical Industry Press, Beijing (2002)
Deng, J., Wang, T.: Tooth profile equation and meshing process of double-circular-arc gear [J]. Coal Mine Machine.
Fu, Z.: Differential Geometry and Principle of Gear Meshing [M]. Petroleum University Press, Shandong (1999)
He, Q.: Run-in characteristics of double-circular-arc gear and analysis of tooth surface contact state [J]. Oil Field Machine. (1982)
Li, Q., et al.: Numerical Method for Solving Nonlinear Equations [M]. Science Press, Beijing (1997)
Litvin, F.L., Lu, X., Gao, Y., Wang S (translated): Principle of Gear meshing [M]. Shanghai Science and Technology Press, Shanghai (1984)
Mei, X., Huang, J.: Differential Geometry [M]. Higher Education Press, Beijing (1998)
Shao, J. (ed.): Circular Arc Gear[M], 2nd edn. China Machine Press, Beijing (1980)
Wang, S.: New Technology Theory of ZC1 Worm Drive [M]. Tianjin University Press, Tianjin (2003)
Wu, D., Luo, J.: Theory of Gear Meshing [M]. Science Press, Beijing (1985)
Wu, D.: Handout on Differential Geometry [M]. People's Education Press, Beijing (1980)
Wu, H. (ed.): Principle of Gear Meshing [M]. Harbin Institute of Technology Press, Harbin (1979)

Wu, X.: Meshing Principle of Gear Drives, 2nd edn. Xi'an Jiaotong University Press, Xi'an (2009)
Xue, D., Chen, Y.: MATLAB For Solving Advanced Applied Mathematics Problems [M], 3rd edn. Tsinghua University Press, Beijing (2013)
Yan, Z.: Mathematical basis of gear meshing (IV)[J]. Acta Mathematicae Applicatae Sinica. (1980)
Zhao, Y., et al.: Novel methods for curvature analysis and their application to TA worm [J]. Mech Mach Theory. (2016)

Chapter 3
Helical Bevel Novikov Gears

Michał Batsch

3.1 Introduction

Bevel gears are widely used in high-power drives in which shaft axes cross at a specific angle. They are implemented, e.g., in drive trains of aircraft propulsion systems (e.g., as the main rotor or tail rotor gears) as well as in the automotive industry (in drive axles). Helical bevel gears are the most commonly used due to their durability. Tooth surface topography largely depends on the machining method. Leading bevel gear machine tool manufacturers such as Gleason and Klingelnberg have developed software (GEMS or KiMOS) to aid in the design of such gears. These software feature modules enabling tooth contact analysis and geometry optimization for the correct contact pattern (VHJ analysis) and strength.

There have been repeated attempts to improve bevel meshing performance by making topological modifications by means of machining with a positive feedback loop [1–3] as well as application of nonconventional tooth profiles. One such profile is the Wildhaber profile, in which circular arcs are present in the normal section [4]. The mathematical framework for modeling this mesh form, generalized over three different gear types, is presented in the study by HouJun et al. [5]. An example of the application of this type of meshing in a bevel gear pair is a nutation drive, in which double circular-arc gearing is used [6]. The bevel gears in the study featured a loxodromic flank pitch line, which offers certain benefits when machined on 5-axis milling centers [7].

Another unconventional profile employed in bevel gear pairs is the eccentric-cycloid profile [8] in which the pinion teeth's profile is a circular arc in the frontal section, whereas the gear teeth's profile is an equidistant curve of an epicycloid. The fundamentals of mathematical modeling and tooth contact analysis in reference to

M. Batsch
Rzeszów University of Technology, Rzeszów, USA

S. P. Radzevich, *Novikov/Conformal Gearing*,
https://doi.org/10.1007/978-3-031-10019-2_3

cylindrical gears are presented in the study by Li et al. [9]. The profile was also applied in bevel gear pairs used in machines and cable railways [10, 11].

Another profile type is the convexo-concave Novikov profile, in which circular arcs appear in the frontal section [12]. The application of the profile in cylindrical gears enables the enhancement of load capacity [13], durability [14], or a reduction of the gear pair's overall dimensions [15]. In the literature, we may find early attempts to implement this type of profile in bevel gears with a circular-arc flank pitch line [16]. Studies such as those by Korotkin et al. [17] and Radzevich [18] provide fundamental information on the geometry and its selection for this type of gear teeth.

This chapter presents a mathematical model of a bevel convexo-concave Novikov-type mesh. Based on this model, tooth mesh simulations were carried out, designed to determine the effect of modifications and the gear axis position error on the contact pattern and transmission error. The results were compared with those of a conventional spiral bevel gear pair of the DUPLEX helical type.

3.2 Mathematical Model of a Gear Mesh

3.2.1 Coordinate Systems

This study analyzes an external bevel gear pair with an axis intersection angle $\mathbf{\Sigma}$ and a helical tooth pitch line in which the pinion has a concave profile and the gear has a convex profile (Fig. 3.1).

The fixed coordinate system f is related to the gear housing, and moving systems 1 and 2 are related to the pinion and the gear, respectively. The convex and concave surfaces of the pinion teeth $\mathbf{\Sigma}_{\mathbf{1cx}}$ and $\mathbf{\Sigma}_{\mathbf{1cv}}$ are represented in system 1 by position vectors $\mathbf{r}_{1\text{cx}}^{(1)}$ and $\mathbf{r}_{1\text{cv}}^{(1)}$, respectively. Similarly, the surfaces of the gear teeth $\mathbf{\Sigma}_{\mathbf{2cx}}$ and $\mathbf{\Sigma}_{\mathbf{2cv}}$ are represented in system 2 by position vectors $\mathbf{r}_{2\text{cx}}^{(2)}$ and $\mathbf{r}_{2\text{cv}}^{(2)}$, respectively. The pinion rotates around axis z_1 by angle φ_1, whereas the gear rotates around axis z_2 by angle φ_2. Ideally, in a deviation-free gear pair, the gear axes intersect at the apex point of their pitch cones A. Tooth flanks may be presented in a fixed coordinate system as (3.1) and (3.2):

$$\mathbf{r}_{1\text{cx,cv}}^{(f)} = \mathbf{Rs}(1 \mapsto f) \cdot \mathbf{r}_{1\text{cx,cv}}^{(1)}, \tag{3.1}$$

$$\mathbf{r}_{2\text{cx,cv}}^{(f)} = \mathbf{Rs}(2 \mapsto f) \cdot \mathbf{r}_{2\text{cx,cv}}^{(2)}, \tag{3.2}$$

where $\mathbf{Rs}(1 \mapsto f)$ is a homogeneous matrix of transformation of system 1 to f and $\mathbf{Rs}(2 \mapsto f)$ is a homogeneous matrix of transformation of system 2 to f. An additional coordinate system h (Fig. 3.2) was introduced in order to allow for gear position errors resulting from, e.g., assembly or manufacturing errors.

Accordingly, the gear will rotate around the new axis z_h, shifted by vectors $\mathbf{V}$, $\mathbf{H}$, $\mathbf{J}$ of the coordinate system h. Considering Figs. 3.1 and 3.2, transformation matrices from systems 1 and 2 to f are given as (3.3) and (3.4), respectively:

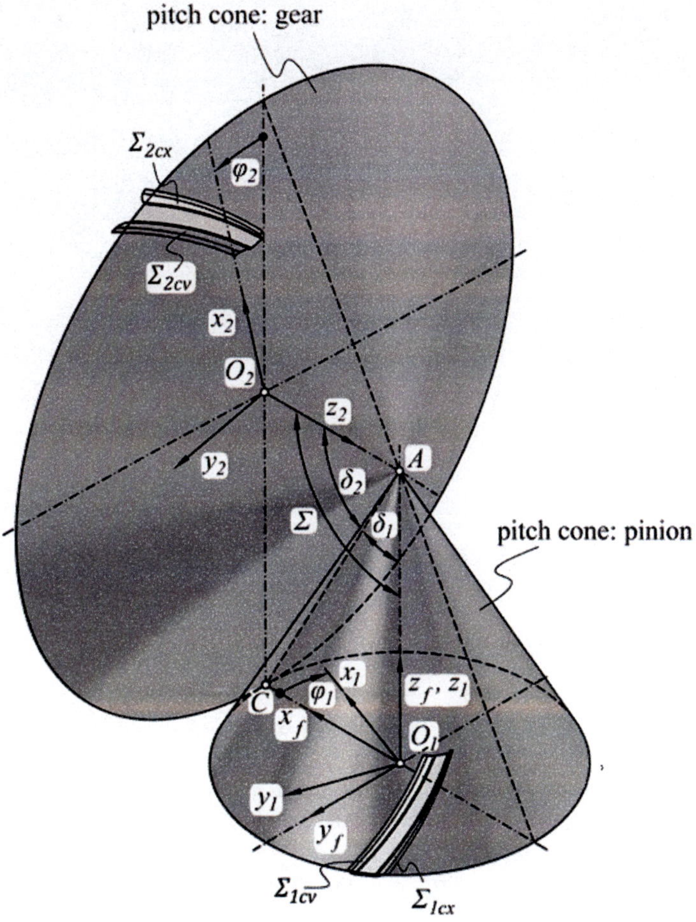

Fig. 3.1 Coordinate systems in bevel meshing analysis

$$\mathbf{Rs}(1 \mapsto f) = \mathbf{Rt}(\varphi_1, Z), \tag{3.3}$$

$$\mathbf{Rs}(2 \mapsto f) = \mathbf{Rs}(h \mapsto f) \cdot \mathbf{Rs}(2 \mapsto h), \tag{3.4}$$

where matrices $\mathbf{Rs}(2 \mapsto h)$ and $\mathbf{Rs}(h \mapsto f)$ are expressed by formulas (3.5), (3.6), respectively:

$$\mathbf{Rs}(2 \mapsto h) = \mathbf{Rt}(-\varphi_2, Z), \tag{3.5}$$

$$\begin{aligned} \mathbf{Rs}(h \mapsto f) = {} & \mathbf{Tr}(H, X) \cdot \mathbf{Tr}(V, Y) \cdot \mathbf{Tr}(J, Z) \cdot \\ & \cdot \mathbf{Tr}(r_{e1} - r_{e2} \cos(\pi - \Sigma), X) \cdot \mathbf{Tr}(r_{e2} \sin(\pi - \Sigma), Z) \cdot \mathbf{Rt}(-\Sigma, Y), \end{aligned} \tag{3.6}$$

and r_{e1}, r_{e2} are the outer pitch radii of the pinion and the gear, respectively. The rotation operator **Rt** and the translation operator **Tr** are defined in Appendix D [18].

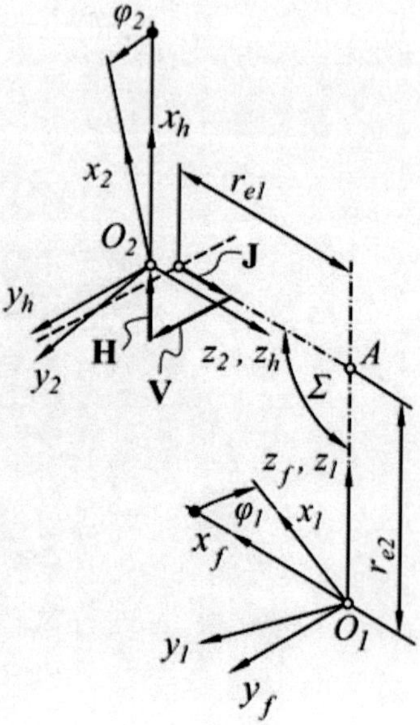

Fig. 3.2 An additional coordinate system

3.2.2 Tooth Surfaces

The tooth profile was defined in the outer transverse section, as shown in Fig. 3.3.

The location of the profile contact point B is defined in the outer transverse section by the external pressure angle α_t. The O' point is the center of the convex profile with radius ρ_{e1}, whereas O is the center of the concave profile with radius ρ_{e2}. Point C is the central point of meshing, which corresponds to the contact point of the back cone pitch circles. Depending on the assumed drive side of the pinion, the contact is made between the convex surface of the pinion tooth and the convex surface of the gear tooth or between the concave surface of the pinion tooth and the convex surface of the gear tooth.

The tooth profile for the concave side of the pinion tooth may be expressed by vector (3.7)

$$\bar{r}^{(1)}_{\mathrm{pr1cv}} = \mathbf{Rt}(-\delta_1, Y) \cdot \begin{bmatrix} \rho_{e1}\cos\theta_1 - d_{CO'}\sin\alpha_t \\ \rho_{e1}\sin\theta_1 + d_{CO'}\cos\alpha_t \\ 0 \\ 1 \end{bmatrix}, \tag{3.7}$$

where θ_1 is the tooth profile parameter. The teeth are tapered. Therefore, the defined profile should be re-scaled so that tooth height and thickness at apex point A equals zero. Furthermore, in order to determine the helical bevel surface, the profile will be

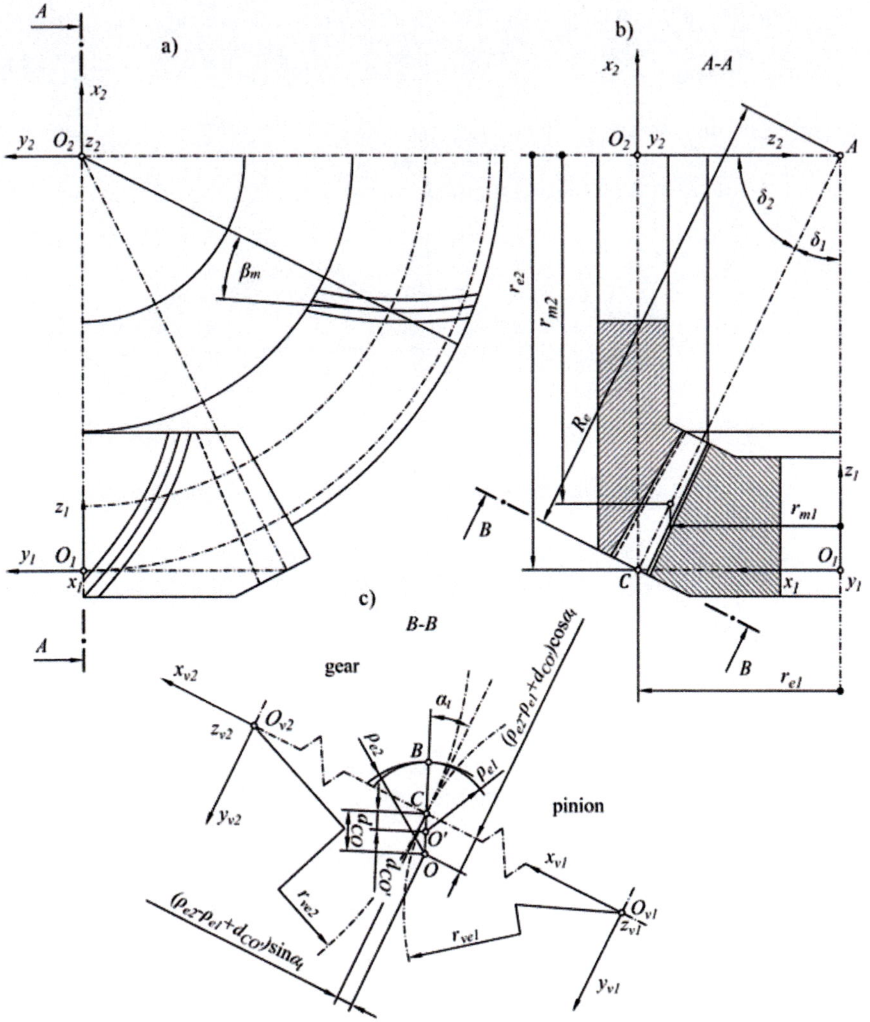

Fig. 3.3 A Bevel Novikov gear in its initial position ($\varphi_1 = \varphi_2 = 0$): (**a**) a view, (**b**) axial section, and (**c**) outer transverse section

shifted along axes x_1, y_1, z_1 by appropriate coordinates of the bevel helix and rotated around axis z_1. In view of the above, the concave surface of the pinion tooth is given as (3.8)

$$\mathbf{r}_{1cv}^{(1)} = \mathbf{Rt}(-\zeta_1, Z) \cdot \mathbf{Tr}(x_{hx1}, X) \cdot \mathbf{Tr}(y_{hx1}, Y) \cdot \mathbf{Tr}(z_{hx1}, Z) \cdot \mathbf{Sca}(r_{z1}/r_{e1}) \cdot \mathbf{r}_{pr1cv}^{(1)}, \tag{3.8}$$

where ζ_1 is the helix parameter, $x_{\mathbf{hx1}}$, $y_{\mathbf{hx1}}$, $z_{\mathbf{hx1}}$ are the coordinates of the helix wrapped around the pitch cone of the pinion (3.9), r_{z1} is the pitch radius in the current section (3.10), and **Sca** is the scale operator (3.11),

$$\begin{cases} x_{hx1} = \left(r_{e1} - \dfrac{H_1 \zeta_1}{2\pi} \tan \delta_1 \right) \cos \zeta_1 \\ y_{hx1} = \left(r_{e1} - \dfrac{H_1 \zeta_1}{2\pi} \tan \delta_1 \right) \sin \zeta_1 \\ z_{hx1} = \dfrac{H_1 \zeta_1}{2\pi} \end{cases}, \tag{3.9}$$

$$r_{z1} = r_{e1} - \frac{H_1 \zeta_1}{2\pi} \tan \delta_1, \tag{3.10}$$

$$\mathbf{Sca}(k) = \begin{bmatrix} k & 0 & 0 & 0 \\ 0 & k & 0 & 0 \\ 0 & 0 & k & 0 \\ 0 & 0 & 0 & 1 \end{bmatrix}, \tag{3.11}$$

where $H_1 = 2\pi r_{m1} \frac{\cos \delta_1}{\tan \beta_m}$ is the spiral lead on the pitch cone, $r_{\mathbf{e1}}$ is the external pitch radius, $r_{\mathbf{m1}}$ is the mean pitch radius, and $\beta_{\mathbf{m}}$ is the mean spiral angle.

The profile of the gear tooth's convex surface is given by vector (3.12)

$$\mathbf{r}_{pr2cx}^{(2)} = \mathbf{Rt}(\delta_2, Y) \cdot \begin{bmatrix} \rho_{e2} \cos \theta_2 - (\rho_2 - \rho_1 + d_{CO'}) \sin \alpha_t \\ \rho_{e2} \sin \theta_2 + (\rho_2 - \rho_1 + d_{CO'}) \cos \alpha_t \\ 0 \\ 1 \end{bmatrix}. \tag{3.12}$$

The helical bevel surface of the gear tooth is determined similarly to that of the pinion (3.13)

$$\mathbf{r}_{2cx}^{(2)} = \mathbf{Rt}(\zeta_2, Z) \cdot \mathbf{Tr}(x_{hx2}, X) \cdot \mathbf{Tr}(y_{hx2}, Y) \cdot \mathbf{Tr}(z_{hx2}, Z) \cdot \mathbf{Sca}(r_{z2}/r_{e2}) \cdot \mathbf{r}_{pr2cx}^{(2)}, \tag{3.13}$$

where ζ_2 is the helix parameter, $x_{\mathbf{hx2}}$, $y_{\mathbf{hx2}}$, $z_{\mathbf{hx2}}$ are the coordinates of the helix wrapped around the pitch cone of gear (3.14), and r_{z2} is the pitch radius in the current section (3.15),

$$\begin{cases} x_{\mathrm{hx2}} = -\left(r_{\mathrm{e2}} - \frac{H_2\zeta_2}{2\pi}\tan\delta_2\right)\cos\zeta_2 \\ y_{\mathrm{hx2}} = \left(r_{\mathrm{e2}} - \frac{H_2\zeta_2}{2\pi}\tan\delta_2\right)\sin\zeta_2 \\ z_{\mathrm{hx2}} = \frac{H_2\zeta_2}{2\pi} \end{cases}, \tag{3.14}$$

$$r_{\mathrm{z2}} = r_{\mathrm{e2}} - \frac{H_2\zeta_2}{2\pi}\tan\delta_2, \tag{3.15}$$

where $H_2 = 2\pi r_{\mathrm{m2}}\frac{\cos\delta_2}{\tan\beta_{\mathrm{m}}}$ is the helix lead on the pitch cone, r_{e2} is the external pitch radius, and r_{m2} is the mean pitch radius.

The profile of the convex side of the pinion tooth may be expressed by vector (3.16)

$$\mathbf{r}^{(1)}_{\mathrm{pr1cx}} = \mathbf{Rt}(-\delta_1, Y)\cdot\mathbf{Rf}_{\mathrm{y}}(x_{\mathrm{v1}}z_{\mathrm{v1}})\cdot\begin{bmatrix}\rho_{\mathrm{e1}}\cos\theta_1 - d_{\mathrm{CO'}}\sin\alpha_{\mathrm{t}} \\ \rho_{\mathrm{e1}}\sin\theta_1 + d_{\mathrm{CO'}}\cos\alpha_{\mathrm{t}} \\ 0 \\ 1\end{bmatrix}, \tag{3.16}$$

where $\mathbf{Rf}_{\mathrm{y}}(X\ Z)$ is the symmetry operator relative to plane XZ (Appendix D, Eq. (D.51) in [18]). Thus, the convex surface of the pinion tooth is represented by vector (3.17)

$$\mathbf{r}^{(1)}_{\mathrm{1cx}} = \mathbf{Rt}(-\zeta_1, Z)\cdot\mathbf{Tr}(x_{\mathrm{hx1}}, X)\cdot\mathbf{Tr}(y_{\mathrm{hx1}}, Y)\cdot\mathbf{Tr}(z_{\mathrm{hx1}}, Z)\cdot\mathbf{Sca}\left({}^{r_{\mathrm{z1}}}/_{r_{\mathrm{e1}}}\right)\cdot\mathbf{r}^{(1)}_{\mathrm{pr1cx}}. \tag{3.17}$$

The gear tooth profile for the concave side is given by the formula (3.18)

$$\mathbf{r}^{(2)}_{\mathrm{pr2cv}} = \mathbf{Rt}(\delta_2, Y)\cdot\mathbf{Rf}_{\mathrm{y}}(x_{\mathrm{v2}}z_{\mathrm{v2}})\cdot\begin{bmatrix}\rho_{\mathrm{e2}}\cos\theta_2 - (\rho_2 - \rho_1 + d_{\mathrm{CO'}})\sin\alpha_{\mathrm{t}} \\ \rho_{\mathrm{e2}}\sin\theta_2 + (\rho_2 - \rho_1 + d_{\mathrm{CO'}})\cos\alpha_{\mathrm{t}} \\ 0 \\ 1\end{bmatrix}. \tag{3.18}$$

The concave surface of the gear tooth is therefore given as (3.19)

$$\mathbf{r}^{(2)}_{\mathrm{2cv}} = \mathbf{Rt}(\zeta_2, Z)\cdot\mathbf{Tr}(x_{\mathrm{hx2}}, X)\cdot\mathbf{Tr}(y_{\mathrm{hx2}}, Y)\cdot\mathbf{Tr}(z_{\mathrm{hx2}}, Z)\cdot\mathbf{Sca}(r_{\mathrm{z2}}/r_{\mathrm{e2}})\cdot\mathbf{r}^{(2)}_{\mathrm{pr2cv}}. \tag{3.19}$$

The unit normal values to the tooth flanks may be determined by Eqs. (3.20) and (3.21):

$$\mathbf{n}_{1\text{cx,cv}}^{(1)} = \frac{\dfrac{\partial \mathbf{r}_{1\text{cx,cv}}^{(1)}}{\partial \zeta_1} \times \dfrac{\partial \mathbf{r}_{\text{cx,cv1}}^{(1)}}{\partial \theta_1}}{\left| \dfrac{\partial \mathbf{r}_{1\text{cx,cv}}^{(1)}}{\partial \zeta_1} \times \dfrac{\partial \mathbf{r}_{1\text{cx,cv}}^{(1)}}{\partial \theta_1} \right|}, \tag{3.20}$$

$$\mathbf{n}_{2\text{cx,cv}}^{(2)} = \frac{\dfrac{\partial \mathbf{r}_{2\text{cx,cv}}^{(2)}}{\partial \theta_2} \times \dfrac{\partial \mathbf{r}_{2\text{cx,cv}}^{(2)}}{\partial \zeta_2}}{\left| \dfrac{\partial \mathbf{r}_{2\text{cx,cv}}^{(2)}}{\partial \theta_2} \times \dfrac{\partial \mathbf{r}_{2\text{cx,cv}}^{(2)}}{\partial \zeta_2} \right|}, \tag{3.21}$$

although they are not shown here in their expanded form due to their complexity.

3.2.3 Modifications

Tooth flank modifications are conveniently defined and presented against a curvilinear coordinate grid, provided they are connected with the tooth profile and flank the pitch line. The modification is introduced in the calculations by means of relationship (3.22)

$$\mathbf{r}_i^{*(i)} = \mathbf{r}_i^{(i)} - C_i \mathbf{n}_i^{(i)}, \tag{3.22}$$

where $\mathbf{r}_i^{*(i)}$ is the vector of the modified surface, $\mathbf{r}_i^{(i)}$ is the vector of the nonmodified surface, $\mathbf{n}_i^{(i)}$ is the unit normal to the nonmodified surface, and C_i is the amount of modification, with $i = 1{,}2$ referring to the pinion and the gear, respectively. In this case, instead of the nonmodified surface, the analyses make use of the surface with modification described by vector (3.22). It should be noted that a positive modification is one that goes deep into the material (toward the tooth), whereas a negative modification is one that is applied to the outside of the material (toward the tooth space). This chapter describes the modification proposed in the study by Batsch [19], shown using the example of the pinion tooth surface in Fig. 3.4.

The amount of modification at any grid point is calculated by parabolic interpolation based on the nodes presented in Fig. 3.4.

3.3 Tooth Contact Analysis

3.3.1 Meshing Equations

To achieve consistent contact between surfaces at a given point, the condition in vector equation system (3.23) must be fulfilled.

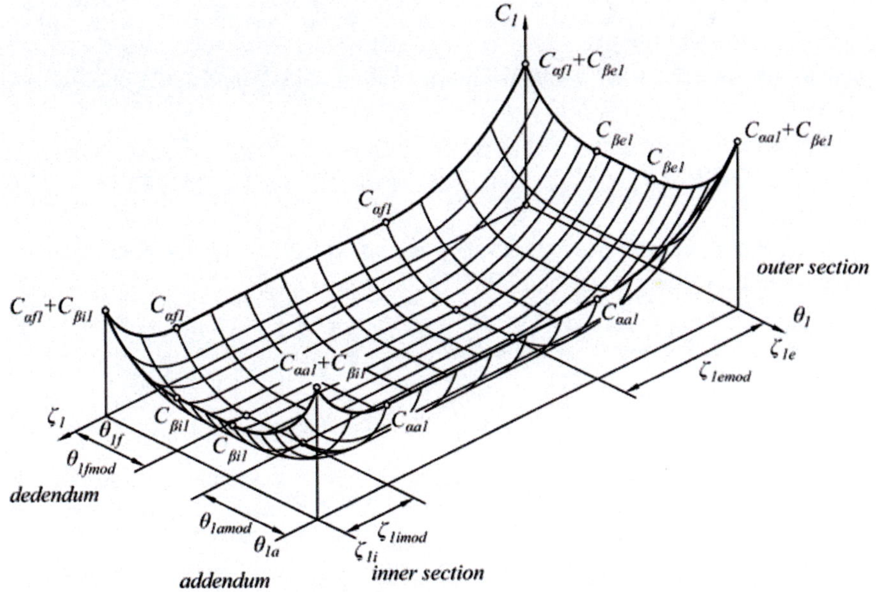

Fig. 3.4 A topological modification of the pinion tooth surface in a bevel Novikov gear

$$\begin{cases} \mathbf{r}_1^{(f)}(\theta_1, \zeta_1, \varphi_1) = \mathbf{r}_2^{(f)}(\theta_2, \zeta_2, \varphi_2) \\ \mathbf{n}_1^{(f)}(\theta_1, \zeta_1, \varphi_1) = \mathbf{n}_2^{(f)}(\theta_2, \zeta_2, \varphi_2) \end{cases}, \tag{3.23}$$

where $\mathbf{n}_1^{(f)}$ and $\mathbf{n}_2^{(f)}$ are the unit normal values to the tooth surface in system *f*. The first equation in the system ensures that the mating surfaces have a common point. The tangency of the surfaces is ensured by the equality of the unit normal values. Considering the identical length of vectors $\mathbf{n}_1^{(f)}$ and $\mathbf{n}_2^{(f)}$, system (3.23) may have the form of a system of five Eq. (3.24)

$$\begin{cases} x_1^{(f)}(\theta_1, \zeta_1, \varphi_1) = x_2^{(f)}(\theta_2, \zeta_2, \varphi_2) \\ y_1^{(f)}(\theta_1, \zeta_1, \varphi_1) = y_2^{(f)}(\theta_2, \zeta_2, \varphi_2) \\ z_1^{(f)}(\theta_1, \zeta_1, \varphi_1) = z_2^{(f)}(\theta_2, \zeta_2, \varphi_2) \\ n_{x1}^{(f)}(\theta_1, \zeta_1, \varphi_1) = n_{x2}^{(f)}(\theta_2, \zeta_2, \varphi_2) \\ n_{y1}^{(f)}(\theta_1, \zeta_1, \varphi_1) = n_{y2}^{(f)}(\theta_2, \zeta_2, \varphi_2) \end{cases}, \tag{3.24}$$

This system of usually nonlinear algebraic equations may be solved for consecutive discrete values of the pinion rotation angle φ_l, numerically by means of, e.g., the Levenberg–Marquardt algorithm [20, 21] or the trust region algorithm [22]. In such a case, we obtain solutions in the form of (3.25)

$$\theta_1(\varphi_1),\zeta_1(\varphi_1),\theta_2(\varphi_1),\zeta_2(\varphi_1),\varphi_2(\varphi_1). \tag{3.25}$$

The use of the above relationships in formulas (3.1) and (3.2) allows us to determine the line of action in the form of (3.26)

$$\mathbf{r}_{\mathrm{LA}}^{(f)} = \mathbf{r}_1^{(f)}(\theta_1(\varphi_1),\, \zeta_1(\varphi_1),\, \varphi_1) = \mathbf{r}_2^{(f)}(\theta_2(\varphi_1),\, \zeta_2(\varphi_1),\, \varphi_2(\varphi_1)). \tag{3.26}$$

Their inclusion in vectors $\mathbf{r}_1^{(1)}(\theta_1, \zeta_1)$ and $\mathbf{r}_2^{(2)}(\theta_2, \zeta_2)$ yields the path of contact on the surface of the pinion tooth (3.27) and the gear tooth (3.28):

$$\mathbf{r}_{\mathrm{PC1}}^{(1)} = \mathbf{r}_1^{(1)}(\theta_1(\varphi_1),\, \zeta_1(\varphi_1)), \tag{3.27}$$

$$\mathbf{r}_{\mathrm{PC2}}^{(2)} = \mathbf{r}_2^{(2)}(\theta_2(\varphi_1),\, \zeta_2(\varphi_1)). \tag{3.28}$$

Moreover, relationship $\varphi_2(\varphi_1)$ makes it possible to determine the transmission error. This error may also be established by other methods. One of them, more conveniently applied to complex surfaces, is described in Sect. 3.3.2.

Differentiation of the first equation in system (3.23) with respect to time t helps determine the kinematics of the meshing. The application of the theorem of the complex total derivative [23] leads to the absolute velocity (observed in the fixed coordinate system f) of contact point (3.29)

$$\begin{aligned}\mathbf{v}_{\mathrm{abs}} &= \frac{\partial \mathbf{r}_1^{(f)}}{\partial t} = \frac{\partial \mathbf{r}_1^{(f)}}{\partial \varphi_1}\cdot\frac{\partial \varphi_1}{\partial t} + \frac{\partial \mathbf{r}_1^{(f)}}{\partial \theta_1}\cdot\frac{\partial \theta_1}{\partial t} + \frac{\partial \mathbf{r}_1^{(f)}}{\partial \zeta_1}\cdot\frac{\partial \zeta_1}{\partial t} = \\ &= \frac{\partial \mathbf{r}_2^{(f)}}{\partial t} = \frac{\partial \mathbf{r}_2^{(f)}}{\partial \varphi_2}\cdot\frac{\partial \varphi_2}{\partial t} + \frac{\partial \mathbf{r}_2^{(f)}}{\partial \theta_2}\cdot\frac{\partial \theta_2}{\partial t} + \frac{\partial \mathbf{r}_2^{(f)}}{\partial \zeta_2}\cdot\frac{\partial \zeta_2}{\partial t}\end{aligned}, \tag{3.29}$$

where $\frac{\partial \varphi_1}{\partial t} = \omega_1$ is the pinion's angular velocity and $\frac{\partial \varphi_2}{\partial t} = \omega_2$ is the gear's angular velocity. Each component of absolute velocity (3.29) is interpreted as relative velocity $\mathbf{v}_{\mathbf{ri}}$ (3.31) connected to the motion relative to the tooth surface and as transfer velocity $\mathbf{v}_{\mathbf{ti}}$ (3.30) connected to the motion of the pinion and tooth coordinate systems:

$$\mathbf{v}_{\mathrm{ti}} = \frac{\partial \mathbf{r}_i^{(f)}}{\partial \varphi_i}\cdot\frac{\partial \varphi_i}{\partial t}, \tag{3.30}$$

$$\mathbf{v}_{\mathrm{ri}} = \frac{\partial \mathbf{r}_i^{(f)}}{\partial \theta_i}\cdot\frac{\partial \theta_i}{\partial t} + \frac{\partial \mathbf{r}_i^{(f)}}{\partial \zeta_i}\cdot\frac{\partial \zeta_i}{\partial t}, \tag{3.31}$$

where index $i = 1{,}2$ refers to the pinion and the gear, respectively. The difference between transfer velocity for individual gears corresponds to sliding velocity (3.32) is

$$\mathbf{v}_{12} = \mathbf{v}_{t1} - \mathbf{v}_{t2}, \tag{3.32}$$

which is the velocity of the contact point assigned to the pinion tooth surface relative to the contact point assigned to the gear tooth surface.

3.3.2 *Ease-Off Topography and Transmission Error*

Determining the transmission error using the method described in Sect. 3.3.1 at times proves problematic, since it involves solving equation system (3.24). The equation system is often complex due to elaborate analytical expressions describing tooth flanks. Therefore, a method was developed for determining the transmission error of surfaces considered discrete. This concept is presented in the study by Klingelnberg [24]. The method was designed to analyze bevel and hypoid gear pairs and was implemented in KiMOS, a computer-aided bevel and hypoid gear pair design system by Klingelnberg. This subsection discusses the author's implementation of the method.

It assumes a perfect mesh between the pinion tooth surface Σ_1 and the gear tooth surface Σ_2, i.e., one in which the gears are rotated by angles resulting from the theoretical gear ratio $i_{12} = z_2/z_1 = \varphi_1/\varphi_2$. Surfaces in the nth position are shown in Fig. 3.5.

It is assumed that the pinion rotation axis z_1 is aligned with axis z_f of the fixed coordinate system f and the surfaces in that system are given by vectors (3.1) and (3.2). In each position n resulting from the discretization of gear rotation angles, between the surfaces there is a certain distance measured along the arc with radius $R_{M2c}^{(f)} = \sqrt{\left(x_1^{(f)}\right)^2 + \left(y_1^{(f)}\right)^2}$ corresponding to the length of arc $k_{kor} = \widehat{M_1 M_2}$, where M_1 is a point on the surface Σ_1 with coordinates $z_1^{(f)}$ and M_2 is a point on the surface

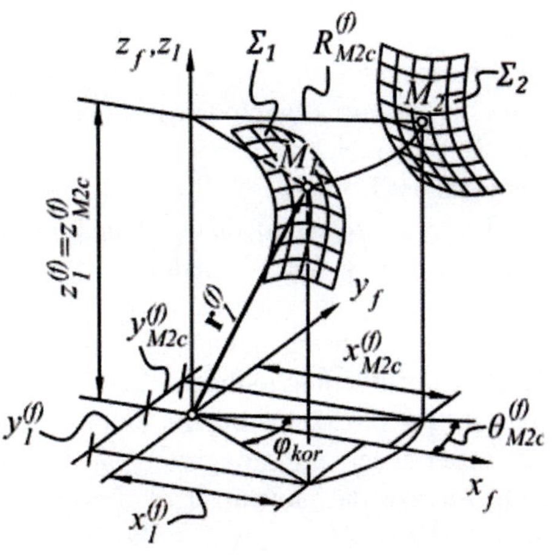

Fig. 3.5 Tooth flanks in the nth position

Σ_2 with coordinates $z_1^{(f)} = z_{M2c}^{(f)}$. Angle φ_{kor}, by which surface Σ_1 would have to be rotated in order for points M_1 and M_2 to overlap, is the central angle corresponding to the length of arc $k_{\text{kor}} = \widehat{M_1 M_2}$ and can be determined based on coordinates as (3.33)

$$\varphi_{\text{kor}} = 2\arcsin\left(\frac{\sqrt{\left(x_1^{(f)} - x_{M2c}^{(f)}\right)^2 + \left(y_1^{(f)} - y_{M2c}^{(f)}\right)^2}}{2R_{M2c}^{(f)}}\right). \tag{3.33}$$

The distance between the flank surfaces is expressed as (3.34)

$$k_{\text{kor}} = R_{M2c}^{(f)} \varphi_{\text{kor}}. \tag{3.34}$$

In each nth position, there is a certain minimum distance $k_{\text{kor}}^{\min}$ with its corresponding angle $\varphi_{\text{kor}}^{\min}$. The relationship between the minimum angle $\varphi_{\text{kor}}^{\min}$ and the gear rotation angle φ_2 determined discretely for all n positions constitutes the transmission error. In addition, in a specific nth position, the distribution of distances k_{kor} on surface Σ_1 is referred to as an instantaneous Ease-Off. The Ease-Off is generated in the form of the lowest-lying envelope of instantaneous Ease-Offs for all n positions. The greatest challenge of the method is to specify the location of point M_2 on surface Σ_2 that corresponds to coordinate $z_1^{(f)} = z_{M2c}^{(f)}$ and radius $R_{M2c}^{(f)}$. The issue may be easily solved by interpolation in the cylindrical coordinate system $\theta_{2c}^{(f)}, R_{2c}^{(f)}, z_2^{(f)}$.

3.3.3 Contact Pattern

In a gear pair, the contact pattern is the area where the contact takes place between the mating tooth surfaces. It is created by the elastic deformation of teeth due to forces exerted on the meshing. A "geometric contact pattern" is defined as an approximation of the actual contact area exclusively based on the geometric properties of the surfaces in contact. In other words, it is determined without taking into account the relationship between the deformation and forces applied or the type of material used.

There are many methods of determining the geometric contact pattern. They include, among others, methods based on mathematical models [3, 25–27] and methods using computer-aided design (CAD) systems [28, 29]. The use of CAD systems involves the mutual penetration of three-dimensional (3D) tooth models to a depth corresponding to the thickness of the gear marking compound applied in gear pair testing or expected deformations. The contact pattern has the form of a flaky construct generated through logical multiplication of thus positioned solid shapes. To visualize the movement of the contact pattern, the above operation must be performed for consecutive discrete positions arising from the gear solids' rotation angles. This method has been detailed in the study by Sobolak [30]. Its equivalent is the method that involves finding the curve of the intersection of tooth surfaces supplied in an analytical manner [3]. Another method is the one proposed in the study by Litvin et al. [26]. The contact pattern is determined as the area for which the distance between the points on the surfaces in contact, measured along the common normal, is lower than the set amount. In order to determine the contact pattern, the method uses an approach similar to determining Hertz contact stress for point contact [13, 31]. It is only based on the main curvatures of surfaces, and the contact pattern resulting from the local approximation of surfaces is always elliptical [27]. The main drawback of the method is the possibility of performing only a local analysis, without taking into account, e.g., the edge contact. Moreover, it is of limited use for nonelliptical contact patterns, which occur, e.g., in gears with an eccentric-cycloid profile [32] or in Novikov gears with a convexo-concave profile [33]. The last method, described in more detail in this chapter, is the one in which the contact pattern is determined as a set of points for which the distance between tooth flank surfaces, measured along the unit normal to the pinion tooth flank, is lower than the set value [25]. The method makes it possible to determine edge and nonelliptical patterns, as experimentally verified for cylindrical Novikov gears [34] and eccentric-cycloid meshing [32].

Let us consider a gear for which the mating tooth surfaces Σ_1 and Σ_2 are given in a fixed coordinate system f by vectors (3.1) and (3.2), respectively. The contact pattern is determined on the basis of vector equation (3.35), resulting directly from Fig. 3.6.

$$\mathbf{r}_1^{(f)}(\theta_1, \zeta_1) + k\mathbf{n}_1^{(f)}(\theta_1, \zeta_1) = \mathbf{r}_2^{(f)}(\theta_2, \zeta_2), \tag{3.35}$$

where k is the distance between the tooth flanks measured along the unit normal to the pinion tooth surface. The above equation can be solved based on the distance k and the gear tooth surface parameters θ_2, ζ_2 for all discrete pinion tooth surface parameters θ_1, ζ_1. The points for which the distance k is smaller than the assumed value form the contact pattern. Assuming that the gears revolve around axis z, the

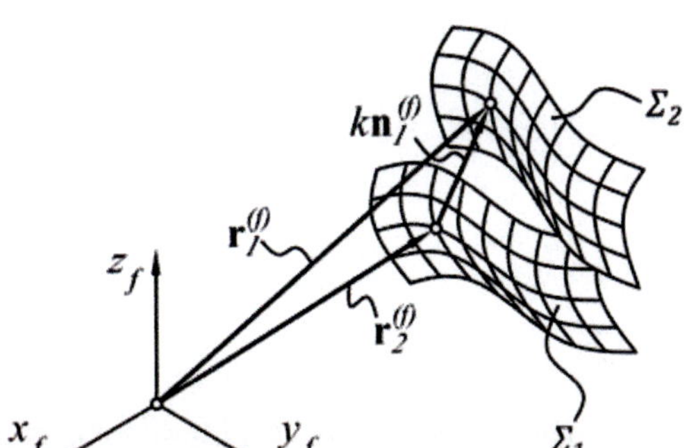

Fig. 3.6 Determining distances between mating tooth surfaces

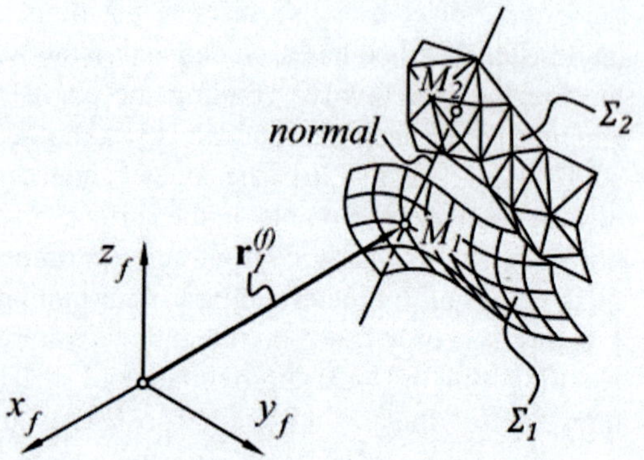

Fig. 3.7 The proposed method for determining distances between mating tooth surfaces

contact pattern is conveniently presented as a projection on plane r_{y1}, z_f where $r_{y1} = \sqrt{\left(x_1^{(f)}\right)^2 + \left(y_1^{(f)}\right)^2}$ is an arbitrary radius of the pinion.

Solving vector equation (3.35), representing a system of three algebraic, usually nonlinear equations, is problematic and time-consuming even if numerical methods are employed [20, 21]. For this reason, it is recommended that the solution should be approximated by means of a geometric method. The author proposes a method (Fig. 3.7) in which the gear tooth surface is presented as a grid of triangles, which can be accomplished using a common method of triangulation by Delaunay [35].

Next, for each point on the pinion tooth surface M_1, we determine point M_2, at which the gear tooth surface triangle intersects with a straight line normal to the pinion tooth surface, which can be accomplished by the algorithm described in the study by Möller and Trumbore [36]. The point for which the distance $|M_1M_2|$ is lower than the set distance is plotted on the chart, forming the gear pair's contact pattern.

3.4 Results

The results of the calculations are presented for a Novikov gear pair with a gear ratio $i_{12} = 1$ and an axis intersection angle $\Sigma = 120°$, the data of which are listed in Table 3.1. Gear pairs of this type are used, e.g., in milling cutters as well as in intermediate tail gearboxes in helicopters.

3.4.1 Ideal Gear Pair

The results of tooth contact analysis for the ideal gear pair (free from axis position errors and flank surface modifications), rotating at pinion's angular velocity $\omega_1 = 104.72$ rad/s (1000 rpm), are shown in Fig. 3.8. They were obtained by solving

Table 3.1 Data of the analyzed Novikov helical bevel gear pair

Parameter	Pinion	Gear
Mean normal module, mm	$m_{nm} = 2.5$	
Number of teeth	$z_1 = 35$	$z_2 = 35$
Shaft angle, °	$\Sigma = 120$	
Overlap ratio	$\varepsilon_\beta = 1.33$	
Face width, mm	$b = 20$	
Mean spiral angle, °	$\beta_m = 30$	
Hand of spiral	right	Left
Normal pressure angle, °	$\alpha_n = 25$	
Translation of the convex tooth profile, mm	$d_{CO'} = 0$	
Profile	convex	concave
Outer profile radius, mm	$\rho_{e1} = 5.32$	$\rho_{e2} = 5.43$
Outer pitch diameter, mm	$d_{e1} = 118.35$	$d_{e2} = 118.35$
Outer tip diameter, mm	$d_{ae1} = 122.58$	$d_{ae2} = 118.52$
Outer root diameter, mm	$d_{fe1} = 117.61$	$d_{fe2} = 113.55$
Pitch angle, °	$\delta_1 = 60$	$\delta_2 = 60$

two equation systems (3.24) for the concave and convex side of the pinion, for all discrete values of its rotation angle. In this manner, the line of action (3.26) and sliding velocity (3.32) were determined (Fig. 3.8).

If the concave surface of the pinion tooth is the drive side (Fig. 3.8a), then the teeth begin to mate at point B for angle $\varphi_1 = 0$, where the sliding velocity is maximum and equals $|\mathbf{v}_{12}| = 0.56$ m/s. As the gears rotate, the point of contact moves along the meshing line to point B' for angle $\varphi_1 = 0.2286$ rad, at which the sliding velocity drops to $|\mathbf{v}_{12}| = 0.40$ m/s and the tooth contact is terminated. Meanwhile, for the convex drive side of the pinion (Fig. 3.8b), the teeth begin to mate on the small modules side at point B' for angle $\varphi_1 = 0.2286$ rad, at which the sliding velocity assumes a minimum value $|\mathbf{v}_{12}| = 0.40$ m/s. Teeth mating ends at point B for angle $\varphi_1 = 0$, at which the sliding velocity increases to $|\mathbf{v}_{12}| = 0.56$ m/s. The meshing line in this type of gear pair is a straight line, the direction of which corresponds to the direction of segment AB, with A being the apex point of the pitch cone. As in a helical Novikov gear pair, the path of the contact point on tooth flanks is the ζ-line of those surfaces, corresponding to the position of point B. The gear pair maintains a constant gear ratio; no transmission error occurs. For Ease-Off charts and contact patterns determined for the thickness of the marking compound $k = 4$ μm, refer to Fig. 3.9.

Contact patterns are elliptical, making an edge contact as they enter and leave the mesh. Moreover, Ease-Off charts allow us to observe areas of potential tooth interference (marked with a red ellipse). The phenomenon is visible in the form of long and narrow contact patterns at the pinion tooth base.

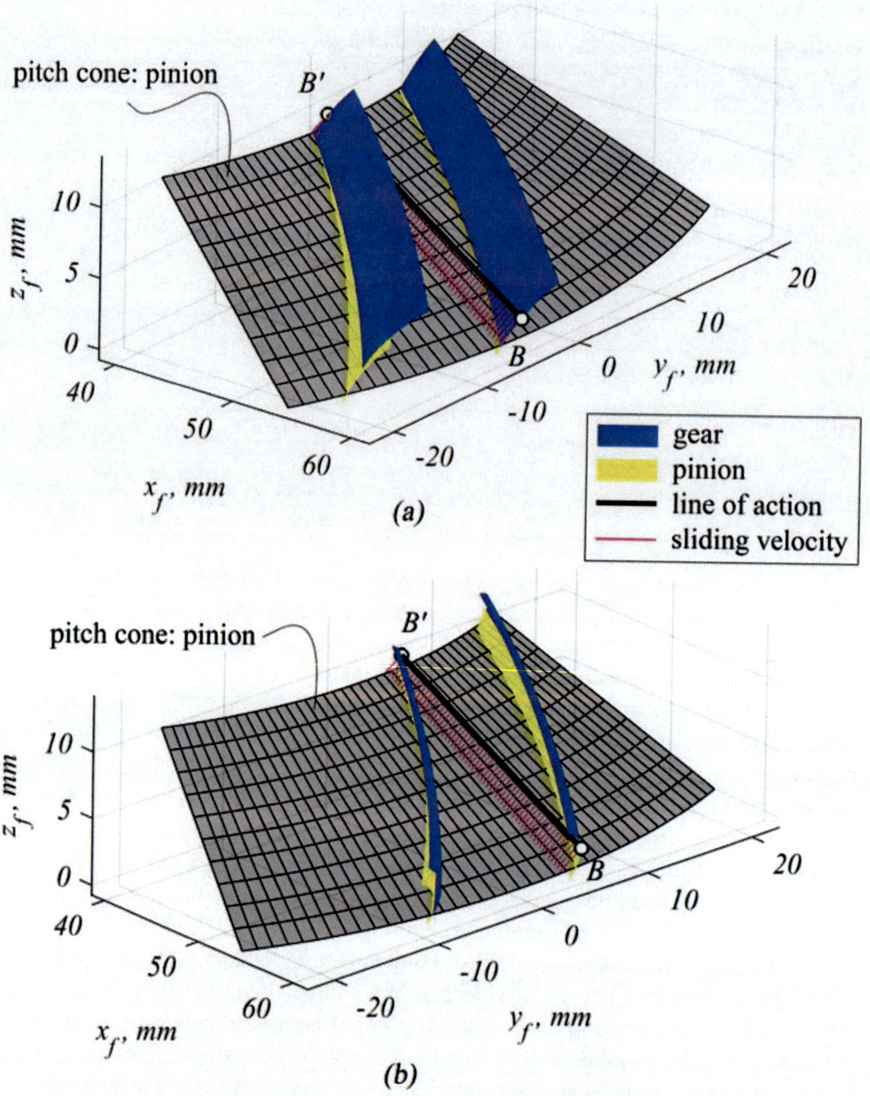

Fig. 3.8 The line of action and sliding velocity in a bevel Novikov gear for the following drive sides: concave side of the pinion (**a**) and convex side of the pinion (**b**)

3.4.2 *Effect of Tooth Surface Modifications*

In order to visualize the effect of tooth flank modifications, tooth contact analyses were carried out for the Novikov gear pair characterized by the data listed in Table 3.1 with the following modification parameters (according to Fig. 3.4):

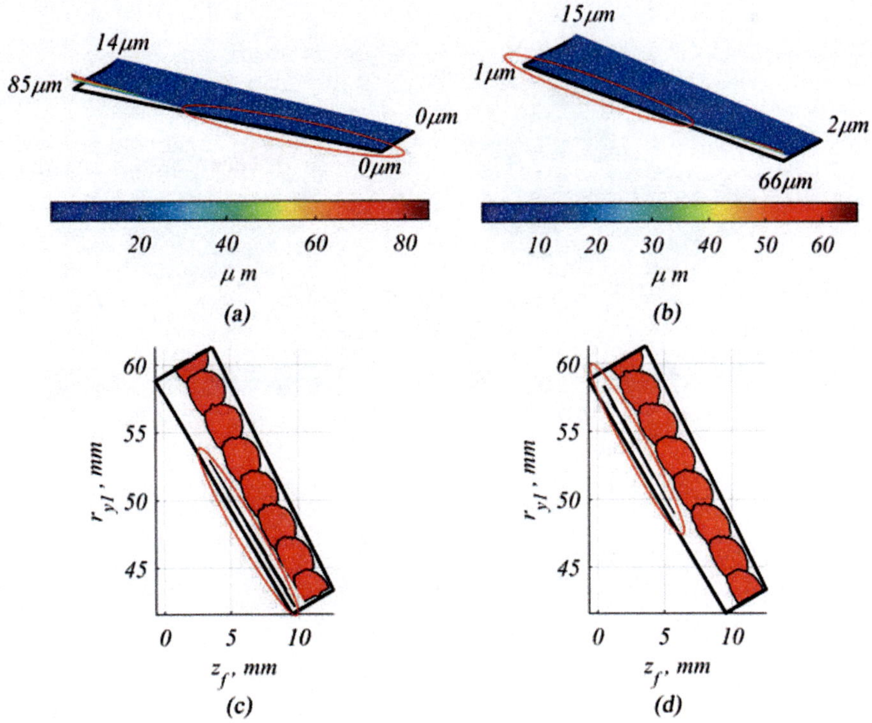

Fig. 3.9 Results of tooth contact analysis for the Novikov gear pair without modification: Ease-Off chart for the drive on the concave side of the pinion (**a**), Ease-Off chart for the drive on the convex side of the pinion (**b**), contact patterns for the drive on the concave side of the pinion (**c**), and contact patterns for the drive on the convex side of the pinion (**d**)

$\zeta_{1\text{imod}} = \zeta_{1\text{emod}} = 0.0171$ mm, $C_{\beta\text{i}1} = C_{\beta\text{e}1} = 0.01$ mm, $\theta_{1\text{amod}} = 0.1745$ rad, $\theta_{1\text{fmod}} = 0.0463$ rad, $C_{\alpha\text{a}1} = 0.01$ mm, $C_{\alpha\text{f}1} = 0.02$ mm, $\zeta_{2\text{imod}} = \zeta_{2\text{emod}} = 0.0171$ mm, $C_{\beta\text{i}2} = C_{\beta\text{e}2} = 0.01$ mm, $\theta_{2\text{amod}} = 0.1745$ rad, $\theta_{2\text{fmod}} = 0.0873$ rad, $C_{\alpha\text{a}2} = 0.05$ mm, $C_{\alpha\text{f}2} = 0.01$ mm. The modification grid is shown in Fig. 3.10.

Applying a tooth flank modification led to an evenly distributed shape of the Ease-Off chart, which assumed minimum values in the areas of the expected tooth contact (Fig. 3.11c, d).

The transmission error does not occur: the gear pair maintains a constant transmission ratio, with transmission charts at the beginning and the end of mesh assuming a shape approximating that of a parabola (Fig. 3.11a, b). Moreover, tooth line modifications helped avoid an edge contact pattern at the beginning and end of the tooth contact (Fig. 3.12).

Due to profile modification, the minor axis of the ellipse is shortened, moving the pattern away from the gear and pinion addendum, thus preventing interference.

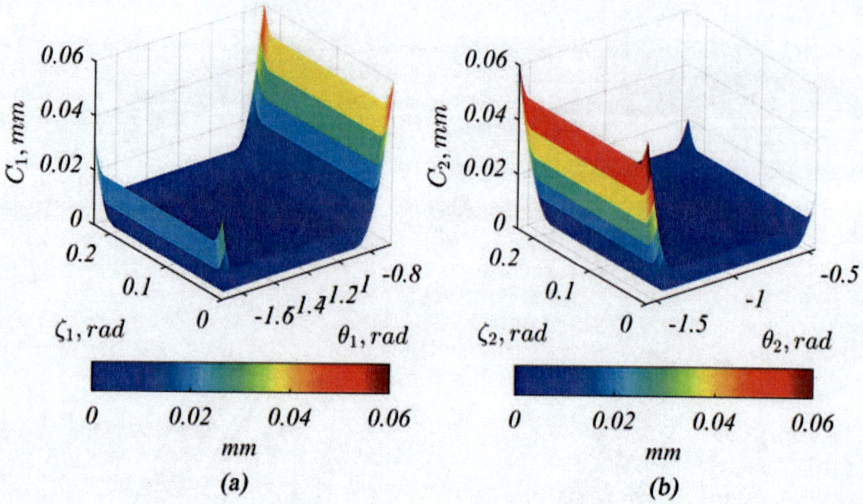

Fig. 3.10 Tooth flank modification in the Novikov a gear pair: pinion (**a**) and gear (**b**)

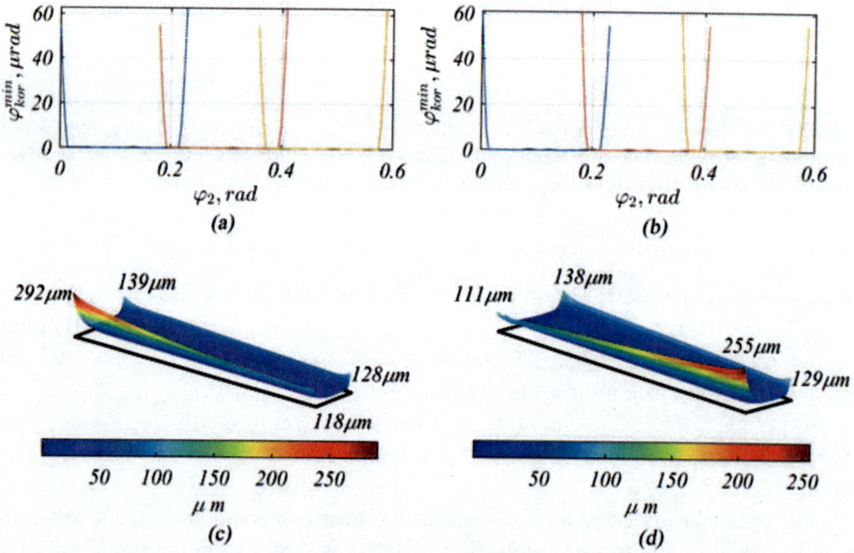

Fig. 3.11 Results of tooth contact analysis for the modified Novikov gear pair: transmission error for the drive on the concave side of the pinion (**a**), transmission error for the drive on the convex side of the pinion (**b**), Ease-Off for the drive on the concave side of the pinion (**c**), and Ease-Off for the drive on the convex side of the pinion (**d**)

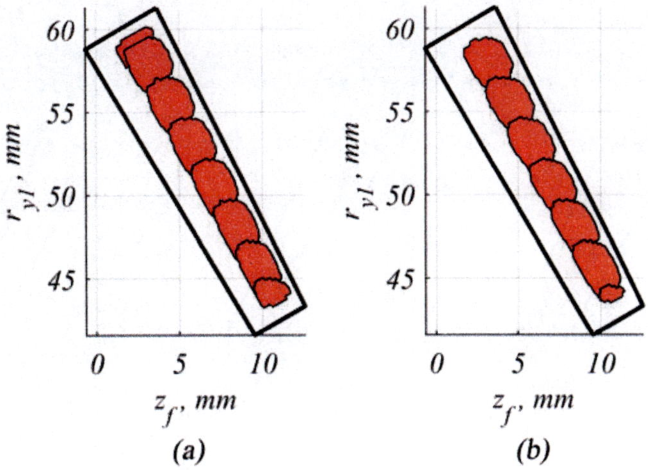

Fig. 3.12 Tooth contact pattern in the modified Novikov gear pair: pinion's concave side (**a**) and pinion's convex side (**b**)

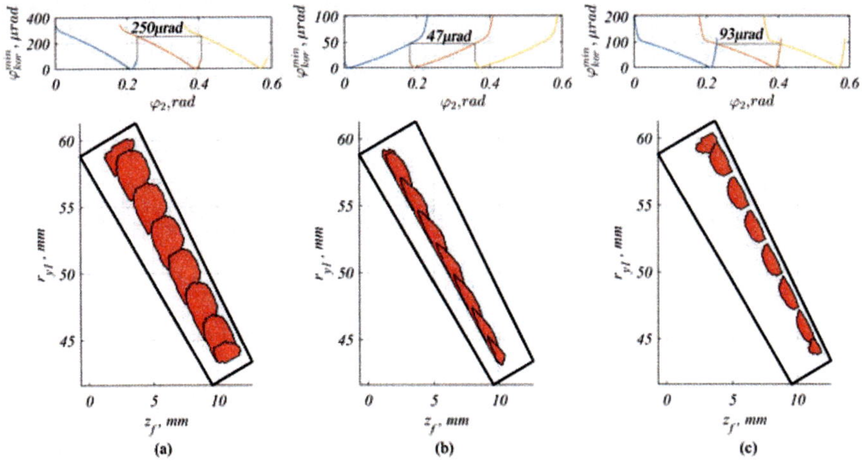

Fig. 3.13 Transmission errors and contact patters in the Novikov gear pair, with axis transmission errors, determined for the concave side of the pinion tooth: $V = 0.05$ mm (**a**), $H = 0.05$ mm (**b**), $J = -0.05$ mm (**c**)

3.4.3 *Effect of Gear Position Errors*

The effect of gear position errors is exemplified by their three values: $V = 0.05$ mm, $H = 0.05$ mm, and $J = -0.05$ mm. The results of the tooth contact analysis for the concave side of the pinion tooth as the drive side are shown in Fig. 3.13.

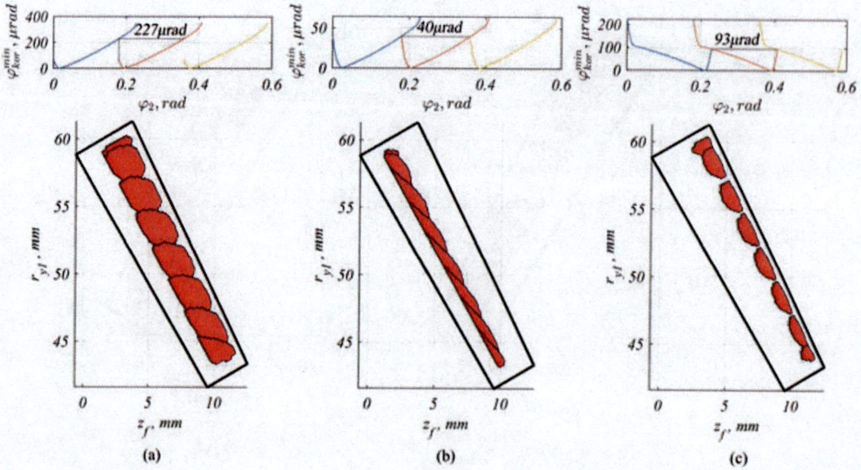

Fig. 3.14 Transmission errors and contact patters in the Novikov gear pair, with axis transmission errors, determined for the convex side of the pinion tooth: $V = 0.05$ mm (**a**), $H = 0.05$ mm (**b**), $J = -0.05$ mm (**c**)

The hypoid offset error V has the greatest impact on motion transmission quality. For $V = 0.05$ mm, the transmission error was 250 μrad. The gear position error $H = 0.05$ mm generates a transmission error of 47 μrad and a reduction in the contact area due to its shift toward the pinion's tooth root. The gear mounting distance error $J = -0.05$ mm causes an inconsistent motion transmission of 93 μrad and shifts the contact pattern toward the pinion's tooth addendum, which also results in a reduction of its surface area. The same relationships are observed with the convex side of the tooth pinion as the drive side (Fig. 3.14).

3.4.4 Comparison with a Conventional Helical Bevel Gear Pair

This section contains a comparison between a Novikov gear pair and a DUPLEX helical spiral bevel gear pair. The macrogeometry of the conventional gear pair corresponds to that of the Novikov gear pair, with data listed in Table 3.2.

The choice of its microgeometry as well as its configuration parameters (Table 3.3) was aided by the KiMOS software.

Tooth flank surfaces were generated with the use of one of many available mathematical models [3]. The surfaces obtained in this manner made it possible to apply the methods described in Sects. 3.3.2 and 3.3.3 to analyze the tooth contact. The results are displayed in Fig. 3.15.

The transmission error was 81.8 μrad for the concave side and 62.5 μrad for the convex side. They correspond to the values obtained with the use of the commercial

Table 3.2 Data of the analyzed DUPLEX helical spiral bevel gear pair

Parameter	Pinion	Gear
Mean normal module, mm	$m_{\mathrm{nm}} = 2.5$	
Number of teeth	$z_1 = 35$	$z_2 = 35$
Shaft angle, °	$\Sigma = 120$	
Face width, mm	$b = 20$	
Mean helix angle, °	$\beta_{\mathrm{m}} = 30$	
Hand of spiral	Right	Left
Outer pitch diameter, mm	$d_{\mathrm{e1}} = 118.35$	$d_{\mathrm{e2}} = 118.35$
Outer tip diameter, mm	$d_{\mathrm{ae1}} = 121.32$	$d_{\mathrm{ae2}} = 121.34$
Outer root diameter, mm	$d_{\mathrm{fe1}} = 114.33$	$d_{\mathrm{fe2}} = 114.63$
Pitch angle, °	$\delta_1 = 60$	$\delta_2 = 60$

Table 3.3 Machine settings for a DUPLEX helical spiral bevel gear pair

Parameter		Pinion	Gear
Radial setting, mm		56.8575	57.3172
Blank offset, mm		0.5612	0
Machine root angle, °		52.7893	58
Sliding base, mm		3.0510	5.7013
Ratio of the gear roll		1.138497	1.153997
Helical motion 1-ord., mm/rad		4.389823	0
Swivel angle, °		24.0198	−23.2722
Tilt angle, °		1.1917	7.0247
Start roll position, °		−48.8120	42.5192
Machine center to cross-point, mm		−0.4465	0
Tool profile angle, °	Concave	14.3330	12.0002
	Convex	29.4913	28.0002
Tool radius, mm	Concave	57.9744	57.9399
	Convex	56.3255	56.3601

software KiMOS: 79.5 μrad and 61.6 μrad, which confirms the correct implementation of the method of determining transmission error as well as surfaces generated by simulated machining. The shapes and locations of the contact patterns (Fig. 3.15) approximate those determined with the help of commercial software.

Figure 3.17 compares the surface areas of instantaneous contact patterns as a function of the pinion rotation angle for the modified Novikov gear pair (Fig. 3.12) and the DUPLEX helical gear (Fig. 3.15).

The Novikov gear is characterized by a larger contact area than that of the conventional gear, regardless of whether the concave side or the convex side is the drive side. For maximum values, the contact area in the Novikov gear pair is approximately 80% greater (for the concave side of the pinion tooth) and approximately 60% greater (for the convex side of the pinion tooth) than the contact area of the DUPLEX helical gear pair. The results are in line with those of earlier analyses for bevel gears of different geometries [19]. Considering the above as well as the fact

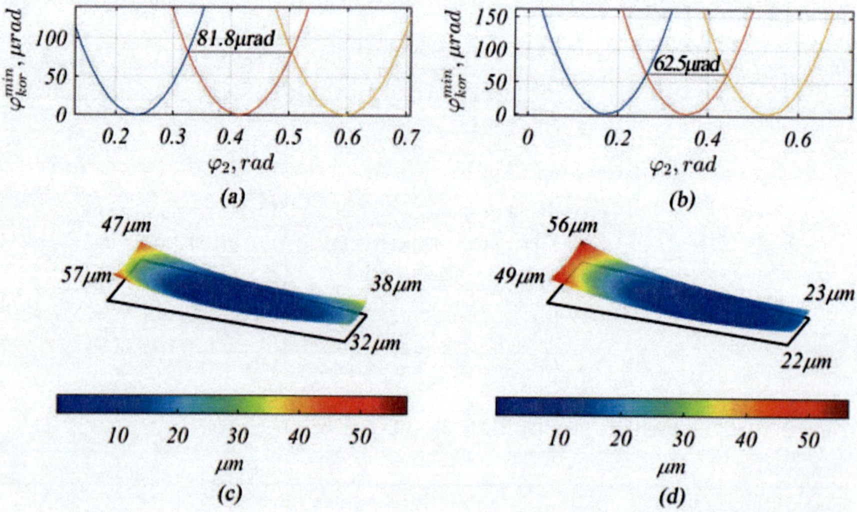

Fig. 3.15 Results of tooth contact analysis for the DUPLEX helical bevel gear pair: transmission error for the drive on the concave side of the pinion (**a**), transmission error for the drive on the convex side of the pinion (**b**), Ease-Off for the drive on the concave side of the pinion (**c**), and Ease-Off for the drive on the convex side of the pinion (**d**)

that the maximum sliding velocity in the DUPLEX helical gear pair, which is equal to 0.803 m/s, is approximately 1.5 times that of the maximum velocity in the Novikov gear (0.557 m/s), we may expect that the gear with the convexo-concave tooth profile offers greater durability or load capacity. This was proved in experimental studies with regard to spur gear teeth [14].

Figure 3.18 lists the transmission error values for comparable gear pairs depending on the values of the gear position error.

An analysis of Fig. 3.16 reveals that the greatest impact on the transmission quality of the bevel Novikov gear pair is exerted by the hypoid offset error V and negative values of errors H and J. Even minimal values of those errors cause the transmission error to rise significantly. For positive errors H and J, the values approximate those of the DUPLEX helical gear pair, which maintains an almost constant value of the transmission error. Consequently, the Novikov gear pair is more sensitive to gear position errors than is the conventional gear pair.

3.5 Conclusion

A series of simulations were carried out on the basis of the mathematical model of Novikov bevel gear teeth and the proposed calculation methods. It was demonstrated that the designed modification of tooth flanks enables the elimination of the edge contact pattern and tooth interference. Moreover, simulations were carried out aimed

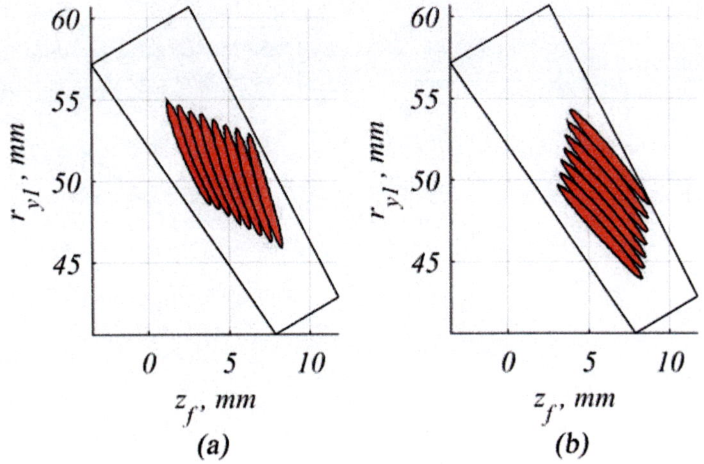

Fig. 3.16 Tooth contact pattern in the DUPLEX helical gear pair: pinion's concave side (**a**) and pinion's convex side (**b**)

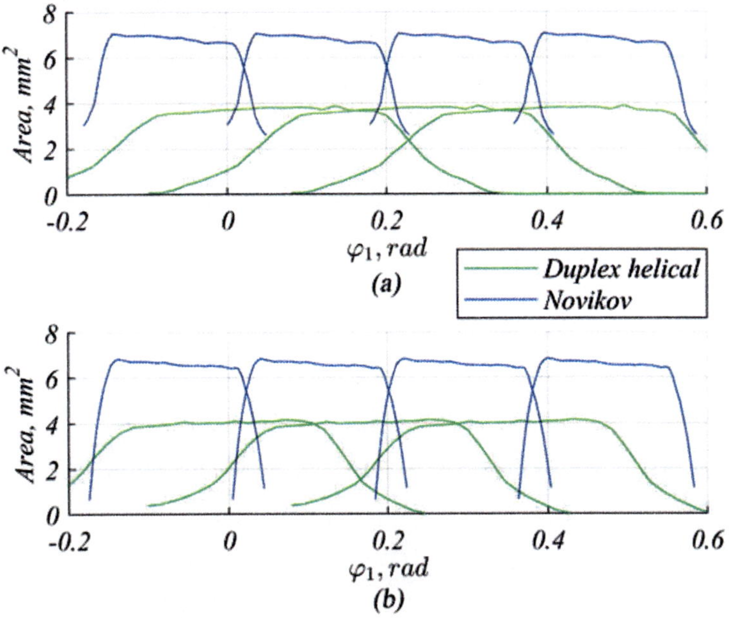

Fig. 3.17 Surface area of the contact pattern as a function of the rotation angle of the pinion in the Novikov gear pair and the DUPLEX helical gear pair: the pinion's concave side (**a**) and the pinion's convex side (**b**)

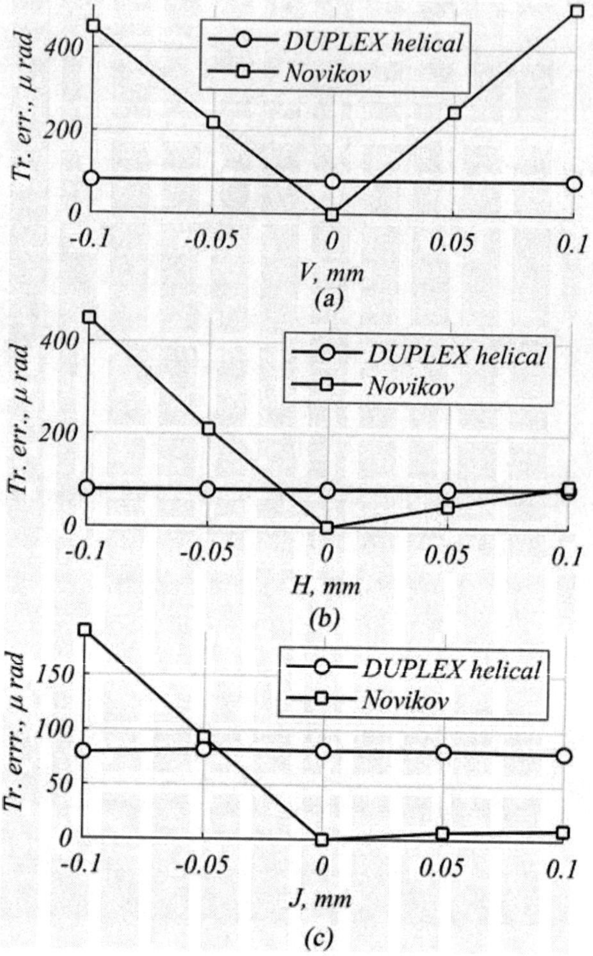

Fig. 3.18 Transmission errors for the Novikov gear pair and the DUPLEX helical gear pair depending on the hypoid offset error V (**a**), gear position error H (**b**), and gear mounting distance error J (**c**)

at a comparison between a bevel Novikov gear pair and a conventional (DUPLEX helical) gear pair. Following an analysis of the results, the following conclusions were formulated:

- Instantaneous contact patterns in the bevel Novikov gear pair are larger
- The maximum sliding velocity is lower in the Novikov gear pair
- The Novikov gear is more sensitive to gear axis position errors

Based on the above conclusions, we may formulate a thesis that the bevel Novikov gear pair will be characterized by a greater surface load capacity (or durability arising from such a load capacity) than the DUPLEX helical gear pair, as long as the following conditions are met:

Fig. 3.19 A sample 3D CAD model of the helical bevel Novikov gear

- The gear case is accurately manufactured
- Gear datum surfaces, as well as gear teeth, are machined with precision
- Bearing rigidity

The above thesis should be experimentally verified in further research. To this end, the mathematical model designed here may be used for developing a 3D CAD model of gear teeth (Fig. 3.19) and subsequently used in finite element method (FEM) simulations and gear tooth machining programming in a computer-aided manufacturing (CAM) system.

References

1. Artoni, A., et al.: Ease-off based compensation of tooth surface deviations for spiral bevel and hypoid gears: only the pinion needs corrections. Mech. Mach. Theory. **61**, 84–101 (2013). https://doi.org/10.1016/j.mechmachtheory.2012.10.005
2. Lin, C.-Y., et al.: Computer-aided manufacturing of spiral bevel and hypoid gears by applying optimization techniques. J. Mater. Process. Technol. **114**(1), 22–35 (2001). https://doi.org/10.1016/S0924-0136(01)00734-8
3. Pisula, J.: An analysis of the effect of the application of helical motion and assembly errors on the meshing of a spiral bevel gear using duplex helical method. Adv. Manuf. Sci. Technol. **40**(1), 19–31 (2016). https://doi.org/10.2478/AMST-2016-0002

4. Wildhaber, E.: US Patent No. 1,601,750, (1926)
5. HouJun, C., et al.: Study on the general principle of normal circular-arc gear transmission. Mech. Mach. Theory. **41**, 1424–1442 (2006). https://doi.org/10.1016/j.mechmachtheory.2006.01.016
6. Yao, L., et al.: Mathematical modeling and simulation of the external and internal double circular-arc spiral bevel gears for the nutation drive. ASME J. Mech. Des. **132**, 1–10 (2010)
7. Zhenyun, D., et al.: Mathematical model and manufacture programming of loxodromic-type normal circular-arc spiral bevel gear. Front. Mech. Eng. **7**(3), 312–321 (2012)
8. Stanovskoy, V.V. et al.: US Patent No. 8,789,437 B2, (2014)
9. Li, X., et al.: Design and investigation of a cycloid helical gear drive. J. Mech. Sci. Technol. **31**(9), 4329–4336 (2017). https://doi.org/10.1007/s12206-017-0831-8
10. Kamchatnyi, S.A., et al.: Mathematical simulation of the formbuilding of the conic driving gear details with EC-gearing. Vestn. Tomsk. Gos. Univ. Mat. Mekh. **2**(28), 5–17 (2014)
11. Stanovskoi, V.V., et al.: Development and implementation of resource-saving technology of a new generation based on eccentric-cycloidal (EC) engagement. Innovations. **12**(182), 117–120 (2013)
12. Novikov, M.L.: USSR Patent 109 750, (1956)
13. Batsch, M., et al.: Cylindrical gears with increased contact area – proposal of application in watercrafts power transmission systems. Solid State Phenom. **236**, 26–30 (2015)
14. Batsch, M., Markowski, T.: Comparative fatigue testing of gears with involute and convexo-concave teeth profiles. Adv. Manuf. Sci. Technol. **40**(2), 5–25 (2016)
15. Dyson, A., et al.: Wildhaber-Novikov circular arc gears: geometry and kinematics. Proc. Royal Soc. A Math. Phys. Eng. Sci. **403**(1825), 313–349 (1986)
16. Goto, J., KojIMA, H.: Study on cutting method of Novikov-type spiral bevel gears. Bullet. JSME. **16**(92), 414–421 (1973)
17. Korotkin, V.I., et al.: Novikov Gearing: Achievements and Development. Nova Science Publishers (2011)
18. Radzevich, S.P.: High-Conformal Gearing: Kinematics and Geometry. CRC Press (2016)
19. Batsch, M.: Mathematical model and tooth contact analysis of convexo-concave helical bevel Novikov gear mesh. Mech. Mach. Theory. **149**, 103842 (2020). https://doi.org/10.1016/j.mechmachtheory.2020.103842
20. Levenberg, K.: A method for the solution of certain problems in least squares. Q. Appl. Math. **2**(2), 164–168 (1944)
21. Marquardt, D.W.: An algorithm for least-squares estimation of nonlinear parameters. J. Soc. Ind. Appl. Math. **11**(2), 431–441 (1963). https://doi.org/10.1137/0111030
22. Nocedal, J., Wright, S.: Numerical Optimization. Springer-Verlag, New York (2006)
23. Bronshtein, I.N., et al.: Handbook of Mathematics. Springer (2015)
24. Klingelnberg, J.: Bevel Gear: Fundamentals and Applications. Springer, Berlin Heidelberg (2016)
25. Kawalec, A., Wiktor, J.: Simulation of generation and tooth contact analysis of helical gears with crowned flanks. Proc. Inst. Mech. Eng. B J. Eng. Manuf. **222**(9), 1147–1160 (2008)
26. Litvin, F.L., et al.: Computerized determination of curvature relations and contact ellipse for conjugate surfaces. Comput. Methods Appl. Mech. Eng. **125**(1), 151–170 (1995). https://doi.org/10.1016/0045-7825(95)00799-7
27. Markowski, T., Batsch, M.: Analityczno-numeryczne metody wyznaczania obszaru styku przekładni wklęsło-wypukłych Nowikowa. Zeszyty Naukowe. Transport/Politechnika Śląska. **82**, 155–165 (2014)
28. Marciniec, A., et al.: Comparative analysis of numerical methods for the determination of contact pattern of spiral bevel gears. Aircr. Eng. Aerosp. Technol. **90**(2), 359–367 (2018). https://doi.org/10.1108/AEAT-08-2016-0133
29. Sobolak, M.: Discrete numerical method of tooth contact analysis (TCA) with α-bufor use. Arch. Mech. Technol. Auto. **27**(2), 153–160 (2007)

30. Sobolak, M.: Analiza i synteza współpracy powierzchni kół zębatych metodami dyskretnymi. Oficyna Wydawnicza Politechniki Rzeszowskiej. (2006)
31. Johnson, K.L.: Contact Mechanics. Cambridge University Press (1987)
32. Batsch, M.: Rapid prototyping and tooth contact analysis of eccentric cycloid gear mesh. J. KONBiN. **49**(1), 369–382 (2019). https://doi.org/10.2478/jok-2019-0019
33. Batsch, M., Markowski, T.: Influence of Novikov convexo-concave gear parameters on contact pattern. Sci. J. Silesian Univ. Technol. Ser. Transport. **89**, 89–99 (2015)
34. Batsch, M., et al.: Measurement and mathematical model of convexo-concave Novikov gear mesh. Measurement. **125**, 516–526 (2018). https://doi.org/10.1016/j.measurement.2018.04.095
35. de Berg, M., et al.: Computational Geometry: Algorithms and Applications. Springer, Berlin Heidelberg (2008)
36. Möller, T., Trumbore, B.: Fast, minimum storage ray-triangle intersection. J. Graphics Tools. **2**(1), 21–28 (1997). https://doi.org/10.1080/10867651.1997.10487468

Chapter 4
Hyperboloidal-Type Normal Circular-Arc Gearing

Houjun Chen, Zhilan Ju, Xiaoping Zhang, and Chang Qu

4.1 Introduction

The conception of a normal circular-arc tooth profile was first presented in Wildhaber's helical gearing, which features a convex-to-concave line contact of tooth flanks in the gear drive with parallel axes. The absence of knowledge in the theory of gearing may be the root cause of why the manufacturing of Wildhaber's helical gearing failed. Fortunately, considering a transverse circular-arc tooth profile, Novikov gearing was proposed and successfully manufactured to achieve convex-to-concave point contact of mating tooth surfaces. Both Wildhaber's helical gearing and Novikov gearing were loosely unified with a new incorrect name, Wildhaber–Novikov gearing.

What are the essential attributes of Novikov gearing? First, is the tooth profile on each a normal circular arc or a transverse circular arc? This is a significant difference between the ideas of these two inventors, and, especially, though the modulus is the same, a set of special hob cutters are required for the generation of the pinion and gear in Novikov gearing with a different number of teeth and different spiral angles, respectively. It increases the difficulty of the production, and, actually, both the design of the special cutters and the machining of Novikov gearing are usually carried out based on a normal circular arc in the manufacturing process. Obviously, the theory is divorced from practice. This would seem to show that a normal circular arc is one of the significant characteristics of Novikov gearing. Second, is it on a line meshing or a point meshing? Wildhaber proposed designing a line meshing gear pair but by applying the principle of mismatching of generating surfaces to generate gears with a localized bearing contact; however, Novikov directly constructed a pair of mating gears with point contact and was the first one to apply mismatched tool surfaces to achieve the generation of helical gears [1]. Actually, both are the same,

H. Chen · Z. Ju · X. Zhang · C. Qu
School of Mechanical Engineering, Nantong University, Nantong, China

S. P. Radzevich, *Novikov/Conformal Gearing*,
https://doi.org/10.1007/978-3-031-10019-2_4

and, viewed from different perspectives and understandings, the two can be unified under the framework of the mismatching principle of conjugation, i.e., the point-contact processing of the line meshing pair.

Following the principle of molding-surface conjugation, hyperboloidal-type normal circular-arc gearing is proposed as a novel improved version of Novikov gearing to transmit motion and power between two orthogonally crossed axes. The basic principle of molding-surface conjugation is introduced in detail, and the general principle of normal circular-arc gearing is systematically discussed to establish the theoretical basis for constructing novel types of gearings. The generation of the pinion and mating gear in HNCGing is presented, mathematical models of conjugate tooth surfaces are established, and the curvature properties of mating tooth surfaces are analyzed. An integrated manufacturing software system for HNCGing is developed, which connects computerized design, generation, simulation of meshing and contact, and the tool path planning and cutting simulation of NC machining in a single application platform.

4.2 Basic Principle of Molding-Surface Conjugation

4.2.1 Molding Surface

Definition 4.1 Suppose that $\mathbf{R}(s)$ is the radius vector of a point on a C^2 continuity spatial curve, $d(s, \theta)$ is the offset distance function describing the distance of a point from a C^1 continuity planar curve to the spatial curve. In this chapter, the spatial curve is called the directrix and the planar curve is called the generatrix. $\mathbf{S}(s, \theta)$ defines a molding surface by the radius vector function:

$$\mathbf{S}(s,\theta) = \mathbf{R}(s) + d(s,\theta)(\sin\theta\mathbf{e}_3(s) + \cos\theta\mathbf{e}_2(s)) \tag{4.1}$$

where

s	is the arc-length parameter of the directrix;
θ	is the profile angle parameter of the generatrix; and
$\mathbf{e}_2(s)$, $\mathbf{e}_3(s)$	are the orthogonal unit vectors that are both perpendicular to the unit tangent vector of the directrix.

As shown in Fig. 4.1, $\{P; \mathbf{e}_1, \mathbf{e}_2, \mathbf{e}_3\}$ is a moving frame of reference set in the directrix C_{d}, in which $\mathbf{e}_1$ is the unit tangent vector of C_{d}; then, the locus of the generatrix C_g in the normal plane $\{P; \mathbf{e}_2, \mathbf{e}_3\}$ of C_{d}, moving with the moving frame along C_{d}, forms the molding surface ψ. The sets $\{\mathbf{S}(s, \theta), s = \text{const}\}$ and $\{\mathbf{S}(s, \theta), \theta = \text{const}\}$ are two families of curves C_{g}, C_{n} on the surface ψ, which constitute the orthogonal system of lines of curvature of the surface.

The geometry of the molding surface is mainly governed by the shapes of the generatrix and directrix, and molding surfaces with various shapes can be constructed by the shape change and relative motion of the two to satisfy different

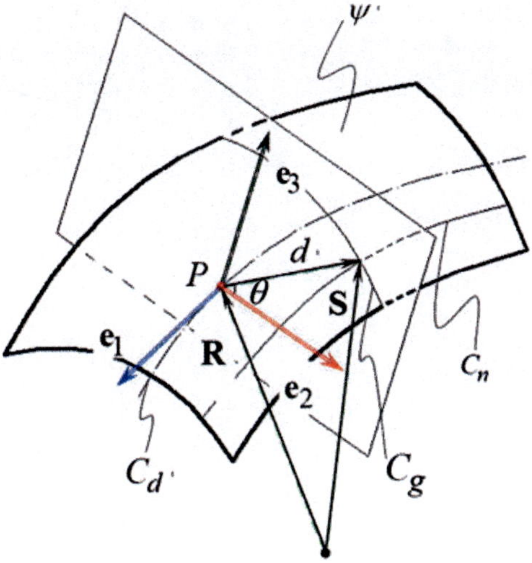

Fig. 4.1 A representation of a molding surface

requirements. In engineering practice, three kinds of familiar structural surfaces can be placed in the same category as molding surfaces, namely,

(i) Surface of revolution: Here, the generatrix is coplanar with a fixed axis of rotation and the directrix is a circle, the center of which is at the axis and lies on the plane perpendicular to the axis.
(ii) Developable surface: Here, the generatrix is a straight line and the molding surface is the envelope of a one-parameter family of planes.
(iii) Normal circular-arc surface (also known as tubular surface): Here, the generatrix is a circle, the center of which is at the directrix, and the molding surface is the envelope of a one-parameter family of spheres.

4.2.2 *Conditional Equation of Molding-Surface Conjugation*

Definition 4.2 The case that a pair of molding surfaces is always in tangential contact along a commonly owned generatrix in their rotating process about two fixed axes is called molding-surface conjugation [2].

As shown in Fig. 4.2, a pair of conjugate molding surfaces $\psi^{(1)}$, $\psi^{(2)}$ is in tangential contact along the generatrix C_{g} and their directrices $C_{\mathrm{d}}^{(1)}, C_{\mathrm{d}}^{(2)}$ are simultaneously tangent at the point P, which is called the conjugate point of the directrices. A shared moving frame, $\{P; \boldsymbol{\alpha}_1, \boldsymbol{\alpha}_2, \boldsymbol{\alpha}_3\}$, is set at point P, and, thereinto, $\boldsymbol{\alpha}_1$ is the common unit tangent vector of $C_{\mathrm{d}}^{(1)}, C_{\mathrm{d}}^{(2)}$. The rotational axes A_1, A_2 of $\psi^{(1)}$, $\psi^{(2)}$ are placed with the center distance A and the shaft angle Σ. A fixed orthogonal right-handed coordinate system, $\{O^{(1)}; \mathbf{i}^{(1)}, \mathbf{j}^{(1)}, \mathbf{k}^{(1)}\}$, is established at the axis A_1, and, thereinto, $\mathbf{i}^{(1)}$ is the unit directional vector of the common normal line $O^{(1)}O^{(2)}$ of the axes A_1, A_2, and $\mathbf{k}^{(1)}$ is the unit directional vector of the axis A_1.

Let $\boldsymbol{\Omega}^{(1)}$, $\boldsymbol{\Omega}^{(2)}$ be the angular velocities of $\psi^{(1)}$, $\psi^{(2)}$ about the axes A_1, A_2, respectively; the following relationships can be obtained from the theory of gearing:

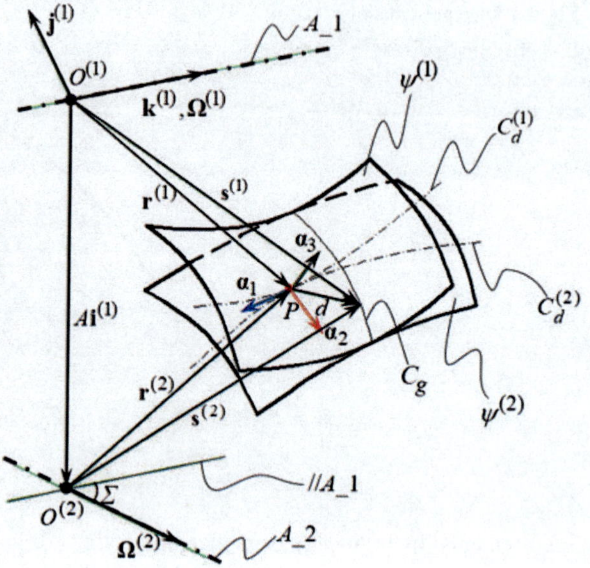

Fig. 4.2 Sketch of molding-surface conjugation

$$\left|\boldsymbol{\Omega}^{(1)}\right|=\frac{d\varphi^{(1)}}{dt},\left|\boldsymbol{\Omega}^{(2)}\right|=\frac{d\varphi^{(2)}}{dt},I=\frac{\left|\boldsymbol{\Omega}^{(2)}\right|}{\left|\boldsymbol{\Omega}^{(1)}\right|},\boldsymbol{\Omega}^{(21)}=\boldsymbol{\Omega}^{(2)}-\boldsymbol{\Omega}^{(1)} \tag{4.2}$$

where

$\varphi^{(i)}$	is the rotational angle of $\psi^{(i)}$ about the axis A_i, and the superscript $i = 1, 2$ denotes the quantities pertaining to $\psi^{(1)}$ and $\psi^{(2)}$ of the pinion and gear, respectively. Similarly, hereinafter;
t	is the time variable;
I	is the speed ratio; and
$\boldsymbol{\Omega}^{(21)}$	is the relative angular velocity between $\psi^{(1)}$ and $\psi^{(2)}$.

Without loss of generality, the following discussions on the kinematics and geometry of conjugation are based on the assumption that I is constant and $|\boldsymbol{\Omega}^{(1)}| = 1$.

According to Definition 4.2, a pair of conjugate molding surfaces is tangentially located along the generatrix, and their directrices are usually in tangential contact at the conjugate point (except developable surface conjugation). Because the moving frame can be flexibly chosen so long as the unit tangent vector of the directrix is one of three coordinate vectors of the frame, a shared moving frame associated with the directrices can be appointed to uniformly describe the kinematics and geometry of conjugation. In this case, the offset distance functions and the profile angles, which define the common generatrix of $\psi^{(1)}$, $\psi^{(2)}$, will be identical in form in this shared frame. According to Eq. (4.1), the families of surfaces $\psi^{(1)}$, $\psi^{(2)}$, which are generated by the rotations about the axes A_1, A_2, respectively, can be described by the radius vector functions:

$$\psi^{(1)}: \mathbf{s}^{(1)}\left(s^{(1)}, \theta, \varphi^{(1)}\right) = B\left(\varphi^{(1)}\right)\mathbf{S}^{(1)}\left(s^{(1)}, \theta\right) = \mathbf{r}^{(1)}\left(s^{(1)}, \varphi^{(1)}\right) + d(\sin\theta\boldsymbol{\alpha}_3 + \cos\theta\boldsymbol{\alpha}_2) \tag{4.3}$$

$$\psi^{(2)}: \mathbf{s}^{(2)}\left(s^{(2)}, \theta, \varphi^{(2)}\right) = B\left(\varphi^{(2)}\right)\mathbf{S}^{(2)}\left(s^{(2)}, \theta\right) = \mathbf{r}^{(2)}\left(s^{(2)}, \varphi^{(2)}\right) + d(\sin\theta\boldsymbol{\alpha}_3 + \cos\theta\boldsymbol{\alpha}_2) \tag{4.4}$$

where

$B(\varphi^{(i)})$	is the rotation group about the axis A_i (i = 1,2), as seen in Chen et al. [2];
$\mathbf{s}^{(i)}$	is the radius vector of the surface $\psi^{(i)}$ in the rotating process;
$\mathbf{r}^{(i)}$	is the radius vector of the directrix $C_{\mathrm{d}}^{(i)}$ in the rotating process, which is equal to $B(\varphi^{(i)})$ $\mathbf{R}^{(i)}$; and
$\boldsymbol{\alpha}_2,\boldsymbol{\alpha}_3$	are the coordinate vectors of the shared frame, which are equal to $B\left(\varphi^{(i)}\right)\mathbf{e}_2^{(i)}, B\left(\varphi^{(i)}\right)\mathbf{e}_3^{(i)}$, respectively.

The significant feature of molding-surface conjugation is that the kinematics and geometry of conjugate surfaces are discussed by means of a pair of directrices of the conjugate surfaces rather than the conjugate surfaces themselves. According to Fig. 4.2, the basic equations of molding-surface conjugation can be established as follows:

$$\begin{cases} \mathbf{r}^{(1)} - \mathbf{r}^{(2)} = A\mathbf{i}^{(1)} \\ \boldsymbol{\alpha}_j = \boldsymbol{\alpha}_j^{(1)} = \boldsymbol{\alpha}_j^{(2)} \quad (j = 1,\ 2,\ 3) \end{cases} \tag{4.5}$$

Differentiating Eq. (4.5) with respect to $\varphi^{(1)}$ and simplifying the expression, the differential equations of molding-surface conjugation are obtained:

$$\begin{cases} \boldsymbol{\alpha}_1\left(\mathrm{d}s^{(1)} - \mathrm{d}s^{(2)}\right) = \mathbf{v}_{\mathrm{P}}^{(21)}\mathrm{d}\varphi^{(1)} \\ \boldsymbol{\omega}^{(1)}\mathrm{d}s^{(1)} - \boldsymbol{\omega}^{(2)}\mathrm{d}s^{(2)} = \boldsymbol{\Omega}^{(21)}\mathrm{d}\varphi^{(1)} \end{cases} \tag{4.6}$$

where

$ds^{(i)}$	is the arc differential of the directrix $C_{\mathrm{d}}^{(i)}$ (i = 1,2);
$\mathbf{v}_{\mathrm{P}}^{(21)}$	is the relative linear velocity of point P, which is equal to $\boldsymbol{\Omega}^{(21)} \times \mathbf{r}^{(1)} - A(\boldsymbol{\Omega}^{(2)} \times \mathbf{i}^{(1)})$; and
$\boldsymbol{\omega}^{(i)}$	is the angular velocity of infinitesimal rotation of the shared frame at $C_{\mathrm{d}}^{(i)}$.

Taking the vector product of the first expression in Eq. (4.6) and $\boldsymbol{\alpha}_1$, as well as the dot product of the second expression and $\boldsymbol{\alpha}_1$, the conditional equations of molding-surface conjugation are obtained:

$$\begin{cases} \mathbf{v}_{\mathrm{P}}^{(21)} \times \boldsymbol{\alpha}_1 = 0 \\ \boldsymbol{\Omega}^{(21)} \cdot \boldsymbol{\alpha}_1 = 0 \end{cases} \tag{4.7}$$

Here, the first expression indicates that the directrices of the conjugate surfaces are tangent at the conjugate point, and their common unit tangent vector must be collinear with the relative linear velocity of the conjugate points of the directrices. The second expression indicates that the common unit tangent vector must be orthogonal to the relative angular velocity of the directrices.

Case 1: Conjugate condition of developable surfaces

Definition 4.3 The case that the line of contact of conjugate molding surfaces is a straight line in the movement is called developable surface conjugation.

In this case, a pair of conjugate molding surfaces is the developable surface, the generatrix of which is the straight line, and their directrices can slide along the line of contact at the conjugate position, while the conjugate surfaces are still tangent without the effect of the sliding, as shown in Fig. 4.3. In the shared moving frame $\{P; \boldsymbol{\alpha}_1, \boldsymbol{\alpha}_2, \boldsymbol{\alpha}_3\}$, $\boldsymbol{\alpha}_1$ is the unit tangent vector of the directrix $C_{\mathrm{d}}^{(1)}$, $\boldsymbol{\alpha}_2$ is the unit directional vector of the straight generatrix C_{g}, and, naturally, $\boldsymbol{\alpha}_3$ is the common normal vector of the conjugate developable surfaces. Combining Figs. 4.3 with 4.2, and referring to Eq. (4.5), the basic equations of developable surface conjugation can be arranged as follows:

$$\begin{cases} \mathbf{r}^{(1)} - \left(\mathbf{r}^{(2)} - l_{\mathrm{s}}\boldsymbol{\alpha}_2\right) = A\mathbf{i}^{(1)} \\ \boldsymbol{\alpha}_j = \boldsymbol{\alpha}_j^{(1)} = \boldsymbol{\alpha}_j^{(2)} \quad (j = 1,\ 2,\ 3) \end{cases} \tag{4.8}$$

where

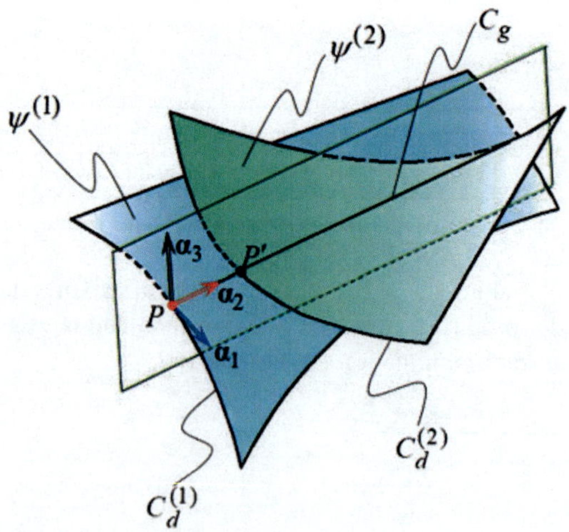

Fig. 4.3 Sketch of developable surface conjugation

l_s	is the relative sliding quantity between $C_d^{(1)}$ and $C_d^{(2)}$ along C_g, i.e., the distance from P to P'.

Differentiating Eq. (4.8) with respect to $\varphi^{(1)}$ and simplifying the expression, the differential equations of developable surface conjugation are obtained as:

$$\begin{cases} \boldsymbol{\alpha}_1\left[\mathrm{d}s^{(1)}-\left(1+l_s k_g^{(2)}\right)\mathrm{d}s^{(2)}\right]+\boldsymbol{\alpha}_2\mathrm{d}l_s=\mathbf{v}_P^{(21)}\mathrm{d}\varphi^{(1)} \\ \boldsymbol{\omega}^{(1)}\mathrm{d}s^{(1)}-\boldsymbol{\omega}^{(2)}\mathrm{d}s^{(2)}=\boldsymbol{\Omega}^{(21)}\mathrm{d}\varphi^{(1)} \end{cases} \tag{4.9}$$

where

$k_g^{(2)}$	is the geodesic curvature of $C_d^{(2)}$ in the shared moving frame $\{P;\boldsymbol{\alpha}_1,\boldsymbol{\alpha}_2,\boldsymbol{\alpha}_3\}$.

Taking the dot products of the first expression in Eq. (4.9) and $\boldsymbol{\alpha}_3$, as well as the second expression and $\boldsymbol{\alpha}_1$, the conditional equations of developable surface conjugation are acquired after the simplification as follows:

$$\begin{cases} \mathbf{v}_P^{(21)}\cdot\boldsymbol{\alpha}_3=0 \\ \boldsymbol{\Omega}^{(21)}\cdot\boldsymbol{\alpha}_1=0 \end{cases} \tag{4.10}$$

Here, the first expression indicates that the common unit normal vector of the conjugate surfaces must be orthogonal to the relative linear velocity of the conjugate points of the directrices. The second expression indicates that the common unit tangent vector of the directrices must be orthogonal to their relative angular velocity.

Case 2: Conjugate condition of normal circular-arc surfaces

Definition 4.4 The case that the line of contact of conjugate molding surfaces is a circular arc in the movement and that the directrices of conjugate surfaces are the loci of the centers of the circular-arc generatrix in the generation process of the surfaces is called normal circular-arc surface conjugation.

In this case, a pair of conjugate molding surfaces is a normal circular-arc surface and they can make the relative rotation about the common tangent line of the directrices at the conjugate point. The basic equations and the differential equations of normal circular-arc surface conjugation are consistent with Eqs. (4.5) and (4.6) in form, respectively; then, the conditional equation of normal circular-arc surface conjugation is acquired after taking the dot product of the first expression in Eq. (4.6) and $\boldsymbol{\alpha}_1$ as follows:

$$\mathbf{v}_P^{(21)}\times\boldsymbol{\alpha}_1=0 \tag{4.11}$$

This indicates that the directrices of conjugate surfaces are tangent at the conjugate point and that their common unit tangent vector must be collinear with the relative linear velocity of the conjugate points of the directrices.

4.2.3 Structural Condition of Molding-Surface Conjugation

The conditional equations of conjugation described above are nonlinear, which means that the condition of molding-surface conjugation is stricter than that of traditional conjugation of surfaces. When the kinematic and positional parameters are given, there may not necessarily be another molding surface, which can conjugate with a known molding surface to comply with the specified requirements of molding-surface conjugation.

Definition 4.5 The restriction on the directrices of conjugate surfaces, which supports the achievement of molding-surface conjugation, is called the structural condition of conjugation.

1. The structural condition of developable surface conjugation

Expanding and rearranging Eq. (4.10), a system of equations can be obtained as follows:

$$\begin{cases} \mathbf{F}\cdot\mathbf{j}^{(1)}\cos\varphi^{(1)} + \mathbf{F}\cdot\mathbf{i}^{(1)}\sin\varphi^{(1)} = \mathbf{G}\cdot\boldsymbol{\Omega}^{(1)} \\ I\sin\Sigma\cos\varphi^{(1)}\mathbf{j}^{(1)}\cdot\mathbf{e}_1^{(1)} + I\sin\Sigma\sin\varphi^{(1)}\mathbf{i}^{(1)}\cdot\mathbf{e}_1^{(1)} = (I\cos\Sigma - 1)\boldsymbol{\Omega}^{(1)}\cdot\mathbf{e}_1^{(1)} \\ \mathbf{F} = I\sin\Sigma\left(\mathbf{R}^{(1)}\times\mathbf{e}_3^{(1)}\right) + AI\cos\Sigma\mathbf{e}_3^{(1)} \\ \mathbf{G} = (I\cos\Sigma - 1)\left(\mathbf{R}^{(1)}\times\mathbf{e}_3^{(1)}\right) - AI\sin\Sigma\mathbf{e}_3^{(1)} \end{cases} \tag{4.12}$$

Here, **F**, **G**, and $\mathbf{e}_1^{(1)}$, $\mathbf{e}_3^{(1)}$ are all one-variable functions with respect to arc-length parameter $s^{(1)}$, and the parameter $\varphi^{(1)}$ is the rotational angle of the directrix $C_d^{(1)}$ with the rotation of $\psi^{(1)}$ about the axis A_1. The function $\varphi^{(1)} = \varphi^{(1)}(s^{(1)})$ can be solved only when the system of equations is compatible, and the compatibility condition would then be the structural condition of conjugation.

Substituting the solutions of $\cos\varphi^{(1)}$ and $\sin\varphi^{(1)}$ in Eq. (4.12) in the identical equation $\cos^2\varphi^{(1)} + \sin^2\varphi^{(1)} = 1$ and rearranging it, the structural condition of developable surface conjugation is obtained:

$$\left[I\sin\Sigma\boldsymbol{\Omega}^{(1)}\cdot\left(\mathbf{F}\times\mathbf{e}_1^{(1)}\right)\right]^2$$
$$=\left\{\boldsymbol{\Omega}^{(1)}\times\left[(I\cos\Sigma-1)\left(\boldsymbol{\Omega}^{(1)}\cdot\mathbf{e}_1^{(1)}\right)\mathbf{F}-I\sin\Sigma\left(\boldsymbol{\Omega}^{(1)}\cdot\mathbf{G}\right)\mathbf{e}_1^{(1)}\right]\right\}^2 \quad (4.13)$$

Obviously, this restricts the kinematic parameter I, the positional parameters A and Σ, and the radius vector $\mathbf{R}^{(1)}$. That is to say, only when the structural condition of conjugation is satisfied, molding-surface conjugation can be achieved in theory.

Case 1: Developable surface conjugation with parallel axes ($\Sigma = 0, \pi$ and $A \neq 0$)

In this case, Eq. (4.13) can be calculated and simplified into:

$$\left(\boldsymbol{\Omega}^{(1)}\cdot\mathbf{e}_1^{(1)}\right)^2\left(\boldsymbol{\Omega}^{(1)}\times\mathbf{e}_3^{(1)}\right)^2=0 \quad (4.14)$$

The condition that Eq. (4.14) holds is $\boldsymbol{\Omega}^{(1)}\times\mathbf{e}_3^{(1)}=0$ or $\boldsymbol{\Omega}^{(1)}\cdot\mathbf{e}_1^{(1)}=0$. The former demands that the normal vector of the developable surface should always be parallel to the axis A_1, which makes no sense to the engineering practice. The latter demands that the directrix $C_{\mathrm{d}}^{(1)}$ should be a planar curve, which lies on the plane perpendicular to the axis A_1. At this time, a line, which passes through $C_{\mathrm{d}}^{(1)}$ and forms a constant inclined angle with the axis A_1, can be appointed as the common straight generatrix of developable surface conjugation.

Case 2: Developable surface conjugation with intersecting axes ($A = 0$ and $\Sigma \neq 0, \pi$)

In this case, Eq. (4.13) can be calculated and simplified into:

$$I^2\sin^2\Sigma\left[\left(\mathbf{R}^{(1)}\cdot\mathbf{e}_1^{(1)}\right)\left(\boldsymbol{\Omega}^{(1)}\cdot\mathbf{e}_3^{(1)}\right)\right]^2$$
$$=(I\cos\Sigma-1)^2\left(\mathbf{R}^{(1)}\cdot\mathbf{e}_1^{(1)}\right)^2\left[1-\left(\boldsymbol{\Omega}^{(1)}\cdot\mathbf{e}_3^{(1)}\right)^2\right] \quad (4.15)$$

The conditions that Eq. (4.15) holds can be summarized into two aspects:

(i) $\mathbf{R}^{(1)}\cdot\mathbf{e}_1^{(1)}=0$ Given that $\mathbf{e}_1^{(1)}=d\mathbf{R}^{(1)}/ds^{(1)}$, where $\mathbf{R}^{(1)}$ is a vector with a constant length, the directrix $C_{\mathrm{d}}^{(1)}$ is a spatial curve on the spherical surface, the center of which is at the intersecting point of the axes A_1 and A_2.

(ii) $\frac{I\sin\Sigma}{I\cos\Sigma-1}=\pm\frac{\sqrt{1-\left(\boldsymbol{\Omega}^{(1)}\cdot\mathbf{e}_3^{(1)}\right)^2}}{\boldsymbol{\Omega}^{(1)}\cdot\mathbf{e}_3^{(1)}}$ It can be found from the theory of gearing that the left side of this equation denotes the tangent of the pitch cone angle of the driving gear in the gear pair with intersecting axes, whereas the absolute value on the right side is the tangent of the included angle between the surface normal and the axis A_1 of the driving gear. Thus, we see that developable surface can be applied as the tooth surfaces of a gear pair when the included angle between the surface normal and the axis A_1 is the pitch cone angle of the driving gear.

Case 3: Developable surface conjugation with orthogonally crossed axes ($\boldsymbol{\Sigma} = \pi/2$ and $A \neq 0$)

In this case, Eq. (4.13) can be calculated and simplified into:

$$
\begin{aligned}
& I^2\left(\mathbf{R}^{(1)} \cdot \mathbf{e}_1^{(1)}\right)^2\left(\boldsymbol{\Omega}^{(1)} \cdot \mathbf{e}_3^{(1)}\right)^2 \\
& = \left\{\left(\mathbf{R}^{(1)} \cdot \mathbf{e}_1^{(1)}\right)\left[\boldsymbol{\Omega}^{(1)} \times \left(\boldsymbol{\Omega}^{(1)} \times \mathbf{e}_3^{(1)}\right)\right] + AI\left(\mathbf{e}_1^{(1)} \times \boldsymbol{\Omega}^{(1)}\right)\left(\boldsymbol{\Omega}^{(1)} \cdot \mathbf{e}_3^{(1)}\right)\right\}^2
\end{aligned} \tag{4.16}
$$

Although it provides the possibility of obtaining developable surface conjugation with orthogonally crossed axes, the decision-making process of the directrix is highly complicated.

2. The structural condition of normal circular-arc surface conjugation

Equation (4.11) can be rewritten in the scalar form, i.e., $\mathbf{v}_P^{(21)} \cdot \boldsymbol{\alpha}_2 = \mathbf{v}_P^{(21)} \cdot \boldsymbol{\alpha}_3 = 0$. Using the method of spreading out and calculating, the system of equations is obtained as follows:

$$
\begin{cases}
\mathbf{P} \cdot \mathbf{e}_2^{(1)} \cos\varphi^{(1)} + \mathbf{Q} \cdot \mathbf{e}_2^{(1)} \sin\varphi^{(1)} = \mathbf{U} \cdot \mathbf{e}_2^{(1)} \\
\mathbf{P} \cdot \mathbf{e}_3^{(1)} \cos\varphi^{(1)} + \mathbf{Q} \cdot \mathbf{e}_3^{(1)} \sin\varphi^{(1)} = \mathbf{U} \cdot \mathbf{e}_3^{(1)} \\
\mathbf{P} = I \sin\boldsymbol{\Sigma}\left(\mathbf{j}^{(1)} \times \mathbf{R}^{(1)}\right) + AI \cos\boldsymbol{\Sigma}\mathbf{j}^{(1)} \\
\mathbf{Q} = I \sin\boldsymbol{\Sigma}\left(\mathbf{i}^{(1)} \times \mathbf{R}^{(1)}\right) + AI \cos\boldsymbol{\Sigma}\mathbf{i}^{(1)} \\
\mathbf{U} = (I\cos\boldsymbol{\Sigma} - 1)\left(\boldsymbol{\Omega}^{(1)} \times \mathbf{R}^{(1)}\right) - AI \sin\boldsymbol{\Sigma}\boldsymbol{\Omega}^{(1)}
\end{cases} \tag{4.17}
$$

Similar to the operation method of Eq. (4.13), the structural condition of normal circular-arc surface conjugation is obtained in the form of Pythagoras' formula:

$$
\left[(\mathbf{U} \times \mathbf{Q}) \cdot \mathbf{e}_1^{(1)}\right]^2 + \left[(\mathbf{P} \times \mathbf{U}) \cdot \mathbf{e}_1^{(1)}\right]^2 = \left[(\mathbf{P} \times \mathbf{Q}) \cdot \mathbf{e}_1^{(1)}\right]^2 \tag{4.18}
$$

Case 1: Normal circular-arc surface conjugation with parallel axes ($\boldsymbol{\Sigma} = 0, \pi$ and $A \neq 0$)

In this case, Eq. (4.18) can be calculated and simplified into:

$$
(1 \mp I)^2\left(\boldsymbol{\Omega}^{(1)} \cdot \mathbf{e}_1^{(1)}\right)^2\left(\boldsymbol{\Omega}^{(1)} \times \mathbf{R}^{(1)}\right)^2 = \left(AI\boldsymbol{\Omega}^{(1)} \cdot \mathbf{e}_1^{(1)}\right)^2 \tag{4.19}
$$

Obviously, the conditions that Eq. (4.19) holds can be summarized into the following aspects:

(i) $\boldsymbol{\Omega}^{(1)} \cdot \mathbf{e}_1^{(1)} = 0$ The directrix $C_d^{(1)}$ is a planar curve, which lies on the plane perpendicular to the rotational axis A_1.

(ii) $\frac{AI}{1 \mp I} = \mp \sqrt{\left(\boldsymbol{\Omega}^{(1)} \times \mathbf{R}^{(1)}\right)^2}$ It can be found from the theory of gearing that the left side of this equation is the pitch radius in parallel-axis gearing, whereas the right side is the projection length of the radius vector $\mathbf{R}^{(1)}$ onto a plane orthogonal to the axis A_1. It is certain that the directrix $C_{\mathrm{d}}^{(1)}$ is a spatial curve wrapped around the pitch cylinder.

Case 2: Normal circular-arc surface conjugation with intersecting axes ($A = 0$ and $\Sigma \neq 0, \pi$)

In this case, Eq. (4.18) can be calculated and simplified into:

$$\left[I \sin \boldsymbol{\Sigma}\left(\boldsymbol{\Omega}^{(1)} \cdot \mathbf{R}^{(1)}\right)\left(\mathbf{R}^{(1)} \cdot \mathbf{e}_1^{(1)}\right)\right]^2 = (I \cos \boldsymbol{\Sigma} - 1)^2 \left(\mathbf{R}^{(1)} \cdot \mathbf{e}_1^{(1)}\right)^2 \left[\boldsymbol{\Omega}^{(1)} \times \left(\boldsymbol{\Omega}^{(1)} \times \mathbf{R}^{(1)}\right)\right]^2 \tag{4.20}$$

Obviously, the conditions that Eq. (4.20) holds can be summarized into two aspects:

(i) $\mathbf{R}^{(1)} \cdot \mathbf{e}_1^{(1)} = 0$ Similar to the previous description, the directrix $C_{\mathrm{d}}^{(1)}$ is a spatial curve on the spherical surface, the center of which is at the intersecting point of the axes A_1 and A_2.

(ii) $\frac{\mathrm{I} \sin \Sigma}{\mathrm{I} \cos \Sigma - 1} = \pm \frac{\sqrt{\left[\boldsymbol{\Omega}^{(1)} \times \left(\boldsymbol{\Omega}^{(1)} \times \mathbf{R}^{(1)}\right)\right]^2}}{\boldsymbol{\Omega}^{(1)} \cdot \mathbf{R}^{(1)}}$ It can be found from the theory of gearing that the left side of this equation denotes the tangent of the pitch cone angle of the driving gear in intersecting-axis gearing, whereas the absolute value on the right side is the tangent of the included angle between the vector $\mathbf{R}^{(1)}$ and the axis A_1. It is certain that the directrix $C_{\mathrm{d}}^{(1)}$ is a spatial curve wrapped around the pitch cone. Its engineering application can be seen in the study by Duan et al. [3].

Case 3: Normal circular-arc surface conjugation with orthogonally crossed axes ($\Sigma = \pi/2$ and $A \neq 0$)

In this case, Eq. (4.18) can be calculated and simplified into:

$$\left[I\left(\boldsymbol{\Omega}^{(1)} \cdot \mathbf{R}^{(1)}\right)\left(\mathbf{R}^{(1)} \cdot \mathbf{e}_1^{(1)}\right)\right]^2 = \left[AI\left(\boldsymbol{\Omega}^{(1)} \cdot \mathbf{R}^{(1)}\right)\boldsymbol{\Omega}^{(1)} \times \mathbf{e}_1^{(1)} + \left(\mathbf{R}^{(1)} \cdot \mathbf{e}_1^{(1)}\right)\boldsymbol{\Omega}^{(1)} \times \left(\mathbf{R}^{(1)} \times \boldsymbol{\Omega}^{(1)}\right)\right]^2 \tag{4.21}$$

Similarly, it provides the possibility of achieving normal circular-arc surface conjugation with orthogonally crossed axes, but the solution of the directrix is also complicated. As a valuable improved version of Novikov gearing, hyperboloidal-type normal circular-arc gearing is proposed to transmit motion and power between two orthogonally crossed axes, and a detailed discussion on how to achieve hyperboloidal-type normal circular-arc gearing is presented in Sect. 4.3.

4.2.4 General Principle of Normal Circular-Arc Gearing

Definition 4.6 The locus of the conjugate points of the directrices of conjugate molding surfaces in a fixed space is called the meshing line. The surface generated by the rotation of the directrix (or the meshing line; the two are identical in the effect on the generation of surface) about the rotational axis is called the datum surface. The outline of the smallest cross section of the datum surface is called the gorge circle.

Definition 4.7 The gearing that complies with the principle of normal circular-arc surface conjugation is named normal circular-arc gearing.

1. Basic relationship of the directrix, the meshing line, and the datum surface

As shown in Fig. 4.4, when the meshing line C_{m} is given, a pair of datum surfaces $\Gamma^{(1)}$, $\Gamma^{(2)}$ can be obtained by the rotations of C_{m} about the axes A_1, A_2, respectively, and the two are tangent at the meshing line. The locus of a movable point P, which moves along C_{m} with a predetermined rule of movement and rotates simultaneously with C_{m}, forms a pair of conjugate directrices $C_{\mathrm{d}}^{(1)}$ and $C_{\mathrm{d}}^{(2)}$ and are always

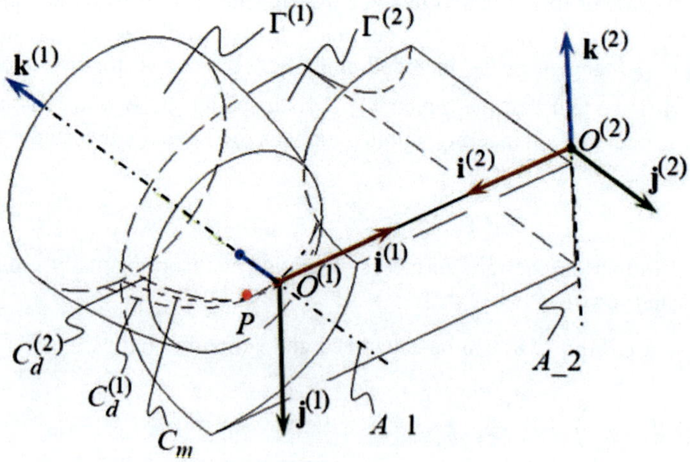

Fig. 4.4 A diagram of the directrix, meshing line, and datum surface

tangent at point P at any instant. A fixed orthogonal right-handed coordinate system, $\{O^{(2)}; \mathbf{i}^{(2)}, \mathbf{j}^{(2)}, \mathbf{k}^{(2)}\}$, is established at the axis A_2, and, thereinto, the coordinate vector $\mathbf{i}^{(2)}$ is in the opposite direction to $\mathbf{i}^{(1)}$. The convention of other symbols is the same as before.

Let $\boldsymbol{\rho}$ be the radius vector of the meshing line $C_{\mathbf{m}}$ and

$$C_{\mathrm{m}}: \quad \boldsymbol{\rho} = \boldsymbol{\rho}(u) = x(u)\mathbf{i}^{(1)} + y(u)\mathbf{j}^{(1)} + z(u)\mathbf{k}^{(1)} \tag{4.22}$$

where

$x(u), y(u), z(u)$	are the coordinate components in $\{O^{(1)}; \mathbf{i}^{(1)}, \mathbf{j}^{(1)}, \mathbf{k}^{(1)}\}$.

According to Definition 4.6, a pair of datum surfaces $\Gamma^{(1)}$, $\Gamma^{(2)}$ can be described with the radius vector equations:

$$\Gamma^{(1)}: \qquad \mathbf{P}^{(1)} = B\left(\lambda^{(1)}\right)\boldsymbol{\rho}(u) \tag{4.23}$$

$$\Gamma^{(2)}: \qquad \mathbf{P}^{(2)} = B\left(\lambda^{(2)}\right)\left(\boldsymbol{\rho}(u) - A\mathbf{i}^{(1)}\right) \tag{4.24}$$

where

$\lambda^{(i)}$	is the rotational angle of the meshing line C_{m} about the axis A_i ($i = 1, 2$).

Referring to Chen et al. [2], Eqs. (4.23) and (4.24) can be rewritten as:

$$\Gamma^{(1)}: \quad \mathbf{P}^{(1)} = x(u)\mathbf{e}\left(\lambda^{(1)}\right) + y(u)\mathbf{e}_1\left(\lambda^{(1)}\right) + z(u)\mathbf{k}^{(1)} \tag{4.25}$$

$$\Gamma^{(2)}: \quad \mathbf{P}^{(2)} = x^*(u)\mathbf{e}\left(\lambda^{(2)}\right) + y^*(u)\mathbf{e}_1\left(\lambda^{(2)}\right) + z^*(u)\mathbf{k}^{(2)} \tag{4.26}$$

where

$\mathbf{e}(\lambda^{(i)})$, $\mathbf{e}_1(\lambda^{(i)})$	are the unit circle vector functions, which denote the rotations of $\mathbf{i}^{(i)}$, $\mathbf{j}^{(i)}$ about $\mathbf{k}^{(i)}$ with the rotational angle $\lambda^{(i)}$ ($i = 1,2$), respectively, as seen in Chen et al. [4] and
$x^*(u), y^*(u), z^*(u)$	are the coordinate components in $\{O^{(2)}; \mathbf{i}^{(2)}, \mathbf{j}^{(2)}, \mathbf{k}^{(2)}\}$.

If the functions $\lambda^{(1)} = \lambda^{(1)}(u)$ and $\lambda^{(2)} = \lambda^{(2)}(u)$ exist and satisfy the predetermined rule of movement for the conjugate point, then a pair of conjugate directrices $C_{\mathrm{d}}^{(1)}$, $C_{\mathrm{d}}^{(2)}$, which are wrapped around the datum surfaces $\Gamma^{(1)}$,$\Gamma^{(2)}$ can be obtained from Eqs. (4.25) and (4.26), respectively. Now, we differentiate Eq. (4.25) with respect to u and let $d\lambda^{(1)}/dt = 1$; then, the tangent vector of $C_{\mathrm{d}}^{(1)}$ can be denoted as:

$$\begin{cases} \dfrac{d\mathbf{P}^{(1)}}{du} = B\left(\lambda^{(1)}\right)\left[\dot{\boldsymbol{\rho}} + \left(\boldsymbol{\Omega}^{(1)} \times \boldsymbol{\rho}\right) \dot{\lambda^{(1)}}\right] \\ \dot{\boldsymbol{\rho}} = \dfrac{d\boldsymbol{\rho}}{du}, \quad \dot{\lambda^{(1)}} = \dfrac{d\lambda^{(1)}}{du} \end{cases} \tag{4.27}$$

2. Conjugate condition and restrictive condition based on the meshing line

In Sect. 4.2.2, both the conjugate condition and the structural condition are directed against a pair of directrices at the conjugate point, whereas the loci of the conjugate points form the meshing line; therefore, it is simple and convenient to discuss the problems on the conjugation by means of the meshing line.

According to Eq. (4.6), the relative linear velocity $\mathbf{v}_{\mathrm{P}}^{(21)}$ of the conjugate point P at the meshing line C_{m} can be rewritten as:

$$\mathbf{v}_{\mathrm{P}}^{(21)} = \mathbf{v}_{\mathrm{P}}^{(2)} - \mathbf{v}_{\mathrm{P}}^{(1)} = \boldsymbol{\Omega}^{(21)} \times \boldsymbol{\rho} - A\left(\boldsymbol{\Omega}^{(2)} \times \mathbf{i}^{(1)}\right) \tag{4.28}$$

where

$\mathbf{v}_{\mathrm{P}}^{(1)}$	is the linear velocity of point P about the axis A_1, which is equal to $\boldsymbol{\Omega}^{(1)} \times \boldsymbol{\rho}$ and
$\mathbf{v}_{\mathrm{P}}^{(2)}$	is the linear velocity of point P about the axis A_2, which is equal to $\boldsymbol{\Omega}^{(2)} \times (\boldsymbol{\rho} - A\mathbf{i}^{(1)})$.

When $\lambda^{(1)} = 0, B(0) = E$ (E is the identical deformation), Eq. (4.27) denotes the tangent vector of $C_{\mathrm{d}}^{(1)}$ at the conjugate point P, i.e.,

$$\frac{d\mathbf{P}^{(1)}}{du}\Big|_{\lambda^{(1)}=0} = \dot{\boldsymbol{\rho}} + \mathbf{v}_{\mathrm{P}}^{(1)} \dot{\lambda^{(1)}} \tag{4.29}$$

As a result, the conditional Eq. (4.11) of normal circular-arc surface conjugation can be rewritten as:

$$\mathbf{v}_{\mathrm{P}}^{(21)} \times \frac{d\mathbf{P}^{(1)}}{du}\Big|_{\lambda^{(1)}=0} = \mathbf{v}_{\mathrm{P}}^{(21)} \times \dot{\boldsymbol{\rho}} - \left(\mathbf{v}_{\mathrm{P}}^{(1)} \times \mathbf{v}_{\mathrm{P}}^{(2)}\right) \dot{\lambda^{(1)}} = 0 \tag{4.30}$$

The prior condition that Eq. (4.30) holds is that the vectors $\mathbf{v}_{\mathrm{P}}^{(21)} \times \dot{\boldsymbol{\rho}}$ and $\mathbf{v}_{\mathrm{P}}^{(1)} \times \mathbf{v}_{\mathrm{P}}^{(2)}$ must be linearly dependent, i.e.,

$$\left(\mathbf{v}_{\mathrm{P}}^{(21)} \times \dot{\boldsymbol{\rho}}\right) \times \left(\mathbf{v}_{\mathrm{P}}^{(1)} \times \mathbf{v}_{\mathrm{P}}^{(2)}\right) = 0 \tag{4.31}$$

By double vector product, it can be simplified into:

$$\left(\mathbf{v}_{\mathrm{P}}^{(1)} \times \mathbf{v}_{\mathrm{P}}^{(2)}\right) \cdot \dot{\boldsymbol{\rho}} = 0 \tag{4.32}$$

It is the restrictive condition of the meshing line and is equivalent to Eq. (4.18). If Eq. (4.32) is not standing, then Eq. (4.30) is certainly false, and, thus, Eq. (4.32) is also the necessary condition for conjugation.

Because the datum surfaces $\Gamma^{(1)}$, $\Gamma^{(2)}$ are the surface of revolution, $\mathbf{v}_P^{(1)}$ and $\mathbf{v}_P^{(2)}$ can be regarded as the tangent vectors of the parallels of $\Gamma^{(1)}$ and $\Gamma^{(2)}$, respectively, at the conjugate point P, and the restrictive condition of the meshing line of normal circular-arc conjugation can be interpreted as follows: the tangent vector of the meshing line must lie on the common tangent plane between the two datum surfaces. Furthermore, Eq. (4.32) can be rewritten in an expanded form as:

$$\mathrm{A}\left(\boldsymbol{\rho}\cdot\mathbf{j}^{(1)}\right)\boldsymbol{\Omega}^{(2)}\cdot\dot{\boldsymbol{\rho}}+\mathrm{I}\sin\Sigma\left(\boldsymbol{\rho}\cdot\mathbf{i}^{(1)}\right)\left(\boldsymbol{\rho}-\mathrm{A}\mathbf{i}^{(1)}\right)\cdot\dot{\boldsymbol{\rho}}=0 \tag{4.33}$$

Case 1: Normal circular-arc gearing with parallel axes ($\Sigma = 0$, π and $A \neq 0$)

In this case, the conditions that Eq. (4.33) holds can be summarized into two aspects:

(i) $\boldsymbol{\Omega}^{(2)}\cdot\dot{\boldsymbol{\rho}}=\boldsymbol{\Omega}^{(1)}\cdot\dot{\boldsymbol{\rho}}=0$ The meshing line C_{m} lies on the plane, which is perpendicular to the axes A_1, A_2.

(ii) $\boldsymbol{\rho}\cdot\mathbf{j}^{(1)}=0$ The meshing line C_{m} lies on the plane, which is determined by the parallel axes A_1, A_2, and it is the instantaneous axis in the gear pair with parallel axes; the datum surfaces are a pair of pitch cylinders.

Case 2: Normal circular-arc gearing with intersecting axes ($A = 0$ and $\Sigma \neq 0$, π)

In this case, the conditions that Eq. (4.33) holds can be summarized into two aspects:

(i) $\boldsymbol{\rho}\cdot\dot{\boldsymbol{\rho}}=0$ The meshing line C_{m} is a spatial curve on the spherical surface, the center of which is at the intersecting point of the axes A_1 and A_2, and the spherical surface is used as a pair of coinciding datum surfaces.

(ii) $\boldsymbol{\rho}\cdot\mathbf{i}^{(1)}=0$ The meshing line C_{m} lies on the plane determined by the intersecting axes A_1, A_2; it is the instantaneous axis in the gear pair with intersecting axes, and the datum surfaces are a pair of pitch cones.

Case 3: Normal circular-arc gearing with orthogonally crossed axes ($\Sigma = \pi/2$ and $A \neq 0$)

In this case, Eq. (4.33) can be simplified into:

$$\begin{cases} (x-A)\dfrac{\mathrm{d}m^2}{\mathrm{d}u}+x\dfrac{\mathrm{d}z^2}{\mathrm{d}u}=0 \\ m^2=x^2+y^2 \end{cases} \tag{4.34}$$

where

m	is the distance between the projection of point P on the plane perpendicular to the axis A_1 and the axis.

Let $u = x$; then, the following expression can be obtained by the transposition and the integral of Eq. (4.34) as:

$$z^2 = \int \left(\frac{A}{x} - 1\right) \frac{dm^2}{dx} dx + C \tag{4.35}$$

where

C	is the integration constant.

$m = m(x)$ can be freely chosen to satisfy the practical requirements, and, subsequently, $z = z(x)$ is solved from Eq. (4.35), and, finally, a spatial curve defined by m, z is the meshing line C_{m}.

3. Determination of the directrix

According to the previous discussion, Eq. (4.11) is equivalent to two independent scalar equations, and, thereinto, one is the restrictive condition of the meshing line, and only when $\dot{\lambda}^{(1)}$ is properly chosen can another scalar condition hold. Only in this manner can the function $\lambda^{(1)} = \lambda^{(1)}(u)$ be obtained, and, then, the directrix $C_{\mathrm{d}}^{(1)}$ is solved. Thus, the condition that $\dot{\lambda}^{(1)}$ meets is sufficient for normal circular-arc surface conjugation.

Taking the scalar product of Eq. (4.30) and $\mathbf{k}^{(1)}$, another scalar condition can be acquired after the simplification as follows:

$$\dot{\lambda}^{(1)} = -\frac{\dot{x}[(I\cos\Sigma - 1)x - AI\cos\Sigma] + \dot{y}[(I\cos\Sigma - 1)y + Iz\sin\Sigma]}{Ixz\sin\Sigma + AIy\cos\Sigma} \tag{4.36}$$

Case 1: Normal circular-arc gearing with parallel axes ($\Sigma = 0$, π and $A \neq 0$)

$$\dot{\lambda}^{(1)} = \frac{(1 \mp I)(x\dot{x} + y\dot{y}) \pm AI\dot{x}}{\pm AIy} \tag{4.37}$$

Case 2: Normal circular-arc gearing with intersecting axes ($A = 0$ and $\Sigma \neq 0$, π)

$$\dot{\lambda}^{(1)} = -\frac{(I\cos\Sigma - 1)(x\dot{x} + y\dot{y}) + Iz\dot{y}\sin\Sigma}{Ixz\sin\Sigma} \tag{4.38}$$

Case 3: Normal circular-arc gearing with orthogonally crossed axes ($\Sigma = \pi/2$ and $A \neq 0$)

$$\dot{\lambda}^{(1)} = \frac{x\dot{x} + y\dot{y} - Iz\dot{y}}{Ixz} \tag{4.39}$$

Theoretically, the function $\lambda^{(1)} = \lambda^{(1)}(u)$ can be obtained by the integral of $\dot{\lambda}^{(1)}$ and by substituting it in Eq. (4.25), we have the directrix $C_{\mathrm{d}}^{(1)}$. Because $\lambda^{(i)}$ is equivalent to $\varphi^{(i)}$, there is a relationship $\lambda^{(2)}/\lambda^{(1)} = I$ between a pair of directrices. By substituting $\lambda^{(2)} = I\lambda^{(1)}(u)$ in Eq. (4.26), the directrix $C_{\mathrm{d}}^{(2)}$ is obtained.

4.3 Geometry of Hyperboloidal-Type Normal Circular-Arc Gears

Definition 4.8 A kind of normal circular-arc gearing in which the datum surfaces are a pair of hyperboloids of one sheet is named hyperboloidal-type normal circular-arc gearing.

4.3.1 Determination of Conjugate Directrices

For normal circular-arc gearing with orthogonally crossed axes, the case $x = x_0$ (const) in Eq. (4.34) is highly valuable. By substituting it in Eq. (4.35), the meshing line C_{m}, which lies on the plane $x = x_0$, is obtained as:

$$\begin{cases} z^2 = y^2 \cot^2\delta^{(1)} + C \\ \tan^2\delta^{(1)} = \dfrac{x_0}{A - x_0} \end{cases} \tag{4.40}$$

As shown in Fig. 4.5, Eq. (4.40) can represent three kinds of meshing lines: when $C = 0, z = \pm y \cot \delta^{(1)}$, the meshing lines are a pair of straight lines with the included angle $\delta^{(1)}$ relative to the $\mathbf{k}^{(1)}$ axis (i.e., the axis A_1); when $C = C_0^2 > 0$, $z^2 = y^2 \cot^2\delta^{(1)} + C_0^2$, the meshing lines are a pair of hyperbolas, which are symmetrical with respect to the $\mathbf{j}^{(1)}$ axis; and when $C = -C_0^2 < 0$, $z^2 = y^2 \cot^2\delta^{(1)} - C_0^2$, the meshing lines are a pair of hyperbolas, which are symmetrical with respect to the $\mathbf{k}^{(1)}$ axis. It should be noted that the vectors $\mathbf{j}^{(1)}$, $\mathbf{k}^{(1)}$ in Fig. 4.5 originate from the translation of $\mathbf{j}^{(1)}$, $\mathbf{k}^{(1)}$ from the plane $x = 0$ to the plane $x = x_0$.

These three kinds of meshing lines can be uniformly described with the following equation:

$$\begin{cases} z = z(y) = \pm\sqrt{y^2 \cot^2\delta^{(1)} + C} \\ x = x_0 \end{cases} \tag{4.41}$$

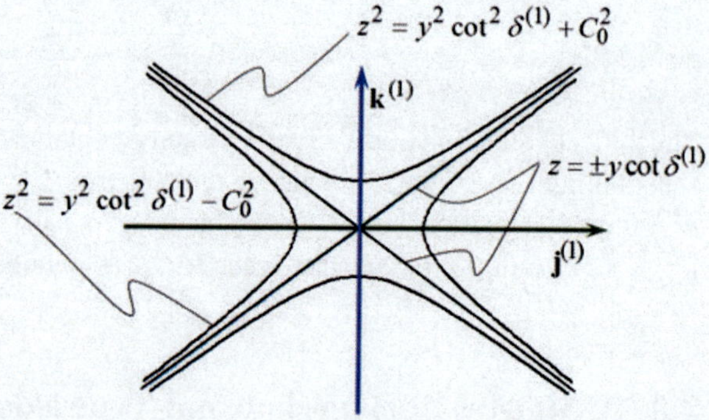

Fig. 4.5 Three forms of meshing lines

When the meshing line is a straight line, i.e., $C = 0$, the surfaces of revolution generated by the meshing line about the axes A_1, A_2 are a pair of hyperboloids of one sheet and hyperboloidal-type normal circular-arc gearing is named after it. In this case, the vector equation of the meshing line C_{m} can be represented in $\{O^{(1)}; \mathbf{i}^{(1)}, \mathbf{j}^{(1)}, \mathbf{k}^{(1)}\}$ as:

$$C_{\mathrm{m}}: \quad \boldsymbol{\rho} = \boldsymbol{\rho}(u) = x_0 \mathbf{i}^{(1)} + u\left(\sin \delta^{(1)} \mathbf{j}^{(1)} - \cos \delta^{(1)} \mathbf{k}^{(1)}\right) \tag{4.42}$$

where

u	is the length variable of the meshing line C_{m} and
x_0	is the interception of the meshing line at the $\mathbf{i}^{(1)}$ axis.

Referring to the above description, the datum surface $\boldsymbol{\Gamma}^{(1)}$ of the pinion can be described by the vector equation:

$$\begin{aligned} \boldsymbol{\Gamma}^{(1)}: \mathbf{P}^{(1)} = B(-\lambda^{(1)})\boldsymbol{\rho}(u) = R^{(1)}\mathbf{e}(-\lambda^{(1)}) \\ +u(\sin \delta^{(1)} \mathbf{e}_1(-\lambda^{(1)}) - \cos \delta^{(1)} \mathbf{k}^{(1)}) \end{aligned} \tag{4.43}$$

where

$\mathrm{R}^{(1)}$	is the gorge radius of the datum surface $\boldsymbol{\Gamma}^{(1)}$, which is equal to the interception x_0.

According to Eqs. (4.22), (4.42), and (4.43), Eq. (4.39) can be simplified and arranged into:

$$\frac{\mathrm{d}u}{\mathrm{d}\lambda^{(1)}} = \frac{IR^{(1)}}{(\sin \delta^{(1)} + I \cos \delta^{(1)}) \tan \delta^{(1)}} \tag{4.44}$$

By the integral, we can obtain the function $u = u(\lambda^{(1)})$ defining the motion of the conjugate point, i.e.,

$$u = p^{(1)}\left(\lambda^{(1)} + \lambda^{(01)}\right) \qquad p^{(1)} = \frac{IR^{(1)}}{\left(\sin\delta^{(1)} + I\cos\delta^{(1)}\right)\tan\delta^{(1)}} \tag{4.45}$$

where

$\lambda^{(01)}$	is an integration constant, generally zero.

Similarly, we can also obtain the function $u = u(\lambda^{(2)})$, i.e.,

$$u = p^{(2)}\left(\lambda^{(2)} + \lambda^{(02)}\right) \qquad p^{(2)} = \frac{1}{I}p^{(1)} = \frac{R^{(2)}\tan\delta^{(1)}}{\sin\delta^{(1)} + I\cos\delta^{(1)}} \tag{4.46}$$

where

$\lambda^{(02)}$	is an integration constant, generally zero and
$R^{(2)}$	is the gorge radius of the datum surface $\Gamma^{(2)}$, which is equal to $A - x_0$.

Because $p^{(1)}$, $p^{(2)}$ are constant, the functions $u = u(\lambda^{(1)})$, $u = u(\lambda^{(2)})$ are linear and the directrices $C_{\mathrm{d}}^{(1)}$, $C_{\mathrm{d}}^{(2)}$ are a pair of spirals wrapped around the datum surfaces $\Gamma^{(1)}$, $\Gamma^{(2)}$ with the pitches $p^{(1)}$, $p^{(2)}$, respectively.

According to Chen et al. [4], the noninterference condition of hyperboloidal-type normal circular-arc gearing is $\tan\delta^{(1)} = \sqrt[3]{I}$, and, therefore, the following relationships between the kinematic parameters and the positional parameters can be summarized from the above discussions:

$$\begin{aligned} &R^{(1)} = A\sin^2\delta^{(1)},\ R^{(2)} = A\cos^2\delta^{(1)}, \\ &p^{(1)} = R^{(1)}\sin\delta^{(1)},\ p^{(2)} = R^{(2)}\cos\delta^{(1)} \end{aligned} \tag{4.47}$$

Then, the rule of movement of the conjugate point of the directrices is $u = \lambda^{(1)}R^{(1)}\sin\delta^{(1)} = \lambda^{(2)}R^{(2)}\cos\delta^{(1)}$. Substituting it in Eq. (4.43), the directrix $C_{\mathrm{d}}^{(1)}$ of the pinion tooth surface in HNCGing with line contact is obtained by the vector equation:

$$\begin{aligned} \mathbf{C}_{\mathrm{d}}^{(1)} : \mathbf{R}^{(1)} = {} & R^{(1)}\mathbf{e}\left(-\lambda^{(1)}\right) \\ & + \lambda^{(1)}R^{(1)}\sin\delta^{(1)}\left(\sin\delta^{(1)}\mathbf{e}_1\left(-\lambda^{(1)}\right) - \cos\delta^{(1)}\mathbf{k}^{(1)}\right) \end{aligned} \tag{4.48}$$

To reduce the effect of manufacturing errors and the misalignment of mating members, the principle of mismatching of generating surfaces is used to achieve the point of -contact of the gear pair, in which the profile radius of the gear is slightly greater than that of the pinion. In such a case, the center of a normal cross section of the gear tooth surface at any position will depart from the directrix $C_{\mathrm{d}}^{(1)}$ so long as the pinion and gear tooth surfaces are in contact.

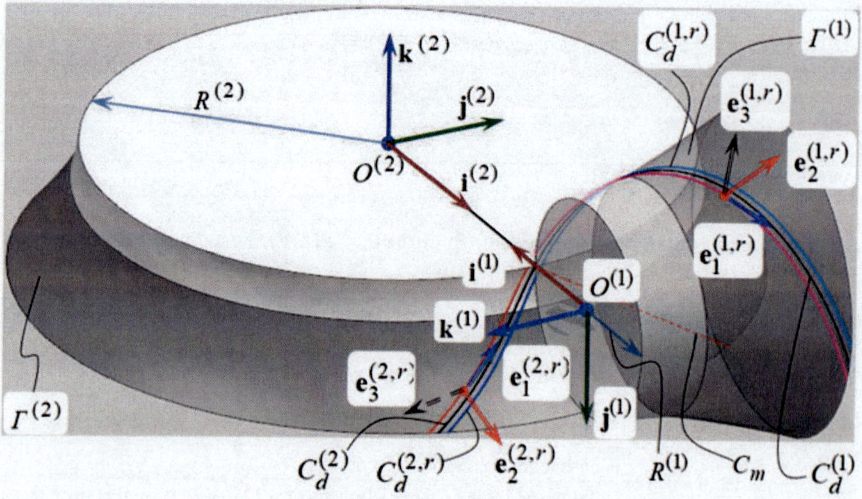

Fig. 4.6 Diagram of real directrices applied to HNCGing with point contact

To comply with the basic principle of molding-surface conjugation, the directrices of the pinion and gear tooth surfaces in HNCGing with point contact must be adjusted in accordance with some rules. As shown in Fig. 4.6, the real directrix $C_{\mathrm{d}}^{(1,\,r)}$ of the pinion is obtained by rotating the theoretical directrix $C_{\mathrm{d}}^{(1)}$ about the axis A_1 with the angle $\Delta\lambda^{(1)}$, i.e.,

$$\begin{aligned}\mathrm{C}_{\mathrm{d}}^{(1,\,r)}:\ \mathbf{R}^{(1,\,r)} = {} & B\left(\pm\Delta\lambda^{(1)}\right)\mathbf{R}^{(1)} = R^{(1)}\mathbf{e}\left(-\lambda^{(1)}\pm\Delta\lambda^{(1)}\right) \\ & +\lambda^{(1)}R^{(1)}\sin\delta^{(1)}\left(\sin\delta^{(1)}\mathbf{e}_1\left(-\lambda^{(1)}\pm\Delta\lambda^{(1)}\right)-\cos\delta^{(1)}\mathbf{k}^{(1)}\right)\end{aligned} \tag{4.49}$$

where

r	denotes the quantities pertaining to the real tooth surface in HNCGing with point contact;
$\Delta\lambda^{(1)}$	is the modification angle, which is equal to $l_p/(R^{(1)}\sin\delta^{(1)})$. Here, l_p is the transverse offset of the tooth profile of the pinion; and
$\pm$	denotes the counter-clockwise (or clockwise) rotation viewed against the direction of $\mathbf{k}^{(1)}$ and is used to describe the quantities pertaining to the right (or left) tooth flank of the tooth in the next derivation.

HNCGing is planned to transmit motion and power between orthogonally crossed axes, and, thus, the inclined angle of the meshing line with respect to the axis A_2 of the gear is $90-\delta^{(1)}$. Similar to the operation of the pinion, the real directrix $C_{\mathrm{d}}^{(2,\,r)}$ of the gear tooth surface can be arranged into:

$$\mathrm{C}_{\mathrm{d}}^{(2,\,r)}:\mathbf{R}^{(2,\,r)}= R^{(2)}\mathbf{e}\left(-\lambda^{(2)}\pm\Delta\lambda^{(2)}\right) \\ +\lambda^{(2)}R^{(2)}\cos\delta^{(1)}\left(\cos\delta^{(1)}\mathbf{e}_1\left(-\lambda^{(2)}\pm\Delta\lambda^{(2)}\right)-\sin\delta^{(1)}\mathbf{k}^{(2)}\right) \tag{4.50}$$

Here, the modification angle $\Delta\lambda^{(2)}$ is determined by the magnitude of the backlash and is less than $I\ \Delta\ \lambda^{(1)}$. The upper (lower) sign of "±" is used to describe the quantities pertaining to the right (left) flank of the gear tooth space (viewed against the direction of $\mathbf{k}^{(2)}$). As pointed out earlier, to comply with the meshing relationships of conjugate directrices, the real directrix $C_{\mathrm{d}}^{(2,\,r)}$ is not the set of the arc centers of all cross sections of the working tooth surface of the gear but is the inner equidistant curve of this set, the offset distance of which is the radius difference of the tooth profiles of the mating pinion and gear.

4.3.2 Mathematical Models of Conjugate Tooth Surfaces

Taking the derivative of Eq. (4.49) with respect to $\lambda^{(1)}$, the tangent vector $\mathbf{t}^{(1,r)}$ of the directrix $C_{\mathrm{d}}^{(1,\,r)}$ can be represented as:

$$\mathbf{t}^{(1,\,r)}=\frac{d\mathbf{R}^{(1,\,r)}}{d\lambda^{(1)}} \\ =R^{(1)}\left[\lambda^{(1)}\sin^2\delta^{(1)}\mathbf{e}\left(\lambda^{(1)}\pm\Delta\lambda^{(1)}\right)+\cos\delta^{(1)}\left(\cos\delta^{(1)}\mathbf{e}_1\left(\lambda^{(1)}\pm\Delta\lambda^{(1)}\right)+\sin\delta^{(1)}\mathbf{k}^{(1)}\right)\right] \tag{4.51}$$

Because the meshing line rotates with the rotation of the directrix $C_{\mathrm{d}}^{(1,\,r)}$, according to Eq. (4.42), the tangent vector $\mathbf{m}^{(r)}$ of the real meshing line $\mathrm{C}_{\mathrm{m}}^{(r)}$ can be denoted as:

$$\mathbf{m}^{(r)}=\frac{d\left(B\left(\lambda^{(1)}\pm\Delta\lambda^{(1)}\right)\boldsymbol{\rho}(u)\right)}{du}=\sin\delta^{(1)}\mathbf{e}_1\left(\lambda^{(1)}\pm\Delta\lambda^{(1)}\right)-\cos\delta^{(1)}\mathbf{k}^{(1)} \tag{4.52}$$

Taking a dot product between Eqs. (4.51) and (4.52) and referring to Chen et al. [4], we can find that:

$$\mathbf{t}^{(1,\,r)}\cdot\mathbf{m}^{(r)}=0 \tag{4.53}$$

This shows that the family of the directrix and the family of the meshing line constitute the orthogonal system of lines of the datum surface. On the basis of this, the moving frame $\left\{P;\mathbf{e}_1^{(1,\,r)},\mathbf{e}_2^{(1,\,r)},\mathbf{e}_3^{(1,\,r)}\right\}$ set along the directrix $C_{\mathrm{d}}^{(1,\,r)}$ can be constituted by:

$$\left\{\begin{aligned} \mathbf{e}_1^{(1,\,r)} &= \frac{\mathbf{t}^{(1,\,r)}}{\left|\mathbf{t}^{(1,\,r)}\right|} \\ &= \frac{\lambda^{(1)}\sin^2\delta^{(1)}\mathbf{e}\left(-\lambda^{(1)}\pm\Delta\lambda^{(1)}\right)-\cos\delta^{(1)}\left(\cos\delta^{(1)}\mathbf{e}_1\left(-\lambda^{(1)}\pm\Delta\lambda^{(1)}\right)+\sin\delta^{(1)}\mathbf{k}^{(1)}\right)}{\sqrt{\cos^2\delta^{(1)}+\left(\lambda^{(1)}\sin^2\delta^{(1)}\right)^2}} \\ \mathbf{e}_2^{(1,\,r)} &= \frac{\mathbf{m}^{(r)}}{\left|\mathbf{m}^{(r)}\right|} = \sin\delta^{(1)}\mathbf{e}_1\left(-\lambda^{(1)}\pm\Delta\lambda^{(1)}\right)-\cos\delta^{(1)}\mathbf{k}^{(1)} \\ \mathbf{e}_3^{(1,\,r)} &= \mathbf{e}_1^{(1,\,r)}\times\mathbf{e}_2^{(1,\,r)} \\ &= \frac{\cos\delta^{(1)}\mathbf{e}\left(-\lambda^{(1)}\pm\Delta\lambda^{(1)}\right)+\lambda^{(1)}\sin^2\delta^{(1)}\left(\cos\delta^{(1)}\mathbf{e}_1\left(-\lambda^{(1)}\pm\Delta\lambda^{(1)}\right)+\sin\delta^{(1)}\mathbf{k}^{(1)}\right)}{\sqrt{\cos^2\delta^{(1)}+\left(\lambda^{(1)}\sin^2\delta^{(1)}\right)^2}} \end{aligned}\right. \tag{4.54}$$

Obviously, $\mathbf{e}_3^{(1,\,r)}$ is the unit normal vector of the datum surface $\Gamma^{(1)}$ along the real meshing line $C_{\mathrm{m}}^{(r)}$ and is directed toward the outside of the entity of the pinion.

According to previous descriptions, a pair of conjugate directrices is tangent at the conjugate point, which means that the two directrices have the common tangent line at that point, and, additionally, a pair of datum surfaces of HNCGing is tangent at the straight meshing line, which shows that the two datum surfaces share the common normal line along the meshing line. These two characteristics provide a chance to set a shared moving frame for describing the kinematics and geometry of conjugation, and, thus $\left\{P;\mathbf{e}_1^{(1,\,r)},\,\mathbf{e}_2^{(1,\,r)},\,\mathbf{e}_3^{(1,\,r)}\right\}$ is used as the base of the shared moving frame $\{P;\boldsymbol{\alpha}_1,\boldsymbol{\alpha}_2,\boldsymbol{\alpha}_3\}$ in the discussion.

Similar to the above operation, the moving frame $\left\{P;\mathbf{e}_1^{(2,\,r)},\,\mathbf{e}_2^{(2,\,r)},\,\mathbf{e}_3^{(2,\,r)}\right\}$ set at the directrix $C_{\mathrm{d}}^{(2,\,r)}$ can be constituted by:

$$\left\{\begin{aligned} \mathrm{e}_1^{(2,\,r)} &= -\frac{\lambda^{(2)}\cos^2\delta^{(1)}\mathrm{e}\left(-\lambda^{(2)}\pm\Delta\lambda^{(2)}\right)-\sin\delta^{(1)}\left(\sin\delta\mathrm{e}_1\left(-\lambda^{(2)}\pm\Delta\lambda^{(2)}\right)+\cos\delta^{(1)}\mathrm{k}^{(2)}\right)}{\sqrt{\sin^2\delta^{(1)}+\left(\lambda^{(2)}\cos^2\delta^{(1)}\right)^2}} \\ \mathrm{e}_2^{(2,\,r)} &= \cos\delta^{(1)}\mathrm{e}_1\left(-\lambda^{(2)}\pm\Delta\lambda^{(2)}\right)-\sin\delta^{(1)}\mathrm{k}^{(2)} \\ \mathrm{e}_3^{(2,\,r)} &= -\frac{\sin\delta^{(1)}\mathrm{e}\left(-\lambda^{(2)}\pm\Delta\lambda^{(2)}\right)+\lambda^{(2)}\cos^2\delta^{(1)}\left(\sin\delta\mathrm{e}_1\left(-\lambda^{(2)}\pm\Delta\lambda^{(2)}\right)+\cos\delta^{(1)}\mathrm{k}^{(2)}\right)}{\sqrt{\sin^2\delta^{(1)}+\left(\lambda^{(2)}\cos^2\delta^{(1)}\right)^2}} \end{aligned}\right. \tag{4.55}$$

It should be noted that $\mathbf{e}_3^{(2,\,r)}$ is the unit normal vector of the datum surface $\Gamma^{(2)}$ along the real meshing line and is directed toward the inside of the entity of the gear, but $\mathbf{e}_2^{(2,\,r)}$ keeps the direction with $\mathbf{e}_2^{(1,\,r)}$ at the conjugate position. This operation ensures the uniformity of $\left\{P;\mathbf{e}_1^{(1,\,r)},\,\mathbf{e}_2^{(1,\,r)},\,\mathbf{e}_3^{(1,\,r)}\right\}$ and $\left\{P;\mathbf{e}_1^{(2,\,r)},\,\mathbf{e}_2^{(2,\,r)},\,\mathbf{e}_3^{(2,\,r)}\right\}$, and, thus we can choose the former as the base of the shared moving frame

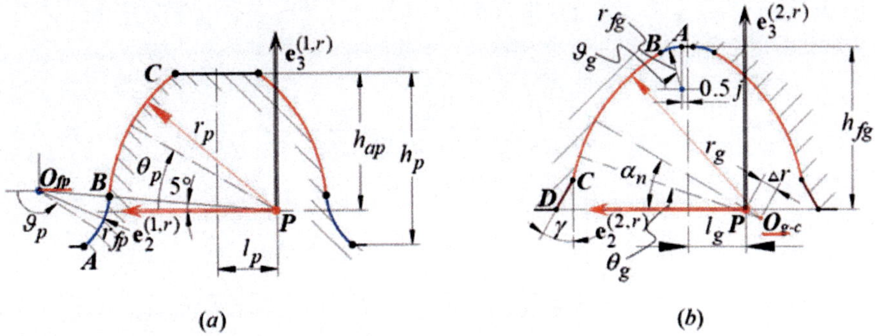

Fig. 4.7 Schematic illustration of basic tooth profiles

$\{P; \boldsymbol{\alpha}_1, \boldsymbol{\alpha}_2, \boldsymbol{\alpha}_3\}$, and the profile angles defining the tooth profiles of the pinion and the gear are consistent in the shared frame. As shown in Fig. 4.7, when the pinion tooth engages with the gear tooth, the working tooth profiles $\overset{\frown}{BC}$ of the pinion and the gear contact at a predetermined point (location) by the pressure angle, and, then, the two moving frames coincide at the conjugate point P, which is the contacting point of the real directrices $C_{\mathrm{d}}^{(1,\,r)}$, $C_{\mathrm{d}}^{(2,\,r)}$ at that position.

It must be pointed out that Fig. 4.7 is achieved by viewing the tooth of the pinion and the tooth space of the gear against the direction of $\mathbf{k}^{(1)}$ and $\mathbf{k}^{(2)}$ and that the left sides of the tooth of pinion and the tooth space of the gear are considered to be the driving side and the driven side, respectively, in the process of conjugation. Now, the sign "-" is chosen to describe the quantities pertaining to the sign "±." According to Eqs. (4.1), (4.49), and (4.50), the contacting tooth surfaces $\psi^{(1)}$, $\psi^{(2)}$ of the pinion and the gear can be described as:

$$\psi^{(1)} : \mathbf{S}^{(1)} = \mathbf{R}^{(1,\,r)} + r_{\mathrm{p}}\left(\sin\theta_{\mathrm{p}}\mathbf{e}_3^{(1,\,r)} + \cos\theta_{\mathrm{p}}\mathbf{e}_2^{(1,\,r)}\right) \tag{4.56}$$

$$\begin{aligned}\psi^{(2)} : \ \mathbf{S}^{(2)} = \ & \mathbf{R}^{(2,\,r)} - \Delta r\left(\sin\alpha_{n}\mathbf{e}_3^{(2,\,r)} + \cos\alpha_{n}\mathbf{e}_2^{(2,\,r)}\right) \\ & + r_{\mathrm{g}}\left(\sin\theta_{\mathrm{g}}\mathbf{e}_3^{(2,\,r)} + \cos\theta_{\mathrm{g}}\mathbf{e}_2^{(2,\,r)}\right)\end{aligned} \tag{4.57}$$

where

$r_{\mathrm{p}}, r_{\mathrm{g}}$	are the radii of the working tooth profiles of the pinion and the gear, respectively.
$\theta_{\mathrm{p}}, \theta_{\mathrm{g}}$	are the profile angles of the working tooth profiles of the pinion and the gear, respectively.
Δr	is the radius difference between the working tooth profiles and $\Delta r = r_{\mathrm{g}} - r_{\mathrm{p}}$.
α_n	is the pressure angle.

According to the conclusion drawn from Eq. (4.53), the directrices of the adjacent teeth of the gear (or the pinion) must be perpendicular to the same meshing line, and the length of the intervals of the adjacent teeth are equal; thus, the normal pitch can be defined as $P_n = \pi m_n$ (m_n is the normal module of the gear pair). On the basis of this, the following parameter relationships can be obtained as follows:

$$R^{(1)} = \frac{z_1 m_n}{2\sin\delta^{(1)}},\ R_2 = \frac{z_2 m_n}{2\cos\delta^{(1)}},\ A = \frac{m_n}{2}\left(\frac{z_1}{\sin\delta^{(1)}} + \frac{z_2}{\cos\delta^{(1)}}\right),\ \tan\delta^{(1)} = \sqrt[3]{\frac{z_1}{z_2}} \tag{4.58}$$

where

z_1 and z_2	are the number of teeth of the pinion and the gear, respectively.

Obviously, the normal module m_n is an important parameter that standardizes the design and machining of HNCGing. When m_n is given, the basic parameters of HNCGing can be acquired by Eq. (4.58), and, then, the rest of the structural parameters can be determined by means of the current standards of Wildhaber–Novikov gearing.

Figure 4.8 is a three-dimensional (3D) representation of a pair of right-handed pinion and gear in HNCGing with point contact. When the pinion makes a positive rotation about the $\mathbf{k}^{(1)}$ axis as a driving member, the gear is driven to rotate positively about the $\mathbf{k}^{(2)}$ axis through the engagement of the tooth surfaces $\psi^{(1)}$, $\psi^{(2)}$.

4.3.3 *Induced Curvatures of Mating Tooth Surfaces*

According to Sect. 4.2.1, the tangent direction of the generatrix and its orthogonal direction (i.e., the tangent direction of the directrix) are the principal directions of the tooth surfaces of HNCGing and the normal curvatures at the two directions are the principal curvatures. On the basis of this property, two principal directions and the surface normal direction (pointing to the center of the circular-arc generatrix) can be used to define a moving frame, $\left\{\mathbf{S}^{(1)}; \mathbf{e}_{s1}^{(1)}, \mathbf{e}_{s2}^{(1)}, \mathbf{e}_{s3}^{(1)}\right\}$, of the pinion tooth surface $\psi^{(1)}$. According to differential geometry, the principal curvatures $k_{s1}^{(1)}$, $k_{s2}^{(1)}$ of the tooth surface $\psi^{(1)}$ can be denoted as:

$$\begin{cases} k_{s1}^{(1)} = -\dfrac{k_g^{(1)}\cos\theta_p + k_n^{(1)}\sin\theta_p}{1 - r_p\left(k_g^{(1)}\cos\theta_p + k_n^{(1)}\sin\theta_p\right)} \\ k_{s2}^{(1)} = \dfrac{1}{r_p} \end{cases} \tag{4.59}$$

where

$k_{s2}^{(1)}, k_{s1}^{(1)}$	

(continued)

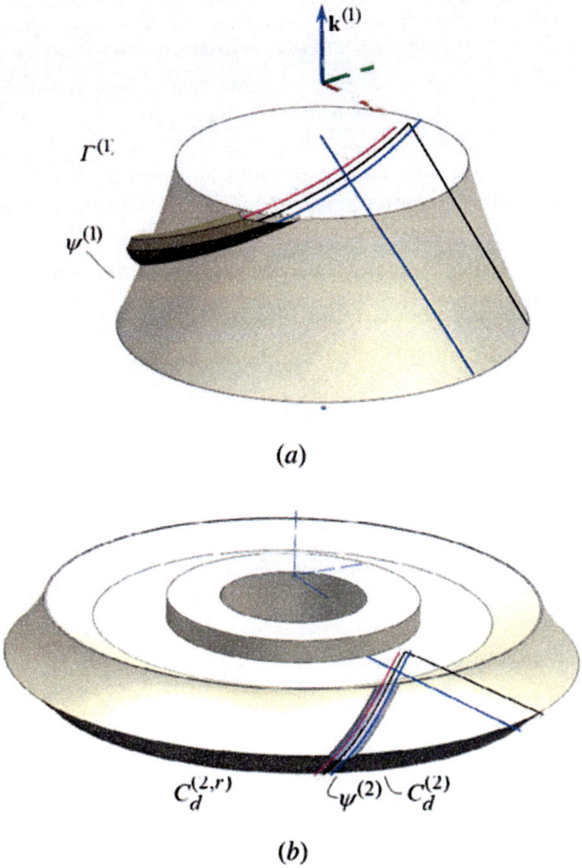

Fig. 4.8 Schematic illustration of one tooth of the pinion and one tooth space of the gear in HNCGing

	are the principal curvatures of $\psi^{(1)}$ in the tangent direction of the generatrix and its orthogonal direction, respectively.
$k_g^{(1)}, k_n^{(1)}$	are the geodesic curvature and the normal curvature of the directrix $C_d^{(1,r)}$ in the shared moving frame $\{P; \boldsymbol{\alpha}_1, \boldsymbol{\alpha}_2, \boldsymbol{\alpha}_3\}$, respectively.

Similarly, according to Eq. (4.59), the principal curvatures $k_{s1}^{(2)}$, $k_{s2}^{(2)}$ of the tooth surface $\psi^{(2)}$ can be denoted as:

$$\begin{cases} k_{s1}^{(2)} = -\dfrac{k_g^{(2)}\cos\theta_g + k_n^{(2)}\sin\theta_g}{1 - r_g\left(k_g^{(2)}\cos\theta_g + k_n^{(2)}\sin\theta_g\right) + \Delta r\left(k_g^{(2)}\cos\alpha_n + k_n^{(2)}\sin\alpha_n\right)} \\ k_{s2}^{(2)} = \dfrac{1}{r_g} \end{cases} \tag{4.60}$$

where

$k_{s2}^{(2)}, k_{s1}^{(2)}$	are the principal curvatures of $\psi^{(2)}$ in the tangent direction of the generatrix and its orthogonal direction, respectively.
$k_{g}^{(2)}, k_{n}^{(2)}$	are the geodesic curvature and the normal curvature of the directrix $C_d^{(2,\,r)}$ in the shared moving frame $\{P; \boldsymbol{\alpha}_1, \boldsymbol{\alpha}_2, \boldsymbol{\alpha}_3\}$, respectively.

For a pair of conjugate tooth surfaces $\psi^{(1)}$,$\psi^{(2)}$, two circular-arc tooth profiles with constant radii r_p,r_g are in tangential contact at the direction of the pressure angle in the common normal plane of the tooth surfaces, and, thus, the induced principal curvature $k_{s2}^{(21)}$ of mating tooth surfaces in the tangent direction of the generatrix is constant, i.e.,

$$k_{s2}^{(21)} = k_{s2}^{(1)} - k_{s2}^{(2)} = \frac{1}{r_p} - \frac{1}{r_g} = \frac{r_g - r_p}{r_p r_g} = \frac{\Delta r}{r_p r_g} \tag{4.61}$$

In the shared moving frame $\{P; \boldsymbol{\alpha}_1, \boldsymbol{\alpha}_2, \boldsymbol{\alpha}_3\}$, the profile angles are $\theta_p = \theta_g = \alpha_n$ when they are used to define the contact position of the tooth surfaces. According to Eqs. (4.59) and (4.60), the induced principal curvature $k_{s1}^{(21)}$ of the conjugate tooth surfaces $\psi^{(1)}$, $\psi^{(2)}$ at that contact position in the common tangent direction of the directrices $C_d^{(1,\,r)}$, $C_d^{(2,\,r)}$ can be denoted as:

$$\begin{cases} k_{s1}^{(21)} = k_{s1}^{(1)} - k_{s1}^{(2)} \\ = -\dfrac{k_g^{(1)}\cos\alpha_n + k_n^{(1)}\sin\alpha_n}{1 - r_p\left(k_g^{(1)}\cos\alpha_n + k_n^{(1)}\sin\alpha_n\right)} + \dfrac{k_g^{(2)}\cos\alpha_n + k_n^{(2)}\sin\alpha_n}{1 - (r_g - \Delta r)\left(k_g^{(2)}\cos\alpha_n + k_n^{(2)}\sin\alpha_n\right)} \\ = -\dfrac{k_g^{(21)}\cos\alpha_n + k_n^{(21)}\sin\alpha_n}{\left[1 - r_p\left(k_g^{(1)}\cos\alpha_n + k_n^{(1)}\sin\alpha_n\right)\right]\left[1 - r_p\left(k_g^{(2)}\cos\alpha_n + k_n^{(2)}\sin\alpha_n\right)\right]} \\ k_g^{(21)} = k_g^{(1)} - k_g^{(2)} \\ k_n^{(21)} = k_n^{(1)} - k_n^{(2)} \end{cases} \tag{4.62}$$

where

$k_g^{(21)}, k_n^{(21)}$	are the induced geodesic curvature and the induced normal curvature of $C_d^{(1,\,r)}$, $C_d^{(2,\,r)}$, respectively.

4.4 An Integrated Manufacturing Software System for HNCGing

Design, engineering, and manufacturing are undergoing a digital transformation, and the need for a collaborative product development environment is becoming an ever-growing requirement. The integrated manufacturing software system of HNCGing meets this need by connecting CAD, CAM, and CAE in a single application platform and can be used as a primary expert system for the design and machining of HNCGing [5].

4.4.1 *Functional Framework*

The idea of an integrated manufacturing software system of HNCGing builds upon digital manufacturing trends, which connect CAD, CAM, and CAE together through 3D modeling, simulation of meshing and contact, and adaptive tool path programming, as shown in Fig. 4.9. In CAD systems, according to the mathematical models of HNCGing and parameter relationships, a 3D model generation program that outputs 3D models of pinion and gear is developed by means of computer graphics and commonly used programming languages. The 3D models of pinion and gear will be used as the criterion model for error analysis, tooth contact analysis, and machining simulation. In CAM systems, selection of the blank, design of the cutting tool, and determination of the cutting regime are brought into the tool path planning, and a file of NC code can be automatically generated with the change of the specification parameters and cutting parameters. This system provides the machining simulation and the human–computer interaction of the machining parameters. In CAE systems, the measurement data of real tooth surfaces, achieved by CAM or 3D scanning, can be imported and fitted and the surface deviation can be calculated by comparing between the fitting model of the real tooth surface and the criterion model. On the basis of this, tooth contact analysis can provide the effect of errors on the path of contact and the area of contact. This system is extendible and reserves the interface to enhance functions for the dynamics and redesign of HNCGing.

4.4.2 *Three-Dimensional Modeling*

The basic method of computer representation of the pinion and gear of HNCGing is presented in the study by Zhang et al. [6].

According to the structural characteristics of teeth arrangement, the tooth element, which is the key substructure extracted from the gear (the pinion), is used as the foundation for constructing a model of the gear (the pinion). The main steps comprise:

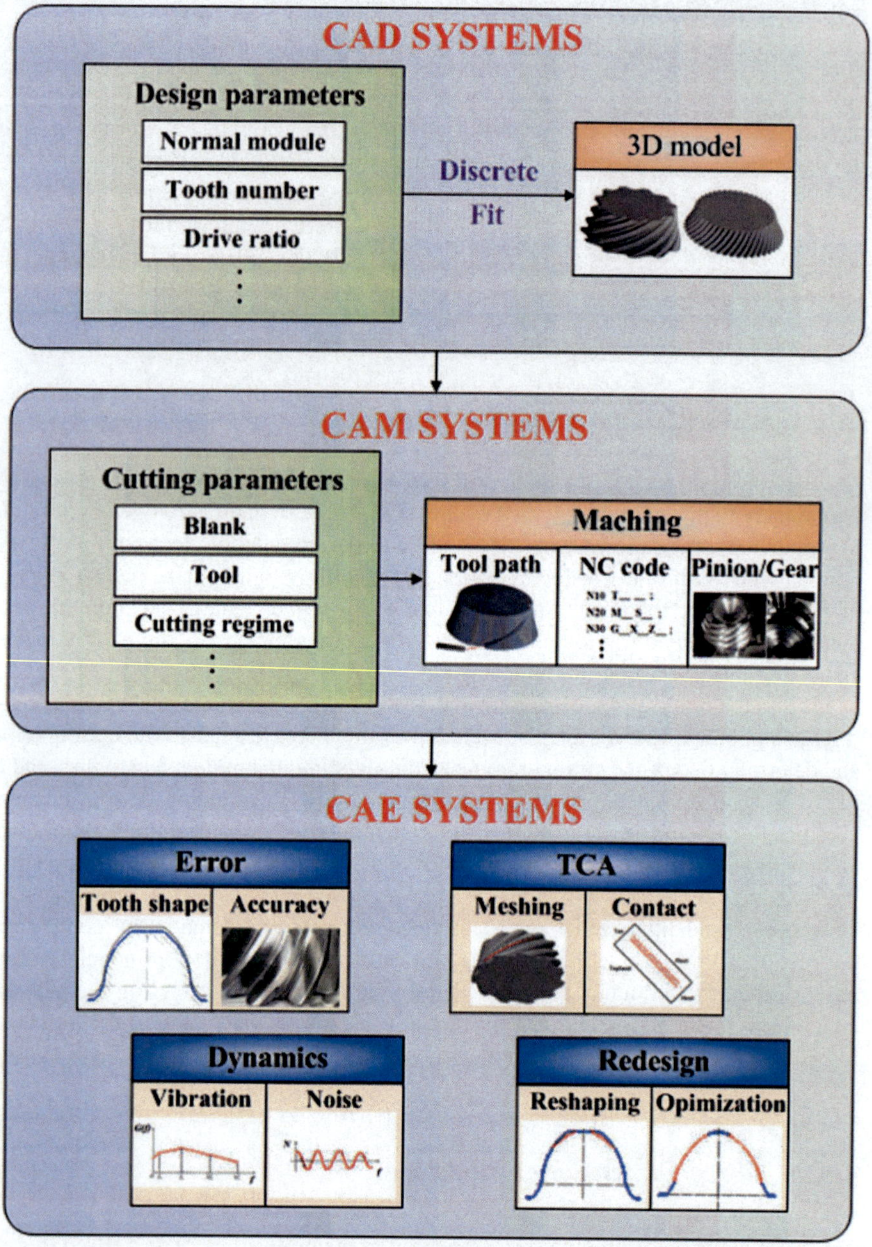

Fig. 4.9 Functional framework of an integrated manufacturing software system of HNCGing

(1) Determining the boundary condition of the tooth element; (2) dividing the tooth element into five parts, i.e., the left fillet arc at the root of the tooth, the left working tooth profile, the addendum circle, the right working tooth profile, and the

Fig. 4.10 Three-dimensional models of HNCGing

right fillet arc, and calculating the connecting condition between the five parts; (3) solving discrete points on the surfaces of the five parts, and fitting these points with triangular or quadrilateral patches to generate the tooth element; and (4) establishing a three-dimensional solid model of HNCGing by arraying the tooth element according to the number of teeth and merging it with the dedendum circle. The three-dimensional models of a pair of pinion and gear with point contact, generated by CAD systems embedded in the integrated manufacturing software system of HNCGing, are shown in Fig. 4.10.

4.4.3 Adaptive Tool Path Programming

The basic idea of adaptive tool path programming for NC machining of HNCGing is presented in the study by Yan et al. [7].

To quickly respond to the changes of design and machining parameters, an adaptive tool path programming is applied to automatically compute the tool path, conduct the post-processing, and generate the G-codes for the 5-axis form milling of HNCGing with different specifications, as shown in Fig. 4.11. The CAM systems embedded in the integrated manufacturing software system of HNCGing support the selection of the blank, the design of the cutting tool, and the tool path planning. On one hand, the tooth width and the maximum radius of the addendum are calculated from the input design parameters and the specifications of the blank can be selected, and, then, the machining parameters, set by the rough and finish machining process of the addendum surface, can be input into the CAM system to calculate the tool path and generate the G-codes of NC turning. On the other hand, according to the input design parameters, the size of the tooth space can be calculated, the forming cutter profiles can be designed, and the machining parameters can be input to meet the production requirements; the tool paths of the rough and finish machining of the tooth surface can be calculated, and, based on the coordinate transformation of the tool paths, they are rearranged in accordance with the NC file template to generate the G-codes. In adaptive tool path programming, the cutting cycle parameters, determined by the feed rate and the finish allowance, are considered to be the cycle indices of program execution.

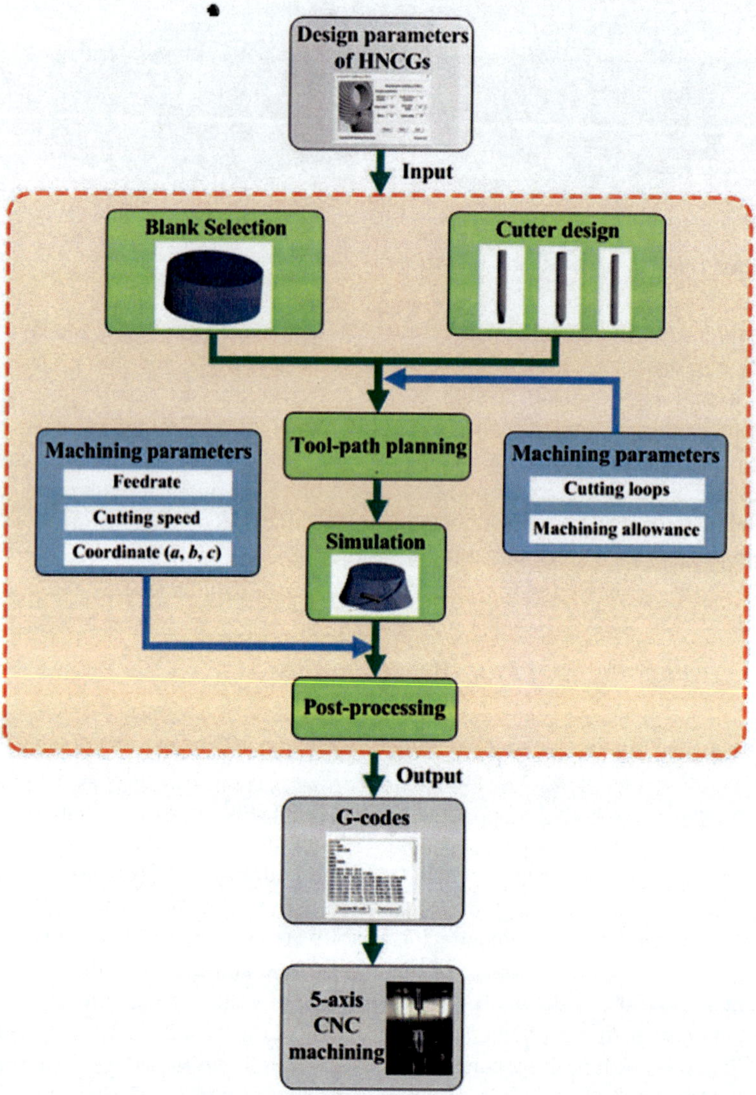

Fig. 4.11 Flowchart of adaptive tool path programming in the integrated manufacturing system of HNCGing

In the program implementation of machining simulation, the blank of the workpiece is regarded as a positive geometric solid and the swept volume of the cutter is regarded as a negative geometric solid; the point set of the blank on the path the cutter passes is removed, and the boundary surface of the blank is replaced by the envelope of the cutter, as shown in Fig. 4.12. Let $\{P_{\text{blank}}\}$ be the point set of the blank, $\{P_{\text{CSB}}\}$ be the point set of the swept volume of the cutter, and $\{P_{\text{CES}}\}$ be the

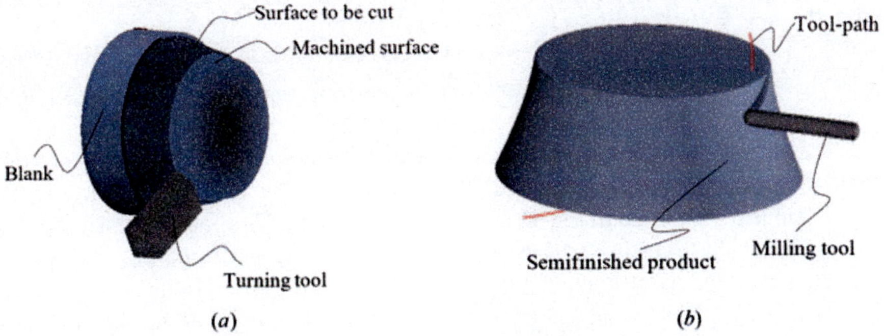

Fig. 4.12 Machining simulation in the integrated manufacturing software system of HNCGing

Fig. 4.13 Photograph of the prototype machine of hyperboloidal-type normal circular-arc gearing

point set of the envelope of the cutter, then the point set $\{P_{\mathbf{RE}}\}$ of the reconstruction model satisfies the following equation:

$$\{P_{\mathrm{Re}}\} \in (\{P_{\mathrm{blank}}\} - \{P_{\mathrm{blank}}\} \cap \{P_{\mathrm{CSB}}\}) \cup \{P_{\mathrm{CES}}\} \quad (4.63)$$

Referring to the Chinese standards on circular-arc gears, a prototype machine of hyperboloidal-type normal circular-arc gearing was designed with the main design parameters as follows: the normal module $m_n = 3$ mm, the number of teeth of the pinion $z_1 = 9$, the number of teeth of the gear $z_2 = 23$, the radius of the working tooth profile of the pinion $r_{\mathbf{p}} = 4.5$ mm, and the radius of the working tooth profile of the gear $r_{\mathbf{g}} = 4.95$ mm. The generation of the pinion and gear tooth surfaces were completed on a five-axis computer numerical control (CNC) machine, and the prototype machine of HNCGing was achieved by assembling all the components, as shown in Fig. 4.13.

4.4.4 *Simulation of Meshing and Contact*

The basic idea of computer simulation of meshing and contact of HNCGing is presented in the study by Chen et al. [8].

The error is the difference between the actual position and the theoretical position that the gear pair would occupy if they were perfectly conjugated, such as the misalignment error, shape error, and kinematic error. The variation denotes the deviation of the transmission process and the meshing performance, such as the position variation of the contact point, the variation of the speed ratio, etc. If the error is considered to be the input disturbance, then the variation can be regarded as a response to this disturbance. Using the moving frame of conjugated tooth surfaces, the errors and variations were quantified, and, then, the response relationship between the variations and errors was established by means of the conditional equation of conjugation. According to the Hertz contact hypothesis, the instantaneous contact area of conjugate tooth surfaces is calculated on the basis of induced curvatures of the conjugate tooth surfaces presented in Sect. 4.3.3 and is represented on the common tangent plane of the conjugate surfaces. Applying the method of center projection, the instantaneous contact area is projected onto the tooth surface, and, thus, a relationship is established between the two-dimensional (2D) contact area and the three-dimensional (3D) contact area. In CAE systems embedded in the integrated manufacturing software system of HNCGing, tooth contact analysis is an important component to directly reflect the contact performance of the gear pair. Because the errors cannot be avoided absolutely, the path of contact may deflect to the tooth root or the tooth top land, and, then, the loading capacity of the teeth will be impacted. Here, the simulation of meshing and contact is achieved by application of the TCA computer program developed by the authors and is used to demonstrate the shift of the bearing contact caused by the misalignment. Figure 4.14 presents the result of the influence of the angle error of the $\mathbf{k}^{(2)}$ axis (i.e., the axis A_2) on the path of contact.

4.5 Conclusion

Under the guidance of the basic principle of molding-surface conjugation, hyperboloidal-type normal circular-arc gearing (HNCGing) is proposed to be the latest improved version of Novikov gearing to transform motions between orthogonally crossed axes. The tooth geometry of the pinion and mating gear in HNCGing is generated, and the curvature characteristics of conjugate tooth surfaces are explored. This helps in calculating the Hertzian contact between the two mating teeth and implementing tool path planning for NC machining. An integrated manufacturing software system of HNCGing is developed, which connects 3D computer-aided design, 3D computer-aided manufacturing, and tooth contact analysis in a single application platform.

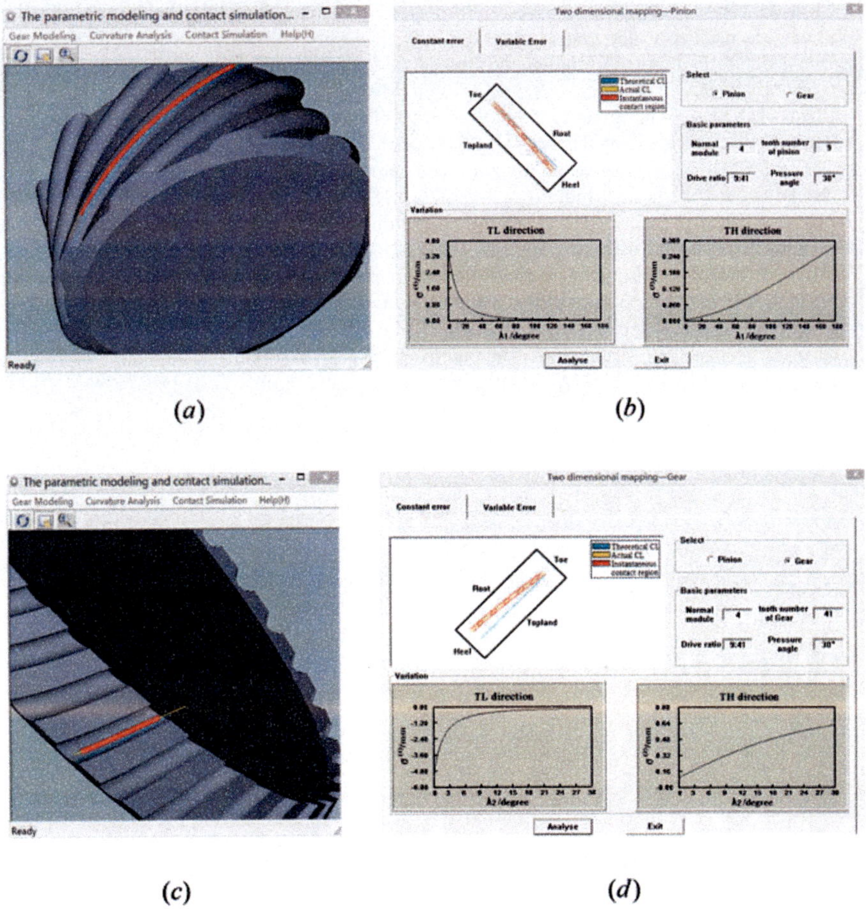

Fig. 4.14 Influence of the angle error of the $\mathbf{k}^{(2)}$ axis on the path of contact: (**a**) 3D contact of the pinion, (**b**) 2D contact of the pinion, 3D contact of the gear, and 2D contact of the gear

Acknowledgments This work was supported by the National Natural Science Foundation of China under Grant No.51105210.

All the authors express their most sincere appreciation to Prof. Jian Liu of Dalian University of Technology of China.

All the authors thank Xiong Cai, Guangwen, Yan and Lingling Zhang of Nantong University of China for their contributions to the integrated manufacturing software system of HNCGing.

References

1. Litvin, F.L., Fuentes, A., Gonzalez-Perez, I., et al.: New version of Novikov–Wildhaber helical gears: computerized design, simulation of meshing and stress analysis. Comput. Methods Appl. Mech. Eng. **191**(49–50), 5707–5740 (2002)

2. Chen, H., Duan, Liu, J., et al.: Research on basic principle of molding-surface conjugation. Mech. Mach. Theory. **43**, 791–811 (2008)
3. Duan, Chen, H., Ju, et al.: Mathematical model and manufacture programming of loxodromic-type normal circular-arc spiral bevel gear. Front. Mech. Eng. **7**, 312–321 (2012)
4. Chen, H., Duan, Wu, H., et al.: Study on the general principle of normal circular-arc gear transmission. Mech. Mach. Theory. **41**(12), 1424–1442 (2006)
5. Yan, G., Chen, H., Zhang, X., et al.: An integrated CAD/CAM system for hyperboloidal-type normal circular-arc gear. J. Ind. Prod. Eng. **37**(4), 186–193 (2020)
6. Zhang, X., Cai, X., Chen, H., et al.: A parametric modeling method for hyperboloidal-type normal circular-arc gear [J]. J. Appl. Sci. **14**(3), 229–236 (2014)
7. Yan, G., Chen, H., Zhang, X., et al.: A dimension-driven adaptive programming for tool-path planning and post-processing in 5-axis form milling of hyperboloidal-type normal circular-arc gears. Int. J. Adv. Manuf. Technol. **106**, 2735–2746 (2020)
8. Chen, H., Zhang, X., Cai, X., et al.: Computerized design, generation and simulation of meshing and contact of hyperboloidal-type normal circular-arc gears. Mech. Mach. Theory. **96**, 127–145 (2016)

Chapter 5
Modern Methods of Estimating and Increasing the Load-Bearing Capacity of Novikov Gearing

Analysis of the performance comparison between involute gears and Novikov gears [1] convinces of the effectiveness of the widespread use of the latter in both general and special purpose drive, in the design of transmission for oil pump in particular.

Thus, it is known that Novikov gears with standard basic profiles are widely used in gearboxes of general and special purposes. At the same time, Novikov gears with chemically and thermally hardened, cemented, and nitrocarburized tooth surfaces are developed based on the standard basic profile BP1 [2] and those with thermally improved (sometimes with additional nitriding) teeth are made on the basis of the standard basic profile BP2 [3].

Currently, the industry is facing a situation in which there are no uniform (standard) methods and computational programs for the geometric and strength calculations of Novikov gearings, which forces consumers to use outdated imperfect methods, often contradicting each other. This makes designing gearings and drives with rational parameters that ensure both high quality and competitiveness of products difficult.

Briefly, the shortcomings of the previous methods (including the current GOST 17744-72 for the calculation of geometry and the methodological recommendation MR 221-86 for the calculation of strength) are as follows:

- There is no unified methodological approach to the calculation of Novikov gears with any hardness of the tooth surfaces.
- Gears with nonzero displacement of the basic profile, which are widespread in gearbox construction, are not covered.
- The calculation of the contact endurance of the tooth surfaces is based on the definition of not effective but normal contact stresses, that is, the shape of the contact pad is not taken into account, which leads to contradictory results.

Financial support was provided to Southern Federal University by the Russian Ministry of Science and Higher Education (project VnGr/2020-04-IM).

S. P. Radzevich, *Novikov/Conformal Gearing*,
https://doi.org/10.1007/978-3-031-10019-2_5

- Changes in the geometry of the teeth in the process of a run-in are not taken into account.
- There is no calculation of the deep contact strength of surface-hardened teeth.
- There is no accounting of the "edge effects" when the contact spot comes out on the edges of the teeth and their connection with the adaptive ability of the real gearing is made with given errors.
- Gears with a new modern basic profile [2], providing a high load-bearing capacity of the drive, as well as modified gears are not covered.
- There are no calculations of the geometric limiting factors and quality of engagement of gears, which determine the rational scope of applicability of the latter.
- Many of the most important design parameters (coefficients of stress concentration, unevenness of load distribution, gearing dynamics, etc.), which must take into account the real conditions of the manufacture and operation of gears, are adopted from the methods of calculations for involute gears, which completely ignore the specifics of Novikov engagement, and, as a result, the reliability of calculations is significantly reduced.
- Permissible stresses and safety factors are not consistent with the structure of the formulas for calculating the active stresses and with the results of tests of Novikov gears, which often leads to unreliable conclusions.

Other shortcomings could be noted, but the above, in our opinion, is enough to judge the unacceptability of the widespread use of the indicated methods.

In this chapter, on the basis of a new approach to eliminate the above shortcomings, a generalization of the results of determining the main parameters necessary for calculating the strength, load-bearing capacity, and service life of Novikov gears with the basic profile BP1 is provided and the method developed by us to radically increase the load-bearing capacity of Novikov gears with basic profile BP2 is described.

It should be noted that the basic profiles of BP1 and BP2 are fundamentally different from each other: the first near-pole zone is turned off from work, which prevents increased contact stresses in the gear, but, at the same time, there is an additional concentrator of bending stresses. The second has a weak contact relation in the near-polar involute-forming zone, but there is no additional stress concentrator.

It should be emphasized that the teeth of Novikov gears are in a complex volumetric stress state, which, moreover, when hooked, constantly changes from one phase of engagement of teeth to another. The task is to find the "dangerous" aspects of the strength phases with maximum contact and bending stresses. These stresses, which we will later call permissible (critical), ultimately determine the load-bearing capacity and durability of the gear transmission and, often, the service life of a gearbox as a whole.

In order to find the permissible stresses, it is necessary:

1. To three dimensionally simulate the solution of the stress-strain state of the contact problem with previously unknown shapes and sizes of the contact spots, provided that the theoretical point (spot) of contact is located anywhere

along the width of the gear rim (along the length of the tooth) up to the end of the tooth at the time of axial switching of the teeth
2. To simulate the process of a real engagement, in which the rigidity is taken into account for the contacting teeth and accompanying parts of the drive (shafts, bearings, housing) as well as the inevitable technological errors (in the probabilistic aspect) of the manufacture and assembly of gears, depending on the specified accuracy

When solving the first problem, the so-called basic parameters such as contact, bending stresses, and stiffness of the teeth when the theoretical point (spot) of contact is located in the middle of the gear rim where the influence of the end of the gear rim is absent, the phase values of these parameters and their relationship with the base ones, and the coefficients of influence of the end of the gear rim, when the contact spot shifts from the middle of the gear rim to its end, are determined. The stresses must be brought to a form that allows them to be compared with the permissible stresses established by the standards [4] for involute transmissions, that is, for theoretically linear contact. When solving the second problem, partial loads (fractions of the total input load) are in various phases of engagement of gears calculated, and, thus, permissible stresses and "dangerous" phases of engagement are defined.

Let us first look at the basic parameters.

5.1 Bending Stresses

Here, it was necessary to determine the volume coefficient Y_V of the tooth shape under the action of the concentrated force on it as well as the parameter Y_a that takes into account the influence of the longitudinal length of the contact spot on bending stresses.

The approximation dependence obtained during statistical processing of modeling results in the ANSYS system and has the Y_V form:

$$Y_V = Az_v^a \pm B|x|^b z_v^c, \tag{5.1}$$

where the coefficients are accepted according to Table 5.1 depending on the equivalent number z_v of the teeth of the gear and the relative (to the module m) value x^* of the displacement of the basic profile when cutting the teeth.

To determine the parameter Y_a, the simulation was based on the contact of the studied equivalent straight tooth with an elastic indenter representing the equivalent straight tooth of a paired wheel, the surface of which consists of theoretical points of contact of the hyperbolic type, in which the radius of the concave part of the surface is equal to the profile radius ρ_f of the tooth leg and the radius of the convex part of the surface is equal to the main longitudinal radius ρ_β of curvature of the contacting surfaces of the tooth with the indenter.

Table 5.1 Values of coefficients in the formula (Eq. 5.1)

Range x^*	Z_v	A	B	a	b	c
$-0.6 \leq x^* \leq 0$	$Z_v \leq 18$	1.17	17.5	−0.312	1.268	−1.262
	$Z_v \geq 18$	0.436	37.7	0.03	1.268	−1.522
$0 \leq x^* \leq 0.6$	$Z_v \leq 18$	1.17	−29	−0.312	0.766	−2.07
	$Z_v \geq 18$	0.436	0.024	0.03	1.387	0

In the process of modeling, the number z of the wheel teeth, the main radii of curvature – profile ρ_α and longitudinal ρ_β [5], the tangential force F_t, and the value a_H of the length of the semi-major axis of the conditional ellipse of contact, determined from Viktor et al. [5] were adopted. The parameter Y_a was calculated as the ratio of the bending stress of the distributed load to the stress of the concentrated force.

The approximation formula obtained from the results of processing a wide range of variations in the geometric and force parameters of the simulation data is as follows (the asterisk means the ratio of magnitude to modulo):

$$Y_a = 1 - \frac{0.0544\left(a_H^*\right)^2 \cdot z_V^{0.12}}{1 + 0.14a_H^* + 0.07\left(a_H^*\right)^2}, \tag{5.2}$$

Taking into account the so-called stress factor λ [6], it is possible to write a formula for determining the basic bending stress σ_F:

$$\sigma_F = \frac{Y_V Y_a F_t}{m^2} \tag{5.3}$$

5.2 Effective Contact Voltages

When calculating gears with theoretically point (practically local) contact, which include Novikov gearing, the contact strength is estimated not by normal σ_H but by effective (equivalent) stresses σ_{He}, determined in accordance with the fourth theory of strength. Of course, the permissible normal contact voltages [4] must be converted into effective ones.

As it is known [7], effective contact stresses are calculated by the formula (with the Poisson coefficient $\mu = 0.3$):

$$\sigma_H = \frac{0.6F_n\sqrt{1-\beta+\beta^2}}{\pi a_H b_H(1+\beta)}, \tag{5.4}$$

where F_n is the normal force, $\beta = b_H/a_H$, a_H is the ellipticity coefficient, and b_H is the value of the small semi-axle of the conditional contact ellipse.

It should be borne in mind that in the process of operation, the teeth of the Novikov gear wheels undergo run-ins, in which the contact spot grows in width to a certain value l depending on the hardness of the contacting surfaces, while the main profile's radius ρ_α of curvature increases. Taking into account this circumstance, the basic effective contact stress can be recorded in the form [8]:

$$\sigma_{He} = 19.93 l^{-1.074} F_n^{0.69} \rho_\beta^{-0.31}. \tag{5.5}$$

The results of modeling the contact of two surfaces are well-consistent with the theoretical dependencies, (Eqs. 5.4 and 5.5). However, when simulating the contact of a real tooth with an indenter, the values σ_{He} differ between (Eqs. 5.4 and 5.5). This is due to the fact that the contact is also accompanied by a bend of the tooth, which makes some adjustments, taken into account by the correction factor K_ψ included as a multiplier in (Eq. 5.5):

$$K_\psi = -933.33\psi^4 + 706.67\psi^3 - 189.67\psi^2 + 20.63\psi + 0.13, \tag{5.6}$$

where $\psi = a_H/b_w$ (b_w is the tooth length).

It should be emphasized that for gearings with teeth hardened to high hardness, it is necessary to take into account the characteristics of the surface layer as well as, in some cases, to assess the deep contact stresses.

5.3 Stiffness of the Teeth

The stiffness of the teeth in our studies is conveniently characterized by the value W of the total (contact and shear) displacement of the center of the contact spot under load, determined in the process of modeling the solution of the contact problem. Knowledge of these movements is necessary to get a complete picture of the distribution of the load and stresses on the contact sites when modeling a real multi-pair engagement.

Statistical processing of a large amount of information has made it possible to obtain dependencies for determining the base value W:

$$W = 0.359 F_t^{0.77} m^{-0.585} \left(\rho_\beta^* + 100\right)^{-0.294} (0.111 z_v)^\gamma, \tag{5.7}$$

where $\gamma = -0.294(0.111 z_v)^{-0.423}$.

Let us now proceed to the report on the results of the study of the phase values of the considered parameters obtained by shifting the contact spots along the length of the teeth from the middle part of the gear rim to one of its ends in the process of

Table 5.2 Values of coefficients in the formula (Eq. 5.8)

D_{00}	D_{01}	D_{02}	D_{03}	D_{04}	
1.481304	−0.923358	20.93405	−38.70766	19.43515	
D_{10}	D_{11}	D_{12}	D_{13}	D_{14}	D_{15}
−0.392466	0.487538	−10.61498	−2.137028	99.47423	−128.8717
D_{20}	D_{21}	D_{22}	D_{23}	D_{24}	D_{25}
0.0896951	−0.177547	3.738188	−8.708485	−2.723311	16.54621
D_{30}	D_{31}	D_{32}	D_{32}	D_{34}	
−0.00625611	0.0136604	−0.308048	1.041986	−1.045351	

Table 5.3 Values of coefficients in the formula (Eq. 5.9)

B_{00}	B_{01}	B_{02}	B_{03}	B_{04}	
1.004276	18.10111	71.25582	−744.2869	1201.738	
B_{10}	B_{11}	B_{12}	B_{13}	B_{14}	B_{15}
−0.0913205	−40.02436	−123.5507	1167.931	−310.1259	−3255.421
B_{20}	B_{21}	B_{22}	B_{23}	B_{24}	B_{25}
0.101049	21.95774	46.85895	−335.5927	−1309.902	3864.673

engagement. The closer the contact spot is to the end, the stronger is the influence of the latter on the stress and stiffness of the teeth. The designations of the phase parameters obtained during modeling, in further presentation, will be provided with an additional index "s," and the influence of the ends of the teeth on them will be evaluated through the so-called coefficients of influence representing the ratios of the phase parameters to the corresponding baseline.

In previous works [6, 9] typical computer graphs of bending stresses and effective contact stresses and contact spots for their basic positions and in the zone of the ends of the teeth are provided. Processing of the obtained data in the entire range of displacements of the contact spot from the base position to the end of the teeth has made it possible to obtain the values of the coefficients of influence of the ends expressed by the developed polynomials for software calculations and Tables 5.2, 5.3, and 5.4:

- Coefficient K_{Fs} of influence of the tooth ends on phase bending stresses

$$K_{\mathrm{Fs}}=\sum_0^4 D_{0q}S^q+(\psi+0.1)^{-1}\sum_0^5 D_{1q}S^q+(\psi+0.1)^{-2}\sum_0^5 D_{2q}S^q+(\psi+0.1)^{-3}\sum_0^4 D_{3q}S^q \tag{5.8}$$

- Coefficient K_{Ws} of influence of the tooth ends on shift in different phases

$$K_{\mathrm{Ws}}=\sum_0^4 B_{0q}(S-0.05)^q+(\psi+1)^{-1}\sum_0^5 B_{1q}(S-0.05)^q+(\psi+1)^{-2}\sum_0^5 B_{2q}(S-0.05)^q \tag{5.9}$$

Table 5.4 Coefficient K_{Hes} of influence of tooth ends on phase effective contact stresses

S	ψ				
	0.05	0.10	0.15	0.20	0.25
0.500	3.15	2.45	2.51	2.63	2.65
0.475	2.46	2.20	2.37	2.53	2.57
0.463	1.66	1.97	2.21	2.43	2.49
0.450	1.52	1.72	2.05	2.32	2.39
0.425	1.31	1.26	1.82	2.09	2.22
0.400	1.19	1.18	1.41	2.03	2.05
0.375	1.12	1.11	1.17	1.72	1.85
0.350	1.06	1.08	1.12	1.34	1.80
0.325	1.02	1.04	1.09	1.13	1.54
0.300	1.01	1.04	1.06	1.08	1.26
0.275	1.00	1.03	1.05	1.05	1.13
0.250	1.00	1.02	1.03	1.03	1.09
0.200	1.00	1.01	1.02	1.02	1.05
0.150	1.00	1.01	1.01	1.02	1.02
0	1.00	1.00	1.00	1.00	1.00

Based on the above, it is possible to present the final formulas of the main parameters of the strength calculation of Novikov gearing:

- Phase bending stress

$$\sigma_{\mathrm{Fs}} = Y_{\mathrm{V}} Y_{\mathrm{as}} K_{\mathrm{Fs}} F_{\mathrm{ts}} / m^2 \tag{5.10}$$

- Phase effective contact stress

$$\sigma_{\mathrm{Hes}} = 19.93 l^{-1.074} K_{\psi \mathrm{s}} K_{\mathrm{Hes}} F_{\mathrm{ns}}^{0.69} \rho_{\beta}^{-0.31}, \tag{5.11}$$

- Phase total (contact shear) displacement

$$\begin{aligned} W_{\mathrm{s}} &= 0.359 K_{\mathrm{Ws}} F_{\mathrm{ts}}^{0.77} m^{-0.585} \left(\rho_{\beta}^{*} + 100\right)^{-0.294} (0.111 z_{\mathrm{v}})^{\gamma}, \\ \gamma &= -0.294 (0.111 z_{\mathrm{v}})^{-0.423}, \end{aligned} \tag{5.12}$$

where F_{ts}, F_{ns} are the tangential and normal partial forces, respectively, on the contact spots in different phases of engagement, determined in the process of modeling the real multi-pair gear engagement. In polynomials and tables S– the distance of the theoretical point of contact to the middle of the gear rim (of base position) is expressed in fractions of the length of the tooth.

Thus, the solution of the spatial contact problem for Novikov gear teeth with the standard basic profile BP1 at the position of the contact spot in any zone along the width of the gear rim makes it possible to obtain working formulas for determining bending and effective contact stresses, as well as the stiffness of the teeth, including the coefficients of the influence of the ends of the gear rim on the main parameters of the contact. The data obtained are the basis for finding permissible stresses and for the definition of the operability of the gear and the resource of the gearbox as a whole.

Using the working formulas (Eqs. 5.10–5.12) for determining phase stresses and displacements, it is possible to proceed to modeling the process of a real multi-pair gear engagement to find permissible stresses.

For this purpose, a system of equations containing equations of the elastic balance and shared displacement of the contact teeth has been compiled:

$$\begin{cases} W_1 + \delta u_1 + \delta l_1 = W_2 = \delta u_2 + \delta l_2 = \cdots = W_n + \delta u_n + \delta l_n \\ \sum_{i=1}^{n} F_{ti} = F_{t\Sigma}; \quad i = 1\ldots n. \end{cases} \quad (5.13)$$

Here, $F_{t\Sigma}$ is total tangential force, F_{ti} is the partial tangential force on the i the contact spot, and δu is a set of technological errors in the manufacture and assembly of gears of the pair, expressed in the probabilistic aspect, which also includes errors caused by deformations of the related parts (shafts, bearings, etc.) of the drive. The solution is implemented by a specially developed program in the MAPLE-17 system. As a result, the tensest sections of the teeth and, accordingly, the load-bearing capacity of the gearbox are determined.

Figure 5.1 shows two-stage gearboxes: serial 1Ts2U-200 with nitro-cemented involute wheels and experimental 6Ts2-160 with nitro-cemented Novikov wheels, calculated according to the developed technique. With the same torque $T = 2500$ Nm, the experimental gearbox is almost 1.5 times lighter than the serial one. With the same specific gravity G/T, which is one of the main indicators of the technical and economic characteristics, an experienced gearbox transmits torque 35% higher than a production gearbox.

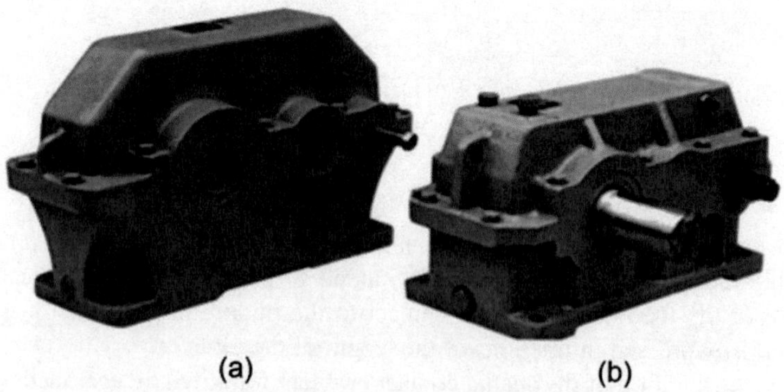

Fig. 5.1 (**a**) Gearbox 1Ts2U-200 (involute gearing): $T = 2500$ Nm, $G = 170$ kg; $G/T = 0.068$ kg/Nm; (**b**) Gearbox 6Ts2-160 (Novikov gearing): $T = 2500$ Nm, $G = 120$ kg; $G/T = 0.048$ kg/Nm

The gearbox 6Ts2-160, with our active participation, has passed many years of factory tests. A large experimental series has been manufactured, which has been successfully operated for many years in the drives of machines and mechanisms for various purposes.

The technique described above was applied to the assessment of the technical characteristics of special purpose gearboxes, in particular, drive gearboxes of oil pumping units equipped with thermally improved Novikov gears with the initial BP2 circuit, which are widely produced in the industry. At the same time, a study of the effectiveness of the longitudinal modification of the working surfaces of the teeth, created by us to radically increase the load-bearing capacity of Novikov gears, was carried out.

The technique described above was applied to the assessment of the technical characteristics of special purpose gearboxes, in particular, drive gearboxes of oil pumping units equipped with thermally improved Novikov gears with the initial BP2 circuit, which are widely produced in the industry. At the same time, a study of the effectiveness of the longitudinal modification of the working surfaces of the teeth, created by us to radically increase the load-bearing capacity of Novikov gears, was carried out.

The task of improving the technical characteristics of the drive system of oilfield equipment is highly relevant. One of the solutions to this problem is to reduce the stresses of the teeth wheels in the drive gearboxes. This makes it possible to increase the load-bearing capacity of the gearboxes and the resource of their operation. At the same time, it is extremely important to preserve the radial and axial dimensions, as well as the general gear ratios of the gears, which makes it relatively easy to achieve the goal without performing any significant design and technological additional works in the production of manufactured products. In the same sense, it is important to fundamentally preserve the type of tooth-cutting and control tools for the manufacture of gears.

Oil pumping units (see Fig. 5.2), produced by the domestic industry by major series, are widespread in the oil industry, working throughout the country under a variety of climatic conditions.

It should be noted that the authoritative American company DARCO (USA), which produces oil pumping machines and exports them to many countries of the world, also uses Novikov gears, which the company considers the best variant, confirmed by tests in China and by the National Aeronautics and Space Administration (NASA), of equipping gearboxes serving oilfield equipment.

In order to modernize the gearbox of an oil pumping unit, we first considered it expedient to replace the serial chevron gears with helical gears with special thrust rings [10] and perceived axial loads, which allows the use of radial bearings without requiring axial adjustment. Such a design has successfully passed a multi-year test while operating the gearbox model TsNK250 and the series Ts3NK, producing highly positive results [11]. In addition, it was necessary to preserve the basic profile [3] as well as to not increase the radial and axial dimensions of the gearbox structure while maintaining its total gear ratio, which, as mentioned above, contributes to a significant facilitation of the use of the proposed modernization in production.

As the object of the study, the widespread three-stage gearbox Ts3NSH-450-40 was chosen (see Fig. 5.3).

Fig. 5.2 General view of an oil pumping unit

Fig. 5.3 Serial chevron gearboxTs3NSH-450-40

At the initial stage of the study, designing Novikov gears with a module exceeding that used in a serial variant was proposed. In comparison with the existing serial version, this led to a decrease in the stresses of the teeth along the contact strength by approximately 1.12 times, an increase in the load-bearing capacity of the gearbox by 1.45 times, an increase in the service life by 2 times, and an increase in the bending strength from 1.3 to 1.6 times, 1.74 times, and 4 times. Along the way, it was possible to reduce the working width of the gear rims at various stages of the gearbox from 1.26 to 1.64 times, which helped improve the weight and size characteristics of the gearbox [12]. Thus, only the increase in the module significantly increased the technical characteristics of the gearbox, which under the above conditions, as shown in the study by Korotkin and Onishkov [12], could not be achieved in principle if involute gears were used instead of Novikov gears.

Our research has revealed another significant reserve for reducing the stresses of Novikov gear teeth, which is advisable to use together with the increase in the module – this is a longitudinal modification of the barrel-shaped type of the working surfaces of the teeth [5].

Figure 5.4a conditionally depicts a pair of the basic gear racks of gear 1 and the wheel 2, which do not have errors according to the contact norms ($\delta a = 0$) and are in contact with each other in a straight line, and, in Fig. 5.4b, the same pair with errors according to the norms of contact, leading to a kinematic gap $\delta a \neq 0$ at one of the ends of the gear rim, is displayed.

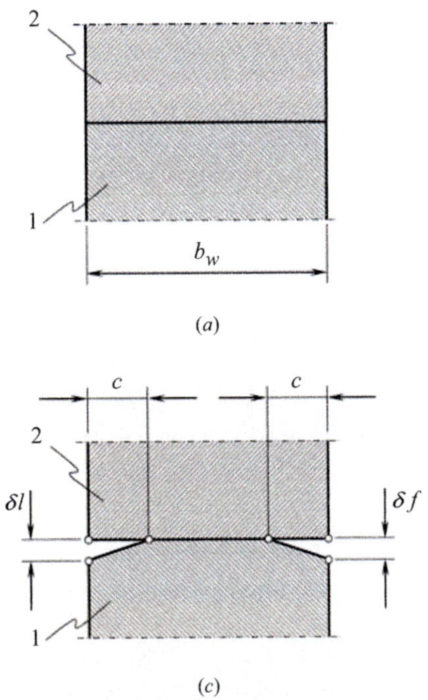

Fig. 5.4 Diagram of tooth contact variants

Figure 5.4c shows the longitudinal flanking of the teeth. With this method, at one of the elements of the pair (more convenient at the gear), the flanks at the ends of the teeth with a depth of δf and a length of c are removed. There are various methods of longitudinal flanking of Novikov gear teeth, in which the line of flank is straight or close to a straight line. With this method, the end of the teeth, as expected, is completely turned off from work, but the main engagement cycle (on the nonflanking part) remains the same, and the alignment of stresses along the contact spots does not occur, which sometimes leads to not only a decrease but also an increase in stresses [13]. Achieving significant stress equalization with the simultaneous exclusion of the end of the teeth from the engagement, as studies have shown, is possible by performing a longitudinal modification, giving the surfaces of the teeth a slightly barrel-shaped shape (see Fig. 5.4d).

A longitudinal modification is deviation of the contacting surfaces of the gears from each other by a certain value δl, called the modification parameter, which, according to the parabolic law $\delta l = b_{\mathrm{w}}^2/8\rho$, decreases along the length of the tooth up to zero in the middle part of the gear rim (ρ is the longitudinal radius of curvature of the tooth surface). The removal δl is best carried out on one of the elements of the toothed pair. As known, a similar method has been successfully used for involute gearings.

Technologically, longitudinal modification of teeth does not present difficulties and can be carried out in production in the same manner as it is performed to obtain barrel-shaped teeth of involute gears, that is, either by diagonal gear milling [14] or with the help of removable copiers to serial gear cutting machines or on tooth-shaving machines with a swinging table [15], in which the radius of the copier is $r_{\mathrm{k}} = \rho\, tg\alpha_{\mathrm{k}}$ (α_{k} is the angle of the basic profile at the nominal theoretical point of contact). The latter method is simpler and more preferable.

The results of the study presented below are obtained when the system (Eq. 5.13) is solved with the addition of a modification parameter, δl.

Table 5.5 shows a comparison of contact (σ_{H}) and bending (σ_{F}) permissible phase stresses of the modified teeth of Novikov helical gears with an enlarged module and the teeth of serial wheels on the stages of the gearbox Ts3NSH-450-40, at a rated torque $T_{\text{вых}} = 40,000$ Nm on the output shaft of the gearbox. On the left part of the table are the values for the serial (unmodified) teeth and on the right are the values obtained for the modified teeth. All gears in the compared gears are shown variants are made without displacements of the basic profile when cutting teeth, from the same material, with the same heat treatment and have really achievable in practice technological deviations corresponding to approximately 9–10 of accuracy grade (if you focus on GOST 1643-81).

Longitudinal modification in combination with an increased module leads to a significant reduction in the stresses of the teeth, and, for the input stage, it was also possible to further reduce the width of the gear rim by 1.27 times.

Similarly, the efficiency of longitudinal modification of the teeth is shown in Fig. 5.5, where λ_{σ} is the ratio of the permissible stress of the serial gearing to the permissible stress of the modified teeth at the same torque $T = 40{,}000$ Nm (curves A) and λ_T is the ratio of the load-bearing capacity of the modified gearing to the load-

Table 5.5 Strength characteristics of Novikov gears of three-stage gearboxes of the Ts3Sh-450-40 model with unmodified (serial version) and modified teeth with definition of the optimal ranges of parameters δl

	Parameter	Serial variant				Modified variant			
Stage	Gear ratio	40	63	90	125	40	63	90	125
Input (1)	m, mm	6.3				8			
	σ_H, MPa	476	442	441	449	268	250	253	262
	σ_F, MPa	258	220	201	189	99	90	85	81
	δl, mm	0	0	0	0	0.06 ... 0.07			
Intermediate (2)	m, mm	7.1				9.0			
	σ_H, MPa	759				299			
	σ_F, MPa	399				165			
	δl, mm	0				0.09 ... 0.11			
Output (3)	m, mm	8.0				10.0			
	σ_H, MPa	1023				420 ... 455			
	σ_F, MPa	662				325 ... 350			
	δl, mm	0				0.1 ... 0.3			

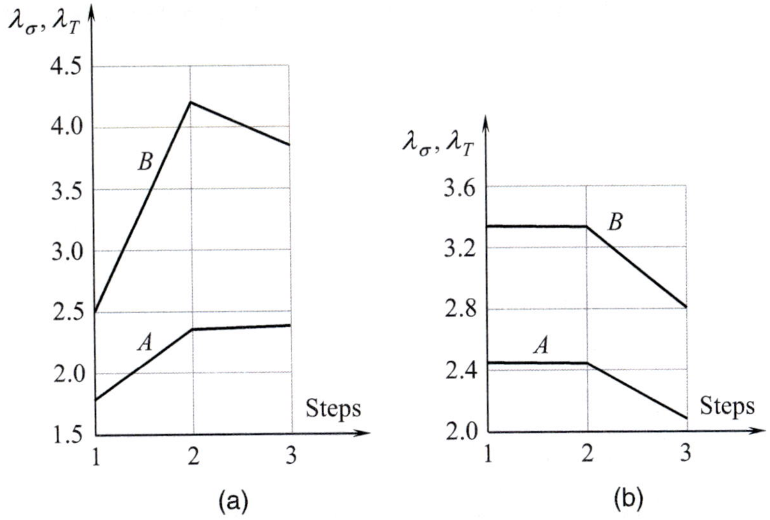

Fig. 5.5 Schedules of the ratio λ_σ, λ_T on the stages of the gearbox Ts3NSH-450-40

bearing capacity of the serial gearing at the equality of the permissible stresses (curves B). Figure 5.5a shows the contact strength and Fig. 5.5b demonstartes bending.

The high efficiency of the modification is clearly visible. So, for the most loaded (output) stage of the gearbox, we have $\lambda_\sigma = 2.25 \ldots 2.44$, $\lambda_T = 3.9$ by contact and $\lambda_\sigma = 1.89 \ldots 2.04$, $\lambda_T = 2.8$ by bend. A similar effect is visible for other gear stages. Of course, it is possible to implement it, apparently, only partially, otherwise it will be necessary to modernize the shafts and bearings. However, it is safe to say that modified Novikov gearing will not be a limiting link at all stages of the gearbox Ts3NSH-450-40 and also in the rest of the gearboxes of the oil pumping units produced.

The effect of reducing stress at the request of consumers can be obtained in another manner. If the gearbox is operated with a serial load (40,000 Nm on the output shaft), then with the obtained reduced stresses of the modified teeth, the resource of the gearbox in accordance with the fatigue curves [4] will increase significantly, based on both contact and bending strength.

Considering the question of increasing the resource of the gearbox, it is necessary to use the well-known universal dependence of the type $\sigma^q N = \text{const}$ [16], arising from the Wohler fatigue curve, where σ is the stress, q is an indicator of the degree of the fatigue curve, and N is the number of cycles of the gear wheel.

According to GOST 21354-87 [4], the contact fatigue curve has two inclined sections with a degree index of $q = q_{\mathrm{H}} = 6$ and $q = q_{\mathrm{H}} = 20$ and the bending endurance curve, in our case, has one inclined section with a degree index of $q = q_{\mathrm{F}} = 6$, before passing into a horizontal section with long durability. Thus, reducing the contact and bending stresses at constant load will increase durability by at least 1–2 orders of magnitude.

The developed computational program determines not only the permissible stresses in the areas but also looks for the optimal modification parameters at which these voltages will be minimal. Such parameters are indicated in Table 5.5. It should be noted that it is not desirable to markedly deviate from the recommended ranges of the modification parameters specified in the table, within which the reduced voltages are obtained because, otherwise, the permissible stresses will begin to grow, as can be seen from Fig. 5.6.

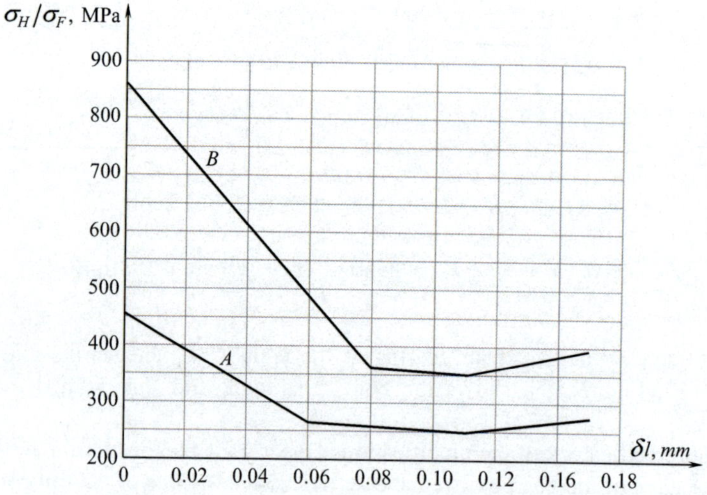

Fig. 5.6 Graphs of changes in bending stresses (A) and effective contact stresses (B) of the modified teeth of Novikov gears at the output stage of the gearbox Ts3NSH-450-40 depending on the values of the modification parameter δl

Analysis of the results shows that an increase in the module of the teeth basically leads to a decrease in the bending tension of the teeth and that the longitudinal modification dramatically reduces their contact tension.

In summary, it can be argued that the longitudinal modification of the teeth of Novikov gears in combination with an enlarged module is a highly effective and fundamentally technologically simple way to radically improve the technical and economic characteristics of drive gearboxes produced by the domestic industry of oil pumping machines, ensuring their high competitiveness in the world market.

References

1. Korotkin, V.I.: The Comparative Effect of Certain Provisions Concerning the Increasing Load Capacity of Novikov Gearing and Involute Gearing. Physics, Mechanics of New Materials and Their Applications (PHENMA-2016) Chapter 78, pp. 571–581. Nova Science Publishers, New York (2017) (ISBN: 978-1-53611-052-4)
2. GOST 30224-96: Novikov Gears Cylindrical with a Hardness of the Tooth Surface of at Least 35 HRC_3. The Basic Profile. Interstate Standard, 5p. Mezhgos. Sovet po standartizatsii, metrologii i certificatsii (1996)
3. GOST 15023-76: Novikov Gears Cylindrical with Two Engagement Lines. Basic Profile, 3p. Izd-vo standartov, Moscow (1976)
4. GOST 21354-87: Gears Cylindrical Involute. Strength Calculation, 125p. Izd-vo standardov, Moscow (1988)
5. Korotkin, V.I., Onishkov, N.P., Kharitonov, Y.D.: Novikov Gearing: Achievements and Development, 249p. Nova Science Publishers, Inc, Hauppauge, New York (2011) (ISBN 978-1-61761-193-3)
6. Korotkin, V.I., Gazzaev, D.A.: Bending intensity of teeth under the influence of a distributed load in different phases of Novikov gearing. J. Mach. Manuf. Reliab. **43**(2), 104–111 (2014) © Allerton Press, Inc., 2014. (ISSN 1052_6188)
7. Kovalsky, B.S.: Calculation Parts on Local Compression, 233p. HVKIU, Kharkov (1967)
8. Korotkin, V.I., Gazzaev, D.A.: Modeling of contact interaction of teeth of wheels of Novikov gearing. Vestnik mashinostroeniya. (11), 31–35 (2014)
9. Korotkin, V.I.: Permissible contact stress, phase displacements and stiffness of teeth of Novikov gears made on the basis of the basic profile according to GOST 15023-76. Spravochnik. Inzhenerniy zhurnal. (6), 14–20 (2016)
10. Pat.2191937 RF: MPK^7F16H1/08: Helical cylindrical gearing with thrust rings
11. Ognev, M.E.: Helical cylindrical gearing with thrust rings. Vestnik mashinostroeniya. (12), 14–17 (2007)
12. Korotkin, V.I., Onishkov, N.P.: Modernization of gearboxes of oilfield equipment drives. Spravochnik. Inzhenerniy zhurnal. (1), 14–19 (2018)
13. Pavlenko, A.V., Fedyakin, R.V., Chesnokov, V.A.: Gears with Novikov Engagement, 144p. Technika, Kiev (1978)
14. Shunaev, B.K., Efimenko, V.F.: Diagonal gear milling of wheels with barrel-shaped teeth. Stanki i instrument. (4), 15–18 (1967)
15. Groman, M.B., Zak, P.S.: Barrel-shaped tooth. Vestnik mashinostroeniya. (4), 23–25 (1976)
16. Chasovnikov, L.D.: Gearing (Toothed and Worm). 2nd ed., rev. and add, 486p. Mashinostroenie, Moscow (1969)

Chapter 6
Some Features of the Contact Strength of *Novikov* Gearing

Viktor I. Korotkin and Nikolay P. Onishkov

Theoretically, the contact point of the interacting teeth of *Novikov* gears under load turns into local contact with an instantaneous contact pad (ICP) of an elliptical, or close to it, form provided during the design. At the same time, effective contact stresses determined in accordance with the Fourth theory strength, σ_e (which in some sources are called stress intensity), regardless of the hardness of the working surfaces of the teeth, have traditionally been determined in the center of the ICP and were considered as permissible stresses—that is, determining the load-bearing capacity of the gearing [1, 2].

For the plastic material of steel teeth of relatively low hardness (i.e., normalized, thermally improved) with a coefficient ellipticity, $\overline{\beta} \leq 0.45$, which is the ratio of the sizes of the small (b) to the large (a) semi-axle of the ICP; this approach is quite justified. When $\overline{\beta} > 0.45$, the maximum effective stresses σ_e are shifted to the periphery side of the contact pad [2], and when $\overline{\beta} = 1$ (i.e., circular contact pad) one has on the periphery the largest excess over stress σ_e in the center the; however, such cases in the gears practically do not occur.

As for teeth hardened by chemical heat treatment (CHT), then, as will be shown in the following, permissible contact stresses, $\sigma_{cr,}$ can occur at the end of the large or small semi-axel of the ICP at some value, $\overline{\beta}$. This is because of the fact that stretching stresses appear at the boundaries of the ICP, leading to pure shift; and a rather fragile hardened CHT material, unlike plastic, is much more sensitive to tensile stresses than to compressive ones, which is usually characterized by the plasticity parameter $\chi = \sigma_+/\sigma_-$, where σ_+, σ_- —uniaxial stress of destruction, respectively, stretching and compression. Obviously: for absolutely plastic materials, $\chi = 1$; for absolutely fragile, $\chi = 0$.

Financial support was provided to Southern Federal University by the Russian Ministry of Science and Higher Education (Project VnGr/2020-04-IM).

V. I. Korotkin · N. P. Onishkov
Southern Federal University, Institute for Mechanics, Rostov-on-Don, Russia

S. P. Radzevich, *Novikov/Conformal Gearing*,
https://doi.org/10.1007/978-3-031-10019-2_6

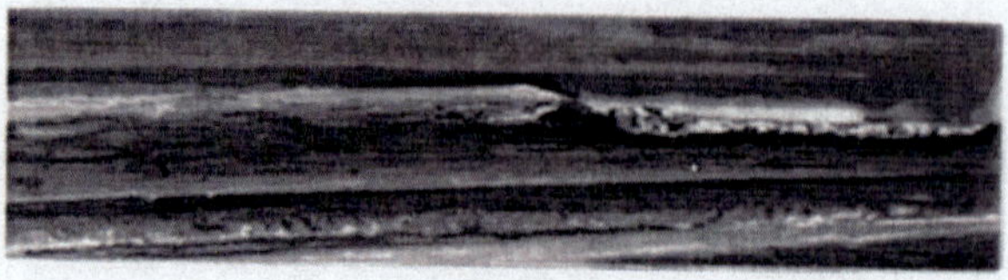

Fig. 6.1 Example of destruction of nitrocarburized teeth of *Novikov* gears

Note that for *Novikov* gearing, contact calculations for the strength of the teeth's working surfaces with convex–concave local contact reliability was considered to be secondary. Surface contact strength was supposed to be a priori provided, and failures, as it was believed, were associated with fractures and chips from excessive bending stresses.

Still, analysis of some published data [3] on the occurrence at the border of the ICP (especially in materials prone to embrittlement) of radial microcracks growing with repeated (even insignificant) stresses, makes one pay serious attention to the detected phenomenon. An additional impetus to the need to study the stress–strain state (SSS) of the teeth in the area of the boundaries of contact spots was given by the following circumstance.

During the tests of *Novikov*'s nitrocemented gears [1], no cases of pitting corrosion were recorded, the destruction of *Novikov*'s teeth was in the nature of end breakage and in most cases the primary surface defects were localized in the active zone near the boundaries of the contact spot, and the fracture surface was close in some cases to a flat surface (see Fig. 6.1). This implies a pure shift under the action of tangential stresses.

The purpose of this work is to substantiate the possibility of destruction of gear teeth in the area of the boundaries of the ICP and to assess the influence of the contact conditions of high-stressed surface-hardened gears with the teeth's local contact on the initiation of alternative (i.e., noncontact) types of their destruction.

It is known that for heat-improved steels there is a close to linear dependence of tensile strength on the σ_{B+} hardness *H* according to *Brinell*: $\sigma_{B+} \approx 0,34HB$ and the conditional yield strengths under tension ($\sigma_{0,\,2+}$) and compression ($\sigma_{0,\,2-}$) are taken to be equal. Yet, the increase in hardness over (450...500)*HB* leads to a violation of these ratios—a clear display of differences in tensile strength (σ_{B+}) and compression (σ_{B-}): $\sigma_{B-} >> \sigma_{B+}$, $\sigma_{0,\,2+}$approaching σ_{B+}, $\sigma_{0,\,2-} << \sigma_{B-}$ [4, 5]. The role of tangential stresses in the process of destruction with the growth of the fragility of the material decreases, destruction will be brittle or mixed (i.e., brittle-plastic) depending on the stiffness of the load. Under these conditions, the use of plasticity criteria (i.e., *Treck*, *Mises*) in assessing the strength leads to incorrect results. Studies have shown that the use of the *Lebedev–Pisarenko* limit state criterion for structurally inhomogeneous material [5] was more objective, in which the current permissible (i.e., critical) stresses are σ_{cr} determined by the formulas:

$$\sigma_{cr} = \chi\sigma_e + (1-\chi)\sigma_1 A^{1-(\sigma_1+\sigma_2+\sigma_3)/\sigma_e} \leq [\sigma] \tag{6.1}$$

Here σ_1, σ_2, σ_3 = the main stresses; A = 0,7...0,8 = statistical parameter of defectiveness; [σ] = some allowable stress for a uniaxial stress state—of stretch = $[\sigma]_+$ or of compression $[\sigma] = [\sigma]_- \approx [\sigma]_+/\chi$. For plastic materials $\chi = 1$, $\sigma_{cr} = \sigma_e$, and evaluation of strength assessment can be made according to the criterion of plasticity of *Treck, Mises.*

The limit state criterion allows one to consider the effect of load stiffness on the decrease or increase [σ] depending on the sign σ_1. Within the framework of the proposed model, the change in these properties is considered by a change in the plasticity parameter. The processing of available experimental data, summarized in Reference [1], made it possible to accept $\chi = 1$ for thermal improvement of steels in the first approximation, and for a high-quality alloyed tool and bearing steels at $H = (62 \ldots 64)$ *HRC*e to accept $\chi = 0.6\ldots0.7$. For hardened structural carbon steels, it is possible to reduce the parameter χ to 0.45...0,55.

The use of calculation models based on the criterion of the limit state made it possible to make significant clarifications in the assessment of the deep contact strength of *Novikov*'s surface-hardened gears) [1, 6] and to substantiate the inexpediency of overestimating the hardness of working surfaces H_0 over (58...60) *HRC*e; whereas, the recommendations of GOST 21354-87 allow up H_0 to 65*HRC*e. Consider, for example, the features of SSS in the field of theoretical point contact "sphere-to-plane" with a circular contact pad.

The stresses on the entire surface of the contact area are compressive, except for the area of its boundary. At the ends of the radius $r_,$ there are equal voltages modulo—radial (tensile) and circumferential (compressive). The effective stresses $\sigma_{e,}$ according to the *Mises* criterion at the *Poisson* coefficient of 0.3 in this case for the circular region of the ICP, are equal $\sigma_{e0} = 0, 2p_0$ in the center of the IPC and $\sigma_{er} = 0.231p_0$ at the border of the ICP (p_0 = pressure in the center of the ICP, the lower index "0" refers to the center of the IPC, the index "*r*" = the border of the ICP). Consequently, the congestion of the ICP boundary relative to the center is 15.5%. For hardened alloy structural steels at $H_0 \approx (60\ldots62)$ *HRC*e and at the parameter of plasticity $\chi \approx 0.7$, according to the *Lebedev–Pisarenko* permissible stresses, will be: in the center $\sigma_{cr0} = 0, 143p_0$ of the ICP and at the border $\sigma_{cr0} = 0, 186p_0$ of the ICP—that is, the relative tension of the border zone increased approximately 12–13%. But in addition to the increase in relative tension, it is important to keep in mind that the level of permissible stresses in the center of the ICP is because of the compressive strength of the material, while at the edge of the ICP it is tensile strength (in the first approximation, the corresponding permissible stress can be considered proportional χ) [7, 8].

The shear stress is maximum at the circular contact area and decreases as the elliptic coefficient, $\overline{\beta}_,$ decreases. Accordingly, the permissible stresses at the boundaries of the ICP decrease and increase in its center. Comparative graphics of changes in tension [σ] (stresses are normalized by p_0) depending on $\overline{\beta}$ on the elliptical ICP shown in Fig. 6.2. (The components of the stress tensor are defined by [6, 7]). From Fig. 6.2 follows:

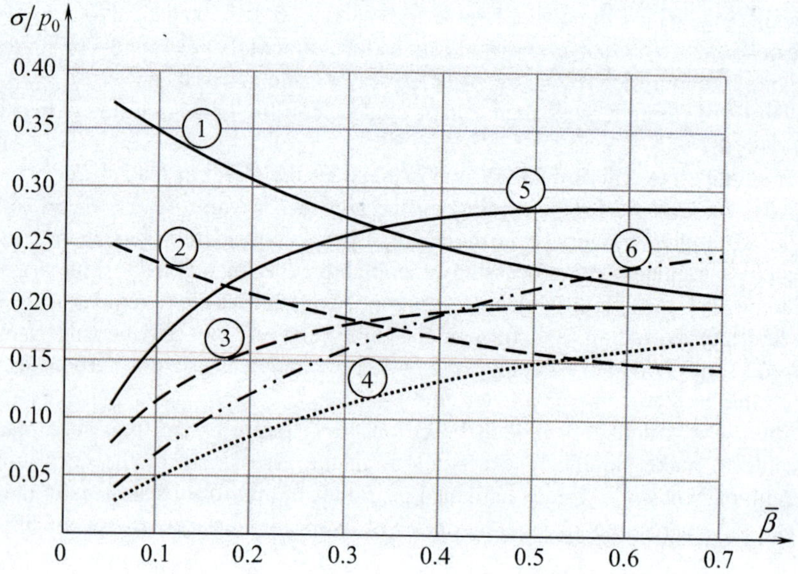

Fig. 6.2 The tension on the elliptical ICP depending on the elliptic coefficient $\bar{\beta}$; 1.2 – σ_{e0}, σ_{cr0}, respectively, the effective stresses and permissible stresses at the plasticity parameter $\chi = 0.7$ in the center of IPC; 3,4 – σ_{cra}, σ_{crb}- permissible stresses at borders respectively of large and small semi-axles of contact ellipse; 5 – σ_{cra}/χ; 6 –σ_{crb}/χ

(a) for all values $\bar{\beta}$ – $\sigma_{cr0} < \sigma_{e0}$
(b) when $\bar{\beta} \geq 0{,}2$ one has $\sigma_{cra}/\chi > \sigma_{cr0}$, which increases the danger of end breakage
(c) at $\bar{\beta} \geq 0.35$ the permissible stresses at both boundaries of the semi-axles of the ICP become determinative of the danger of the slice

Studies conducted for *Novikov* gearings in a wide range of parameters of the basic profile and geometry of the gears showed that by varying these parameters, it is possible, by changing the elliptic coefficient, to control the ratio of permissible stresses in the center and at the borders of the contact pad and thereby identify the teeth's dangerous areas .

The increase in hardness determines the increase in the contact strength of steel parts, but at the same time there is a decrease in the plastic properties of the material with the opposite tendency. Consequently, there is, as mentioned earlier, some threshold value of the hardness of steel, after which its further increase will give a negative effect. As part of the study of deep contact strength of chemically–thermally hardened gears based on the *Lebedev–Pisarenko* criterion, the acting stresses were determined by formula (6.1), and allowable stress by the formula in Reference [9]:

$$[\sigma] = [\sigma]_{-} = \chi(\chi - 0,\ 11128)H_{HV}D \tag{6.2}$$

where $D = Z_{NK}K_1K_2K_3K_4K_5$: Z_{NK} = longevity coefficient, $K_1 - K_5$ = coefficients taking into account the nature of contact, the number of potentially dangerous zones, the effect of tangential load, the quality of the material and CHT, and the dispersion of material properties in the layer.

The dependence shown in formula (6.2) was obtained in the study of deep contact strength and resolved with respect to destroying compression stresses. On the surface, at the border of the ICP, there is a shift and destructive stresses (i.e., stretching). Hardened steels have various resistances to tension and compression, therefore the allowable stresses for the boundary zone are, taken in the first approximation, reduced by proportionally, χ:

$$[\sigma] = [\sigma]_+ = \chi^2(\chi - 0,\ 11128)H_{HV}D \tag{6.3}$$

The approximation of the dependence (6.3) is explained by the fact that the parameter D does not consider the possible decrease in the mechanical characteristics of the material caused by inevitable surface defects, estimated by several authors as 10–15%. Let's analyze the effect of hardness on the bearing capacity of the contact on the example of the case discussed previously—the sphere–plane contact.

Source data:

- In the center of the IPC: $\sigma_{10} = \sigma_{20} = -0,\ 8p_0$, $\sigma_{30} = -p_0$, $\sigma_{e0} = 0,\ 2p_0$,
- On the border of the IPC: $\sigma_{1r} = 0,\ 133p_0$, $\sigma_{2r} = 0$, $\sigma_{3r} = -0,\ 133p_0$, $\sigma_{er} = 0,\ 231p_0$
- Considered interval of change of hardness: $H_0 = (650 - 800)$HV

As the calculations showed, and with the equality of the operating and permissible stresses (6.1) and (6.2) for the center of the ICP and (6.1) and (6.3) for the boundary of the ICP, one understands that in the central zone of the ICP the change in hardness does not lead to a noticeable fluctuation in pressures. On the border of the same spot, where there is a $\sigma_{1r} > 0$ decrease (with increasing hardness) of the plasticity parameter, one gets a significant decrease, p_0 (up to 30%). Thus, increasing the hardness above 700HV (58HRCe) is impractical.

6.1 Conclusions

1. For the first time it was found that stress for a hardened gear of the CHT at local elliptical contact on its the periphery can exceed the stress in its center and initiate alternative types of failures (e.g., the observed fracture of the teeth).
2. It is necessary to conduct additional research and clarify the levels of permissible stresses within the contact area.
3. To prevent a possible decrease in the contact strength of the teeth, especially along the boundary area of the ICP, it is desirable to limit the hardness of their working surfaces to the level of 58HRCe.

References

1. Korotkin, V.I., Onishkov, N.P., Kharitonov, Y.D.: Novikov Gearing: Achievements and Development. Nova Science Publishers, Inc., Hauppauge, New York. 249p (2011)
2. Kovalsky, B.S.: Calculation Parts on Local Compression, p. 233. HVKIU, Kharkov (1967)
3. Morozov, E.M.Z., M.V.: Contact problems of fracture mechanics. M., Mashinostroenie. 544 p. (1999)
4. Geller Y.A.: Instrumental steels. M., Mashinostroenie. 525p (1983)
5. Pisarenko, G.S., Lebedev, A.A.: Deformation and strength of materials in a complex stress state. Kyiv: Naukova Dumka.. 415 p. (1976)
6. Onishkov, N.P., Korotkin, V.I.: To the assessment of contact-fatigue longevity of chemical-heat-strengthened gears. Vestnik DGTU., T.17. №3(90). S.5-13 (2017)
7. Johnson, K.: Mechanics of contact interaction. M.: Mir.. 432p. (1989)
8. Beskopylny, A., Onishkov, N., Korotkin, V.: Bending strength assessment of chemically-heat-strengthened Novikov gearing. MATEC Web of Conferences 224, 02033. (2018) Volume 224, Number 6
9. KorotkinV, I., Onishkov, N.P., Goltsev, A.V.: To the estimation of deep contact endurance of involute gears with superficially hardened teeth. Vestnik mashinostroeniya, No5. S.9–No5. S14 (2008)

Chapter 7
Tooth Relieving of Worm Hobs for Cutting Novikov Gears with Double Lines of Action

Aleksandr Sandler

7.1 Introduction

The peculiarity of the formation of screw and relieved surfaces by disk grinding wheels lies in the inevitable deviation of the profile of the ground surface from the profile of the wheel, which is called an organic error. For a straight basic rack, this organic error is minimized as much as possible, or, at the attained degree of minimization, an obtained profile is taken as the basic rack [1–3]. For screw and relieved surfaces with a profile of substantial and variable curvature, this organic error must be eliminated to the maximum degree.

A well-known example of a solution to this problem is the method of forming a profile of an unruled worm of the ZT2 type, proposed by F.L. Litvin in 1961 (later, in the monograph by Litvin and Fuentes [4], such worms were considered as worms of the ZF-II type). In this example, the axial profile of a grinding wheel was set in the form of a circular arc. Then, based on the condition that this arc is the contact line of the wheel and the ground surface, the angle of installation of the grinding wheel axis and the axial profile of the worm thread were sought. Obviously, to obtain a worm hob with the relieved surfaces of the teeth close in profile to the found worm threads, a solution for the inverse problem is required.

In the studies by Sandler et al. [2] and Sandler and Lagutin [5], the authors proposed a solution for a similar problem with respect to the convex thread profile of the worm and the hob for cutting the teeth of gears with liquid friction. They specified a substantially curvilinear axial profile of the working worm (or the hob's generating worm) and also took the profile of the grinding wheel in the axial section of the worm. The setting angle of the grinding wheel axis is determined on the basis of the condition that this axis intersects two normal lines to the thread axial profile and lies on a plane parallel to the axis of the worm. The proposed method

A. Sandler
LLC "SELLACTION", Moscow, Russia

S. P. Radzevich, *Novikov/Conformal Gearing*,
https://doi.org/10.1007/978-3-031-10019-2_7

ensures minimization of the organic error in the profiling of the worm thread and, taking into account a similar organic error in tool profiling, the necessary identity of the profile of the hob's generating worm.

In the work by Sandler and Lagutin [6], the authors made the first attempt to apply the general principles of this method to the study of the relieving process of worm hobs for cutting Novikov gears. In this case, two significant factors were taken into account. First, the curvature of the profile of the generating rack of these worm hobs is a variable in terms of not only its radius of curvature but also its sign. Second, the possibilities of the relieving machine allow setting the axis of the grinding wheel on a plane not parallel to the axis of the ground product that improves the grinding conditions for a number of parameters.

In this work, this research is developed and continued. This has solved the main issues of the relieving technique, which are functionally oriented to solving the following problems: providing the rear angles of the teeth necessary for the wear resistance of the cutting edges of the hob, minimizing the organic error of the profile of the generating rack, and determining the main parameters for setting up the relieving machine and the profile of the grinding wheel.

7.2 Novikov Gearing: Parameters of the Basic Rack

For the first time, helical gears with an initial point tangency of the circular-helical surfaces of the teeth were proposed by the eminent American inventor E. Wildhaber in 1926 with a US patent no. 1,601,750.[1] However, at that time, this invention went unnoticed, and such gears underwent rapid development only after the Soviet engineer M.L. Novikov formulated the general principle of their formation and showed that they will have an increased load capacity, primarily in contact endurance [7]. The fundamental novelty of Novikov's invention was analyzed in the works by S.P. Radzevich [8, 9].

Initially, M.L. Novikov proposed a version of helical overcentrode gears, in which tooth profiles, convex on the pinion and concave on the wheel, were delineated with the circular arcs in the face section of the gears. Later on, V.N. Kudryavtsev showed that the same effect can be achieved in gears synthesized on the basis of two incongruent basic racks with the circular-arc profiles and cut by two worm hobs: one for the pinion with convex teeth and the other for the wheel with concave ones [10].

[1] A comment from the editor: It is proven [8] that the invention by E. Wildhaber (1926) is not workable at all, as the fundamental laws of gearing are violated in this invention. It is also proven [8] that the fundamental laws of gearing are fulfilled in Novikov gearing, and, therefore, Novikov gearing is workable. It is a wrong practice to combine Novikov gearing with the helical gearing by E. Wildhaber in a common term – the terms "Wildhaber–Novikov gearing" and "W-N gearing" are meaningless by nature.

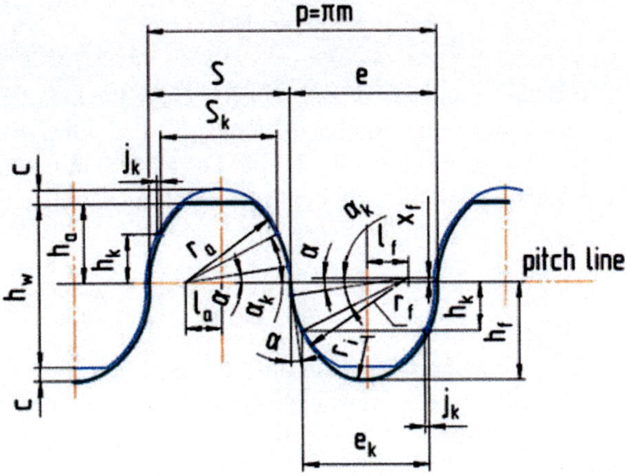

Fig. 7.1 Basic rack for Novikov gears according to GOST 15023-76

However, at present, basic racks are standardized and widely used in production with a full tooth profile, which provides, in gears, two lines of action and allows for cutting the pinion and the wheel by one worm hob [11]. In particular, according to the Russian Standard GOST 15023-76, a basic rack consists of a convex addendum, a concave dedendum, and a short straight section between them. The geometry of Novikov gears with two lines of action is calculated according to the Russian Standard GOST 17744-72. Figure 7.1 shows the main parameters of a basic rack, the values of which depend on a range of modules.

The pressure angle α_k at the predetermined contact points on the tooth addendum and dedendum is equal to 27° for any case. These points are also characterized by parameters such as the distance h_k from the pitch line, the hob tooth width e_k at its addendum, and the space width S_k at its dedendum; the difference $j_k = e_k - S_k$ provides a backlash between the teeth of the wheels to be cut. The pressure angle α in the straight section of the tooth depends on the module and is equal to ≈8°.

The hob tooth in the normal section is formed according to the dimensions of the space of the basic rack. Similarly, the space between the hob teeth forms the tooth of the wheel to be cut.

The convex profile of the hob tooth is outlined by the radius r_f, the center of curvature of which is located at a distance l_f from the axis of symmetry of the head and is shifted relative to the pitch line to the hob axis by x_f. The center of the concave arc radius r_a of the hob tooth space is located on the pitch line, at a distance l_a from the axis of symmetry of the space. Dimensions along the pitch line are the thickness e of the hob tooth and the width S of the hob space. The height h_f of the hob tooth head is equal to the dedendum of the gear to be cut. The depth of the space is equal to the sum of the addendum h_a of the gear and the radial clearance c between the gear and the hob when cutting. Moreover, in Fig. 7.1, the height h_w of the tooth active profile in the gear and the radii r_i of the arc transition curves on the head and in the space of the hob tooth are indicated.

7.3 Worm Hob Parameters

The main parameters of any worm hob are the axial module m, the radius r_F of the pitch cylinder, the number of treads z_0, the screw parameter $p = 0.5m\ z_0$ of the generating surface, and the number of teeth z_f in the face section. From these data, the lead angle γ_1 of the helical surface of the cutting edges on the hob pitch cylinder is determined from the expression:

$$\tan\gamma_1 = p/r_F \tag{7.1}$$

The worm hob design profile (a generating rack) is considered in the normal section of the generating worm, which is tangent to the screw front surface. For single-thread hobs, this section and the front surface practically coincides in the profiling zone.

When cutting the teeth on a gear hobbing machine, the plane of the generating rack is installed perpendicular to the direction of the teeth of the cut wheel, while the hob axis is set, taking into account the lead angle of the hob front surface.

When the teeth of the worm hob are relieved, its axis is installed in the centers of the grinding-relieving machine, that is, in the horizontal plane. For a reliable reproduction of the profile of the generating rack in the normal section of the hob, with the adopted profiling method, it is necessary to recalculate the parameters of the profile of the generating rack into the axial section of the tooth surface to be relieved.

First of all, it is necessary to determine the axial profile of the helical surface, on which the cutting edges of the hob teeth are located. To do this, the prescribed angle α of the normal profile of the straight section should be replaced by an angle α_1 of the axial profile according to the formula:

$$\tan\alpha_1 = \tan\alpha/\cos\gamma_1 \tag{7.2}$$

The curvature radii r_a and r_f of the two parts of the active profile in the normal section are replaced by the corresponding radii r_{a0} and r_{f0} in the axial section. For single-thread hobs, these radii with a sufficient approximation are determined by the Meusnier theorem from the expression:

$$r_{a0,f0} = r_{a,f}/\cos\gamma_1 \tag{7.3}$$

The arc radius r_i on the head of the hob tooth, which is a processed fillet of the wheel tooth, in the normal section is replaced with the corresponding radius r_0 in the axial section of the hob. According to Euler's formula, these radii are related by the expression:

$$r_0 = r_i/\cos^2\gamma_1 \tag{7.4}$$

The thread thickness b in the axial section should be determined on the minimum ground radius $r_{\min}$ of the hob, proceeding from the thickness b_n of the normal section at the end point of machining:

$$b = b_n / \cos \gamma_{(r_{\min})} \tag{7.5}$$

where $\gamma_{(r_{\min})} = \text{atan}\ (p/r_{\min})$ is the lead angle on the hob dedendum cylinder.

The calculated coordinates of the axial profile of the helical surface of the cutting edges are related to the coordinates of the generating rack profile by the dependencies:

- For point A_1 on the tooth dedendum:

$$x_{01} = x_{n1};\, z_{01} = z_{n1} / \cos \gamma_{x01}$$
$$\tan\ \alpha_{01} = \tan\ \alpha_{n1} / \cos \gamma_{x01},$$

where:

$$z_{n1} = 0{,}5\pi\, m_n + l_a - (r_F - x_{n1})\ \cot\ \alpha_{n1}$$
$$\gamma_{x01} = \text{atan}\ (p/x_{01}),$$

- For point A_2 on the tooth addendum

$$x_{02} = x_{n2}$$
$$z_{02} = z_{n2} / \cos \gamma_{x02}$$
$$\tan\ \alpha_{02} = \tan\ \alpha_{n2} / \cos \gamma_{x02},$$

where:

$$z_{n2} = (x_{n2} - r_F + x_f)\ \cot\ \alpha_{n2} - l_f$$
$$\gamma_{x02} = \text{atan}\ (p/x_{02})$$

The found parameters of the axial section of the helical surface of the cutting edges of the hob teeth are also parameters of the axial section of the relieved surfaces of the teeth.

For clarity, let us consider Fig. 7.2, which shows an example of the basic rack for gears with the module $m_n = 12$ mm.

For full-profile Novikov gears, the pitch line of the basic rack divides the tooth height in half. The active parts of the hob tooth profile are limited by the dimensions 10.2 and 10.32 mm from the pitch line. At the same points, there is a joint of the radius sections of the profile. The tooth thickness in the normal section at the minimum machining diameter is $b_n = 31.7$ mm.

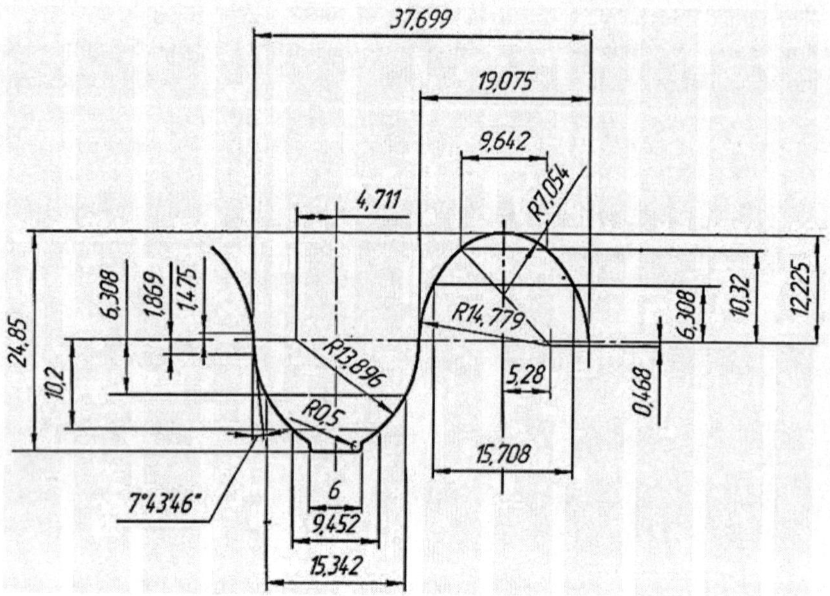

Fig. 7.2 An example of the basic rack of the hob for gears with the module $m_n = 12$ mm

The "predesigned parameters" of the worm cutter are as follows: the outer diameter is 180 mm; the number of hob threads $z_0 = 1$; the number of teeth in the face section $z_f = 10$; the pitch diameter is 155.55 or its radius $r_F = 77.775$ mm; and the fall of the relieving cam on the outer cylinder $K = 10$ mm.

The "calculated parameters" are as follows: the axial module $m = 12.038$; the screw parameter of the generating worm $p = 6.019$; the lead angle on the pitch cylinder $\gamma_F = 4.425°$; the parameter of relieving $k = K\, z_f/2\pi = 15.915$; the minimum radius of grinding on the cylinder of hob spaces $r_{min} = 66.215$; the lead angle on this cylinder is 5.194°; and the maximum tooth thickness $b = 31.83$.

The parameters of the axial profile of the hob generating surface are as follows: the minimum pressure angle on the straight section $\alpha_{1\ min} = 7.7523°$; the concave arc radius on the tooth head $r_{0f} = 14.823$; the convex arc radius in the space $r_{0a} = 13.938$; and the radii of transitive curves on the tooth head $r_{i\ f} = 7.096$ and in the space $r_{i\ a} = 7.002$. The pressure angles at special points of the work area are as follows:

- At a point of conjugation of the radii r_{0f} and r_{if} on the head of the tooth $\alpha_{0f} = 46.95°$
- At the contact point on the head of the tooth $\alpha_{01} = 27.35°$
- At a point of conjugation of the radii r_{0a} and r_{ia} in the space $\alpha_{0a} = 47.34°$
- At the contact point in the space $\alpha_{02} = 27.08°$

7.4 Parameters of Radial–Axial Relieving

The hobs for cutting the involute gears have an almost straight tooth profile with the pressure angle $\alpha_0 \approx 20°$. The relieving of such hobs is usually carried out by the radial method.

The tooth of hobs for cutting Novikov gears has a substantially curved profile with a large difference in pressure angles. The pressure angle in the straight section of the tooth profile near the line is $\alpha_0 \approx 8°$, and, therefore, the rear angles in this section are minimal, which, during cutting, significantly reduces the wear resistance of the cutters as a whole, since it is precisely in this section that they are intensively formed wear sites [12]. In addition to the arc sections with large pressure angles, a large shift of the profile relief surface relative to the pitch line will occur, which will appear at re-sharpening of the hob. So, such hobs cannot be relieved with only a purely radial method, and, therefore, radial–axial relieving must be used.

In the general case of radial–axial relieving of the teeth of the worm hob [1–3], the direction of relieving motion is perpendicular to the hob axis an angle φ_c, which ensures obtaining a necessary rear angle λ_6 near the cutting edges of the hob teeth.

The grinding wheel axis O_w–O_w is turned (in the projection to the horizontal plane) at an angle φ_0 to the hob axis O_1–O_1 (see Fig. 7.5) and is simultaneously inclined to this plane by the angle β_w. Such a method of setting the axis of the grinding wheel eliminates the large differences in the curvature of the grinding surface of the wheel, improves the grinding conditions, and increases the life of the grinding wheel.

During the relieving, the hob rotates around its axis O_1–O_1 with the angular velocity ω_1, the grinding wheel moves along the hob axis at a speed $p\omega_1$ and performs reciprocating motion with a speed $k\omega_1$, where $k = K\, z_f/2\pi$ is the relieving parameter and K is the recession of the Archimedean spiral of the cam of the relieving mechanism on the angular pitch of the hob teeth. The relieving carriage slides are turned perpendicular to the hob axis by an angle φ_c, which allows increasing the rear angle λ_b on the lateral surface of the tooth.

The current radius r_w of the grinding wheel is determined at the point of its contact with the tooth of the hob on its current cylinder. For the initial position of the wheel, it is advisable to take the tangency of its maximum radius r_{wm} with the minimum radius of the work piece surface $r_{i\ min}$.

With radial–axial relieving, the lead angle γ_0 of the relieved surface at each of the two selected points ($i_{1,2}$) is determined from the expression [1, 2]:

$$\tan \gamma_{0i} = [p \pm k \sin (\alpha_i + \varphi_c)/\cos \alpha_i]/x_{0i} \tag{7.6}$$

where the radius x_{0i} of the location of the selected point on the ground surface is determined from the drawing of the product.

The sign "+" in square brackets refers to the teeth side where the lead angle of the relieved surface is greater than the lead angle of the generating surface of the hob, sign "–", respectively, to the opposite side of the teeth.

From formula Eq. (7.6), it follows that the values of the angles γ_{0i} depend on the angle φ_c of the installation of the relieving support, which is determined by the required rear angle near the cutting edge of the tooth. For hobs with variable profile curvatures, this angle is determined in the area with the smallest value of the angle α_{0i}, that is, in the considered example for a straight-line segment near the pitch line, where $\alpha_{0\ \min} = 7.7523°$.

7.5 Calculation of the Angle φ_c Installation of the Relieving Support of the Machine

For worm hobs intended for finish cutting of involute gears with module 12 mm of the thermally hardened steel, the Russian State Standard 9324-80 recommends parameters of radial relieving providing a normal relief angle of ≈4° on a pitch cylinder. Based on this recommendation, for the considered type of hob, we establish the following dependence, which ensures the determination of φ_c with sufficient accuracy for practical calculations:

$$(\tan \gamma_{0F} - \tan \gamma_{xF}) \cos \alpha_{0\min} = \tan 4° \tag{7.7}$$

Expanding the meanings of $\tan \gamma_{0F}$ from Eq. (7.6) and $\tan \gamma_{xF} = p/r_F$, we obtain the following expression for determining the value of φ_c:

$$\sin\left(\alpha_{0\min} + \varphi_{cR,L}\right) = 0.44\, r_F / \left(K z_\Phi\right) \tag{7.8}$$

The subscript "$_{R,L}$" relates to the right and left sides of the hob tooth. For the example considered above with $\alpha_{0\ \min} = 7.7523°$, we obtain $\varphi_{cR,L} = 12.229°$, taking into account the price of the rotation angle scale that we set according to the rounding rules. $\varphi_{cR,L} = 12°$.

Table 7.1 shows the calculated parameters of the profile of the hob tooth and the lead angles of the relieved surface of the teeth at seven key points of the profile $i = 1 \ldots 7$:

1. The boundary of the active section of the profile on the tooth leg
2. The point of contact on the leg of the hob tooth
3. The conjugation of the arc concave section with a rectilinear section
4. The point on the hob pitch cylinder
5. The conjugation of the rectilinear and arc sections on the tooth head
6. The point of contact on the head of the hob tooth
7. The boundary of the active section of the profile on the tooth head

In Figs. 7.3 and 7.4, the graphs of the dependence of the lead angles γ_{0i} of the relieved surface on the radius $x_{0i} = r_i$ of the profile point, taking into account the pressure angle α_{0i} at $\varphi_c = 12°$, are shown for both sides of the tooth.

Table 7.1 Parameters of hob tooth profile in key points (for the given example)

Numbers of profile points	Radius of point, mm $x_i = r_i$	Pressure angle in axial section, α_{0i}°	Lead angle of screw surface of cutting edges, γ_i°	Lead angles of relieved surface on the tooth sides:	
				Right, γ_{0iR}°	Left, γ_{0iL}°
1	67.575	47.338	5.09	21.207	−11.853
2	71.467	27.079	4.814	13.598	−4.200
3	75.906	7.752	4.587	8.576	0.446
4	77.775	7.752	4.425	8.373	0.435
5	79.25	7.752	4.343	8.219	0.427
6	84.083	27.35	4.094	11.679	−3.635
7	88.095	46.951	3.909	16.439	-9.001

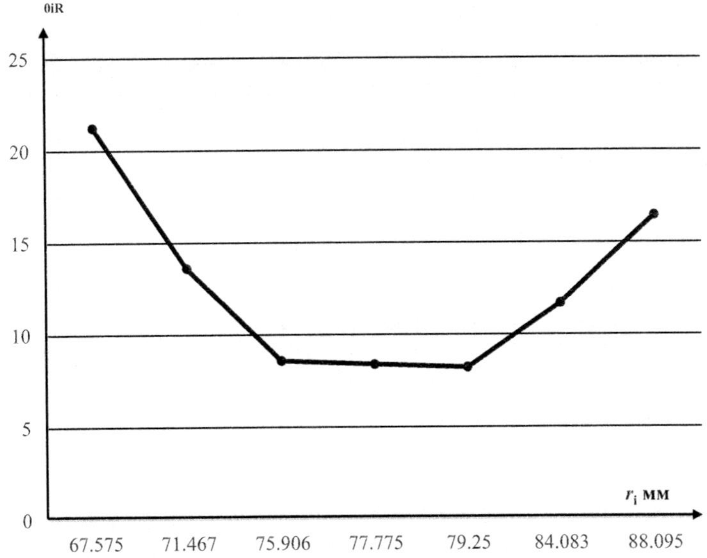

Fig. 7.3 Graph of the function $\gamma_{0iR} = f(r_i)$ – right tooth side

Analysis of the calculated data shows that the lateral surfaces of the teeth of such a hob are organically characterized by the presence of paired points with equal lead angles of the relieved surface, but the values of these angles differ significantly from the lead angle of the helical generating surface. In this regard, still one task when choosing the relieving parameters is to exclude pruning of the cutting edges of the teeth.

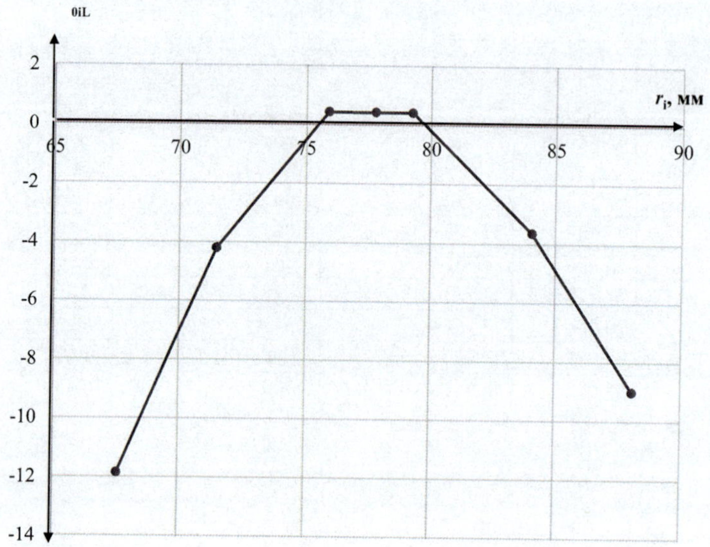

Fig. 7.4 Graph of the function $\gamma_{0iR} = f(r_i)$ – left tooth side

7.6 Choice of the Designed Points to Construct Designed Normals

The choice of two points of the axial profile of the relieved surface with the parameters $A_1(x_{01}, z_{01}, \alpha_{01})$ and $A_2(x_{02}, z_{02}, \alpha_{02})$ must satisfy two conditions. First, as much as possible, the part between the selected points should cover the active part of the tooth profile, including contact points ($\alpha_0 \approx 27°$). Second, undercutting of the tooth cutting edges should be excluded.

To do this, on the right side of the teeth of the hob with right-hand threads, the angle β_w of setting the axis of the grinding wheel in a plane parallel to the hob axis should not significantly exceed the maximum lead angle of the generating surface. When relieving of the opposite side of the tooth, the angle β_w should not be substantially less than the lead angle of the helical surface of the cutting edges on the tooth head.

Point A_1 should be selected on the tooth leg of the hob and point A_2 on its head. A study conducted with concrete examples (see the calculation example below) showed that these conditions are most fully satisfied if, as point A_1, we select the design contact point in the tooth space on a concave arc at $\alpha_{01} \approx 27°$ (point 2 in the Table 7.1). As point A_2, we should choose the conjugation of the convex arc section with the transition curve on the tooth head at $\alpha_{02} \approx 47°$ (point 7 in Table 7.1). In this case, the angle β_w of the inclination of the wheel axis is less than the smaller of the values of the angle γ_{0i} at the selected design points.

7.7 Determination of the Angle β_w of Inclination of the Grinding Wheel Axis

The design scheme for determining the angle β_w of setting the axis of the grinding wheel for radial–axial relieving of surfaces with a variable curvature profile is shown in Fig. 7.5.

In the coordinate system XYZ (see Fig. 7.5), the Z-axis is directed along the axis of the hob, the X-axis is aligned with the axis of symmetry of the thread, XZ is the plane of the axial section of the hob, and the Y-axis is perpendicular to this section. The plane of the axis of the grinding wheel is parallel to the Y-axis and intersects the X-axis at a distance a to the Z-axis under an angle $\varphi_0 \neq 0$.

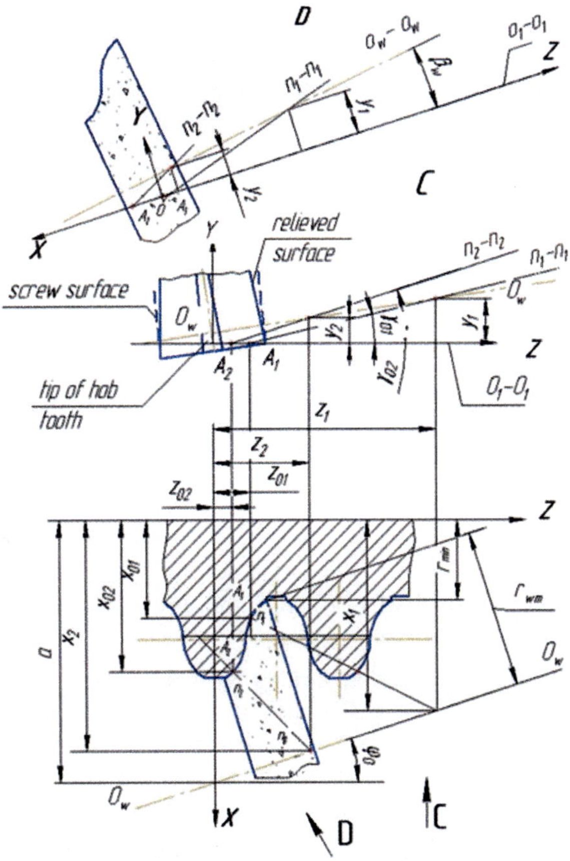

Fig. 7.5 The design scheme for determining the angle β_w of the installation of the axis of the grinding wheel for radially axial occultation of surfaces with a profile of variable curvature

As shown in the lower projection of Fig. 7.5, the value of distance a is determined as:

$$a = r_{i\ \min} + 0.5\mathrm{b}\tan\varphi_0 + r_{\text{нм}}/\cos\varphi_0 \tag{7.9}$$

where b is the thickness of the worm thread being ground with a minimum radius $r_{i\ \min}$.

The equation of the plane containing the axis of the grinding wheel is written in the form:

$$x(z) = a - z\tan\varphi_0 \tag{7.10}$$

The coordinate z_i of the intersection of normals restored from points A_1 and A_2 with the plane of the axis of the grinding wheel is:

$$z_{i\,(1,\,2)} = (a - x_{0i} + z_{0i}\tan\alpha_i)/(\tan\alpha_i + \tan\varphi_0) \tag{7.11}$$

The two other coordinates $x_{i(1.2)}$ and $y_{i(1.2)}$ of each of the intersection points of the normals with this plane are determined through z_i. When determining the coordinate $y_{i(1.2)}$ we should consider the frontal projection (view C):

$$x_{i\,(1.2)} = a - z_i\tan\varphi_0;\ y_{i\,(1.2)} = (z_i - z_{0i})\tan\gamma_{0i} \tag{7.12}$$

The angle β of the inclination of the axis of the grinding wheel is determined in the tangent function (from top view on the plane of the axis of the grinding wheel) by the formula:

$$\tan\beta_{\mathrm{w}} = \frac{y_1 - y_2}{\sqrt{(x_1 - x_2)^2 + (z_1 - z_2)^2}} \tag{7.13}$$

7.8 Choice of the Angle φ_0 for Relieving the Tooth Flanks

As mentioned above, turning the hob axis by the angle φ_0 improves grinding conditions by increasing the curvature of the surface of the wheel at the points of contact with a processed surface and reduces the difference between the angle β_{w} of the inclination of the wheel axis and the lead angles of the helical line of the cutting edges of the hob teeth. The choice of the angle φ_0 is determined by two restrictions. On the one hand, it is necessary to minimize the mentioned difference of angles β_{w} and γ_{0i}. On the other hand, the value φ_0 should not lead to pruning of the cutting edges of the teeth on the opposite side of the thread of the generating surface, i.e., the grinding wheel should enter the space without damaging the adjacent tooth.

Now, we consider the issue of pruning prevention in more detail. The minimum distance between the back (flat) side of the grinding wheel and the opposite side of the thread of the generating surface takes place opposite to the most protruding point on the head of the hob tooth. Such a point is a contact point on the tooth head with a pressure angle $\alpha_{0i} \approx 27°$.

Due to the difference between the angle β_w of the wheel inclination and the lead angle γ_i of the generating surfaces, the dangerous zone of approach of the surfaces is shifted from the horizontal plane (down when processing the right side of the tooth). The dangerous approach reaches its maximum value at the extreme point of intersection of the face sections of the outer wheel cylinder (radius r_{wm}) and the cylinder of the hob radius ($r_F + h_k$) at the contact point on the tooth head. For simple geometric reasons, the distance from this point to the horizontal plane is determined by the formula:

$$l_{\max} = r_{wm} \sin \psi, \tag{7.14}$$

where:

$$\psi = a\cos\left(\frac{r_{wm}^2 + (r_{wm} + r_{\min})^2 - (r_F + h_k)^2}{2r_{wm}(r_{wm} + r_{\min})}\right) \tag{7.15}$$

The linear displacement Δ of the back surface of the grinding wheel to the middle of the space at a length l_{max} in a screw projection onto the axial plane of the cutter to both tooth sides will be:

$$\Delta_{R,L} \approx \pm l_{\max}\left[\tan \beta_{wR,L} \cos \varphi_0 - p/(r_F + h_k)\right] \tag{7.16}$$

where the sign "+" relates to the right side of the right-hand hob and the sign "−" to its left side.

In the angular dimension, this displacement is determined from the expression:

$$\tan \varphi_\beta = \Delta/(r_F + h_k) \tag{7.17}$$

The pruning of the cutting edges of the opposite side of the thread is guaranteed to be excluded to satisfy the condition $\varphi_0 + \varphi_\beta \leq \alpha_{0i}$, where the value α_{0i} is accepted for the key point 6 from Table 7.1. For a preliminary calculation of the relieving parameters, the angle φ_0 should be considered as $\varphi_0 = 18°$. Then, having determined the angle φ_β by formula Eq. (7.17), it is possible to correct the angle φ_0 by increasing it to the maximum permissible value:

$$\varphi_{0\max} = \alpha_{0i} - \varphi_\beta. \tag{7.18}$$

Given that increasing φ_0 reduces the desired value of the angle β, such an adjustment is advisable.

As an example, we will calculate the relieving parameters β_w and φ_0 as applied to the hob considered above and study the features of the resulting side surfaces.

As the initial points for constructing the designed normals, we take the points: on the leg A_1 at the contact point with the pressure angle $\alpha_{01} \approx 27°$ (key point 2 from Table 7.1) and, on head A_2 at the conjugation between the active profile and the transition curve with the angle $\alpha_{02} \approx 47°$ (key point 7). For the coordinates of these points and the exact values of the pressure angles, see Table 7.1.

The maximum radius of the grinding wheel is assumed to be $r_{wm} = 60$ mm, the minimum radius of the relieved surface on the space $r_{min} = 66.215$ mm, and the height of the grinding profile of the hob tooth $H_1 = 23.785$ mm, for both sides of the teeth of the manufacturing surface. The value of the turning angle of the relieving support to provide a rear angle in the middle section of the profile $\varphi_c = 12°$ for both sides of the teeth. The angle φ_0 is preliminarily considered to be equal to 18°.

The lead angles γ_{0i} of the relieved surface at points A_1 and A_2 are shown in Table 7.1 for each side of the tooth of the right-hand hob. On the right side of the tooth, at which the lead angle of the relieved surface is greater than that of the generating worm, they are $\gamma_{01} = 13{,}598°$ and $\gamma_{02} = 6439°$. On the opposite left tooth side, correspondingly, we have $\gamma_{01} = -4200°$ and $\gamma_{02} = -9001°$.

The negative values of the lead angles γ_{0i} of the relieved surface at the points of the arc sections of the profile indicate its transformation in these sections into an analogue of the left helical surface, which is a characteristic for single-thread hobs with a standard lateral rear angle of 4° or more, with the number of modules in the pitch diameter of 12 or more.

Table 7.2 summarizes the main stages of the calculations with a mention of the needed formulas.

For the right side, the difference between the values of the initial φ_0 and the adjusted φ_{0max} angles in the preliminary calculation is less than 1°, and, therefore, a change in φ_0 is not required. For the left side, the final calculation is carried out by increasing the angle φ_0 to 19° and recalculating β_w, which has changed very slightly and, in practice, in both cases, is rounded to the value −0.1°.

In the studies by Lagutin and Sandler [1] and Sandler et al. [2], it was shown that the discrepancy between the profiles of the grinding wheel surface and the hob tooth flank to be ground will be minimal if in the vicinity of the calculated contact point, the line of their contact will not cross the axial section of the hob but touch it.

This condition is satisfied in the case when the axis of the grinding wheel is inclined to the axial plane at an angle β_w, which is determined depending on the pressure angle α_0, the lead angle γ_{0i}, and the radius r_i of the relieved surface, taking into account the installation angle φ_0 of the wheel axis in the projection onto the horizontal plane, by the expression:

$$\tan \beta_w = \tan \gamma_{0i} \cos \alpha_0 / \cos (\alpha_0 + \varphi_0) \tag{7.19}$$

Table 7.2 Calculation of the parameters β_w and φ_0 of the hob teeth

		Estimated values		
		Preliminary		Final
Data	Formula number	Right	Left tooth side	
φ_0°	Entering values	18.0	18.0	19.0
a	(9)	134.47	134.47	135.15
x_1	(12)	107.76	107.76	107.21
z_1	(11)	82.21	82.21	81.14
y_1	(12)	17.17	−5.17	−5.13
x_2	(12)	122.47	122.47	122.44
z_2	(11)	36.94	36.94	36.91
y_2	(12)	9.47	−5.09	−5.08
β_w°	(13)	9.19	−0.095	−0.064
ψ°	(15)	34.64	34.64	34.64
l_{max}	(14)	34.11	34.11	34.11
Δ	(16)	2.81	2.49	2.47
φ_β°	(17)	8.93	7.95	7.89
φ_{0max}°	(18)	18.42	19.40	19.46

Table 7.3 Values of the angle β_w calculated for the different sides of the hob tooth

Point numbers	1	2	3	4	5	6	7
x_{0i}, mm	67.575	71.467	75.906	77.775	79.25	84.083	88.095
α_{0i}°	47.34	27.08	7.75	7.75	7.75	27.35	46.95
γ_{0i}° (right side)	21.207	13.598	8.576	8.373	8.219	11.679	16.439
β_w° (right side)	32.22	16.96	9.42	9.198	9.029	16.641	25.441
γ_{0i}° (left side)	−11.853	−4.20	0.446	0.435	0.427	−3.625	−9.001
β_w° (left side)	−18.882	−5.29	0.491	0.479	0.470	−4.590	−14.325

Table 7.3 shows the values of the angle β_w calculated according to this dependence for seven key points of the axial section of the right and left sides of the hob teeth:

In Fig. 7.6 (the upper part), a graphical interpretation of the dependence Eq. (7.19) of the angle β_w on the hob radius r_1 is displayed for the right side of the teeth. Let us draw two boundary straight lines: one through the starting points A_1 and A_2 of the designed normals and the second horizontal line at a minimum value $\beta_w = 9.19°$ (from Table 7.3) until they mutually intersect at some point C.

The sections of the profile that fell into the sector between the boundary lines have practically no organic error, since any straight line drawn through the intersection point C inside the sector also passes through two points of the axial profile of the relieved surface. Outside the obtained sector, the relieved profile has a small organic deviation from the given profile of the grinding wheel.

In the lower part of Fig. 7.6, a similar dependence of the angle β_w on the hob radius r_1 is shown for the left side of the teeth. We also draw two boundary straight lines. One of them passes through the points of the graph corresponding to the values

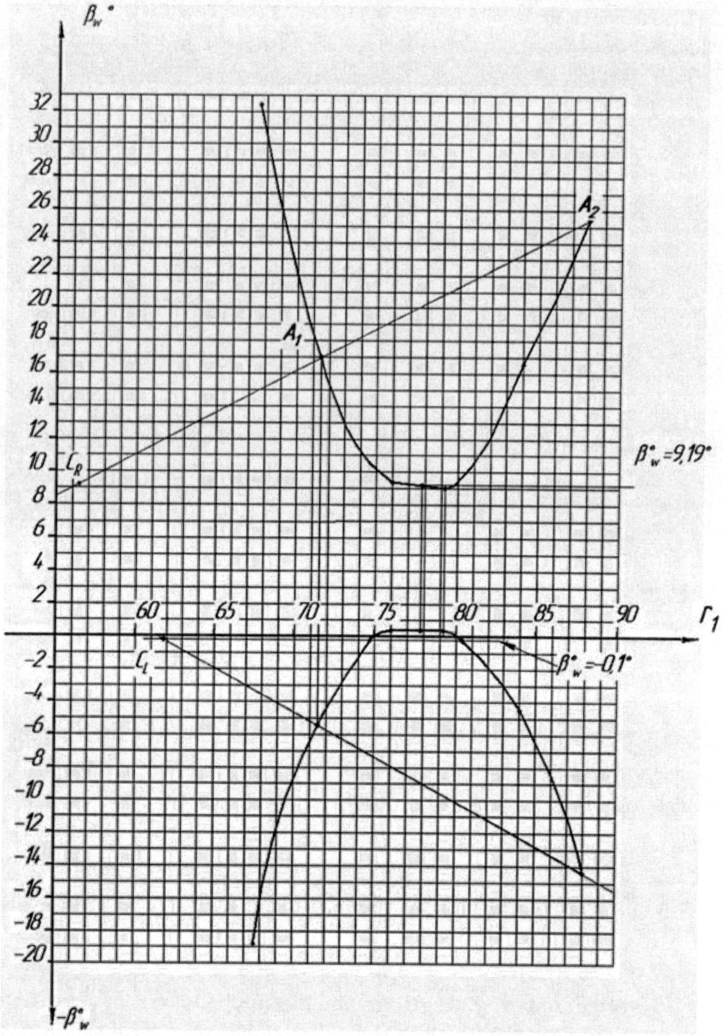

Fig. 7.6 The dependence of the angles β_w from the current radius r_1 of the hob

of the radius at points A_1 and A_2. The second line runs horizontally, touching the graph at its upper point with a value of $\beta_w = -0.1°$. The intersection of these lines forms a sector in which there is practically no organic deviation of the relieved profile.

7.9 Profiling of Grinding Wheels

Significant difficulties in the manufacture of worms and worm hobs with a profile of substantial and variable curvature are due to the fact that the wheel dressing copiers included in the delivery of grinding machines are not suitable for reproducing such a profile [1, 2]. Therefore, it is necessary to design a special knurling roller or create a special ruling device. At present, with the development of computer numerical control (CNC), ruling devices with diamond rollers and programmed control of their movements relative to the working surface of the grinding wheel have gained a certain distribution.

The characteristic of the surface of the grinding wheel, i.e., the line of its contact with the surface to be ground, is oriented along the axial section of the latter. Therefore, profiling of the wheel in its axial section with a sufficient degree of approximation reproduces the section of the surface being ground that is normal to a helical line with an inclination angle equal to β_w. That is, the axial profile of the grinding wheel for single-thread hobs practically repeats the profile of the basic rack. In this case, the value of the angle α_w of the wheel profile in a rectilinear section is determined from the expression:

$$\tan \alpha_{\mathrm{w}} = \tan\left(\alpha_{0\ \min} + \varphi_0\right)\cos\beta_{\mathrm{w}} \tag{7.20}$$

In general, the coordinates of the axial profile of the grinding wheel, based on the coordinates of the axial section of the relieved surface of the hob teeth, the lead angles of this surface, and the found angle of inclination of the wheel axis, are determined by the following expressions:

$$\begin{aligned} z_{\mathrm{w}} &= z_{0i}/\left(\cos\beta_{\mathrm{w}} + \tan\gamma_{0i}\sin\beta_{\mathrm{w}}\right) \\ x_{\mathrm{w}} &= x_{0i} + z_{\mathrm{n}}^2\sin 2\beta_{\mathrm{w}}\tan\gamma_{0i}/(4x_{0i}) \end{aligned} \tag{7.21}$$

7.10 Sensitivity to the Hob Regrinding

If the relieving of surfaces with a variable curvature profile is performed by the radial–axial method, then the axial component of the relieving parameter $k_x = k \sin(\alpha_{0i} + \varphi_c)/\cos\alpha_{0i}$ included in formula Eq. (7.6) is variable along the profile. In this case, strictly speaking, the profile distortion that occurs during regrinding of the hob teeth on the front surface is inevitable. However, the considered profile has points with identical pressure angles α_{0i} (for example, at the conjugate points of large and small radii on the tooth's head and leg) and the lead angles γ_{0i} of the relieved surface, and, therefore, at these points and in the areas between them, the profile deviations during regrinding are minimal.

7.11 Conclusion

1. The technology of relieving of the tooth flanks of the worm hobs for cutting gears with Novikov gearing should provide the necessary rear angles of the hob teeth and the quality of the implementation of the basic rack of gears by the generating worm of the hob.
2. Changing the lead angle of the relieved surface of the tooth hobs with a basic rack for Novikov gears with two lines of action has a parabolic dependence on the radius and pressure angle of the tooth profile, and the relieved tooth surfaces of such hobs are organically characterized by the presence of paired profile points with equal lead angles.
3. A technique has been developed for calculating the parameters of installation, movement, and profile of the grinding disk wheel when grinding the tooth flanks, which allows obtaining at least four contact points between the wheel and the grinding surface in the axial section of the hob. Thus, the organic error of grinding when profiling the grinding wheel is minimized, as close as possible to the basic rack profile.
4. It was determined that the minimum rear angle of the hob teeth takes place near the hob pitch cylinder and that the installation angle φ_c of the relieving support of the grinding-relieving machine is necessary to calculate based on the minimum value of the rear angle of 4°. The technique for calculating this angle is provided.
5. It was revealed that when grinding both the right and left sides of the teeth, it is preferable to assign the angle φ_0 of the installation of the axis of the grinding wheel to 18° and clarify it from the condition to prevent pruning of the cutting edges on the opposite side of the space between the teeth. The technique for refining this angle is provided.
6. When relieving the left side of the teeth of the right-hand cutter, the directions of the lead angles of the grinding surface and the helical surface of the cutting edges have different signs. Accordingly, the angle β_w of inclination of the axis of the grinding wheel to the horizon can also have a sign opposite to the lead angle γ_1 of the generating surface of the hob.
7. The above technique is also proposed for use in the development of control programs for setting up a grinding-relieving machine and the dressing mechanism of a grinding wheel in the manufacture of worm hobs on relieving machines with numerical program control.

References

1. Lagutin, S.A., Sandler, A.I.: Grinding of Helical and Relieved Surfaces, 110 p. Publishing House "Mashinostroyeniye" (in Russian) (1991)
2. Sandler, A.I., Lagutin, S.A., Verhovski, A.V.: Manufacturing of Worm Gears, 272 p. Publishing House "Mashinostroyeniye" (in Russian) (2008)

3. Sandler, A.I., Lagutin, S.A., Gudov, E.A.: Theory and Practice of Manufacturing of General Type Worm Gears, 346 p. Publishing House "Infra-Engineering" Moscow-Vologda (in Russian) (2016)
4. Litvin, F.L., Fuentes, A.: Gear Geometry and Applied Theory, 800 p. Cambridge University Press (2004)
5. Sandler, A.I., Lagutin, S.A.: Features of forming screw and relief surfaces for fluid friction worm gears. In: International Symposium on Theory and Practice of Gearing, Izhevsk, Russia, pp. 375–380 (2014)
6. Sandler, A.I., Lagutin, S.A.: Relieving the teeth of hobs for cutting the Wildhaber-Novikov gears. In: The 10th International Conference KOD 2018 "Machine and Industrial Design in Mechanical Engineering", 6–8 June 2018, Novi Sad, Serbia IOP Conference Series: Materials Science and Engineering, vol. 393, conference 1. Published under licence by IOP Publishing Ltd (2018)
7. Novikov, M.L.: Gears with New Engagement. Publ. VVIA n.a. Zhukovsky (in Russian) (1958)
8. Radzevich, S.P.: High-conformal gearing: a new look at the concept of Novikov gearing. In: Proceedings of International Conference on Gears 2015, Technical University of Munich (TUM), Garching (near Munich), Germany, 5–7 October 2015, pp. 457–470
9. Radzevich, S.P.: About the priority of M.L. Novikov in the development of "Novikov Gearing". Bull. Sci. Tech. Dev., www.vntr.ru. **11**(159) (2020). https://doi.org/10.18411/vntr2020-159-1
10. Kudryavtsev, V.N.: On the issue of Novikov gearing. In: Research and Mastering of Gears with Novikov Gearing, pp. 33–44. Publ. House of the USSR Academy of Sciences (in Russian) (1960)
11. Korotkin, V.I., Onishkov, N.P., Kharitonov Yu, D.: Novikov Gearing: Achievements and Development. Nova Science Publishers (2011)
12. Stebletsov, J.N., Tarapanov, A.S.: Wear resistance of the cutting tool when processing Novikov gears. In: Fundamental and Applied Problems of Engineering and Technology, vol. 4/3, pp. 83–86. Orel: State University – UNPK (2011)

Chapter 8
Mikhail L. Novikov: The Inventor of Novikov Gear System

8.1 Introduction

Novikov gearing has long since firmly entered the practice of mechanical engineering. It has proven itself well in cases in which relatively soft gears with a hardness of the working surfaces of the teeth up to HB 350 are used to transmit rotation. The situation is not so good with the use of Novikov gearing with hardened gears. The existing difficulties with the use of Novikov gears with harder wheels are a consequence of both an insufficiently deeply understood essence of this type of a gear and an unacceptably poor technology for the manufacture of gears. An example of the successful use of Novikov gear transmission in the design of a helicopter transmission from Westland Helicopters, Ltd. is shown in Fig. 8.1.

A new type of gearing, developed by Dr. M.L. Novikov, is often called Wildhaber–Novikov gearing, or W-N gearing, which is incorrect. The author adheres to the point of view that the name Wildhaber–Novikov gearing is incorrect. Novikov gearing is the only correct name for this type of gearing. Below, this point of view is substantiated on the basis of both materials published in an open domain and materials that became known to the author from conversations with specialists who personally knew Dr. M.L. Novikov and Dr. E. Wildhaber.

8.2 Novikov Gear System

Novikov gearing was developed by Dr. M.L. Novikov and systematically presented in his doctoral dissertation [8]. As the author pointed out, the type of gearing proposed by him can be used to design gears "with parallel, intersecting and crossing axes, with external and internal gearing, with constant and variable gear ratios, . . ." [11].

S. P. Radzevich, *Novikov/Conformal Gearing*,
https://doi.org/10.1007/978-3-031-10019-2_8

Fig. 8.1 Use of Novikov gearing in the design of helicopter transmission (Westland Helicopters, Ltd.)

It is appropriate to start the discussion of Novikov gearing with a brief biographical note of the inventor, especially since his 110th anniversary will be celebrated very soon.

8.2.1 A Brief Biographical Sketch of Dr. M.L. Novikov

About Mikhail Leontievich Novikov, Colonel-Engineer, Doctor of Technical Sciences, Professor, Former Head of the Department at the Zhukovsky Military Aviation Engineering Academy (MAEA), winner of the Lenin Prize (1959), most readers know little. Therefore, a summary of the available facts from his biography is of interest, especially for those who are beginning their path in science.

Mikhail Leontievich Novikov was born on March 25, 1915 in the city of Ivanovo into a working-class family [9]. He began working as a locksmith apprentice at the age of 15. In 1934, M.L. Novikov entered the first year of the MVTU – Bauman State Technical University. After completing his second year at the MVTU, he was enrolled as a student of the Zhukovsky Military Aviation Engineering Academy (MAEA), after graduating from which (in 1940) he left to work at one of the leading special departments. In a short time, M.L. Novikov worked his way up from being a junior teacher to the head of this department.

Along with the educational and methodological work assigned by his position, Mikhail Leontievich was engaged in scientific research and invention. More than ten of his inventions are important for the development of aviation technology.

Mikhail Leontievich Novikov (1915–1957)

Novikov, M.L. paid attention to the problem of increasing the bearing capacity of gears. This research topic was of great importance, both for aircraft engine manufacturers and for mechanical engineering as a whole. The result of many years of hard work in this direction was the development of a new type of gearing by M.L. Novikov. The developed type of gearing is defended by a certificate on invention (that is an equivalent of a patent on invention) [11].

Finally, for the development of a new type of gearing, M.L. Novikov was awarded the Lenin Prize[1].

Despite his great fame and wide popularity, Dr. M.L. Novikov remained a modest man. Here is an example from the memoirs shared by Dr. A.S. Yakovlev, (city of Oryol, Russia), who personally knew Dr. M.L. Novikov: "Mikhail Leontyevich was possessed, infinitely devoted to science, and above all a modest, noble and attentive person and an objective interlocutor, without a shadow of arrogance and ambition. Here are some typical examples. Arriving for the first time at MAEA, I brought two variants of the factory transmission with Novikov gearing, and one of the variants was DLZ (which was not known from his dissertation). On the gears of both variants, I calculated the common normals, having seen which Mikhail Leontyevich said that there was no common normal in his gearing. I stood my ground and argued the existence of the normal. His assistants who were present at the same time, amicably and noisily began to prove the absence of a normal.

Mikhail Leontyevich calmly said to me, 'Well, okay, prove it.' After a short discussion, Mikhail Leontyevich thought about it, called the audience to silence and said: 'He is right, the common normal exists.' He said calmly and everyone agreed.

[1] According to Ms. I.M. Averina, Dr. Novikov's daughter, in 1957, The Lenin's Prize Council considered and rejected Novikov's project: 11 votes against, 10 in favor, 0 to reject. In 1959, when the application of Novikov gearing was going on at many factories of the Soviet Union, the decision was revised.

And further. During one of our conversations, the secretary came up to him and said that he was urgently summoned by the Lieutenant General, the head of the Academy. Mikhail Leontyevich told me: 'Sorry, Anatoly Sergeevich, I will try to free myself as soon as possible' and, addressing the assistants: 'Roman Vasilyevich, Viktor Alekseevich, please take Anatoly Sergeevich.' He, at that time the head of the department, Colonel MAEA, Doctor of Technical Sciences, Professor, and I am just a young cocky engineer! Have you met many leaders who treat their interlocutors so many ranks lower?"

Novikov, M.L. suddenly died on August 19, 1957, at the age of only 42 years.

8.2.2 *Kinematics and Geometry of Novikov Gearing*

The concept and theory of a new type of gearing was presented by Dr. M.L. Novikov in his doctoral dissertation [8] and defended by the certificate on invention (the patent on the invention) of the USSR [11]. A monograph [9] was subsequently published based on the materials of the dissertation (without the direct participation of M.L. Novikov).

The title page of the first official certificate issued on the Novikov gear system is depicted in Fig. 8.2.

The USSR author's certificate issued for Novikov gearing [11] is illustrated by the diagram of the contact of the convex profile of the pinion tooth with the concave profile of the tooth of the gear (see Fig. 8.3). In the text of the description of the invention, the following explanation is given: "The essence of the invention is as follows.

> In the space in which the axes of rotation of the gears are fixed, the line of action is set in the form of a straight or smooth curve passing in the immediate vicinity of the instantaneous axis of relative rotation-sliding. Along the line of action, the movement of the point of engagement with a constant, or smoothly changing speed is assigned. A moving point of engagement describes contact lines in spaces associated with rotating gears. A number of some surfaces can be drawn through these lines, which can be conjugate working surfaces of the teeth, if they have a common normal at each current position of the point of engagement and the main theorem of engagement is satisfied if the curvatures of the surfaces satisfy a given gear ratio and if, finally, there is there is no mutual intersection of surfaces within their working shifts.
>
> The proposed types of teeth working surfaces fulfil the stated conditions and provide high contact strength of teeth.
>
> In planes perpendicular to the instantaneous axis of relative rotation-sliding, and passing through the current position of the point of engagement, circular arcs are drawn from the centers located on a straight line passing through the point of engagement and the instantaneous axis, and in the immediate vicinity of this axis. Circular arcs can be thought of as tooth profiles.

Fig. 8.2 The first certificate issued on the Novikov gear system. (Courtesy of Ms. I.M. Averina)

A continuous set of profiles for all current positions of the point of engagement forms mating surfaces of the teeth, while the working surface for one of the gears will be convex, and for the other – concave (in the direction perpendicular to the instantaneous axis). In a particular case, the radii of the profiles of the teeth can be the same in magnitude and equal to the distance from the point of engagement to the instantaneous axis.

When designing teeth, preference should be given to point gearing with a small difference in the values of the profile radii.

The profiles of the teeth can have a shape that differs from the circular arc, however, the curves of the profiles of a different shape (passing through the point of engagement) must be inside the mentioned circular profile with the center at a point located on the instantaneous axis (they must go into the "body" of the teeth).

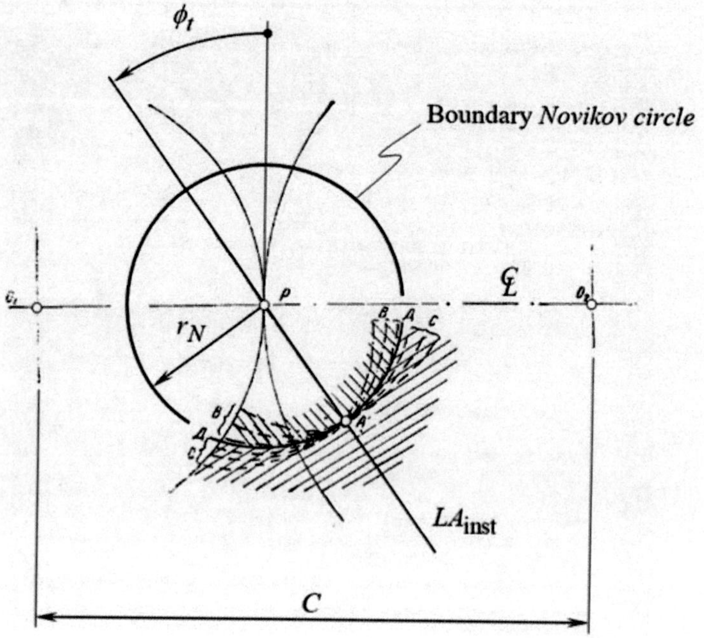

Fig. 8.3 The concept of Novikov gearing (Pat. No. 109,113, 1956 (USSR); the concept of the boundary circle of radius, r_N, was introduced (~2008) by Prof. S.P. Radzevich)

The law of motion of the contact point (namely, the speed and trajectory of its movement) is chosen so that the friction and wear losses are insignificant. Friction and wear losses are proportional to the relative sliding speed in the engagement, therefore, it is necessary to strive to decrease the latter, and for this the engagement line should not be significantly remote from the instantaneous axis. However, an excessive approach of the line of engagement to the instantaneous axis is impractical, since it is associated with a decrease in the contact strength.

The width of the rim of the gear wheels, or the length of the teeth, should be in such a ratio with their pitch, at which the specified overlap ratio would be ensured when the pairs of teeth are re-engaged. The gears can be single-point mesh, that is, only one pair of teeth can participate in the work (except for the re-pairing period), and there can be transmissions with multi-point gearing, when several pairs of teeth are in simultaneous operation.

In the case of gears with parallel axes, it is most convenient, based on design and technological considerations, to choose a line of engagement in the form of a straight line parallel to the axes of the wheels, and the speed of movement of the point of engagement is taken constant. In this case, the radii of the profiles of the teeth in all planes perpendicular to the axes will be the same, the working surfaces of the teeth will be regular helical surfaces.

The description of the invention [11] as well as the diagram illustrating the concept of Novikov gearing are not composed in the best possible manner. It would be better to focus more clearly on the following fundamental differences inherent in Novikov gearing (e.g., a gear train with parallel axes):

(a) Through the pole of engagement P, there is a straight line making up with the center perpendicular O_1O_2 and the face angle of engagement α_t.
(b) At some distance from the pitch point P on the straight line, a point A is chosen.
(c) Through the point A, parallel to the axes O_1 and O_2 (and, therefore, perpendicular to the plane of the drawing in Fig. 8.3), there is a straight line – the line of action.[2]
(d) An arc of some smooth curve is drawn through the point A so that the center of curvature of this curve at the point A is located on a straight line PA.
(e) An arc of some other smooth curve is drawn through the point A so that the center of curvature of this curve at the point A is also located on a straight line PA.
(f) The constructed arcs are used as profiles of the gear and pinion teeth (for this, the curvature of the arcs must exclude the possibility of interference of the teeth both in the differential vicinity of the point A and within the entire height of the gear and pinion teeth).
(g) When transmitting rotation, the contact point A moves along the line of engagement at a constant speed, whereas the projection of the point A in the end plane does not change its position.
(h) The contact ratio m_t in a Novikov gear pair is equal to its axial contact ratio m_F, that is, the relation $m_t = m_F$ holds while the end overlap coefficient is equal to zero ($m_p = 0$) (since there is equality, $m_t = m_p + m_F$; in this case, the equality is simplified to the form $m_p = 0$).

The equality of the component $m_p = 0$ is fundamental to Novikov gearing. If we assume that $m_p \neq 0$, then this completely kills the concept of the gearing developed by M.L. Novikov.

Shown in Fig. 8.4 (see Fig. 62 on page 146 in Novikov [9]) is a more complete (in comparison with Fig. 8.3) idea of the concept of a new type of gearing proposed by Dr. M.L. Novikov.

[2] It was later shown by the author of this chapter [14] that this is not a line of action, as indicated in the description of the invention, but a pseudo-line of action instead. The actual line of action is located in the plane of the drawing in Fig. 8.3 and has zero length.

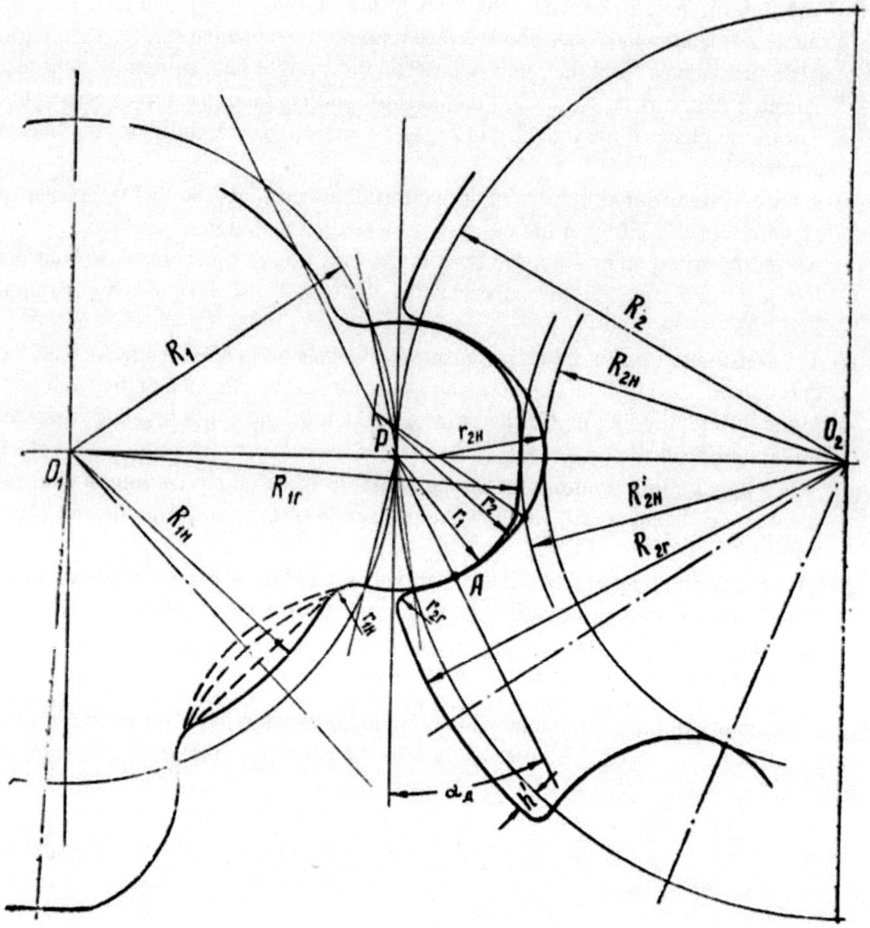

Фиг. 62

Fig. 8.4 Illustration. (Fig. 62 on page 146 in Novikov [9]) explaining the concept of Novikov gearing

8.3 Wildhaber Gearing

Ernst Wildhaber worked on the problem of increasing the bearing capacity of gears. He drew attention to the fact that in the well-known and widely used involute gears, the contacting teeth of the wheels have convex profiles. This contact results in high contact stresses, which limits the power density delivered by the gear train. It was natural to strive to replace the contact of two convex profiles with a contact of profiles, one of which is convex and the other is concave. Ultimately, this led E. Wildhaber to the invention of a new type of gear with a circular tooth profile.[3]

In total, 279 patents for inventions related to the design and manufacture of gear wheels were issued in the name of E. Wildhaber (the total number of patents issued in the name of E. Wildhaber exceeds 320). The design of the hypoid gear proposed by him, as well as the method of cutting spur bevel gears, are known to almost all engineers and researchers working in the field of gear drives.

Wildhaber, E. is also the author of a series of interesting theoretical works[4] published in 1945–46 in the *American Machinist* journal.

8.3.1 A Brief Biographical Sketch of Dr. E. Wildhaber

The information available to the author about Dr. Ernest Wildhaber is extremely scanty. It is known that he was born in 1893 in Switzerland. In 1919, he graduated from the Higher Technical School of the University of Zurich (Technische Hochschule of Zurich University). In 1924, he joined The Gleason Works (Rochester, NY). Many (including the author of this chapter) consider E. Wildhaber to be one of the most prolific specialists who has ever worked in the field of gear design and manufacture of gear wheels.

---

[3] The author of this chapter has strictly proved that the gear transmission proposed by E. Wildhaber is inoperative in principle. It may be that the inventor himself understood this. An indirect confirmation of this point of view is the following.

Suppose that E. Wildhaber, who understood the importance of his invention, succeeded (as he claims in Chironis [1]) to make a workable gear train, made up of wheels with a circular tooth profile. As a talented and advanced engineer, why did he not bring the invention to practical use? Did he not have the opportunity for this? No, he had them, which is convincingly confirmed by the successful implementation of his other less promising inventions. The author of this chapter adheres to the point of view that the only gear pair, which was made by E. Wildhaber, turned out to be inoperative, which cooled the excitement of the inventor. Otherwise, where was he all this time (from 1926 to the late 1950s and early 1960s)? Why he did not raise the question of his importance in the development of a new type of gearing for almost 30 years?

[4] At around 1945–46, E. Wildhaber's theoretical articles were written and published in the *American Machinist* journal, coming out with an interval of about 1 month.

Here, it will not be superfluous to also pay attention to the following interesting fact. In the USSR, great interest was shown in the theoretical works of E. Wildhaber. His articles (1945–46) were translated from English into Russian, after which they were published as a separate book: Wildhaber, E., *Fundamentals of Meshing of Bevel and Hypoid Gears*, Translated from English and the comments of Eng. *A.V. Slepak*. – M.: Mashgiz, 1948, 176 pages. Such a book has never been published in English.

Ernest Wildhaber (1893–1979)

8.3.2 *Kinematics and Geometry of Wildhaber Gearing*

A gear set with parallel axes, for which E. Wildhaber was issued a US patent [10] (see Fig. 8.5), is made up of wheels with a circular tooth profile. In this chapter, there is no need to consider all the details of E. Wildhaber's gear transmission. Let us only dwell on those features of this gear set that are related to Novikov gearing.

At the very end of the first page of the description of the invention, E. Wildhaber states that: "In gearing according to my invention, the contact point between two normal profiles passes over the whole active profile during a turning angle, which corresponds to less than one half the normal pitch, and usually to much less than that ..." The meaning of this statement by E. Wildhaber is that in the process of engagement, the contact point runs through the entire height of the active profile of the teeth. This is confirmed by the analysis of Fig. 8.5, which shows that the contact point moves from position 11 to position $11'$ during engagement. This feature of E. Wildhaber's gear transmission is repeated many times in the description of the invention [10] and is especially reflected in the claims. Thus, it should be concluded that the transverse contact ratio $m_{\rm p}$ in E. Wildhaber's gear transmission is greater than zero ($m_{\rm p} > 0$) and not $m_{\rm p} \equiv 0$ as in Novikov gearing. This difference is fundamental by nature.

8.4 Fundamental Differences Between Novikov Gearing and Wildhaber Gearing

The discussion of this issue should begin with the fact that at first no information about Novikov gearing was published. For a long time, information about this engagement could only be obtained from the source [8]. Serious shortcomings in the preparation of the description of the invention [8] led to a complete

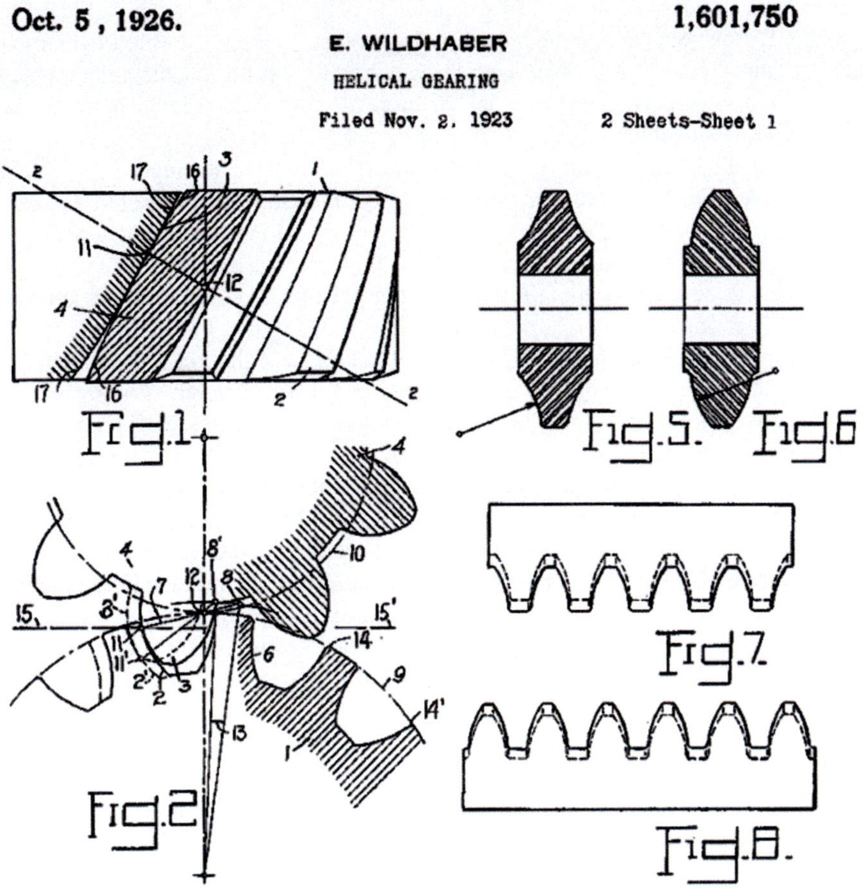

Fig. 8.5 Illustrations from Pat. No. 1.601.750 (USA) [10] explaining the concept of Wildhaber gearing

misunderstanding of the essence of Novikov gearing in the West. In the overwhelming majority of cases, the emphasis, unfortunately, is erroneously placed on the fact that the tooth profiles are circular (and this is not fundamental) while the fundamental differences ($m_p = 0$ and $m_t = m_F > 1$) have faded into the background. This ultimately served as a disservice to the importance of Novikov gearing. In this regard, a single but illustrative example is sufficient. In their study [3], Dyson et al. analyze both Novikov gearing and Wildhaber gearing in sufficient detail. In this case, the authors [3] refer to USSR Pat. No. 109750,[5] which has nothing to do with Novikov's invention (a highly common mistake in the Western literature

[5]Pat. No. 109750, (USSR), *Water spray.*/P.F. Pisulin, Declared January 2, 1957, Cl. 36d 1, 28b, 801.

relating to Novikov gearing). This unambiguously indicates that for a long time, neither Dyson, A. et al. nor the majority of specialists in the West were familiar with the original works of M.L. Novikov; these works were not available and they were forced to use secondary sources of information, the correctness of which, as it turned out, was questionable.

Repeated attempts to understand how different Novikov gearing and Wildhaber gearing are from one another are available in the literature. Chironis [1] came close to understanding this issue. However, he did not reveal the fundamental difference between one program and another.[6]

Here, it is appropriate to return to the above-mentioned characteristic features of Novikov gearing.

Since the point of contact travels along the profile of the teeth of the gear and pinion, the transverse contact ratio m_p in E. Wildhaber's gear transmission is necessarily different from zero ($m_p > 0$). From what has been said, the conclusion unambiguously follows that the existence of a gear train in which the transverse contact ratio m_p (as in Novikov gearing) is necessarily equal to zero ($m_p = 0$) and at the same time (as in Wildhaber gearing) is not equal to zero ($m_p > 0$) is physically impossible. Therefore, the combination of M.L. Novikov's invention with E. Wildhaber's invention is absurd from the outset – it is impossible to combine the non-unified (there cannot be a gear pair that fulfills the conditions $m_p = 0$ (like in Novikov gearing) and $m_p > 0$ (like in Wildhaber gearing) at the same time). It follows that the terms "Wildhaber–Novikov gearing" and "W-N gearing" are incorrect, and, moreover, they are absurd.

The difference between Novikov gearing and Wildhaber gearing has been well-traced from the standpoint of recent advances in the theory of gearing [13]: Novikov gearing fulfills all three fundamental laws of the theory of gearing, whereas Wildhaber gearing does not meet these laws, which is well-illustrated in Fig. 8.6.

In Novikov gearing, the unit normal vector $\mathbf{n}_g$ is always pointed along the instantaneous line of action (in other words, along the instantaneous line of engagement) LA_{inst}, as shown in Fig. 8.6a. In this case, the profile of the tooth of the wheel $\mathcal{G}$ is correctly located in relation to the "boundary $N-$ circle" of radius r_N (in other words, to the "Novikov boundary circle" of radius r_N).

In contrast to Novikov gearing, in Wildhaber gearing, the direction of the unit normal vector $\mathbf{n}_g$ does not coincide with the direction of the instantaneous line of action LA_{inst}, as shown in Fig. 8.6b. In this case, the profile of the tooth of the wheel $\mathcal{G}$ is incorrectly positioned in relation to the "boundary $N-$ circle" of radius r_N.

[6]An interesting fact, on page 135 in the study by Chironis [1], the following reaction of E. Wildhaber to a discussion of the similarities and differences between Wildhaber gearing and Novikov gearing is cited: "According to Ernest Wildhaber, who received the U.S. patent in 1926 on a circular tooth form, 'All the characteristics of the Novikov gearing are completely anticipated by my patent. My gearing never had a real test here, although a pair of gears was made in the 1920s. It would be wrong to dismiss the Russian development as old and known. My gear was never really put to the test here, although a pair of gears were made in the 1920s. It would be wrong to reject Russian developments as old and well-known'"].

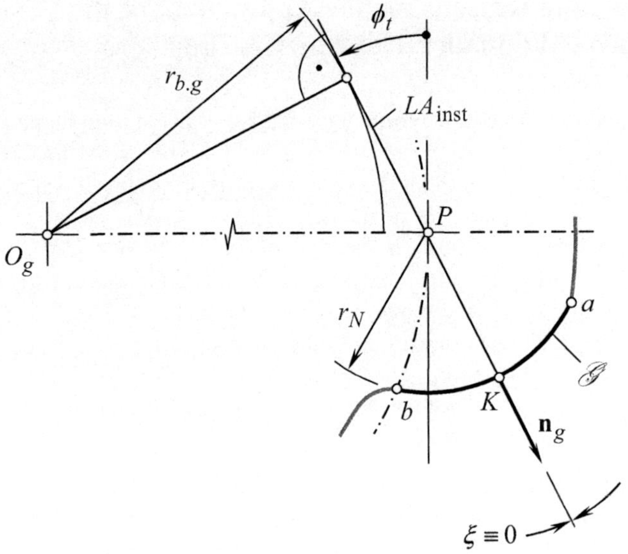

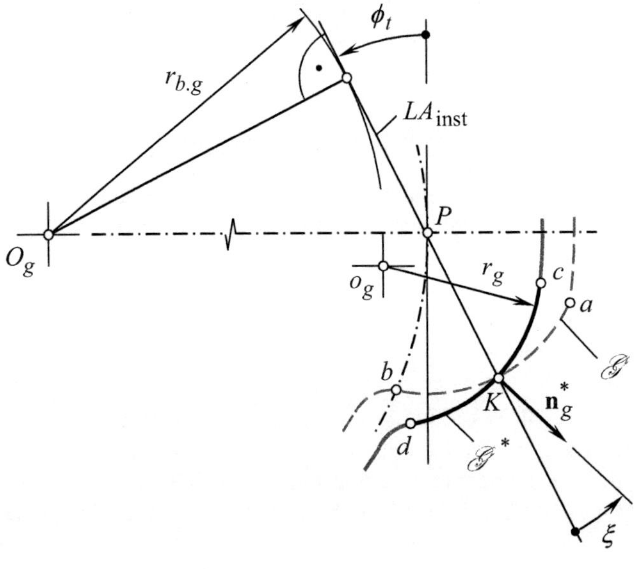

Fig. 8.6 Correct $\mathcal{G}$ (**a**) and incorrect $\mathcal{G}^*$ (**b**) configurations of the circular-arc tooth profile of a gear in Novikov gearing (the common perpendicular $\mathbf{n_g}$ is aligned with the instantaneous line of action $\mathrm{LA_{inst}}$) and in Wildhaber gearing (the common perpendicular $\mathbf{n_g}$ does not align with the instantaneous line of action $\mathrm{LA_{inst}}$)

8.5 Probable Reasons for the Appearance of the Wrong Term "Wildhaber–Novikov Gearing"

Revealing the reasons that led to the fact that Novikov gearing in the West is called Wildhaber–Novikov gearing, or W-N gearing, may be useful for establishing the importance of M.L. Novikov in the new type of gearing developed by him.

A new type of gearing was developed by M.L. Novikov when he served at the Zhukovsky Military Aviation Engineering Academy (MAEA). Zhukovsky MAEA is a regime institution. It is clear that a thesis, completed in a secure institution, could not be accessible to a wide range of specialists in the USSR. The atmosphere of secrecy around this invention was intensified in connection with the consideration of the issue of possible patenting of M.L. Novikov abroad.

For a wide range of specialists outside the USSR, the published works of M.L. Novikov including his:

(a) Dissertation [8]
(b) Descriptions of the invention in the certificate [11]
(c) Monograph [9]

were practically inaccessible. Scientific contacts between specialists from the USSR and the West were extremely limited. Western experts did not have the opportunity to obtain reliable first-hand information about Novikov gearing. Therefore, in the West, transmissions with Novikov gearing were reproduced based on fragmentary and incomplete information. In the overwhelming majority of cases, the emphasis, unfortunately, was placed on the fact that the tooth profiles are circular (and this is not fundamental) while the fundamental differences ($m_{\mathrm{p}} = 0$ and $m_{\mathrm{t}} = m_{\mathrm{F}} > 1$) have faded into the background.

An incomplete/incorrect understanding of the kinematics and geometry of Novikov gearing led to an incorrect comparison of this type of gearing with Wildhaber gearing (such a comparison is incorrect). An incorrect conclusion drawn from such a comparison was "the first reason" that the incorrect term "W-N gearing" came into use in the West. Thus, the excessive "secrecy" regime around Novikov gearing played a cruel joke on this indisputable achievement of M.L. Novikov.

The importance of M.L. Novikov was widely promoted in the USSR. So, on November 13–16, 1957, the Institute of Mechanical Engineering of the Academy of Sciences of the USSR together with the Zhukovsky Military Aviation Engineering Academy (MAEA) held the All-Union Scientific and Technical Conference "Experience of introducing gear drives with Novikov gearing into the industry." To commemorate the great contribution of M.L. Novikov to the development of a new type of gearing, the conference decided to assign the name of their author to the new gears and henceforth refer to them as the Novikov gear system or abbreviated as Novikov gearing [9]. All the specialists, who were even to the slightest degree connected with gears, knew about it. At the same time, the position of

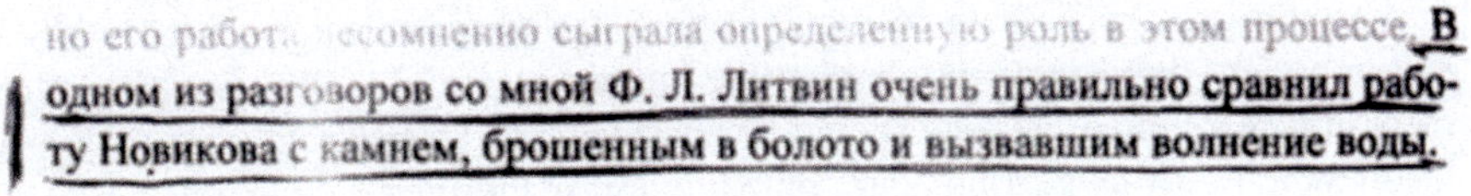
но его работа, несомненно сыграла определенную роль в этом процессе. В одном из разговоров со мной Ф. Л. Литвин очень правильно сравнил работу Новикова с камнем, брошенным в болото и вызвавшим волнение воды.

Fig. 8.7 Fragment from *Memoirs* by Prof. Ya.S. Davydov [2]

scientists and specialists, both in the USSR and in modern Russia in relation to Novikov gearing in many cases was and remains far from what it should have been. Let us give a fairly illustrative example.

A skeptical (if not deeply dismissive) attitude toward Novikov gearing is given in *Memoirs* [2]. The author of *Memoirs* explains his position on this issue with the following remark: "... In one of his conversations with me, F.L. Litvin very correctly compared Novikov's work with a stone thrown into a swamp and caused a wave of water" (see Fig. 8.7). If the scientific achievement of Dr. M.L. Novikov is compared with a "stone," then should not it also be clarified what is the "swamp"? The author of the *Memoirs* is silent about this.

Another example, on page 46 in the journal *Reducers and Gear Drives* (No. 5, 7/2006) published a letter from Prof. F.L. Litvin, in which, in particular, an extremely simplified interpretation of the essence of Novikov gearing is given: "The meaning of Novikov's idea is that he proposed cylindrical gears with point gearing, that is, with a localized working contact of surfaces . . ." and further: "I do not agree that Novikov turned the theory of links upside down." As Dr. M.L. Novikov has stressed in Pat. No. 109,113 [11], gears with a new type of engagement can be made "with *parallel, intersecting* and *crossing* axes, with external and internal engagement, with constant and variable gear ratios, . . ." Does this mean that Prof. F.L. Litvin is not familiar with the works of Novikov [8, 9] and Pat. No. 109,113 [11]?

For what purpose did F.L. Litvin distort the essence of M.L. Novikov and limit his scientific result to only cylindrical gears? Either F.L. Litvin did not understand the idea of Novikov gearing or he deliberately distorted the concept of Novikov gearing – there is simply no other way. This is something to seriously think about here.

There is no reason to doubt that adhering to this position, neither the author of *Memoirs* nor Prof. F.L. Litvin will defend the importance of Novikov gearing. Therefore, it is not surprising that in the well-known books of F.L. Litvin (1960, 1968) [5], Novikov gearing is correctly called "Novikov gearing," but, in the books published later (1994, 2004) [6, 7] this same gearing is called "Wildhaber–Novikov gearing." Litvin, F.L. did not consider it necessary to take advantage of the opportunity and enlighten Western specialists, convincingly showing them the fundamental difference between Novikov gearing and Wildhaber gearing. Was it a lack of professionalism or were there other reasons?

There are many examples of the undeservedly dismissive attitude on the part of some of the specialists toward the work developed by Dr. M.L. Novikov for a new type of gearing.

Therefore, the inadequate position of scientists and specialists, both in the USSR and in present-day Russia in asserting the importance of M.L. Novikov's development of a new type of gearing should be considered as "the second reason" that the incorrect term "W-N gearing" has come into use in the West.

The author of this chapter ok was in a full sense discouraged to learn that even now, not all researchers of Novikov gearing are familiar with the invention of E. Wildhaber. This ignorance is found even among specialists who have devoted decades of their work to the study of Novikov gearing. In other words, along with the lack of awareness of foreign researchers about Novikov gearing, the symmetrical lack of awareness of Soviet/Russian specialists with the invention of E. Wildhaber is widespread [10]. Therefore, insufficient awareness of E. Wildhaber's invention is "the third reason" for the ambiguities encountered in the interpretation of M.L. Novikov's invention.

This list of reasons could be continued.

The author of this chapter adheres to the position that the invention of Novikov gearing is comparable, in its importance, to the involute gearing invented by L. Euler. Euler, L. solved the problem for the case of a spur gear ($m_F = 0$), when the transverse contact ratio m_p in the gear set is greater than one, i.e., $m_p > 1$ (this technical solution was later developed into the field of helical gears). Novikov, M.L. solved a similar problem for the case when the transverse contact ratio m_p in the helical gear is zero, i.e., $m_p = 0$, and the condition of meshing of the teeth is fulfilled due to the axial overlap of the teeth, the face contact ratio of which is greater than one ($m_F > 1$). Here, I will allow myself a pun: "Let us reward Euler with what is attained by Euler, Novikov – with what is attained by Novikov, and Wildhaber – with what is attained by Wildhaber."

8.6 Future Developments in the Realm of Conformal/Novikov Gearing

The potential of the Novikov gear system is far from being exhausted. This system of gearing has a large reserve to increase the power density transmitted by the gear transmission system.

Novikov gearing is an example of the so-called conformal gearing. Conformal gears include those with a small difference in the radii of curvature of the profiles of the teeth at the point of contact: the smaller this difference, the higher is the degree of conformity of the functional surfaces of the gear teeth. So, the degree of conformity of the profiles of the teeth of the internal involute gear transmission is higher (with other conditions being the same) than the transmission with external gearing. Those interested can refer to the studies by Radzevich [12] and [14], in which the questions of conformity of the contacting surfaces are considered in detail.

An increase in the degree of conformity of the functional surfaces of the teeth of the gears makes it possible to increase the power density of the gear transmission system.

Developed by Dr. M.L. Novikov, this gear system allows designing gears, for which the convex surfaces of the pinion teeth contact the concave surfaces of the gear teeth. The difference between the radii of curvature of the interacting profiles is assigned in order to "absorb" the manufacturing errors, mounting errors, and elastic deformations of the entire gear transmission system (taking into account the elastic deformations of the body, shafts, gear fields, etc.) under load. In this setting, the solution to the problem of increasing the power density transmitted by the gear transmission system should be considered as necessary but not sufficient.

While studying the materials on Novikov gearing published over the past ~50 years, it was impossible not to pay attention to the unacceptably simplified idea of the importance of the geometry of contact of the surfaces of the teeth and its effect on the output parameters of the gear transmission system. A serious, if not the most important, omission in the study of Novikov gearing was ignoring the contact geometry of the side surfaces of the teeth of the wheel and gear. The transition from the contact of the convex surfaces of the teeth of the wheels to the contact of the convex and concave surfaces in itself is only a necessary, but not sufficient, condition for increasing the bearing capacity of the functional surfaces of the teeth of the wheels. At the same time, it is important to ensure a certain minimum required degree of conformity (the threshold) of the contacting surfaces of the teeth, starting from which the effect of increasing the bearing capacity begins to manifest itself (see Fig. 8.8). For "soft" wheels, this disadvantage was leveled out by the quick running-in of the teeth and their self-adaptation to the operating conditions. For highly hard wheels, running-in does not save the day. The hardened wheels must initially (a) have the required shape and (b) the correct positioning under load. In the

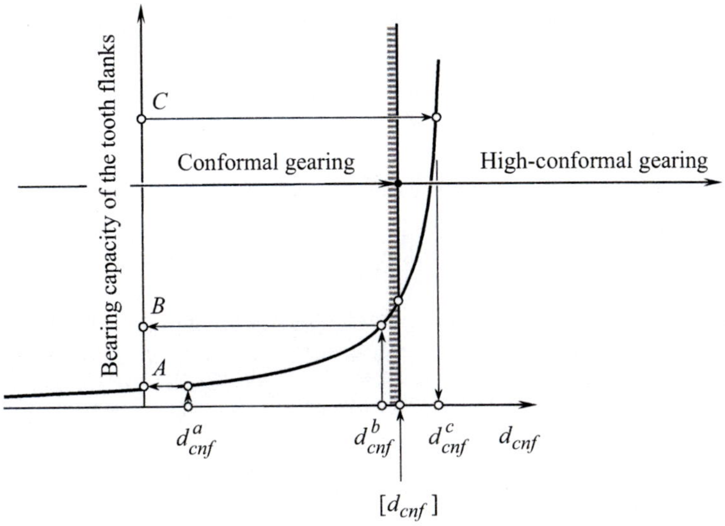

Fig. 8.8 Impact of the degree of conformity δ_{cnf} at the current point of contact of the gear tooth flank $\mathcal{G}$ and of the pinion tooth flank $\mathcal{P}$ on the bearing capacity of the tooth flanks

literature, there is no answer to the question of what the "required form" is. The answer to this question can be obtained on the basis of the results presented in the studies by Radzevich [12, 14], etc.

Back in 2006, the author of this chapter shared with the lead gear experts of Russia my own considerations and some research results on this issue. It was about the conceptual essence of Novikov gearing, more precisely – Novikov gearing using high hardness (for example, carbonized and ground) wheels, that is, when the running-in of wheels is practically impossible: "The main reason for failures with this type of transmission is a lack of understanding of what is happening in the very local contact area of the surfaces of the teeth of the wheels. For me, it is strange why they immediately began to develop the profiles of the basic rack, without understanding thoroughly the physics of what is happening in the contact zone of the surfaces of the teeth. Contact is extremely essential!

The required tooth profiles of the gears are a consequence of the required local geometry of the contacting surfaces, but not vice versa. The geometry of the contact of the surfaces of the teeth of the wheels is the key to solving fundamental issues related to the operability of gears." Furthermore, the following conceptual tasks can be solved analytically:

(a) "For the given transmission conditions, tolerances can be calculated so that a functional transmission can be guaranteed
(b) According to the accuracy of the existing gear-cutting machine and tool, it is possible to answer the question whether it is possible on this machine and with a given tool to machine the wheels for a guaranteed working transmission
(c) With known rigidity . . ., that is, of all that determines the resulting displacement of one wheel in relation to another wheel under load, to answer the question whether it is possible to apply Novikov gearing in principle under these conditions."

It should be noted that not all of what was produced by the industry under the name Novikov gearing was (and still is) such. Under the flag of Novikov gearing, something was often produced that vaguely resembled Novikov gearing but, upon closer examination, was not. In other words, much of what they unsuccessfully tried to introduce into the industry, in fact, had (and has) a remote relation to Novikov gearing. All the gear wheels for Novikov gearing, finish cut in the generating method of gear cutting (namely, cut with hob cutters, shapers, etc.), are not Novikov gear wheels (wheels for Novikov gears cannot be finish cut under generating machining conditions). From this, we can draw a conclusion about the reported successes and failures of the industrial use of gears by Novikov gearing.

Of course, what was done by Dr. M.L. Novikov deserves to be investigated in detail. Based on the results of such research, a monograph can be written. The author of this chapter had extremely limited material at his disposal about Dr. M.L. Novikov and Dr. E. Wildhaber.

8.7 Concise Information About the Private Life of M.L. Novikov

The author of this chapter had an opportunity to communicate with Ms. Irina M. Averina, the daughter of Dr. M.L. Novikov. It is worthy to begin this portion of this chapter with an excerpt from one of the numerous letters the Ms. I.M. Averina sent to me: "I – your son – will try to continue your path to the level of advanced people of human society, to the level of the creators of new technology, to the level of an engineer, not an engineer, "hat", which unfortunately is now, but a real engineer – designer – creator . . . ." (Private communication on March 9, 2020).

Ms. I.M. Averina kindly shared with the author some pictures, those taken of Dr. M.L. Novikov, his family, and his friends. Unfortunately, not many of such pictures are available to the author at this time.

1948 tea party with Alekseyev family. (Courtesy of Ms. *I.M. Averina*)

A 1948 tea party with the Alekseyev family is the oldest available picture of M.L. Novikov. Here, M.L. Novikov is shown third from the left. His wife, Ms. Galina S. Novikova, is sitting next to M.L. Novikov.

In 1969, a memorial plaque was opened in the apartment building where M.L. Novikov used to live for several years. In the center, Ms. Galina S. Novikova is pictured.

1969 Memorial plaque. (Courtesy of Ms. *I.M. Averina*)

1960 Fediakin, Chesnokov, and Shtoda. (Courtesy of Ms. *I.M. Averina*)

In the next picture, Novikov's friends, R.V. Fedyakin, V.A. Chesnokov (both at rear), and Shtoda, A.V. (in front), are pictured. "Andrey Vladimirovich Shtoda served with my father. After the death of my father, he became the head of the department, previously headed by my father. Shtoda, V.A. did a lot to create a special laboratory for Novikov gearing at the Academy, in which R.V. Fedyakin and V.A. Chesnokov worked actively. I note that R.V. Fedyakin and V.A. Chesnokov are talented engineers, extremely devoted to my father, his work and our family."

Much work on the preservation and popularization of the heritage of M.L. Novikov is being held by his daughter, Averina Irina Mikhailovna.

2018 – Averina, I.M. – *Novikov*'s daughter. (After *T. Gogoleva*, [4])

She was 5 years old when her father died and does not have many memories of him. Much more information has been preserved in the funds of various archives and museums of the country. The museum of the Air Force Academy also contains many documents, one way or another, related to the activities of Dr. M.L. Novikov. Recently, Irina Mikhailovna has visited there, and, for several days, she carefully studied the certificates, extracts, reports, etc. – everything that could add information about the great scientist, whose work and life, by the way, was classified "top secret" for a very long time.

Ms. I.M. Averina says that: "Two years ago I wrote a Book about my father, which included some letters from my father, his father, relatives and workmates. Separate documents, family and official photos, memoirs, my comments were given. But since the Book contained many personal, family stories, this Book was published by me in a small edition and donated to the relatives and children, grandchildren (already adults) of my father's comrades."

The author of this chapter had made an offer to Ms. I.M. Averina to join this project and share her memories about her dad M.L. Novikov. The response is immediately below: "I have already received offers from historians, writers, scriptwriters of the Russian Federation to provide information about my father, but I always refused. And now I cannot make a decision to tell strangers from my father" (private communication on March 13, 2020). A few days earlier: "My father's life story is very complex. Many facts in open sources are not true. So far, I cannot make a decision to publicly describe my father's life. I was very young when we were left alone. But the memory of keeping the joyful days when dad was with us" (private communication on March 9, 2020).

I would like to end this portion of this chapter with the following excerpt from Ms. I.M. Averina‘s letter:

"Dear Stepan Pavlovich, good morning!

Thank you for your work on the further development and application of the theory of spatial gearing with point contact, set out in the doctoral dissertation "Basic questions of the geometric theory of point gearing intended for high-power gear wheels" by my father, Mikhail Novikov, and defended in December 1955. Father started working since 1950.

After the tragic death of my father in 1957, the employees of the Zhukovskii Military Aviation Engineering Academy (MAEA), where my father studied and worked, moved ALL scientific documents (manuscripts, drawings, business papers and notes, service letters) that were in our home to the Archives of this Academy.

I am sending you a film made in 1961 about the works of my father – I legally bought this film from the Archive of Documentary Films of the Ministry of Culture of the Russian Federation. The authors of this film were probably the well-known to you V.R. Fedyakin, V.A. Chesnokov, and A.V. Shtoda. When creating the film, the manuscripts of my father were used, which were at that time in the Academy – these documents can be used, since they are officially featured in the film."

8.8 Conclusion

A brief biography about M.I. Novikov and E. Wildhaber is provided.

The features of Novikov gearing are briefly considered in comparison with the cylindrical gearing proposed by E. Wildhaber. The fundamental difference between these two types of gears is shown. The latest advances in gear theory are used for this comparison. The possible reasons for the appearance of the incorrect term "Wildhaber–Novikov gears" are analyzed. It has been convincingly shown that Dr. M.L. Novikov is the only author of a new type of gearing and that the name Novikov gearing is the only correct name for the invented gearing. The possible ways of development of Novikov gearing are shown.

References

1. Chironis, N.: New tooth shape taking over? Design of Novikov gears. In: Chironis, N. (ed.) Gear Design and Application, pp. 124–135, 374p. McGraw-Hill Book Company, New York (1967)
2. Davidov, Y.S.: Memories (manuscript) (March 1991). http://www.reduktor-news.ru/forum/index.php?fid=1&id=11564943501613
3. Dyson, A., Evans, H.P., Snidle, R.W.: Wildhaber-Novikov circular arc gears: geometry and kinematics. Proc. R. Soc. Lond. A Math. Phys. Sci. **403**, 313–340 (1986)
4. Gogoleva, T.: "Novikov Gearing", Patriot, Newspaper of the Military Educational and Scientific Center of the Air Force "Air Force Academy named after Professor N.E. Zhukovsky and Y.A. Gagarin" (Voronezh, Russia), No. 10 (128), p. 4 (October 2018)
5. Litvin, F.L.: Theory of Gearing, 2nd edn, 584p. Nauka, Moscow (1968) (1st edn, Nauka, Moscow, 1960)
6. Litvin, F.L.: Gear Geometry and Applied Theory, 724p. Prentice Hall, Englewood Cliffs (1994)
7. Litvin, F.L., Fuentes, A.: Gear Geometry and Applied Theory, 2nd edn, 800p. Cambridge University Press, Cambridge (2004)
8. Novikov, M.L.: Fundamentals of geometric theory of gearing with point meshing for high power density transmissions. Doctoral thesis, Moscow, Zhukovskii Military Aviation Engineering Academy (MAEA) (1955)
9. Novikov, M.L.: Gearing with a Novel Kind of Teeth Meshing, Moscow, 186p. Zhukovskii Aviation Engineering Academy Zhukovskii Military Aviation Engineering Academy (MAEA) (1958)
10. Pat. No. 1.601.750 (USA), Helical Gearing./Ernest Wildhaber, filed: November 2, 1923
11. Pat. No. 109,113, (USSR). Gear Pairs and Cam Mechanisms Having Point System of Meshing. /M.L. Novikov, National Classification 47h, 6; Filed: April 19, 1956, published in Bull. of Inventions No. 10, 1957
12. Radzevich, S.P.: Fundamentals of Surface Generation, Monograph, 592p. Rastan, Kiev (2001)
13. Radzevich, S.P.: Theory of Gearing: Kinematics, Geometry, and Synthesis, 2nd edn, revised and expanded, 934p. CRC Press, Boca Raton (2018) [1st edn: Radzevich, S.P.: Theory of Gearing: Kinematics, Geometry, and Synthesis, 760p. CRC Press, Boca Raton (2012)]
14. Radzevich, S.P.: High-Conformal Gearing: Kinematics and Geometry, 2nd edn, 506p. Elsevier, Amsterdam (2020) [1st edn: Radzevich, S.P.: High-Conformal Gearing: Kinematics and Geometry, 332p. CRC Press, Boca Raton (2015)]

Bibliography

Radzevich, S.P.: Generation of Surfaces: Kinematic Geometry of Surface Machining, 738p. CRC Press, Boca Raton (2014) [1st edn: Radzevich, S.P.: Kinematic Geometry of Surface Machining, 508p. CRC Press, Boca Raton (2008)

Radzevich, S.P.: High-conforming gearing: kinematics and geometry. In: International Gear Conference 2014, Conference Proceedings, Vol. II, Lyon Villeurbanne, France, 26–28 August 2014, pp. 846–857 (2014)

Radzevich, S.P.: High-Conformal Gearing: Kinematics and Geometry, 332p. CRC Press, Boca Raton (2015) https://www.amazon.com/High-Conformal-Gearing-Kinematics-Stephen-Radzevich-ebook/dp/B016MD4KLI/ref=sr_1_6?dchild=1&keywords=radzevich+stephen+p&qid=1605119111&s=books&sr=1-6

Radzevich, S.P.: High-conformal gearing: a new look at the concept of *Novikov gearing*. In: Proceedings of International Conference on Gears 2015, October 5–7, 2015, Technical University of Munich (TUM), Garching (near Munich), Germany, pp. 457–470 (2015)

Radzevich, S.P.: An examination of high-conformal gearing. Gear Solutions, pp. 31–39 (Feb 2018)

Radzevich, S.P.: Theory of Gearing: Kinematics, Geometry, and Synthesis, 2nd edn, revised and expanded, 934p. CRC Press, Boca Raton (2018) [1st edn: Radzevich, S.P.: Theory of Gearing: Kinematics, Geometry, and Synthesis, 706p. CRC Press, Boca Raton (2012)]

Radzevich, S.P.: A possible way of evolution of *Novikov* gearing. In: Proceedings of the IX International Conference "Complex Assurance of Quality of Technological Processes & Systems", May 14–16, 2019, Chernihiv (Ukraine), vol. 1, pp. 39–40 (2019)

Radzevich, S.P.: High-Conformal Gearing: Kinematics and Geometry, 2nd edn, 506p. Elsevier, Amsterdam (2020) https://www.amazon.com/-/es/Stephen-Radzevich/dp/0128212241?language=en_US

Chapter 9
Poor Understanding of the Scientific Theory of Gearing by the Majority of Gear Scientists and Engineers

In this monograph, the modern scientific theory of gearing is disclosed to the best possible extent [1]. This is a new theory in the realm of gear science and arts. Despite the fact that the origin of the scientific theory of gearing can be traced back to the fundamental research carried out by Leonhard Euler in the eighteenth century, rapid growth of this theory began only in the early twenty-first century. Through the first two decades of the twenty-first century, most (but not all) of the key problems in the proposed theory of gearing have been solved. It is the right time now to study what has already been achieved in this area of mechanical engineering and to implement these results in every-day gear practice. Unfortunately, the latest achievements in the scientific theory of gearing are ignored by the majority of gear scientists and gear engineers. It is likely that a poor understanding of the scientific theory of gearing by the majority of gear scientists, engineers, and practitioners is the root cause of this.

It takes time for the considerable number of gear experts to get familiar with the theory, to study it, and begin to use it in every-day practice. Evidently, this cannot be done instantly.

To properly evaluate the advantages and benefits of the usage of the proposed scientific theory of gearing, an in-depth knowledge of the theory along with a reasonably extensive experience of its implementation in practice are necessary. Otherwise, it is impossible to evaluate the theory properly and to draw benefits from it. Moreover, a toxic environment can be formed in the gear community due to the poor knowledge about the proposed scientific theory of gearing.

Nowadays, some well-known gear experts, who pose as gear theoreticians, commit huge mistakes when trying to evaluate the gear theory, as they are not qualified to conduct such an evaluation.

S. P. Radzevich, *Novikov/Conformal Gearing*,
https://doi.org/10.1007/978-3-031-10019-2_9

Fig. 9.1 Facsimile of the article: Stadtfeld [2]

As an illustrative example, let us consider a recently published article [2] in the *Gear Solutions* magazine. The author of this article made a weak attempt to review some key accomplishments in the scientific theory of gearing, thus demonstrating the absence of an understanding of the cornerstone ideas of gear science (see Fig. 9.1).

Fig. 9.1 (continued)

Gear scientists, gear engineers, and gear manufacturers have worked successfully for many decades on finding the optimal flank forms and the optimal non-conjugate flank surface interaction.

By Dr. HERMANN J. STADTFELD

In 1924, Ernest Wildhaber, a well-known gear scientist, invented hypoid gearing. Compared to spiral bevel gears, hypoid gears provide an offset that allows lowering of the body of rear wheel drive vehicles by 50mm or more. This is possible because the propeller shaft between engine/transmission and the driving axle is not positioned at the center of the drive axle but is lowered by the offset amount (Figure 1). This allows the vehicle designer to lower the floor of the vehicle and subsequently the entire body by the same amount. Lowering the center of gravity of a passenger car by 50mm reduces the inertia responsible for sideways rolling by more than 10 percent, which provides better vehicle handling and more active safety. The lower body also reduces the CV coefficient for air resistance, providing higher gas mileage. Less than five years after the invention of hypoid gearing, all large automotive manufacturers around the world had converted their passenger cars and trucks to hypoid drive axles with a lower vehicle body.

Ernest Wildhaber emigrated from Switzerland to the U.S. in 1919 and was hired by The Gleason Works as a gear theoretician. Wildhaber received 279 patents, many of which changed the world of gearing. The cylindrical gear tooth profile that is today called *Wildhaber-Novikov* gearing was invented by Ernest Wildhaber in 1926. Mikhail Novikov, a Russian scientist with no access to western publications, invented the same tooth profile independent from Wildhaber in 1956. The contributions of both scientists are honored today by calling this system *Wildhaber-Novikov* gears.

Ernest Wildhaber is the father of modern gear theory. His pioneering contributions have been invaluable for the development of today's gear calculation and manufacturing processes.

NEWS ABOUT HYPOID GEARS?

A recently published article by S.P. Radzevich [1] attempts to explain that hypoid gearsets designed and manufactured today lack good performance and are not perfect but approximate. The article seemingly questions the credibility of many scientists and gear engineers who worked on the theory and its improvement over approximately 100 years. However, reading the article carefully quickly reveals some key misperceptions and false assumptions. This write-up is meant to be a response on behalf of established gear theory and manufacturing.

In the following sections, certain statements and mathematical relationships of the published article are discussed and countering responses are presented and explained. Also, some assumptions about manufacturing and application procedures are discussed and compared to established practices today. The topics of this paper are structured accordingly in:

- The three fundamental laws of gearing.
- Perfect conjugacy.
- Real world applications.
- Transmission design.
- Heat treatment, lapping and grinding.

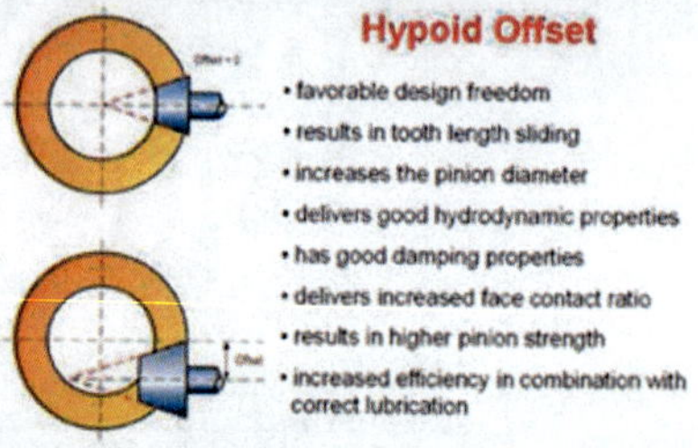

Figure 1: Features of pinion offset.

THE THREE FUNDAMENTAL LAWS OF GEARING

The first fundamental law of gearing quoted in [1] $\mathbf{n}_g \cdot \mathbf{V}_\Sigma = 0$ also implies $|\mathbf{N}_1 \times \mathbf{R}_1| = i \cdot |\mathbf{N}_2 \times \mathbf{R}_2|$, where i is the constant transmission ratio. The three cases in Figure 2 visualize the problem of non-constant ratio and are noncompliant with the first gearing law because of a ratio change from case to case. This problem led Leonard Euler to discover the involute tooth profile. A simplistic mathematical approach teaches that the effective radius vector **R** remains unchanged while the contacting point between two mating flanks moves from $\mathbf{R}_{b2}$ to $\mathbf{R}_{b1}$ as shown to the

May 2019 43

Fig. 9.1 (continued)

right in Figure 3 (movement along the line of action).

The line of action in parallel axes cylindrical gearing is a straight line, connecting the two base circles. If the surface normal vectors N1 and N2 are within the line of action, then the vector product $|N_i \times R_i|$ remains constant during a complete mesh cycle. The consequent application of this principle leads to the construction of an involute, as shown to the right in Figure 3.

The line that forms the tooth surface elements while traveling from position "a" to "f" along the line of engagement (line of action) is always perpendicular to the line of action. This principle implies that a tool, simply with straight cutting edges as shown in Figure 4, can be used to form the complex involute profile. Figure 4 also demonstrates the principle of profile shift while maintaining the first fundamental law of gearing.

The second fundamental law of gearing, which was proposed in 2017, $p_{in} \times V_m + n_g = 0$ [1], is a redundant relationship to the first gearing law and it is limited to cylindrical gears with parallel axes and straight bevel gears without hypoid offset. In this case, it adds nothing to the first gearing law; conjugacy is already given by the relationships required in the first gearing law.

The third fundamental gearing law is quoted in [1] in two different notations. The first notation covers only the special case of ratio = 1:

$\varphi_{b.g} = \varphi_{b.p} = \varphi_{b.op}$

However, the first notation is incorrect for all cases with different base pitch diameters because the circular pitch and not the angular pitch has to be equal between pinion, gear, and the operating base pitch. The second notation:

$p_{b.g} = p_{b.p} = p_{b.op}$

is consistent with the requirement of equal circular pitch, but the article makes no mention of the different notations.

PERFECT CONJUGACY IN STRAIGHT BEVEL GEARSETS

Bevel gears with intersecting axes is the topic of a series of three papers published between October 2014 and January 2015 [2]. A straight bevel gearset with skew teeth was modeled, and a sample was manufactured. This publication addressed two points: the design of a gearset with low tooth count and the solution for perfect conjugacy. There is no mention in the article that the Coniflex® straight bevel gear system has been used since the 1940s for low tooth count gearsets such as differential gears.

Straight bevel gears such as Gleason Coniflex have tapered depth teeth, where the pitch cones roll on each other and the pitch apexes of pinion and gear match the crossing point of the axes. In the standard case, the face and root cone apexes match the crossing point. In such a standard Coniflex design, the base elements are also cones with cone apexes that match the crossing point of the axes. The involute development in Figure 3 can be applied to an infinite number of normal sections along the face width of a straight bevel gear which allows an involute development similar to that for cylindrical gears. The conical base elements of both members can be connected with a straight line (the line of action) in each section along the face width, whereas the plurality of all lines of action forms a plane (the plane of action).

This principle is shown in Figure 5. The two cones in Figure 5 are base cones of a straight bevel or a spiral bevel gearset. In the right two graphics, the view is directed such that the plane of action appears as a line that is tangential to the two cone-enveloping surfaces. The left side graphic in Figure 5 shows the plane of action three-dimensionally and how it connects the two base elements.

The plane of action cannot be extended beyond its tangential contacting line with the base element as shown incorrectly in

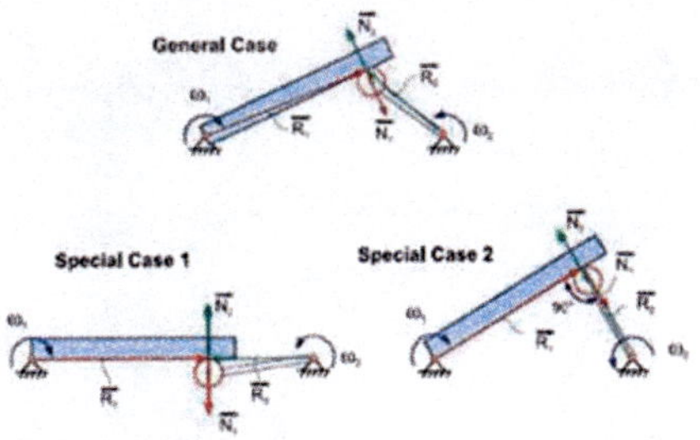

Figure 2: Three general cases of motion transmission with a bar and a crank.

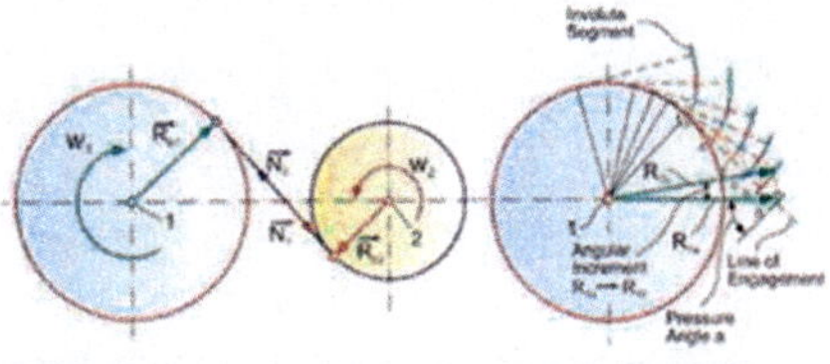
Figure 3: Line of action between base circles (left) and subsequent involute development.

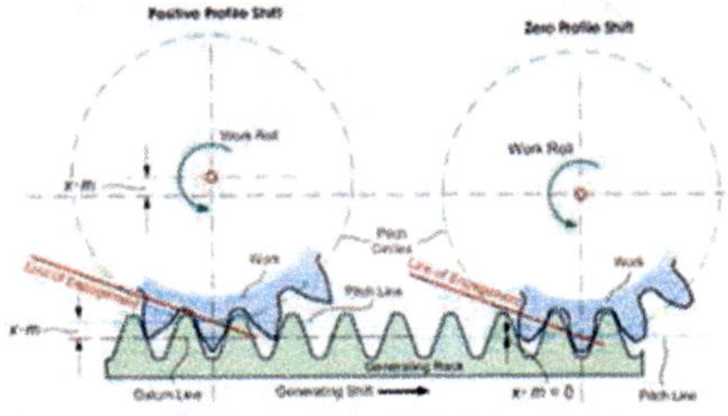
Figure 4: Generating rack with straight profile forming involute profiles.

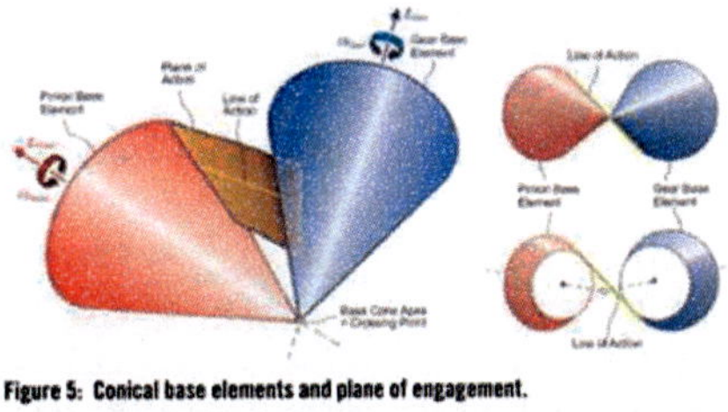
Figure 5: Conical base elements and plane of engagement.

Figure 1 in [2] and Figure 6 in [1]. The plane of action only exists where tooth engagement is possible and it is different than the generating gear plane (more specifically explained later).

There is, however, one difference from the true involute of cylin-

44 gearsolutions.com

Fig. 9.1 (continued)

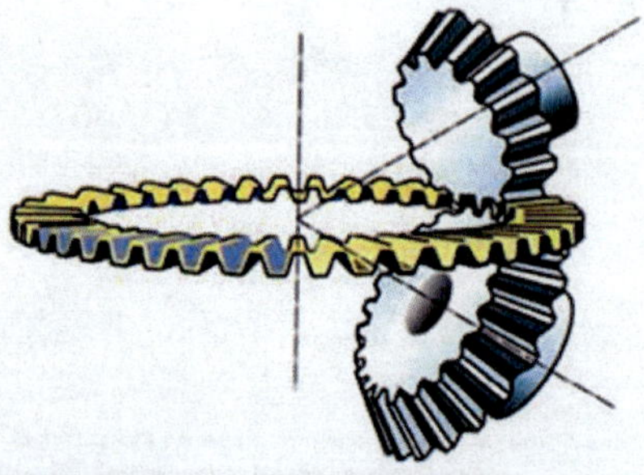

Figure 6: Ring shaped generating rack with trapezoidal profile.

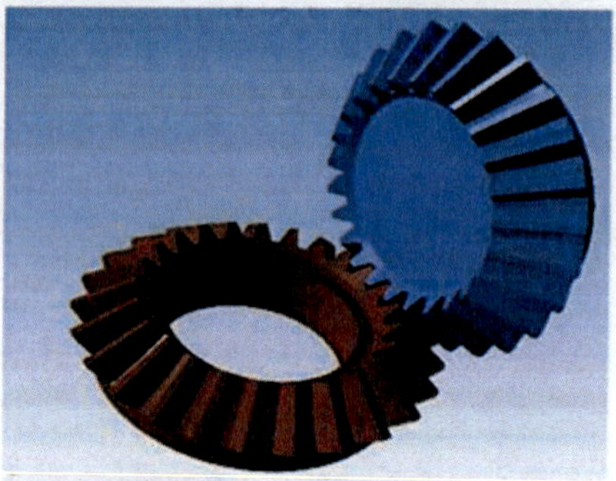

Figure 7: Solid model computer graphic of a perfectly conjugate straight bevel gearset.

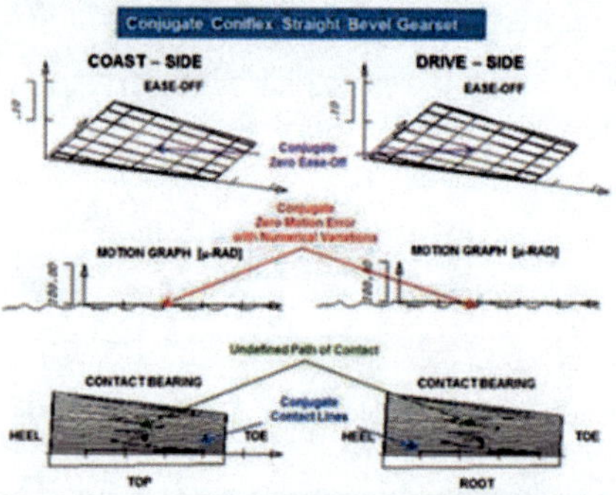

Figure 8: Tooth contact analysis of the perfectly conjugate straight bevel gearset.

drical gears. The rotation of pinion and gear does not occur in the normal plane but in the transverse plane. Because of this difference, the flank profile of straight bevel gears (and all other bevel gear types) is called Octoide. The Octoide is the analog function of an involute, and it provides to bevel gears the same advantages as an involute provides to cylindrical gears. Those advantages are constant ratio, center distance insensitivity, and ease of manufacturing.

Like cylindrical gears, bevel gears also have a trapezoidal generating profile. The straight rack of cylindrical gears becomes a ring, as shown in Figure 6. It is required to establish certain conditions in order to make the ring rack the generating gear for a pinion and a ring gear that will mesh perfectly together with zero transmission error and line contact identical to cylindrical gears. Those conditions are postulated in the kinematic coupling requirements:

- The flank surfaces of the generating gears of the two mating bevel gears are congruent (same shape but mirror images, as given in the example of Figure 6).
- The generating gears of the two mating bevel gears require identical axes of rotation (Top and bottom side of the generating gear in Figure 6 form the same generating gear, which rotates in both cases around the same axis and therefore satisfies condition 2).
- The surface of engagement of pinion and ring gear must be identical to the surface of engagement between pinion and generating gear and also to the one between ring gear and generating gear (without detailed knowledge of the surfaces of engagement, the global condition in Figure 6 seems to satisfy this requirement).

The generating gear principle has to be understood as the ultimate vehicle to form the teeth of two mating gears. The first fundamental law of gearing is fully executed by choosing trapezoidal profiles and by applying the kinematic coupling requirements. Gears are designed and manufactured in order to mesh with each other. What better way to manufacture them by way of a generating gear? The generating gear is represented by the manufacturing machine; it forms the teeth of the gear (at the bottom in Figure 6) while meshing with this gear perfectly. If the pinion is manufactured with the same generating gear but on the opposite side (at the top in Figure 6) and if the generating gear is imagined infinitely thin, then the result is a pinion that perfectly meshes with the gear having zero motion error. It is also given, in such a case, that line contact between the pinion and gear flank surfaces exists along the entire face width. The extraordinary effort shown in [2] to calculate straight bevel gear tooth surfaces that are skew with asymmetric teeth is based on the falsely assumed conjugate requirement. The sample gear pair shown in [2] was manufactured on a machining center with a ball nose endmill in a milling operation likely taking many hours to complete. It is rather simple and straightforward to manufacture a significantly better, perfectly conjugate straight bevel gearset in only a few minutes by using the generating gear method and a standard bevel gear manufacturing machine.

An example straight bevel gearset computer model is shown in Figure 7. The solid model in Figure 7 has been generated by using standard Gleason basic settings, based on the generating gear approach and by applying a standard Coniflex® Plus cutter head as used on Phoenix bevel gear manufacturing machines. The Coniflex straight bevel gear calculation for conjugate contact has to be conducted with a dish angle of 0° and no root angle correction (ΔGammaM = 0.). The dish angle is creating the length crowning and a ΔGammaM creates the profile crowning. With a dish angle of 0° and a ΔGammaM of 0° flank lead lines are straight lines and the profile is a true Octoide (involute equivalent).

A contact analysis of the gearset in Figure 7 is shown in Figure 8. The top of the figure shows the Ease-Offs of left and right flank (called coast and drive side in the graphic). The center of the figure shows the motion transmission errors of the pinion and gear flank pairs. The two bottom graphics are the representation of the tooth contact pattern. The contact pattern graphics are axial

May 2019 45

Fig. 9.1 (continued)

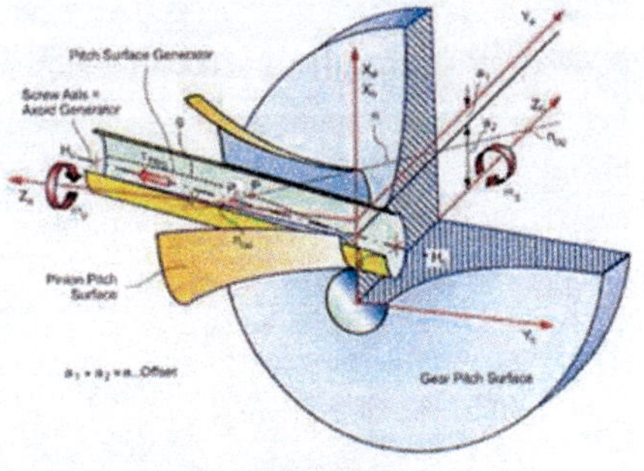

Figure 9: The correct pinion and gear pitch surfaces.

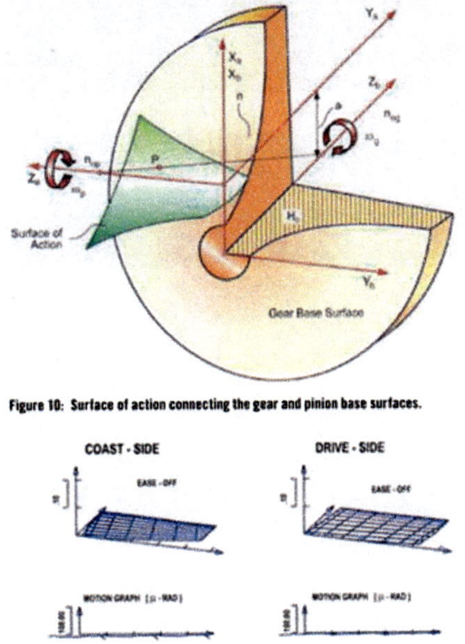

Figure 10: Surface of action connecting the gear and pinion base surfaces.

Figure 11: Ease-Offs, motion graphs, and contact lines of a real conjugate hypoid gearset.

projections of the flank surfaces and the contact lines in the same plane a two-dimensional part print would show the tooth area.

The contact analysis in Figure 8 confirms the full line contact in each roll position (lower graphics) as well as the zero motion error (only numerical static) in the center graphic which makes this example the perfectly conjugate straight bevel gearset. In the lower graphic the path of contact was calculated as a zig-zag line which indicates an undefined contact path. This means that due to the conjugacy, every point along each contact line is a path of contact point which makes the analysis program pick random points.

The Ease-Off base plane (top graphics in Figure 8) defines the conjugate state of a flank surface pair. Because the Ease-Off graph of the calculated flank pairs matches the presentation plane (base plane) precisely, that is the proof that a conjugate and precisely rolling gearset was the input of this contact analysis calculation.

The above experiment creating a conjugate straight bevel gearset is strictly academic. Conjugacy is the basis of all gearsets manufactured in high volume on dedicated manufacturing machines. A conjugate bevel gearset cannot be used for a power transmission because manufacturing tolerances and load affected deflections as well as material expansions and deformations under high operating temperatures will result in edge contact and high load concentrations. The load concentrations already start with moderate load and cause material damage and considerable noise emission. Although conjugacy is used as a reference for each design, predetermined amounts of length and profile crowning are applied. The right amount of crowning makes a gearset quiet and gives it the required load carrying capacity. The crowning is shown in the Ease-Off graphics with the conjugate reference always being present as the Ease-Off base plane. Several Ease-Off examples of a gearset with length and profile crowning are shown in the proceedings of this paper.

PERFECT CONJUGACY IN HYPOID GEARSETS

It begins to become more problematic for hypoid gears. In [1] the pitch elements of crossed axes hypoid gears are drawn as cones. Even though the face cones of hypoid gears and pinions are machined conical, the pitch elements are hyperboloids.

Ernest Wildhaber and Arthur Stewart described their invention of hypoid gearing in 1926 [3]. Boris Shtipelman published in 1978 the relationships and derivations required to understand hypoid gears and their hyperbolic pitch elements [4]. Figure 9 offers a graphical interpretation of the hyperbolic pitch elements and their generator. The pitch surface generator is a line which winds on the surface of a cylinder beginning at the crossing point of the axes, equal to the first contact point of the pinion and gear pitch surfaces. The pitch surface generator is developed by the connecting line between the pinion and gear pitch surfaces (n_{op}- n_{og}) by shifting the connecting line along the pinion and gear axes. The connecting line (n_{op}- n_{og}) is normal to the pitch elements. Point P is one point of the pitch surface generator. If the division of the vector products of the distances between the axes and point P and their respective axis direction equals the ratio i of the hypoid gearset then one point of the pitch surface generator is found with:

$$\{(n_{op} - P) \times Z_a\} / \{(P - n_{og}) \times Z_b\} = i$$

Although the pitch elements are hyperbolic and not conical, it is possible to use conical faces for the blanks of pinion and gear. If point P in Figure 9 was chosen at the center of the face width, then line (n_{op}- n_{og}) can be used as a normal vector to define the face angle a blank with straight face cones if the hypoid set was manufactured by face hobbing, which implies parallel depth teeth.

Straight face cones will merely influence the top root clearance of the gearset in the range of 30 to 60 microns. Using straight face cones will not change the form of the pitch surface nor will it influence the base surface. Those functional surfaces are given by the

Fig. 9.1 (continued)

kinematical relationships and have to be considered when thinking about the shape of the surface of action. No plane of action can exist between two hyperbolic base elements. The correct surface of action is curved and warped, as shown in Figure 10.

The conclusion is that the second fundamental law of gearing according to [1] does not exist and is not related in any way to the conjugacy of gears (as explained and demonstrated in the following section).

CONJUGACY BETWEEN MESHING FLANKS

The term conjugate is used in mathematics for two or more surfaces that contact each other along a line. Since the 1970s, the term conjugate has also been employed in gear technology literature to define the "exact" gear pair which presents a triple plurality of line contact between two gear flanks during the meshing process [5].

DEFINITION OF THE CONJUGATE GEAR PAIR

1: The flanks contact along a line (contact line), which is only limited by the boundaries of the teeth i.e. the working area.

2: The line contact between the flanks exists within the entire area of engagement in every mesh position.

3: Line contact is maintained in the entire area of engagement if pinion and ring gear are rotated by angular increments where: (angular pinion increment) / (angular ring gear increment) = transmission ratio.

The Ease-Off is a three-dimensional graphic of the flank deviations from a conjugate pair. It is calculated by transformation of a pinion flank "into" the gear coordinate system according to the first gearing law, resulting in a virtual gear flank that is conjugate to the actual pinion flank. This conjugate gear flank will then be compared to the present gear flank, where all differences in arc length are plotted point by point in ordinate direction into the Ease-Off graphic.

If both mating bevel gears have conjugate manufacturing data, then the Ease-Off graphic has no deviations in ordinate direction. Also, if the pinion flanks and the gear flanks have spiral angle and pressure angle errors of equal amounts, the Ease-Off graphic will not show any deviation. Although the individual gears are considered incorrect in this case, they will roll conjugate with each other, which subsequently leads to an Ease-Off without any ordinate values. Figure 11 shows the analysis results of a typical conjugate hypoid gearset. The Ease-Off graphics have zero crowning magnitudes in the ordinate direction. The motion graph has, next to some numerical entrance and exit variation, zero motion error. The contact bearings show line contact within the entire working area. The coast side contact ends at a toe root undercut (section ε).

Each spiral bevel and hypoid gearset with uniform tooth depth has a conjugate base design. This applies to all face hobbed and some face milled gearsets. A simple explanation of hypoid gearset conjugacy is possible with a non-generated gear that meshes with a generated pinion. For the calculation and manufacturing process, the hyperbolic pitch elements are calculated for the gear first. Then a suitable blade profile (gear cutter in Figure 12) is chosen and positioned in a face cutter head. The cutter head is positioned to create the desired spiral angle. With this procedure, a non-generated gear can be created by computer simulation and it can be manufactured with a bevel gear cutting machine [6].

A pinion cutter (see Figure 12) is positioned in a mathematical model or in a bevel gear cutting machine such that it represents one tooth of the non-generated gear by rotating around its axis. An additional simultaneous rotation around the pinion generating gear axis results in this pinion cutter becoming the generat-

Fig. 9.1 (continued)

ing gear of a conjugate pinion. If the pinion is positioned with the same offset that was used to determine the pitch surfaces (Figure 9), then the cutter rotation around the pinion generating gear axis will form a pinion that is perfectly conjugate to the non-generated gear. The tooth contact analysis in Figure 11 has been obtained from such a non-generated hypoid gearset and therefore shows perfect conjugacy.

More complicated is the generation of a conjugate bevel or hypoid gearset with tapered depth teeth (see Figure 13). If the generating gear axes have identical axes of rotation that are perpendicular to the pitch line, the rotating cutter heads and their blades will not follow the root line of a tapered depth tooth. Tilting the cutter head in order to follow the root line would violate the first kinematic coupling requirement for teeth that are congruent to the slots of the mating member. The solution was developed in the 1940s [7].

If the cutting edges are adjusted in the cutting machine such that tooth reference profile and depth is matched at midface, and if an axial motion of the cradle is introduced that guides the blades along the tapered root line while the generating roll progresses along the face width, then the requirements of congruent teeth and slots are fulfilled with the result of perfect conjugacy.

The process configuration and kinematics in Figure 13 is called *Duplex Completing*. Today, all face-milled and ground spiral bevel and hypoid gears are manufactured with the Duplex Completing process. The axial cradle movement in this process is called Helical Motion and was first introduced with mechanical bevel gear machines in the late 1940s. The *Helical Motion* of the days of mechanical machines required an additional change gear box which actuated a cam that moved the sliding base during the generation process. Today's CNC controlled Phoenix® free form machines have the Helical Motion capability automatically by their interpolated axes movement.

WHY IS CONJUGACY NOT DESIRABLE FOR REAL WORLD APPLICATIONS?

In 1926, Ernest Wildhaber [3, 8] was the first to propose applying surface crowning on hypoid gears. Wildhaber acknowledged that the slightest deviations in the gear housing and in the building position, as well as deflections affected by load and heat, will cause edge contact with peak stress levels of a multiple of the allowable values the gearset had been designed for. The conjugate gearset used for the TCA in Figure 11 was repeated with realistic displacement values of 50μm offset, 50μm pinion cone and 30' of shaft angle change. The results in Figure 14 show warped and tilted Ease-Offs and severe edge contact on heel and top. This edge contact will cause noisy operation followed by pittings and tooth fracture.

The theoretically conjugate hypoid set will not fulfill any of the fundamental gearing laws in case of the smallest gearbox inaccuracies or deflections. As mentioned above, already small deflections at moderate loads lead to load concentrations on the edges of conjugate flank pairs and can cause material damage and considerable noise emission. As such, the conjugate gear pair is not suitable for any task in a power transmission.

The introduction of 80μm length crowning and 15μm profile crowning to the conjugate hypoid design delivers the analysis results shown in Figure 15. The crowning makes the gearset insensitive to expected inaccuracies in the gear housing and load and heat affected deflections. Applying the same amounts of shaft displacements then used for the TCA of the conjugate hypoid set in Figure 14 moves the mean point slightly out of the initial position (see Figure 16), but a large contact area within the tooth boundaries is still maintained.

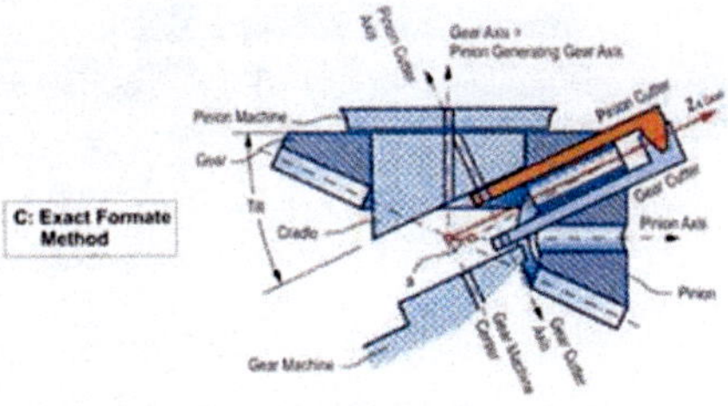

Figure 12: Generating of perfect conjugate hypoid pair.

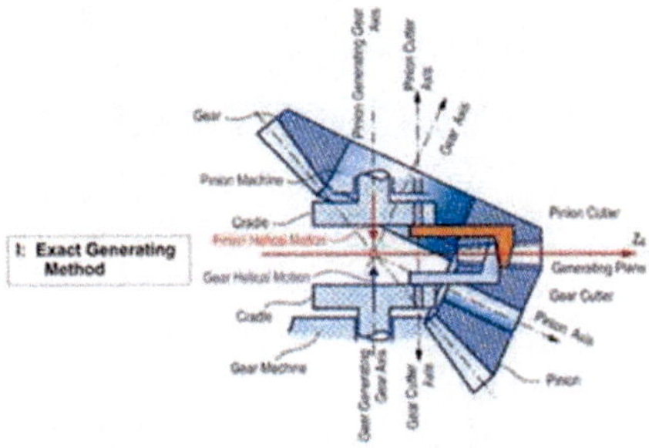

Figure 13: Generating of a conjugate tapered depth bevel gearset.

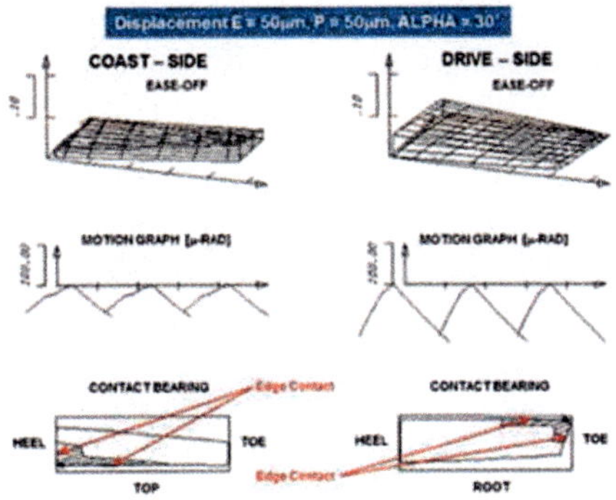

Figure 14: TCA of conjugate hypoid set from Figure 11 with shaft displacements.

It was demonstrated that the hypoid gearset with length and profile crowning in Figure 15 was developed based on a conjugate design. The first and third fundamental gearing law, mentioned in this paper, applies to the hypoid set in Figure 15 at the mean point roll position, if the load is zero. The first and third fundamental gearing law will apply in the area of contact as the load increases and the Hertzian contact spreads in contact line direction as well as in path of contact direction. This ideal condition can only be

Fig. 9.1 (continued)

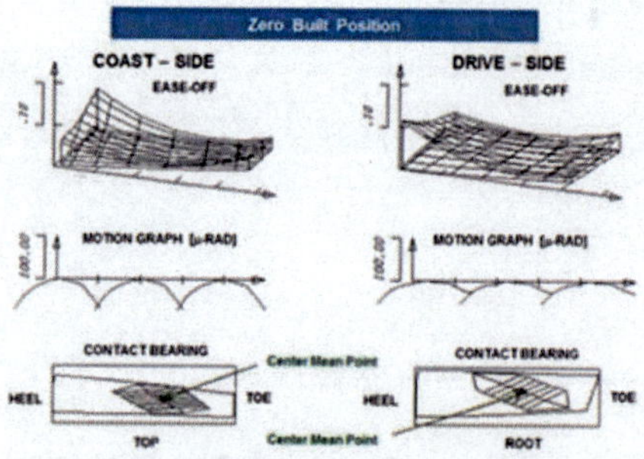

Figure 15: TCA of hypoid set from Figure 11 without any displacements.

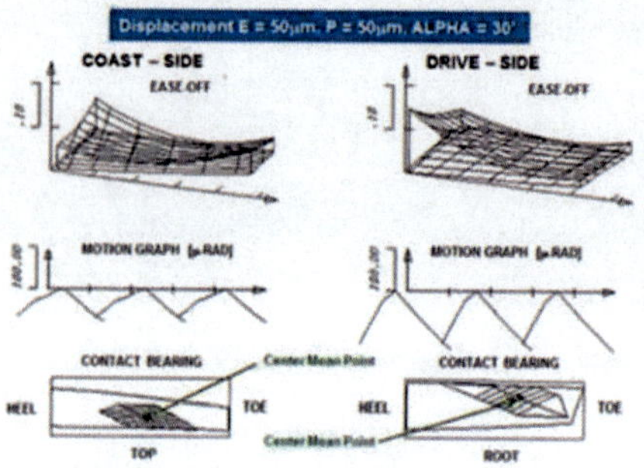

Figure 16: TCA of hypoid set from Figure 11 with added crowning and shaft displacement.

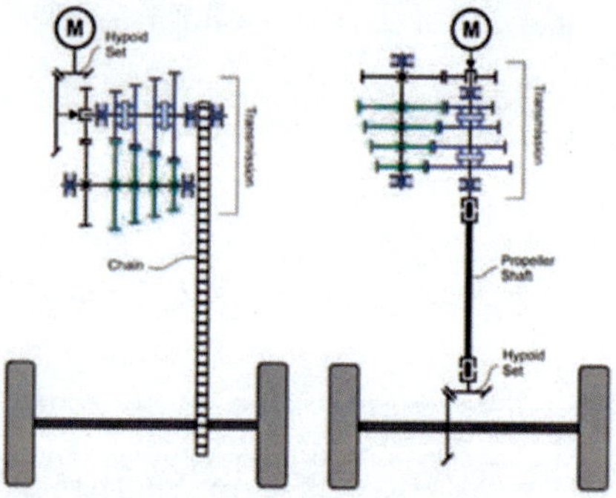

Figure 17: Hypoid gearset locations in vehicles with rear wheel drive.

achieved with correct amounts of crowning, adjusted to the operating displacements. It is interesting to mention that a hypoid gearset with crowning will fulfill the first and third fundamental gearing laws even in case of gearbox inaccuracies.

TRANSMISSION ORIENTATIONS

The proposal in [1] to place the hypoid gearset between motor and transmission is not practical. Hypoid gears are not used as simple reducers, but their purpose is to redirect rotation and torque by a certain angle, commonly 90°. In case of hypoids, the second purpose of lowering the center of gravity of a vehicle body has become very important in the automotive and truck industry. The redirection of rotation and torque has to be done at the driving axle of a vehicle. The engine orientation of vehicles with a rear wheel drive is longitudinal. Because engines of cars and trucks are commonly in the front, a longitudinal oriented propeller shaft transmits the engine rotation to the rear drive axle (Figure 17, right side). The transmission, which is located between engine and propeller shaft, needs to have a gear shaft orientation identical to the direction of the engine crankshaft. The hypoid gearset redirects rotation and torque and provides the final reduction at the drive axle. The advantage of this concept is the lower torque in the complex shift or automatic transmission and the high torque only at the ring gear at the drive axle.

A concept as shown in the left side graphic in Figure 17 would require, for example, a chain connection from the transmission output to the drive axle. This solution has low efficiency, causes high noise, provides an unacceptable packaging, and is not very reliable.

IS LAPPING AN ATTEMPT TO MAKE HYPOID GEARS CONJUGATE?

In [1, 2] it is mentioned that lapping of bevel and hypoid gearsets can be omitted if the gear geometry consists of a conjugate design. This reveals a misperception about the reason for lapping (and grinding). Lapping and grinding are hard-finishing operations. The soft manufactured bevel and hypoid gearsets have to be heat-treated, which in the most common case begins with a case carburizing of standard gear steels steel such as AISI 8620 or 16MnCr5. In order to give the low carbon steel a surface hardness in the 60 HRc range, a layer of carbon enrichment below the surface of 0.8 to 1.5mm depth is placed by a diffusion process. After the carburizing a quenching in oil and an additional tempering takes place. The result is a surface hardness that is commonly close to 60HRc and a core hardness in the 30HRc range. Case hardening provides an ideal transition between surface and core hardness that makes gears on the surface hard and wear resistant and in the core ductile. This makes shock loads and certain small plastic deformations tolerable without failure of the gearset. One major side effect of the heat-treatment process is the distortion of the gears that is caused by the carburizing, the re-crystallization of the steel, and the quenching. In order to make a gearset after heat-treatment suitable for power transmissions, for example in cars and trucks, a hard-finishing operation is required. The hard-finishing operating eliminates the heat-treatment distortions hereby providing the flank surfaces with the correct geometry from before the heat-treatment back. In addition, hard-finishing improves the surface finish to a low roughness and waviness, which enhances the hydrodynamic lubrication and reduces noise. Grinding and skiving are the preferred hard-finishing methods, creating a defined surface form that duplicates the original designed surfaces in the single micron range.

In case of face-hobbed bevel and hypoid gearsets, grinding is not possible, because of the epicyclical flank lead function. Skiving can generate epicyclical lead functions, but is not yet accepted for the high production volumes in the automotive and truck industry. This leaves only the lapping process for the hard-finishing of face-hobbed angular gearsets. However, the face-hobbed surface texture and the relative sliding between the flanks of hypoid gears make lapping an ideal alternative. Lapping can remove the surface scale

May 2019 49

Fig. 9.1 (continued)

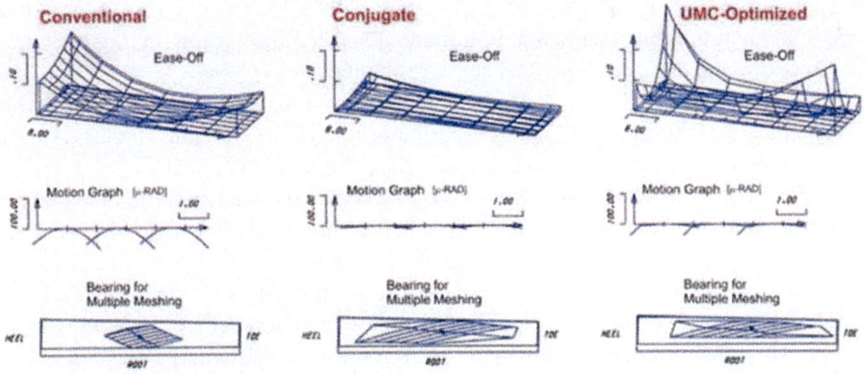

Figure 18: From conventional crowning via conjugate to UMC-optimized.

left from heat-treatment, and it re-matches two mating members by removing some runout and flank form distortions. Lapping can reduce the transmission error in many cases due to the fact that the major material removal is in the center region of the teeth where the tooth contact under light load is expected. In order for the lapping to work well, more crowning than required in the hard-finished gearset is used in the gearset design for the soft cutting. Lapping removes about 30 percent of this crowning, such that the length and profile crowning is just right after the lapping. Soft cutting of parts that are lapped after heat treatment considers a stock allowance of 0.03mm in the pinion and 0.01mm in the gear. If grinding is the hard finishing process (for face milled gearsets) then the design crowning is identical to the desired crowning after hard finishing. Between soft cutting and grinding, a uniform stock allowance of 0.10mm to 0.15mm is applied to the pinion and gear flanks.

SUMMARY

Conjugacy between the members of straight bevel, spiral bevel, and hypoid gears was only the first step and goes back more than 100 years. Quickly, the early scientists and engineers found out that conjugacy only gives us an important basis, but not a solution for power transmissions. Angular gearsets under load experience deflections that move them away from their theoretical position by half a millimeter and more. Well-designed and manufactured bevel and hypoid gearsets today can live up to those requirements and still maintain a power density that is four times higher than it was 50 years ago.

Transmission errors of 50 to 150 micro-radiant that were normal in the 1970s are in today's high-power density gearsets only between 5 and 15 micro-radiant. All this was achieved by converting a global length and profile crowning (Figure 18, left) first back to conjugacy (Figure 18, center) and then into a UMC™-optimized selective crowning, which is limited to particular regions of the teeth as shown in the right graphic in Figure 18. It is notable that the flank center of the UMC-optimized Ease-Off is conjugate, and the transmission error is next to zero. In lapping, similar effects as in grinding are achieved by using low inertia spindles with rotational compliance and high-speed torque control (SmartLap™).

The dream of conjugate angular gearsets turned out to be a false objective. Gear scientists, gear engineers, and gear manufacturers worked successfully for many decades on finding the optimal flank forms and the optimal non-conjugate flank surface interaction. The conjugate tooth design is today considered simple compared to a sophisticated higher order surface modulation. There is still room for improvement, but this cannot be achieved by going back to antiquated conjugate designs.

REFERENCES

[1] Radzevich, S.P.; Design Features of Perfect Gears for Crossed-Axes Gear Pairs; Gear Solutions Magazine, February 2019, Pages 36 to 43.

[2] Radzevich, S.P.; Preliminary Results of Testing of Low Tooth-Count Bevel Gears of a Novel Design, Part I, II and III, Gear Solutions Magazine, October 2014, December 2014 and January 2015.

[3] Wildhaber, E.; Stewart, A.J.; Design, production and application of the hypoid rear axle gear; The journal of the Society of Automotive Engineers Vol. 18, Issue 6, June 1926, Pages 575 to 589.

[4] Shtipelman, B.; Design and Manufacture of Hypoid Gears; John Wiley & Sons, Inc., New York, 1978.

[5] Stadtfeld, H.J.; Gleason Bevel Gear Technology; Expert Publishing, Renningen, Germany, May 2017.

[6] Schriefer, H.; Verzahnungsgeometrie und Laufverhalten bogenverzahnter; Kegelradgetriebe Dissertation RWTH Aachen, 1983.

[7] Wildhaber, E.; Relationships of Bevel Gears – Design for Duplex Cutting; American Machinist, October 1946.

[8] Wildhaber, E.; Relationships of Bevel Gears -Tooth Contact; American Machinist, February 1946.

ABOUT THE AUTHOR Dr. Hermann J. Stadtfeld is vice president of Bevel Gear Technology at The Gleason Works.

Fig. 9.1 (continued)

The article [2] is instructive, as it clearly shows the poor knowledge of gear experts when they make an attempt to participate in a discussion on the theoretical aspects of the gear art in a general sense of the term.

More than a year later, another article was published in the *Gear Solutions* magazine [3]. The author's attention in the article [3] was mainly focused on the analysis of the principal mistakes committed in the article [2].

This article appeared in the August 2020 issue of the *Gear Solutions* magazine. Shortly thereafter, the article was removed from the magazine. The August 2020 issue of the *Gear Solutions* magazine does not contain the earlier published article anymore [3]. This is unacceptable, but the already published article was withdrawn from the issue of the magazine[1] (evidence of this can be found on page 6 of this issue of the *Gear Solutions* magazine – for details, see Fig. 9.2, where the name of the author of this article is mistakenly retained). Scientific problems cannot be solved in this manner.

The article [3] is important, as it reports the numerous common mistakes made by other gear experts. However, because the article [3] was removed from the August 2020 issue of the *Gear Solutions* magazine and, as such, it is no longer available in the public domain, a facsimile of this article can be found in Fig. 9.3 immediately below.

In this second article [3], the author's comments about all the mistakes and inconsistencies in the article [2] pertaining to gear science are concisely outlined.

9.1 Conclusion

In this monograph, it has been outlined that the scientific theory of gearing is an efficient means for solving sophisticated problems in gear design, gear production, gear inspection, and gear application.

Only proficient, experienced, and knowledgeable gear experts are allowed to use the proposed theory of gearing. This theory has not been developed for amateurs or for inexperienced gear experts.

[1] The original version of the entire August 2020 issue of the *Gear Solutions* magazine (that contains the article [3]) is available upon request.

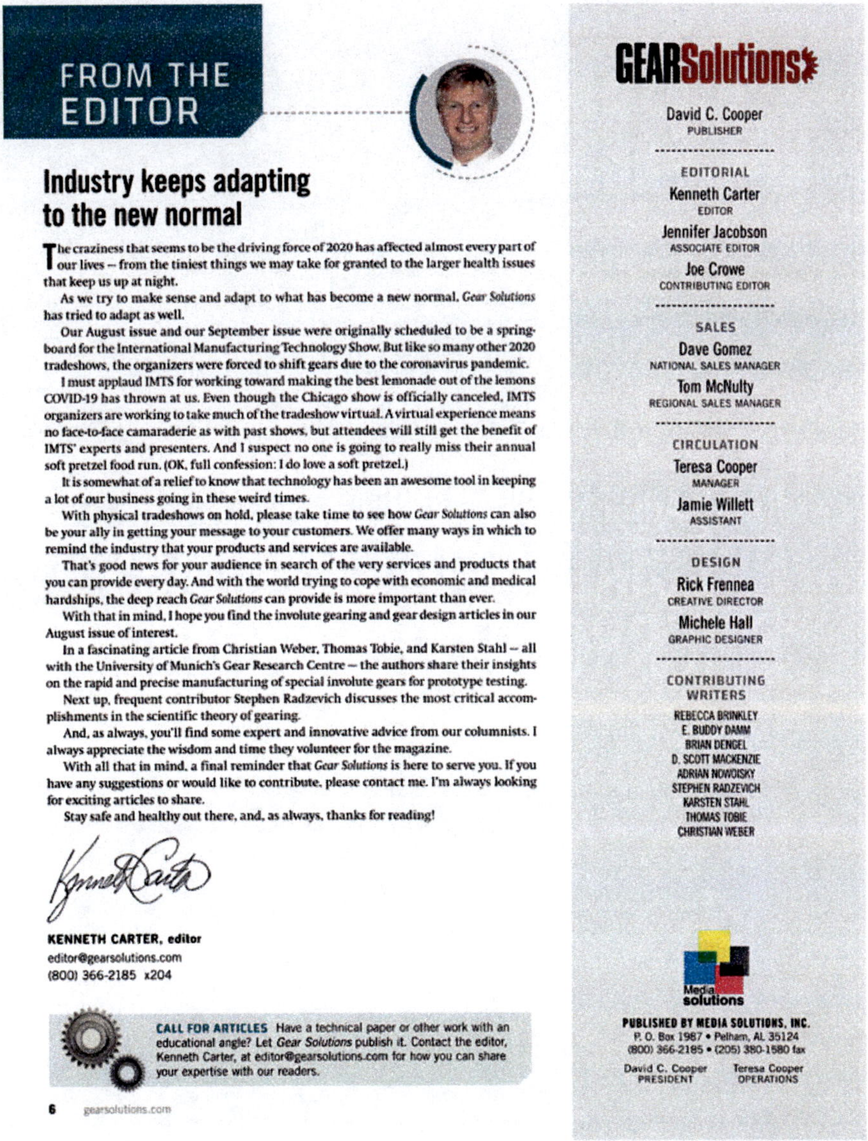

FROM THE EDITOR

Industry keeps adapting to the new normal

The craziness that seems to be the driving force of 2020 has affected almost every part of our lives – from the tiniest things we may take for granted to the larger health issues that keep us up at night.

As we try to make sense and adapt to what has become a new normal, *Gear Solutions* has tried to adapt as well.

Our August issue and our September issue were originally scheduled to be a springboard for the International Manufacturing Technology Show. But like so many other 2020 tradeshows, the organizers were forced to shift gears due to the coronavirus pandemic.

I must applaud IMTS for working toward making the best lemonade out of the lemons COVID-19 has thrown at us. Even though the Chicago show is officially canceled, IMTS organizers are working to take much of the tradeshow virtual. A virtual experience means no face-to-face camaraderie as with past shows, but attendees will still get the benefit of IMTS' experts and presenters. And I suspect no one is going to really miss their annual soft pretzel food run. (OK, full confession: I do love a soft pretzel.)

It is somewhat of a relief to know that technology has been an awesome tool in keeping a lot of our business going in these weird times.

With physical tradeshows on hold, please take time to see how *Gear Solutions* can also be your ally in getting your message to your customers. We offer many ways in which to remind the industry that your products and services are available.

That's good news for your audience in search of the very services and products that you can provide every day. And with the world trying to cope with economic and medical hardships, the deep reach *Gear Solutions* can provide is more important than ever.

With that in mind, I hope you find the involute gearing and gear design articles in our August issue of interest.

In a fascinating article from Christian Weber, Thomas Tobie, and Karsten Stahl – all with the University of Munich's Gear Research Centre – the authors share their insights on the rapid and precise manufacturing of special involute gears for prototype testing.

Next up, frequent contributor Stephen Radzevich discusses the most critical accomplishments in the scientific theory of gearing.

And, as always, you'll find some expert and innovative advice from our columnists. I always appreciate the wisdom and time they volunteer for the magazine.

With all that in mind, a final reminder that *Gear Solutions* is here to serve you. If you have any suggestions or would like to contribute, please contact me. I'm always looking for exciting articles to share.

Stay safe and healthy out there, and, as always, thanks for reading!

KENNETH CARTER, editor
editor@gearsolutions.com
(800) 366-2185 x204

CALL FOR ARTICLES Have a technical paper or other work with an educational angle? Let *Gear Solutions* publish it. Contact the editor, Kenneth Carter, at editor@gearsolutions.com for how you can share your expertise with our readers.

6 gearsolutions.com

GEARSolutions

David C. Cooper
PUBLISHER

EDITORIAL
Kenneth Carter
EDITOR
Jennifer Jacobson
ASSOCIATE EDITOR
Joe Crowe
CONTRIBUTING EDITOR

SALES
Dave Gomez
NATIONAL SALES MANAGER
Tom McNulty
REGIONAL SALES MANAGER

CIRCULATION
Teresa Cooper
MANAGER
Jamie Willett
ASSISTANT

DESIGN
Rick Frennea
CREATIVE DIRECTOR
Michele Hall
GRAPHIC DESIGNER

CONTRIBUTING WRITERS
REBECCA BRINKLEY
E. BUDDY DAMM
BRIAN DENGEL
D. SCOTT MACKENZIE
ADRIAN NOWOISKY
STEPHEN RADZEVICH
KARSTEN STAHL
THOMAS TOBIE
CHRISTIAN WEBER

Media solutions
PUBLISHED BY MEDIA SOLUTIONS, INC.
P. O. Box 1987 • Pelham, AL 35124
(800) 366-2185 • (205) 380-1580 fax
David C. Cooper PRESIDENT
Teresa Cooper OPERATIONS

Fig. 9.2 Facsimile of page 6 in the August 2020 issue of the *Gear Solutions* magazine, where the marks of the removed article [3] are still available

Fig. 9.3 Facsimile of the article: Radzevich [3]

Fig. 9.3 (continued)

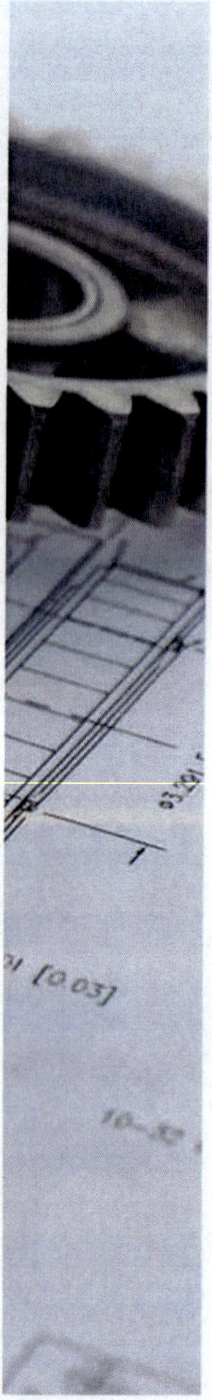

In this article, a few of the most critical accomplishments in the scientific theory of gearing will be discussed.

By STEPHEN P. RADZEVICH

The now-famous equation, "knowledge is power" ("scientia potestas est"), was coined by Francis Bacon (January 22, 1561-April 9, 1626). Formulated as early as 1597, this famous equation was proven to be correct for centuries. Knowledge in any and all areas of human activity gives one a strong advantage over those less knowledgeable ones. Gearing is no exception. Knowledgeable and skilled gear engineers always have a greater chance for success doing business, rather than less knowledgeable and less skilled ones. This article is aimed at the promotion of knowledge (namely, the promotion of the "scientific theory of gearing") to gear practitioners, and, in this way, to equip them with a means to achieving success running a business in the gear industry.

INTRODUCTION

The necessity of scientific foundations of gearing for design and manufacturing has long been realized by gear practitioners. Gears that operate on parallel axes of rotation of a gear and a mating pinion are the first to attract attention of mathematicians and gear theoreticians. It was Leonhard Euler who, in the mid-18th century, proved that an involute tooth profile fits the best gears that operate on parallel axes of rotation of a gear and a mating pinion.

Having started his own career in gearing in 1975, the author had many opportunities to get familiar with state-of-the-art innovations in both gear design and gear production. Shortly after graduating from college, the author was faced with: (a) numerous specific problems in gearing, and (b) the absence of an adequate means to solve these problems. The decades of research in the field of gearing theory by the author has resulted in a 2018 monograph titled "Theory of Gearing: Kinematics, Geometry, and Synthesis" [10]. It definitely makes sense to the gear community to get familiar with the capabilities that the proposed theory of gearing provides gear practitioners. In this article, a few of the most critical accomplishments in the scientific theory of gearing will be briefly discussed. The author concludes many in the gear community may be unfamiliar with the key accomplishments in the scientific theory of gearing. A recently published (2019) *Gear Solutions* article [15] (and not this article only) is taken as a representative example that proves this conclusion. Unfortunately, there are plenty of publications of this sort in current journals and magazines for gear engineers and scientists. Moreover, the article [15] (as well as other similar publications) may mislead gear practitioners, offering wrong approaches in the field of gear design and gear manufacturing. All the ambiguities in the practice of gear design and gear production should be eliminated. An old Chinese proverb, "The beginning of wisdom is to call things by their right names," is a good place to start.

MOTIVATION

This article was necessitated by the recently published article: Stadtfeld, H.J., "Why are Today's Hypoids the Perfect Crossed-Axes Gear Pairs?" *Gear Solutions* magazine, May 2019, pp. 42-50 [15]. These comments pertain to the key accomplishments in the scientific theory of gearing. The comments could be confusing to inexperienced gear community members, especially those who attempt to study the gear theory on their own and apply the theory in their practical work.

Here and below in this text, the article [15], is used solely as a representative example of an improper understanding of the scientific theory of gearing by a majority of gear practitioners, including those who may call themselves "gear theoreticians." An additional reason to reference the article [15] is that it claims to be written on behalf of established gear theory and manufacturing (see page 43 in [15]): "This write-up is meant to be a response on behalf of established gear theory and manufacturing.") No reference is provided in [15] as to whether this "established gear theory" is available to the reader in the public domain.

Recommendation to the readers of the current article: Except for section 1, the rest of the sections in this article intentionally reflect the article [15]. This is to help the reader better understand what these two articles are about, their differences, and why the approach outlined in the current article is advantageous over the approach recommended in the article [15]. When reading the current article, it is recommended to have the article [15] for comparison.

1 THE INTRODUCTORY SECTION: NOVIKOV GEARING (CONFORMAL GEARING)

The evolution of gear art is heavily associated with Ernest Wildhaber. No doubt, Wildhaber is a gifted inventor and a famous gear engineer. Despite that, Wildhaber has never been recognized as a gear scientist, as is loosely claimed in [15] (page 43). Because of that, Wildhaber cannot be granted with the title of "the father of modern gear theory" (see page 43 in [15]). Both these statements are incorrect: Wildhaber was NOT a gear scientist at all, and, therefore, it is incorrect to refer to him as the father of modern gear theory. Wildhaber is credited with NO key accomplishments in the scientific theory of gearing (see [10] for details). A difference between the terms "a

August 2020 39

Fig. 9.3 (continued)

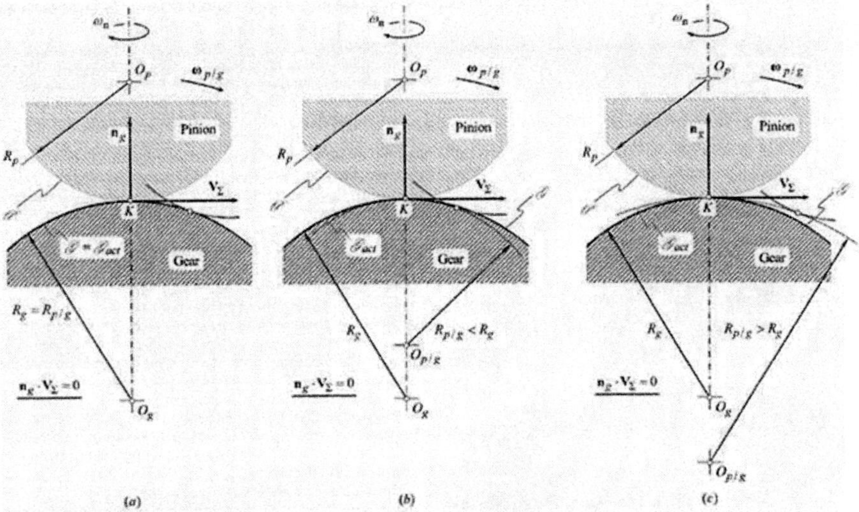

Figure 1: Examples of (a) fulfillment, and violation: (b) interference, and (c) separation, of a gear, $\mathcal{G}$, and its mating pinion, $\mathcal{P}$, tooth flanks: the "Shishkov equation of contact, $\mathbf{n}_g \bullet \mathbf{V}_\Sigma = 0$" is fulfilled in all three cases (a) through (c), while the law of conjugacy is violated in the cases (b) and (v).

gifted inventor and a famous gear engineer" and "a gear scientist" must be noted.

The earlier performed comparison of "Novikov gearing" [2], and Wildhaber "Helical gearing" [1] reveals these two gear systems are completely different, and, moreover, these gear systems are not compatible with one another [10, 7, 9]. On top of that, the term "Wildhaber-Novikov gearing" is incorrect by nature, as these two gear systems (namely, Wildhaber "Helical gearing" [1] and "Novikov gearing" [2]) cannot be combined into a common gear system (see [10, 7, 9], and others, for details). It is a huge mistake to adopt the term "Wildhaber-Novikov gearing."

"Novikov gearing" can be viewed as a reduced case of a parallel-axes involute gear pair, while Wildhaber "Helical gearing" cannot ([3, 6] and numerous other sources).

Today's gear professionals should know the difference between "Novikov gearing" [2] and Wildhaber "Helical gearing" [1].

2 NEWS ABOUT HYPOID GEARS?

The article [15] proves today's hypoid gearsets are approximate by nature, and, thus, they are not geometrically-accurate gearsets. No alterations can be even anticipated.

Further, no attempt is taken in the article [15] to "question the creditability of the scientists and gear engineers who worked on the theory and its improvement over approximately 100 years." No gear theory was developed during these 100 years, and no key scientific accomplishments are attained at this period of time. "Trial-and-error method" is the main – and perhaps only – tool used for "over approximately 100 years." But this is not a theory of gearing. Instead, this is a collection of more-or-less reasonable solutions to separate gear problems. Combined together, these solutions do not form a gear theory.

3 THE THREE FUNDAMENTAL LAWS OF GEARING

Leonhard Euler can be credited for originating geometrically-accurate gears (even though the term "geometrically-accurate gears" was introduced much later [10]). It is proven [10] that, in order to be referred to as "geometrically-accurate," the gears have to fulfill three fundamental laws of gearing. The fulfillment of three fundamental laws of gearing is necessary and sufficient to refer to a gearset as a "geometrically-accurate gearset." To the best possible extent, all three fundamental laws of gearing are discussed in [10]. These laws of gearing are briefly outlined here:

The first fundamental law of gearing: "At every point of contact of tooth flanks of a gear and a mating pinion, the vector of their instantaneous relative motion has to be orthogonal to the common perpendicular at every instant of time."

In this form, the first fundamental law of gearing is valid for gearing of all designs that operate on parallel axes, intersecting axes, and crossing axes of rotation of a gear and its mating pinion.

Numerous interpretations of the first fundamental law of gearing are known. The most practical and extensively adopted way of interpretation of the first fundamental law of gearing is to analytically describe this law of gearing in the form of the "Shishkov equation of contact, $\mathbf{n}_g \bullet \mathbf{V}_\Sigma = 0$." The "Shishkov equation of contact, $\mathbf{n}_g \bullet \mathbf{V}_\Sigma = 0$" is known to practically every gear engineer. Professor V.A. Shishkov was the first to express (~1948) the first fundamental law of gearing in the form of the dot product of a unit vector of a common perpendicular, $\mathbf{n}_g$, by the vector of the resultant linear velocity, $\mathbf{V}_\Sigma$, of the tooth flanks in relation to one another. Both the vectors, $\mathbf{n}_g$ and $\mathbf{V}_\Sigma$, are calculated at the contact point of a gear and its mating pinion tooth flanks.

The second fundamental law of gearing is often referred to as the "conjugate action law." When the "Shishkov equation of contact,

40 gearsolutions.com

Fig. 9.3 (continued)

Figure 2: Rotationally-positive external parallel-axes gear pair (Pat. No. 163857, USSR, A Helical Gearing./B.V. Shitikov, N.A. Bayazitov, Int. Cl. F06h, Filed: February 25, 1963, Published: July 22, 1964.).

$\mathbf{n}_g \bullet \mathbf{V}_\Sigma = 0$" is fulfilled, the "conjugate action law" can either be fulfilled, or it can be violated. In Figure 1, an example of interaction of local patches of the tooth flanks, $\mathcal{G}$ and $\mathcal{P}$, is schematically illustrated. The tooth flanks, $\mathcal{G}$ and $\mathcal{P}$, make contact at point, K. The radii of curvature of the interacting tooth flanks at contact point, K, equal to R_g and R_p, correspondingly (see Figure 1a). The centers of curvature of the tooth profiles, $\mathcal{G}$ and $\mathcal{P}$, are denoted by O_g and O_p, correspondingly. In the instantaneous motion of the pinion, $\mathcal{P}$, in relation to the gear, $\mathcal{G}$, the pinion performs an instantaneous rotation, $\omega_{p/g}$, about the point O_g. The radius of curvature of the generated actual gear tooth flank, $\mathcal{G}_{act}$, equals to $R_{p/g} = R_g$. In this scenario, the second fundamental law of gear is fulfilled, and the actual tooth flank, $\mathcal{G}_{act}$, is identical to the desirable gear tooth flank, $\mathcal{G}$, as shown in Figure 1a.

Note: In Figure 1a, the rotation, ω_n, of the pinion, $\mathcal{P}$, in relation to the gear, $\mathcal{G}$, about the contact perpendicular, $\mathbf{n}_g$, is not prohibited by the second fundamental law of gearing.

If the instantaneous rotation is performed either about the center $O_{p/g}$ (when $R_{p/g} < R_g$, see Figure 1b), or about the center $O_{p/g}$ (when $R_{p/g} > R_g$, see Figure 1c), the second fundamental law of gear is violated, and the actual tooth flank, $\mathcal{G}_{act}$, differs from the desirable gear tooth flank, $\mathcal{G}$.

Reminder: The first fundamental law of gearing (as well as the "Shishkov equation of contact, $\mathbf{n}_g \bullet \mathbf{V}_\Sigma = 0$") is fulfilled in all three cases shown in Figure 1, while the second fundamental law of gearing is fulfilled only in the first case illustrated in Figure 1a. The schematic in Figure 1 is helpful for understanding the difference between the first and second fundamental laws of gearing and prevents the making of unsubstantiated conclusions in this regard.

Important: When the second fundamental law of gearing is fulfilled, the first fundamental law of gearing is always fulfilled, and not vice versa.

With that said, it makes sense to go back to the article [15] and compare the results illustrated in Figure 1 with a loosely made statement (see page 44 in [15]): "The second fundamental law of gearing, ... , is a redundant relationship to the first gearing law, and it is limited to cylindrical gears with parallel axes and straight bevel gears without hypoid offset. In this case, it adds nothing to the first gearing law; conjugacy is already given by the relationships required in the first gearing law." It is correct to question whether a gear practitioner has a chance for success in the design and production of today's sophisticated gear systems if he or she strictly follows the afore mentioned quote (see page 44 in [15]).

In the most general case of gearing (namely, in the case of "crossed-axes gearing", or "$C_a - gearing$", for simplicity) the second fundamental law of gearing is formulated as:

The second fundamental law of gearing: "In order to smoothly transmit a uniform rotary motion from a driving shaft to a driven shaft by means of gear teeth, perpendiculars to the tooth flanks of the interacting teeth at all points of their contact must intersect the axis of instantaneous rotation in the gear pair."

In a reduced case of gears that operate on parallel axes of rotation of a gear and its mating pinion (namely, in the case of "parallel-axes gearing", or "$P_a - gearing$", for simplicity) this fundamental law of gearing is formulated as:

The second fundamental law of (parallel-axes) gearing: "In parallel-axes gearing, in order to smoothly transmit a uniform rotary motion from a driving shaft to a driven shaft by means of gear teeth, perpendiculars to the tooth flanks of the teeth at all points of their contact must pass through a stationary point located on the line of centers, namely, the pitch point P; the pitch point subdivides the center-distance reciprocal to the angular velocities of the gear and the pinion." (Robert Willis, 1841 [16])

These days, the second fundamental law of gearing (in a case of parallel-axes gearing) is commonly referred to as "Camus-Euler-Savari fundamental law of gearing" (or " fundamental law of gearing," for simplicity).

The second fundamental law of gearing is also often referred to as "conjugate action law" — and this is also correct.

The design of "Alpha Worm Gearing" (New Venture Gear, Syracuse, NY; US Pat. No. 6,148,683, Nov. 21, 2000) is a good example of the violation of the second fundamental law of gearing. Violation of the second fundamental law of gearing is the main reason for this design fail, and it has caused major financial damage to the gear industry.

The third fundamental law of gearing relates to the distribution of the gear teeth over the periphery of a gear and a mating pinion. In the most general case of gearing (in the case of crossed-axes gearing), the third fundamental law of gearing is formulated as:

The third fundamental law of gearing: "In order to smoothly transmit a uniform rotary motion from a driving shaft to a driven shaft by means of gear teeth, at every instant of time the angular base pitch of a gear, and that of a mating pinion, both have to be equal to the operating angular base pitch in the gear pair."

In a reduced case of "parallel-axes gearing," this fundamental law of gearing is formulated as:

The third fundamental law of gearing (in parallel-axes gearing): "In parallel-axes gearing, in order to smoothly transmit a uniform rotary motion from a driving shaft to a driven shaft by means of gear teeth, at every instant of time base pitch of a gear and that of a mating pinion, both must be equal to the operating base pitch in the gear pair."

August 2020 **41**

Fig. 9.3 (continued)

All the geometrically-accurate gears (that is, all the precision gears – with no exception) meet all three fundamental laws of gearing – this is a must.

4 PERFECT CONJUGACY IN STRAIGHT BEVEL GEARSETS

The title of the section "Perfect Conjugacy in Straight Bevel Gearsets" in the article [15] may be confusing to readers, as "conjugacy" by nature cannot be "perfect" or "imperfect." Gearsets of all kinds are either conjugate, or they are not conjugate. The term "perfect conjugacy" could be considered analogous to saying "a bit pregnant."

It is loosely claimed (see page 44 in [15]) that: "The plane of action cannot be extended beyond its tangential contacting line with the base element as shown incorrectly .... The plane of action exists where tooth engagement is possible." Gearsets that feature the plane of action that extends far beyond its tangential contacting line with the base element have been known for decades. As an example, a parallel-axes gear set of this particular kind is depicted in Figure 2 [14]. Later on, this concept was evolved (~2008) by Prof. S.P. Radzevich to the cases of intersected-axes gearing, as well as to crossed-axes gearing [10].

In a set of articles [11], [12], and [13], the preliminary results of testing of low-tooth-count spiral bevel gears of a novel design are discussed. Without going into details of design and manufacture of spiral bevel gears of this design (proposed by Prof. S.P. Radzevich, ~2008, and designed using the earlier developed scientific theory of gearing [10]), it is right to stress the following:

1. As shown in Figure 3, the predicted contact patch is perfect (see Figure 7a on page 21 in [13]).

2. As shown in Figure 4, the predicted contact pattern is perfect (see Figure 9 on page 23 in [13]).

3. As shown in Figure 5, the contact pattern in the roll test is also perfect (see Figure 10 on page 23 in [13]), and it indicates excellent correlation with the predicted one.

4. No lapping was used to finish the gears.

5. No adjusting for the axial configuration of the gear and the mating pinion is required, and the gears are ready to run.

In today's production of precision gears, the use of the lapping process is inevitable. Those knowledgeable in the scientific theory of gearing can eliminate dirty and obsolete lapping processes from finishing tooth flanks of gears for precision gearsets.

Concerning the statements on page 45 in [15] that state: "The Octoid is the analog function of an involute, and it provides to bevel gears the same advantage as an involute provides to cylindrical gears. ... bevel gears have a trapezoidal generating profile. The straight rack ... becomes a ring ... as shown in Figure 6." It has been known for a while that the tooth profile of a crown rack for generating a geometrically-accurate bevel gear has to be shaped in the form of a curve illustrated in (see Figure 6) [10], and NOT in the form of a straight-sided crown rack. Only approximate bevel gears can be generated by the just-mentioned straight-sided crown rack "shown in Figure 6" [15]. No gearing with the so-called "octoidal" path of contact is possible at all: If one assumes the path of contact shaped in a form of an octoid curve, this inevitably entails violation of the second fundamental law of gearing – it is important to bear this in mind. The so-called "octoid gearing" is a mistake that often travels from one publication to another.

Concerning the statement on page 46 in [15] that states: "Conjugacy is the basis of all gearsets manufactured in high volume on dedicated manufacturing machines." All the bevel gears manufactured in high volume on dedicated manufacturing machines are generated by means of straight-sided crown rack (shown in Figure 6, page 45 in [15]), and, thus, all of them are approximate gears.

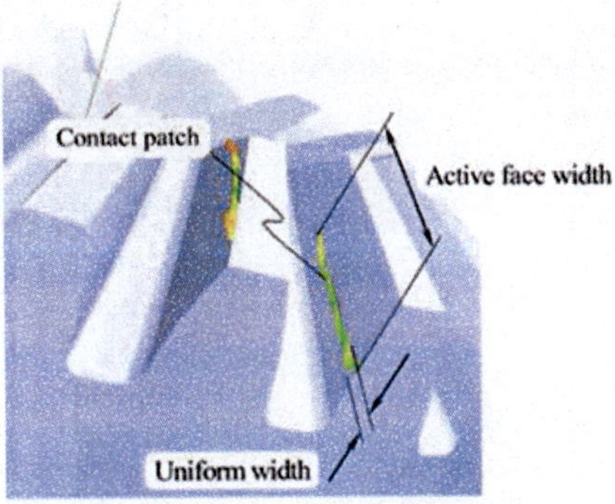

Figure 3: Predicted contact patch in geometrically-accurate intersected-axes spiral bevel gears.

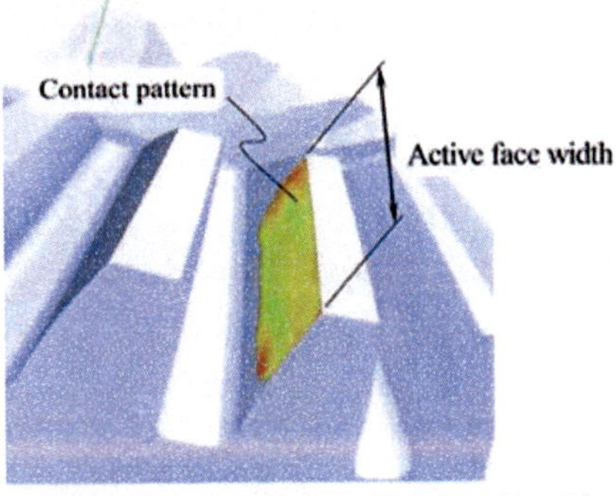

Figure 4: The predicted contact pattern in geometrically-accurate intersected-axes gearing.

Only approximate bevel gears can be generated by just mentioned straight-sided crown rack "shown in Figure 6" [15]. As the generated bevel gears are approximate, no conjugate bevel gear pairs can be composed of these gears.

Concerning the statement on page 46 in [15] that states: "Conjugate bevel gearsets cannot be used for a power transmission because manufacturing tolerances and ..." Today's gear industry is capable of manufacturing precision helical involute gears that are extensively used, for example, in the design of turbine reducers. The gear industry is NOT capable of manufacturing precision bevel gears, and this is the main reason for precision bevel gears NOT being extensively used in the industry. The bottom-line is: Precision bevel gears are NOT extensively used in the industry solely because the gear industry is not capable of producing (for a reasonable cost) these kinds of gears.

Concerning the statement on page 46 in [15] that states: "The right amount of crowning makes a gearset quiet and gives it the required load carrying capacity". How can the "right amount" be determined for a particular bevel gear application? The desirable amount of crowning can be determined by means of a "trial and error method" – this is the only tool developed by the "established

Fig. 9.3 (continued)

Figure 5: Contact pattern in geometrically-accurate spiral bevel gears in roll test.

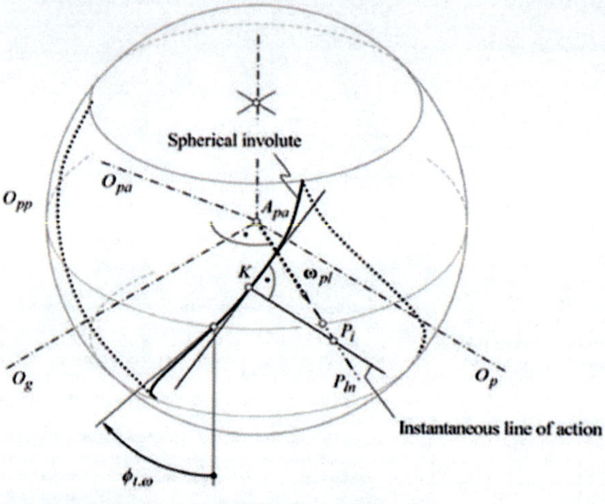

Figure 6: Trace of contact point in geometrically-accurate intersected-axes gear pair.

gear theory" in [15]. Today's scientific theory of gearing not only gives an answer to the question of how the "right amount" can be determined, but it also provides an in-detail specification of the entire modified tooth flank. Properly engineered gears (both design and production) do not need to use a "trial and error method."

5 PERFECT CONJUGACY IN HYPOID GEARSETS

The discussion on "perfect conjugacy in hypoid gearsets" in [15] is similar to "perfect conjugacy in straight bevel gearsets" discussed earlier (see Section 4). Again, the title of the section "Perfect Conjugacy in Straight Bevel Gearsets" in the article [15] may be confusing to readers as "conjugacy," by nature, can neither be "perfect"' nor "imperfect." Gearsets of all kinds are either conjugate, or they are not conjugate.

Concerning the statement on page 46 in [15] that states: "... the pitch elements of crossed axes hypoid gears are drawn as cones." It fails to describe one of the key features of the modern scientific theory of gearing [10]: In order to design gears for hypoid gearsets, no pitch elements are required at all. The term: "... the pitch elements are hyperboloids" is obsolete and is no longer used in the modern scientific theory of gearing [10].

Concerning the statement on page 47 in [15] that states: "No plane of action exists between two hyperbolic base elements." It is correct, as it proves there is no place for hyperboloids in the kinematics and the geometry of gears with crossing axes of rotation of a gear and its mating pinion.

Concerning the statement on page 47 in [15] that says: "The correct surface of action is curved and wrapped, as shown in Figure 10." Surface of action in hypoid gearing (as well as in gearing of all other kinds) is a plane through the axis of instantaneous rotation, or it is a plane through the common perpendiculars constructed at all points of the line of contact between the interacting tooth flanks of a gear and its mating pinion. The normal forces of interaction in gearing are commonly described by means of vectors. A vector is a quantity possessing both magnitude and direction. Vectors are described by means of straight-line segments with a specified direction. This leaves no room for "curved surface of action."

5.1 CONJUGACY BETWEEN MESHING FLANKS

Concerning the statement on page 47 in [15] that says: "The term conjugate is used in mathematics for two or more surfaces that contact each other along a line. Since the 1970s, the term conjugate has also been employed in gear technology literature to define..." Line contact between two surfaces (namely, between tooth flanks of two mating gears) is not sufficient (and is not always necessary) to refer to the pair of surfaces as to "conjugate surfaces." In addition, the second fundamental law of gearing has to be fulfilled for conjugate surfaces, which is a must. Moreover, it has been discovered over the last 10 years that two surfaces (two tooth flanks) can be conjugate to one another even when they are in point contact.

5.2 DEFINITION OF THE CONJUGATE GEAR PAIR

Concerning the statement on page 47 in [15] that says: "... hypoid gearset conjugacy is possible with a non-generated gear that meshes with a generated pinion." This reveals a poor understanding of the term "conjugacy," as no conjugate action is possible between non-generated tooth flank and generated tooth flank.

6 WHY IS CONJUGACY NOT DESIRABLE FOR REAL WORLD APPLICATIONS?

Most gear engineers don't doubt whether the conjugacy of inter-

Fig. 9.3 (continued)

acting tooth flanks is desirable. Conjugacy is always desirable – especially in high-power-density transmissions and in high RPM-gearsets. Unfortunately, with no lapping process in finishing the tooth flanks, the gear industry is not capable of producing precision gears for real-world gearsets. For over a decade, a concept of crossed-axes gearsets that are insensitive to the mating gears axes displacements (both, the linear displacements, as well as the angular displacements) were introduced to the public [10]. Gears of this system are commonly referred to as "S_{pr} – *gearing*." The "S_{pr} – *gearing*" is a kind of point contact conjugate gearing capable of absorbing all the linear and angular displacements of a gear and its mating pinion axes of rotation, as long as the actual values of these displacements are within the prespecified tolerances for the accuracy of the gearset. At the same time, the "S_{pr} – *gearing*" features the highest possible power density being transmitting by the gearset.

7 TRANSMISSION OPERATIONS

Only time will tell if it is practical "... to place the hypoid gearset between motor and transmission ...?" (see page 49 in [15]). Looking to the future, one can conclude that the vehicle powertrain is not the only potential application of gears with crossing axes of rotation of a gear and a mating pinion.

8 IS LAPPING AN ATTEMPT TO MAKE HYPOID GEARS CONJUGATE?

Eventually, the lapping process in the production of gears for bevel and hypoid gearsets will be eliminated [6]. The gear-grinding process is a No. 1 candidate for the future of gear-finishing operations. The sooner the key accomplishments in the scientific theory of gearing reaches gear practitioners, the sooner gear lapping will be replaced by gear grinding. This negates the statement on page 49 in [15] that says: "This reveals a misperception about the reason for lapping."

An in-depth familiarity with the key accomplishments in the scientific theory of gearing will make it possible to eliminate the gear-lapping process and replace it with the gear-grinding process. This is a reliable evolution of the gear design practice and the gear-finishing processes in particular.

9 SUMMARY

It is important to note that over the last 100 years, the gear industry has failed to:

- Eliminate lapping process.
- Avoid the necessity of adjustment of bevel/hypoid gears in pairs (that makes the gears not-self-replaceable).
- Eliminate the necessity of pairing of gears that operate on intersected, as well as on crossing axes of rotation of a gear and its mating pinion.
- Solve the problem of excessive gear noise excitation and vibration generation.

CONCLUSION

In conclusion, it is likely the gear industry could end up wasting funds and time following the direction of evolution outlined in the article [15].

In the article [15] there is no evidence of:

- Whether the third fundamental law of gearing is understood by the author.
- The importance of the concept of the "operating base pitch" in a gear pair is realized.
- The concept of a gear, of a mating pinion base pitch in intersected-axes gearing, and in crossed-axes gearing, along with the concept of the "operating base pitch" in intersected-axes, and in crossed-axes gearsets as realized by the author.

Not all the inconsistencies in [15] are addressed here – only some of the larger issues. However, a more detailed analysis (if necessary) can be undertaken.

REFERENCES

[1] Pat. USA, No. 1,601,750, Helical Gearing, /E. Wildhaber, Patented: October 5, 1926, Filed: November 2, 1923.

[2] Pat. USSR, No. 109,113, Gear Pairs and Cam Mechanisms Having Point System of Meshing. /M.L. Novikov, National Classification 47h, 6; Filed: April 19, 1956, published in Bull. of Inventions No.10, 1957.

[3] Radzevich, S.P., "An Examination of High-Conformal Gearing", Gear Solutions, February, 2018, pages 31-39.

[4] Radzevich, S.P., "Conjugate Action Law in Intersected-Axes Gear Pairs and in Crossed-Axes Gear Pairs", Gear Solutions magazine, June 2020, pages 42-48.

[5] Radzevich, S.P., "Design Features of Perfect Gears for Crossed-Axes Gear Pairs", Gear Solutions, February, 2019, pp. 36-43.

[6] Radzevich, S.P., Gear Cutting Tools: Science and Engineering, 2nd Edition, CRC Press, Boca Raton, FL, 2017, 606 pages.

[7] Radzevich, S.P., Dudley's Handbook of Practical Gear Design and Manufacture, Third Edition, CRC Press, Boca Raton, FL, 2016, 629 pages.

[8] Radzevich, S.P., High-Conformal Gearing: Kinematics and Geometry, 2nd edition, Elsevier, Amsterdam, 2020, 506 pages.

[9] Radzevich, S.P., "On the Inconsistency of the Term "Wildhaber-Novikov Gearing": A New Look at the Concept of "Novikov Gearing"", Appendix A, pp. 487-501, in: Radzevich, S.P., (Editor), Advances in Gear Design and Manufacture, CRC Press, Boca Raton, Florida, 2019, 549 pages.

[10] Radzevich, S.P., Theory of Gearing: Kinematics, Geometry, and Synthesis, 2nd edition, revised and expanded, CRC Press, Boca Raton, FL, 2018, 934 pages.

[11] Radzevich, S.P., et al, "Preliminary Results of Testing of Low-Tooth-Count Bevel Gears of a Novel Design. Part 1", Gear Solutions, October 2014, pp. 25-26.

[12] Radzevich, S.P., et al, "Preliminary Results of Testing of Low-Tooth-Count Bevel Gears of a Novel Design. Part 2", Gear Solutions, December 2014, pp. 20-21.

[13] Radzevich, S.P., et al, "Preliminary Results of Testing of Low-Tooth-Count Bevel Gears of a Novel Design. Part 3", Gear Solutions, January 2015, pp. 20-23.

[14] Shitikov, B.V., Bayazitov, N.A., A Helical Gearing, Pat. No. 163857, USSR, Int. Cl. F06h, Filed: February 25, 1963, Published: July 22, 1964. [see also: Bayazitov, N.A., Helical Gears with a New Type of Gearing, Ph.D. Thesis, Kazan', Kazan' Technological & Chemical Institute, 1964.].

[15] Stadtfeld, H.J., "Why are Today's Hypoids the Perfect Crossed-Axes Gear Pairs?", Gear Solutions magazine, May 2019, pp. 42-50.

[16] Willis, R., Principles of Mechanisms, Designed for the Use of Students in the Universities and for Engineering Students Generally, London, John W. Parker, West Stand, Cambridge: J. & J.J. Deighton, 1841, 446 pages.

ABOUT THE AUTHOR Stephen P. Radzevich, Dr.(Eng.)Sci, can be reached at radzevich@usa.com or at (586) 292-7209.

Fig. 9.3 (continued)

References

1. Radzevich, S.P.: Theory of Gearing: Kinematics, Geometry, and Synthesis, 2nd edn, revised and expanded, 934p. CRC Press, Boca Raton (2018) [1st edn: Radzevich, S.P.: Theory of Gearing: Kinematics, Geometry, and Synthesis, 760p. CRC Press, Boca Raton (2012)]
2. Stadtfeld, H.J.: Why are today's hypoids the perfect crossed-axes gear pairs? Gear Solutions, pp. 42–50 (May 2019)
3. Radzevich, S.P.: Knowledge (of gear theory) is power in the design, production, and application of gears. Gear Solutions, pp. 38–44 (August 2020)

Chapter 10
Gear Manufacturing Accuracy Prediction, Control, and Management

Mykola E. Terniuk, Anatolii V. Kryvosheia, Alexander M. Krasnoshtan, Pavlo M. Tkach, and Sergey V. Lutskii

10.1 Introduction

The difference between real shaping and nominal (ideal) shaping is the presence of the geometrical deviations of the real profile compared with the nominal ones.

The main final problems, to be solved in the research of the regularities of real shaping, are those of forecasting with a given reliability of arising errors as well as the problems of control, including the optimal one, by these errors.

Controlling is impossible without control. Therefore, the tasks of optimizing processes and control schemes should also be studied.

In accordance with the current standards for the methods of standardizing the errors of a gear, the conditions for ensuring accuracy, not lower than the required one, can be modeled by the following inequalities:

$$\begin{aligned} &[\Pi_{ni}]^{+} \geq f_{1i}\left(\overline{\Delta}_{\mathrm{np}}, \overline{\Pi}_{\mathrm{np}}, P\right) \geq [\Pi_{ni}]^{-};\\ &[\Pi_{ci}]^{+} \geq f_{2i}\left(\overline{\Delta}_{\mathrm{cp}}, \overline{\Pi}_{\mathrm{cp}}, P\right) \geq [\Pi_{ci}]^{-};\\ &i \in J, \end{aligned} \tag{10.1}$$

where

Π_{ni}, Π_{ci} are the limiting and average values, respectively, of the ith standardized error.

M. E. Terniuk · S. V. Lutskii
International Academy of Sciences and Innovation Technologies, Kyiv, Ukraine

A. V. Kryvosheia
Bakul' Institue for Superhard Materials, Kyiv, Ukraine

A. M. Krasnoshtan
National Transport University, Kyiv, Ukraine

P. M. Tkach
Paton Electric Welding Institute, Kyiv, Ukraine

S. P. Radzevich, *Novikov/Conformal Gearing*,
https://doi.org/10.1007/978-3-031-10019-2_10

f_{1i}, f_{2i} are the analytical or algorithmic functions.

$\overline{\Pi}_{\text{np}}$, $\overline{\Pi}_{\text{cp}}$ are the vectors of the limiting and average values, respectively, of primary errors and process factors that determine the errors of the gear.

P is the given forecast reliability.

J is the number of standardized errors that determine the kinematic accuracy, smoothness of operation, contact pattern, and backlash in the transmission.

Section 10.2 contains general equations of real tooth profiles, dependencies of the f_{1i} and f_{2i} types are obtained, and methods of information support of the obtained dependencies are developed. Sections 10.3 and 10.4 present the possible methods and technical means of control and consider the possible ways to enforce conditions (10.1).

The resulting dependencies are common to all methods of shaping teeth by machining. The generality is necessary for implementation of the principle of completeness of information when solving problems of optimization synthesis.

The prerequisites for carrying out general studies are the proposed thesis on the isomorphism of the processes of real shaping as well as the previously obtained fundamental results in the field of geometry of gears, the theory of shaping of surfaces [5, 6, 21, 22, 26, 27, 44], and the theory of accuracy [2, 3, 36].

10.2 Modeling the Main Regularities of the Processes of Real Shaping of Teeth and Development of the Methods for Analytical Forecasting of Errors

10.2.1 Development of a System Model of the Process of Real Shaping of Teeth

The model of the real shaping process should reflect the factors causing errors, taking into account the connections between them.

The methodology of the general approach adopted in this work requires, at the first stage of research, the advancement of the assumption that all elements of a technological system affect the processing accuracy. Therefore, potentially valid initial objects are established: blanks, technical means, environment, performers, and the details themselves.

It is obvious that errors occur at different stages of the life cycle of a technological system, but they usually come about at the stage of work. Moreover, a number of primary errors of technical means and blanks materialize at the stages of manufacture (restoration) and adjustment. Therefore, it can be argued that the process of real shaping is influenced by operators and the workpiece, the manufacture (restoration) and adjustment of technical means, the environment, parts, and the process of the system itself, which directly affects the environment, the workpiece, and the part.

From the above-listed factors, we will select those that significantly affect the geometric deviations.

Based on the analysis of the physical processes occurring in technological systems, carried out in the study by Bazrov [3], it can be considered that they significantly affect the geometric deviations: the initial geometric characteristics and the thermal, power, kinematic, and wear phenomena.

These phenomena are the sources of possible geometric inaccuracies, internal stresses, forces, wear, temperature fluctuations, and distortions of the trajectories of the working parts of systems. As a result, the so-called primary errors of a system arise: geometric, power, thermal, kinematic, and wear errors.

All of these types of primary errors, as a result, materialize initially in the form of geometric deviations of the machine–fixture–tool–part (MFTP) system and then in the form of geometric deviations of the profiles of the teeth of a machined gear.

Geometric deviations of an MFTP system can be emphasized with varying degrees of detail.

For practical use, a system for detecting the geometric errors of an MFTP system is desirable, which would make it possible to separately identify the degree of influence of the main elements of this system on the total error of gears. In addition, it is advisable to isolate errors that would make it possible to reveal their physical meaning and make a direct measurement.

The above conditions are reached by a system that distinguishes the following:

(a) Generalized errors of the machine are errors of doubling of the final links; there may be six such errors according to the number of the possible degrees of freedom of the final links in relative motion, like any material body.
(b) Errors in setting the tool and parts as the displacement of the latter in relation to the end links of the machine; each of the errors, in the general case, contains six linearly independent components according to the number of the possible degrees of freedom of solid parts.
(c) Errors of the instrument itself as deviations of real (actual) coordinates in relation to the nominal ones; there can be three such linearly independent errors in accordance with the number of coordinates.
(d) Errors of the part itself as recoverable deformations (elastic and thermal), which always occur during processing; there can be three errors, like the errors of the tool, in accordance with the number of coordinates.

Thus, the geometric errors of an MFTP system can be reduced to a full set of errors, consisting of 24 components. These 24 components, due to the fact that the entire multitude of any error of an MFTP system can be reduced to them, are called cited primary errors.

By being aware of the abovementioned primary errors, it is possible, by developing a system for their accounting, to calculate the normalized errors of a ring gear: indicators of kinematic accuracy, smoothness of operation, contact patch, and side clearance.

If we take quaternions of errors as elements of reduction, then, on the basis of the above, the general system model of the occurrence and accounting of errors, as the main model of the process of real shaping, can be represented in the form of a block diagram as shown in Fig. 10.1.

The resulting model reflects the general patterns of the processes of real shaping of teeth and allows one to proceed to drawing up the general equations of their profiles.

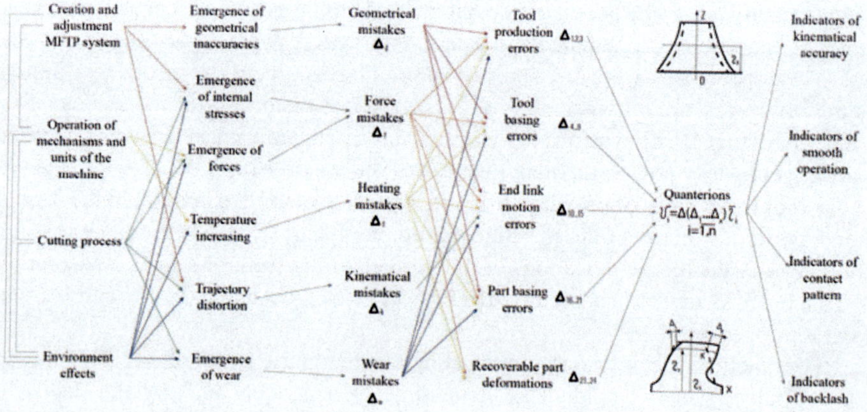

Fig. 10.1 System model for the occurrence and accounting of errors

10.2.1.1 Derivation of the General Equations of the Real Profiles of the Teeth of Machined Gears

To obtain the general equations of real tooth profiles, we associate the coordinate system with the nominal end links of a machine tool

$$S'_{1H}\{X_{1H},\ \ Y_{1H},\ \ Z_{1H}\}\ \text{И}\ S'_{2H}\{X_{2H},\ \ Y_{2H},\ \ Z_{2H}\}.$$

Let us give these systems a relative motion that realizes all six degrees of freedom and coordinate the components of the motion with the parameters, as shown in Fig. 10.2.

The nominal axis *I–I*, around which a tool can rotate with the parameter ϕ_1, is crossed with the nominal axis II–II of the workpiece rotation with the parameter ϕ_2 at an angle γ. The center-to-center distance as well as the displacements of the *I–I* axis in a plane parallel to the plane $Y'_{1H}O'_{1H}Z'_{1H}$ are indicated by the parameters a_w, ψ_y and ψ_z, respectively. The quantities γ, a_w, ψ_y, ψ_z, and ϕ_2 are functions of the envelope parameters and, in turn, ϕ_1 and ψ are the functions of time, i.e.,

$$\gamma=\gamma(\phi_1,\psi),\ a_w=a_w(\phi_1,\psi),\ \psi_y=\psi_y(\phi_1,\psi),\ \psi_z=\psi_z(\phi_1,\psi),\ \text{and}\ \phi_2=\phi_2(\phi_1,\psi)$$

The relationship between the coordinate systems S'_{1H} and S'_{2H} (Fig. 10.2) excluding the rotations of the end links can be expressed by a matrix compiled by the method described in the study by Litvin and Fuentes [16]:

$$\widetilde{M}_H=\begin{pmatrix} 0 & \cos\xi & -\sin\xi & a_w \\ -\sin\gamma & \cos\gamma\sin\xi & \cos\gamma\cos\xi & \psi_y \\ \cos\gamma & \sin\gamma\sin\xi & \sin\gamma\cos\xi & \psi_z \\ 0 & 0 & 0 & 1 \end{pmatrix} \tag{10.2}$$

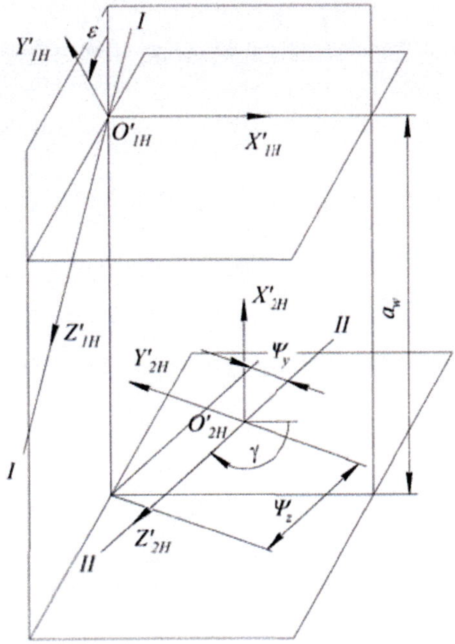

Fig. 10.2 Nominal position of coordinate systems

In accordance with the general model (Fig. 10.1), the errors of a machined gear can be presented as a result of the disturbance of the shaping movement [3] as well as a result of the transfer of the primary errors of the tool to the workpiece and the manifestation of recoverable deformations of the part.

Violation of the shaping movement occurs due to deviations from the nominal positions of the axes of the end links of the machine as well as the mismatch of the geometric axes of the tool and workpiece with these axes.

In accordance with the obtained results [3], we describe such violations by the matrices: $\widetilde{M}_{\Delta H}$,reflecting the possible displacements of the real axes of the end links of the machine in relation to the nominal ones, and $\widetilde{M}_{\Delta 1}$ and $\widetilde{M}_{\Delta 2}$, taking into account the installation errors, namely, the displacement of the base surfaces of the tool and part, respectively, in relation to the real axes of the end links of the machine, on which the tool and part are installed.

It is possible to find the structure and values of the matrix elements $\widetilde{M}_{\Delta 1}$, $\widetilde{M}_{\Delta 2}$, and $\widetilde{M}_{\Delta H}$ due to the fact that these matrices describe the possible displacements of one body with respect to another, and, thus, they can realize all six degrees of freedom.

Based on the study by Litvin and Fuentes [16], we have

$$\widetilde{M}_{\Delta i} = \widetilde{M}_{\Delta iz} \cdot \widetilde{M}_{\Delta iy} \cdot \widetilde{M}_{\Delta ix}; \quad i \in \{H, 1, 2\}, \tag{10.3}$$

where

$$\widetilde{M}_{\Delta iz} = \begin{pmatrix} \cos \Delta\phi_{zi} & \sin \Delta\phi_{zi} & 0 & 0 \\ -\sin \Delta\phi_{zi} & \cos \Delta\phi_{zi} & 0 & 0 \\ 0 & 0 & 1 & \Delta z_i \\ 0 & 0 & 0 & 1 \end{pmatrix};$$

$$\widetilde{M}_{\Delta iy} = \begin{pmatrix} \cos \Delta\phi_{yi} & 0 & -\sin \Delta\phi_{yi} & 0 \\ 0 & 1 & 0 & \Delta y_i \\ \sin \Delta\phi_{yi} & 0 & \cos \Delta\phi_{yi} & 0 \\ 0 & 0 & 0 & 1 \end{pmatrix};$$

$$\widetilde{M}_{\Delta ix} = \begin{pmatrix} 1 & 0 & 0 & \Delta x_i \\ 0 & \cos \Delta\phi_{xi} & \sin \Delta\phi_{xi} & 0 \\ 0 & -\sin \Delta\phi_{xi} & \cos \Delta\phi_{xi} & 0 \\ 0 & 0 & 0 & 1 \end{pmatrix}.$$

Here, $\Delta\phi_{zi}$, $\Delta\phi_{yi}$, and $\Delta\phi_{xi}$ are excessive turns, and Δz_i, Δy_i, and Δx_i are excessive displacements of real axes (bases) with respect to the nominal ones. The accepted sequence of the offsets of the coordinate systems is illustrated in Fig. 10.3, where the subscripts r and p indicate the coordinate axes of the auxiliary systems.

Taking into account the possible rotations of the tool and the workpiece, and the resulting matrices (10.2) and (10.3), the relation between the technological axes is expressed by the matrix:

$$\widetilde{M}_{21} = \widetilde{M}_{2H} \cdot \widetilde{M}_{\Delta H} \cdot \widetilde{M}_{H} \cdot \widetilde{M}_{H1}, \tag{10.4}$$

where

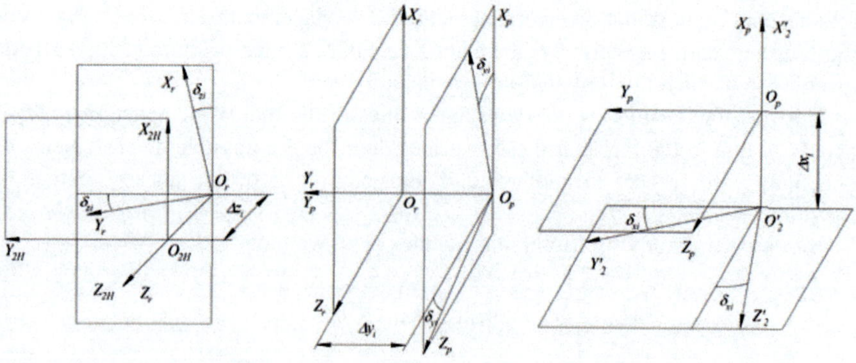

Fig. 10.3 Intermediate coordinate systems

Fig. 10.4 Structural element diagram of the simplest MFTP system

$$\widetilde{M}_{2H}=\begin{pmatrix}\cos\phi_2 & \sin\phi_2 & 0 & 0\\ -\sin\phi_2 & \cos\phi_2 & 0 & 0\\ 0 & 0 & 1 & 0\\ 0 & 0 & 0 & 1\end{pmatrix};\ \widetilde{M}_{H1}=\begin{pmatrix}\cos\phi_1 & -\sin\phi_1 & 0 & 0\\ \sin\phi_1 & \cos\phi_1 & 0 & 0\\ 0 & 0 & 1 & 0\\ 0 & 0 & 0 & 1\end{pmatrix}.$$

Here, ϕ_1 and ϕ_2 are the angles of rotation of the tool and the gear around the axes Z_1' and Z_2', respectively.

Matrix (10.4) does not reflect all the factors that violate the required law of relative motion of the elements of the AIDS system, since it does not take into account the possible displacements of the base surfaces of the tool and part with respect to the technological axes. However, it is necessary for further derivation of the equation of the real surfaces of the teeth of machined gears. This matrix was compiled for a case in which the structure of the MFTP system is the simplest and corresponds to Fig. 10.4.

With regard to this case, the general matrix of the relationship between coordinate systems that are rigidly connected to the tool surface and the structural base of the machined gear, reflecting the violation of the shaping movement in full, has the form:

$$\widetilde{M}_{\Sigma}=\widetilde{M}_{\Delta 2}\cdot\widetilde{M}_{2H}\cdot\widetilde{M}_{\Delta H}\cdot\widetilde{M}_{H}\cdot\widetilde{M}_{H1}\cdot\widetilde{M}_{\Delta 1}. \tag{10.5}$$

Due to the fact that the matrices $\widetilde{M}_{\Delta 2}$, $\widetilde{M}_{\Delta 1}$, and $\widetilde{M}_{\Delta H}$ contain six linearly independent (albeit correlated in modulus) elements, the total number of parameters reflecting the violation of the shaping movement is 18.

The resulting matrices (10.5), jointly with expressions for determining the radius vector of the tool surface, make finding the equations of the real profiles of the teeth of the gear being machined possible. However, these equations will not highlight those parameter changes that can be accepted for active control of machining accuracy.

If we take into account the above evidence that a full set of the given primary errors contains 24 components, 18 of which reflect the violation of the shaping movement and the remaining 6 are the errors of the instrumental surface itself and the recoverable deformations of the gear, then it can be argued that a full set of all possible methods of error compensation should contain 3 groups of techniques. One is to compensate for errors in shaping movements, the second is to compensate for errors on the instrumental surface itself, and the third one is to compensate for recoverable deformations of the gear. Error compensation issues are discussed in

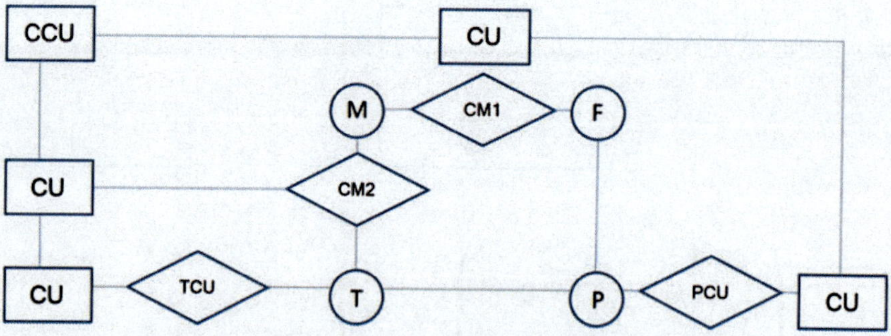

Fig. 10.5 Block diagram of an MFTP system containing a full set of devices for compensating errors

detail below. Here, on the basis of the aforementioned, we note that an MFTP system, in the general case, may contain devices for changing the movements of the machine during processing – compensation mechanisms for small displacements – as well as devices for controlling the size and shape of the tool surface and devices for controlling the recoverable deformations of the part, in particular, the deformations of its teeth. The structural diagram of such an MFTP system is shown in Fig. 10.5.

In Fig. 10.5,

CM is the compensation mechanism, which, in the general case, can be two – one at each of the output links of the machine.
CU is the control unit.
CCU is the common control unit.
TCU is the tool dimensions and shape controlling unit – a device for controlling the size and shape of the tool.
PCU is the part recoverable strain controlling unit – a device for managing the recoverable strain of the part.
M is the machine.
F is the fixture.
T is the tool.
P is the part.

In general, each of the compensation mechanisms can have six linearly independent controlled movements. These movements can be described by a matrix having a matrix structure $\widetilde{M}_{\Delta i}$, the elements of which are indicated by the indices K_1 and K_2.

The function of devices for controlling the dimensions and shape of the tool, as well as the recoverable deformations of the part, can be described by the corresponding three-component increments of the radius vectors $\Delta\bar{r}_{1K}$ and $\Delta\bar{r}_{2K}$.

The above allows one to proceed to the derivation of the general equation of real tooth profiles, taking into account the possible actions to manage errors by compensating for errors. In this case, we additionally take into account that due to the appearance of processing absence zones and large initial errors (for example, when shaving or grinding), adjustment errors for size, the phenomenon of "edge loss" [12, 36], and other reasons, the case in which only a part of the surface is processed (and remains "black") can be considered common. From this, as a specific case, the case of complete surface treatment of the teeth follows.

In the general case, the processing of toothed ears can be carried out according to a two-parameter scheme. The one-parameter processing scheme is a special one. Formally, multi-parameter schemes, by replacing parameters, can be reduced to two-parameter schemes [22].

The estimation of the errors of gears should be made not from a technological but from an operational (constructive) base, although the processing is carried out from a technological one. The relationship between these bases is expressed by a matrix $\widetilde{M}_{\Delta 2}$.

Let the nominal tool surface $S_1\{X_1,\ Y_1,\ Z_1\}$ be specified in the coordinate system

$$\overline{r}_{1H}=\overline{r}_{1H}(u,\, v). \tag{10.6}$$

Then, the coordinates of the real instrumental surface on the ith pass, taking into account the errors described by the increment of the radius vector $\Delta\overline{r}_{1i}$, and the compensation of errors due to the input $\Delta\overline{r}_{1K}$, taking into account (10.6) will be

$$\overline{r}_{1i}=\overline{r}_{1H}+\Delta\overline{r}_{1i}+\Delta\overline{r}_{1K}. \tag{10.7}$$

The envelope (if it exists) of the family of surfaces formed by the instrumental surface described by (10.7), when it moves with parameters ϕ_1 and ψ, can be set by the equations:

$$\overline{r}_{2i}=\overline{r}_{2i}(u,\, v,\, \Delta\overline{r}_{1i},\, \Delta\overline{r}_{1K},\, \phi_1,\, \psi); \tag{10.8}$$

$$\left[\frac{\partial\overline{r}_{2i}}{\partial u},\, \frac{\partial\overline{r}_{2i}}{\partial v},\, \frac{\partial\overline{r}_{2i}}{\partial\phi_1}\right]=0; \tag{10.9}$$

$$\left[\frac{\partial\overline{r}_{2i}}{\partial u},\, \frac{\partial\overline{r}_{2i}}{\partial v},\, \frac{\partial\overline{r}_{2i}}{\partial\psi}\right]=0. \tag{10.10}$$

Equations (10.9) and (10.10) can be transformed. To do this, we will rewrite them in the form:

$$\left(\frac{\partial\overline{r}_{2i}}{\partial u}\times\frac{\partial\overline{r}_{2i}}{\partial v}\right)\frac{\partial\overline{r}_{2i}}{\partial\phi_1}=0; \tag{10.11}$$

$$\left(\frac{\partial\overline{r}_{2i}}{\partial u}\times\frac{\partial\overline{r}_{2i}}{\partial v}\right)\frac{\partial\overline{r}_{2i}}{\partial\psi}=0. \tag{10.12}$$

For the case of reference $\overline{r}_{2i}$ relative to the technological axis, we write:

$$\overline{r}_{2i} = \widetilde{M}_{\Sigma H}\overline{r}_{1i}, \tag{10.13}$$

where $\widetilde{M}_{\Sigma H} = \widetilde{M}_{2H} \cdot \widetilde{M}_{\Delta H} \cdot \widetilde{M}_{H} \cdot \widetilde{M}_{H1} \cdot \widetilde{M}_{\Delta K} \cdot \widetilde{M}_{\Delta 1}$.

In accordance with the structure of the matrix $\widetilde{M}_{\Sigma H}$, its elements are functions ϕ_1 and ψ; therefore, using (10.13) we find:

$$\frac{\partial \overline{r}_{2i}}{\partial u} = \widetilde{M}_{\Sigma H} \cdot \frac{\partial \overline{r}_{1i}}{\partial u} \tag{10.14}$$

$$\frac{\partial \overline{r}_{2i}}{\partial v} = \widetilde{M}_{\Sigma H} \cdot \frac{\partial \overline{r}_{1i}}{\partial v} \tag{10.15}$$

$$\frac{\partial \overline{r}_{2i}}{\partial \phi_1} = \frac{\partial \widetilde{M}_{\Sigma H}}{\partial \phi_1} \cdot \overline{r}_{1i} + \widetilde{M}_{\Sigma H} \cdot \frac{\partial \overline{r}_{1i}}{\partial \phi_1} \tag{10.16}$$

$$\frac{\partial \overline{r}_{2i}}{\partial \psi} = \frac{\partial \widetilde{M}_{\Sigma H}}{\partial \psi} \cdot \overline{r}_{1i} + \widetilde{M}_{\Sigma H} \cdot \frac{\partial \overline{r}_{1i}}{\partial \psi} \tag{10.17}$$

it is known [5, 14], that

$$\frac{\partial \overline{r}_{2i}}{\partial u} \times \frac{\partial \overline{r}_{2i}}{\partial v} = \overline{n}_2 = \left(\widetilde{M}_{\Sigma H}^{-1}\right)^T \cdot \overline{n}_1, \tag{10.18}$$

where

$$\overline{n}_1 = \frac{\partial \overline{r}_{1i}}{\partial u} \times \frac{\partial \overline{r}_{1i}}{\partial v}; \tag{10.19}$$

«T» is the transposition sign and the superscript "-1" means an inverse matrix.

Substituting (10.18) and (10.19) in (10.11) and (10.12) and taking into account (10.14), (10.15), (10.16), and (10.17), like it was done in the study by Erikhov [5], we obtain:

$$\left(\frac{\partial \overline{r}_{1i}}{\partial u} \times \frac{\partial \overline{r}_{1i}}{\partial v}\right)\left(\widetilde{M}_{\Sigma H}^{-1} \frac{\partial \widetilde{M}_{\Sigma}}{\partial \phi_1} \cdot \overline{r}_{1i} + \frac{\partial \overline{r}_{1i}}{\partial \phi_1}\right) = 0; \tag{10.20}$$

$$\left(\frac{\partial \overline{r}_{1i}}{\partial u} \times \frac{\partial \overline{r}_{1i}}{\partial v}\right)\left(\widetilde{M}_{\Sigma H}^{-1} \frac{\partial \widetilde{M}_{\Sigma}}{\partial \psi} \cdot \overline{r}_{1i} + \frac{\partial \overline{r}_{1i}}{\partial \psi}\right) = 0; \tag{10.21}$$

Thus, the envelope of the surfaces of the teeth relative to the technological axis will be described by Eqs. (10.8), (10.20), and (10.21).

Let us introduce and consider the definite unit functions of the asymmetrically positive $U_+(x)$ and negative $U_-(x)$. Their graphs are shown in Fig. 10.6.

Then, if these functions are considered as an argument $x = |\overline{r}_{2i}| - |\overline{r}_{2(i-1)}|$, then one can write:

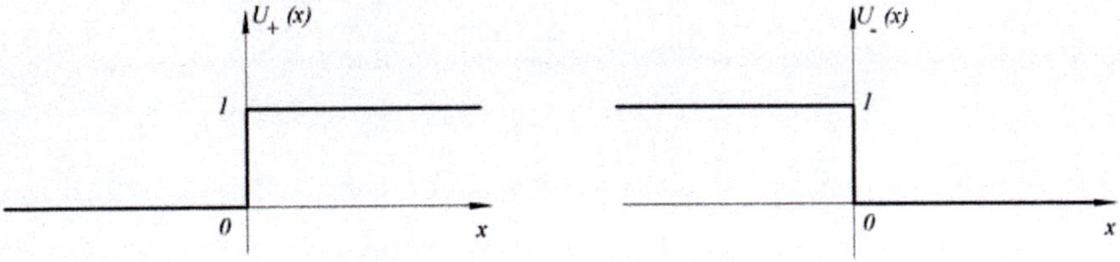

Fig. 10.6 Asymmetric unit function plot

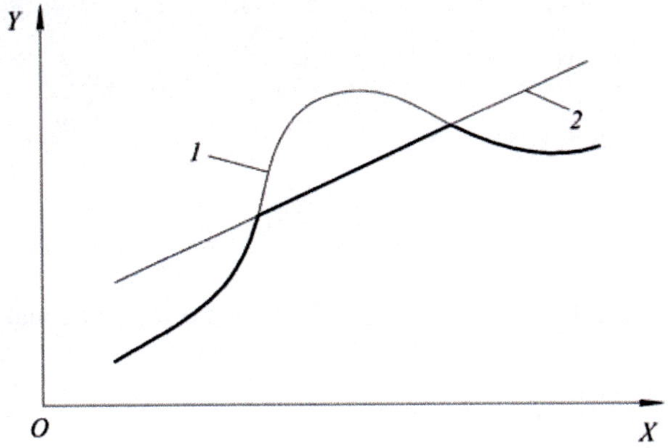

Fig. 10.7 Real surface formation diagram

$$\bar{r}_{2\Phi} = \bar{r}_{2i} \cdot U \pm \left(|\bar{r}_{2i}| - |\bar{r}_{2(i-1)}|\right) + \bar{r}_{2(i-1)} \cdot U \pm \left(|\bar{r}_{2(i-1)}| - |\bar{r}_{2i}|\right). \tag{10.22}$$

Here, the "+" sign refers to the processing of the external engagement profiles, and the "–" sign refers to the internal engagement profiles.

Equation (10.22), which is essentially nonlinear, reflects the presence of "blacks," i.e., areas of the surface described by the radius vector $\bar{r}_{2(i-1)}$. Thus, this equation describes the actual surface obtained, taking into account the areas of nontreatment.

This is illustrated in Fig. 10.7, where line 1 specifies the original surface, i.e., the surface described by the radius vector $\bar{r}_{2(i-1)}$; line 2 is the surface obtained on the *i*th considered passage, i.e., the surface described by the radius vector $\bar{r}_{2i}$; and the contour line marks the actual surface formed on the *i*th pass and is described by Eq. (10.22).

Thus, taking into account the change in the bases noted above, the general equations of the real surfaces of the teeth can be written in the following forms:

$$\bar{r}_{2p} = \widetilde{M}_{\Delta 2}(\bar{r}_{2\Phi} - \Delta\bar{r}_{2YT} + \Delta\bar{r}_{2K}); \tag{10.23}$$

$$\bar{r}_{2\Phi} = \bar{r}_{2i} \cdot U \pm \left(|\bar{r}_{2i}| - |\bar{r}_{2(i-1)}|\right) + \bar{r}_{2(i-1)} \cdot U \pm \left(|\bar{r}_{2(i-1)}| - |\bar{r}_{2i}|\right); \tag{10.24}$$

$$\bar{r}_{2i} = \widetilde{M}_{\Sigma H}\bar{r}_{1i}; \tag{10.25}$$

$$\left(\frac{\partial \overline{r}_{1i}}{\partial u}\times\frac{\partial \overline{r}_{1i}}{\partial v}\right)\left(\widetilde{M}_{\Sigma H}^{-1}\frac{\partial \widetilde{M}_{\Sigma}}{\partial \phi_1}\cdot\overline{r}_{1i}+\frac{\partial \overline{r}_{1i}}{\partial \phi_1}\right)=0; \tag{10.26}$$

$$\left(\frac{\partial \overline{r}_{1i}}{\partial u}\times\frac{\partial \overline{r}_{1i}}{\partial v}\right)\left(\widetilde{M}_{\Sigma H}^{-1}\frac{\partial \widetilde{M}_{\Sigma}}{\partial \psi}\cdot\overline{r}_{1i}+\frac{\partial \overline{r}_{1i}}{\partial \psi}\right)=0; \tag{10.27}$$

$$\widetilde{M}_{\Sigma H}=\widetilde{M}_{2H}\cdot\widetilde{M}_{\Delta H}\cdot\widetilde{M}_{H}\cdot\widetilde{M}_{H1}\cdot\widetilde{M}_{\Delta K}\cdot\widetilde{M}_{\Delta 1}; \tag{10.28}$$

$$\overline{r}_{1i}=\overline{r}_{1H}+\Delta\overline{r}_{1i}+\Delta\overline{r}_{1K}. \tag{10.29}$$

Considering the independence of $\overline{r}_{1H}$ from ϕ_1 and ψ, we represent

$$\frac{\partial \overline{r}_{1i}}{\partial \phi_1}=\frac{\partial \Delta\overline{r}_{1i}}{\partial \phi_1}+\frac{\partial \Delta\overline{r}_{1K}}{\partial \phi_1}; \tag{10.30}$$

$$\frac{\partial \overline{r}_{1i}}{\partial \psi}=\frac{\partial \Delta\overline{r}_{1i}}{\partial \psi}+\frac{\partial \Delta\overline{r}_{1K}}{\partial \psi}. \tag{10.31}$$

From the obtained general Eqs. (10.23), (10.24), (10.25), (10.26), (10.26), (10.27), (10.28), (10.29), (10.30), and (10.31), the equations of the real teeth profiles of all existing processing schemes are derived as special cases. For example, when shaped by the one-parameter envelope method in the reduced system of equations, Eqs. (10.26) or (10.27) turns into an identity and is dropped from consideration. When processing by the copying method, both of these equations become identities.

With complex movements of the machine, it is difficult to "manually" perform calculations of coordinates according to the reduced system of equations. It is advisable to use a computer. At the same time, the use of the method of structural modeling of shaping processes, first proposed by Prof. B.A. Perepelitsa, is especially effective [21, 22], since all matrices and radius vectors can be modeled by a set of unified operators. The differences that should be taken into account in the programs are the account of nonlinearities in dependence (10.26) and strictly specified types of matrices.

The resulting system of equations for the real surfaces of the teeth allows determining the actual coordinates of these surfaces:

$$\begin{aligned} X_{2p}&=X_{2p}\left(\phi_1,\psi,\overline{\Delta}_n,\overline{\Pi},\overline{\Delta}_K\right);\\ Y_{2p}&=Y_{2p}\left(\phi_1,\psi,\overline{\Delta}_n,\overline{\Pi},\overline{\Delta}_K\right);\\ Z_{2p}&=Z_{2p}\left(\phi_1,\psi,\overline{\Delta}_n,\overline{\Pi},\overline{\Delta}_K\right). \end{aligned} \tag{10.32}$$

where X_{2p}, Y_{2p}, Z_{2p} are the Cartesian coordinates of the system associated with the design center distance such that the axle O_{2p} Z_{2p} coincides with the direction of the geometric axis of the gear and the axis O_{2p} X_{2p} и O_{2p} Y_{2p} located in the base end plane.

$\overline{\Delta}_n$ is the 24-component vector of reduced primary errors.
$\overline{\Pi}$ is the vector nominal geometric parameter of the MFTP system.

$\overline{\Delta}_K$ is the vector that takes into account the parameter increments in order to control the accuracy by the above methods.

From (10.32), one can obtain the equations for the nominal surfaces of the teeth by setting $\overline{\Delta}_n = 0$ and $\overline{\Delta}_K = 0$:

$$X_{2H} = X_{2H}(\phi_1, \psi, \overline{\Pi}); \quad Y_{2H} = Y_{2H}(\phi_1, \psi, \overline{\Pi}); \quad Z_{2H} = Z_{2H}(\phi_1, \psi, \overline{\Pi}). \quad (10.33)$$

Based on Eqs. (10.32) and (10.33), the increment coordinates ΔX_{2p}, ΔY_{2p}, and ΔZ_{2p} of the real profiles are calculated in relation to the nominal:

$$\Delta X_{2p} = X_{2p} - X_{2H}; \quad \Delta Y_{2p} = Y_{2p} - Y_{2H}; \quad \Delta Z_{2p} = Z_{2p} - Z_{2H}. \quad (10.34)$$

Being aware of (10.34), one can proceed to determining the relationship between these increments and the normalized errors of the gears.

However, before doing this, we will note the features of the obtained equations of real tooth profiles.

The features, in addition to generality, will include the following:

- Accounting for precision control in all possible ways, the total set of which has cardinality equal to three
- Dependence on the full set of reduced primary errors, the cardinality of which is 24
- Taking into account the functional nature of the given errors (i.e., the approach to considering the formation of errors as a process and not as a phenomenon)
- Nonlinearity of the equations obtained, which excludes the application of the principle of independence of the action of primary errors and agrees with the results obtained in the nonlinear theory of the accuracy of mechanisms
- Taking into account the simultaneous impact of all factors and not each separately, and the ability to take into account the correlation of error modules

These features allow one to proceed to the consideration of qualitatively new problems in the theory of the accuracy of gears:

- Development of methods for optimal accuracy assurance of any kind of gear
- Development of methods for predicting the accuracy of gears at any stage of production with the required forecast reliability
- Creation of systems for optimal regulation of the accuracy of the elements of an MFTP system: machine tools, fixtures, tools, and blanks of machined gears
- Identifying methods for diagnosing the accuracy parameters of the elements of an MFTP system
- Systematic presentation of the theory of control over the accuracy of gears (and other details) on the basis of the revealed methods of compensatory control of system errors

- Systematization of research in the field of accuracy of gears by replacing the study of the entire set of specific errors (more than 200 [4]) with a one-time study of the 24 given primary errors and the extension of the results obtained for all specific cases
- Synthesis of real engagement with given quality indices, taking into account possible engagement errors and others

Most of these tasks will be discussed below.

10.2.1.2 Establishing the Relationship Between Coordinate Increments and Standardized Accuracy Indicators of Gears

The solution to this problem can be obtained in several ways. The difference between the methods lies in the choice of the base accuracy indicator, which, on the one hand, is connected by certain dependencies with the increment of the radius vector

$$\Delta\bar{r}_{2p} = \Delta X_{2p}\bar{i} + \Delta Y_{2p}\bar{j} + \Delta Z_{2p}\bar{k},$$

and, on the other hand, with standardized values for the accuracy of the teething.

There are four known ways to identify these dependencies. The first one provides a direct calculation of standardized accuracy indicators by increments of the radius vector $\Delta\bar{r}_{2p}$ [4]. The second one is based on the use of the increment of the line of action ΔF [36], the third one is based on the use of the "increment of the radius of curvature"$\Delta\rho_K$ [2], and the fourth one is based on the arc deviation Δg [13, 39].

To ensure the effective use of the already developed dependencies applying each of the named indicators, we will find the relationship between these indicators, based on the definitions introduced in the studies by Alikulov [2], Khlus and Ternyuk, [13], and Taits [36].

The nature of the increment ΔF of the line of action is illustrated in Fig. 10.8, where index 1 denotes the real gear, index 2 is the measuring (reference), and L–L is the line of action.

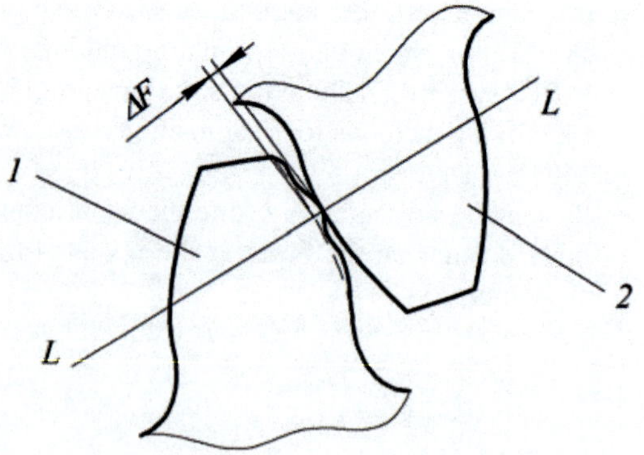

Fig. 10.8 Scheme for identifying the line of action increment

Fig. 10.9 Scheme for identifying the "curvature radius increment"

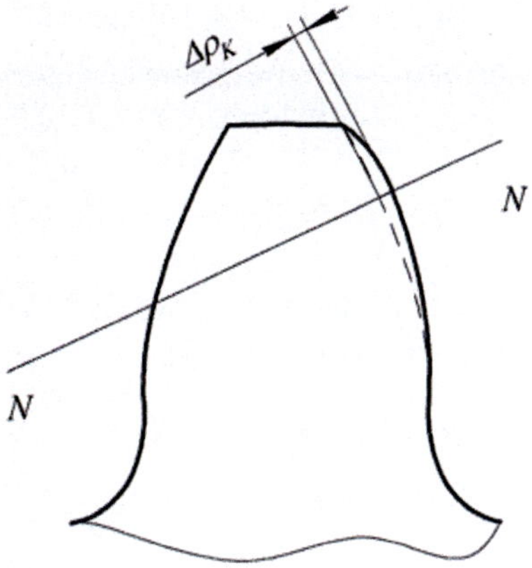

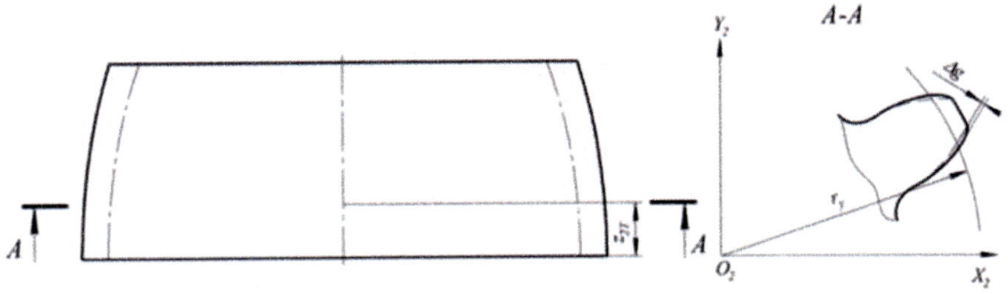

Fig. 10.10 Scheme for identifying arc deviation

In the study by Alikulov [2], the parameter "increment of the radius of curvature" actually represents the deviation of the point of the real profile from the nominal one in the direction of the normal N–N (Fig. 10.9) to the nominal profile.

The arc deviation Δg (Fig. 10.10) means the distance measured along the arc of a circle radius r_y in the considered end plane A–A of the gear at a distance Z_{2T} from the origin (base), enclosed between the points of intersection of this circle with the actual (solid line) and nominal (contour line) tooth profiles [13].

Considering that the contact of surfaces is always carried out at the point through the common normal, the following relationship between ΔF and $\Delta \rho_K$ can be written:

$$\Delta F = \max_{bN} \{\Delta \rho_K\} \cos \xi, \tag{10.35}$$

or for flat gearing:

$$\Delta F = \max_{bN} \{\Delta \rho_K\}, \tag{10.36}$$

where ξ is the angle between the normal to the nominal profile and the end plane and the effective (maximum within the limits of the contact line bN) value $\Delta \rho_K$ is marked with a sign $\max\limits_{bN}$.

Based on the consideration of the directions of the increment of the line of action and arc deviation, an extreme nature of ΔF, and taking into account the smallness of the error moduli, it is possible to write:

$$\Delta F = \max_{bN} \{\Delta g\} \sin \xi_1 \cos \xi, \tag{10.37}$$

where ξ_1 is the angle between the normal and the direction of the radius vector at the point under consideration.

Similarly, taking into account (10.35), we obtain:

$$\Delta \rho_K = \Delta g \cos \alpha, \tag{10.38}$$

The presence of extrema in (10.35), (10.36), (10.37), and (10.38) does not allow establishing a single-value inverse function among $\Delta \rho_K$, Δg, and ΔF.

A relationship between Δg and $\Delta \bar{r}_{2p}$ can be established on the basis of Fig. 10.11, where it is indicated that M is a point on the nominal profile, for which the arc deviation is identified as Δg, and τ–τ is the tangent to the actual (real) profile at point $K\,(x_2 + \Delta x_{2p};\, y_2 + \Delta y_{2p})$. The designations of the remaining quantities can be seen in the figure.

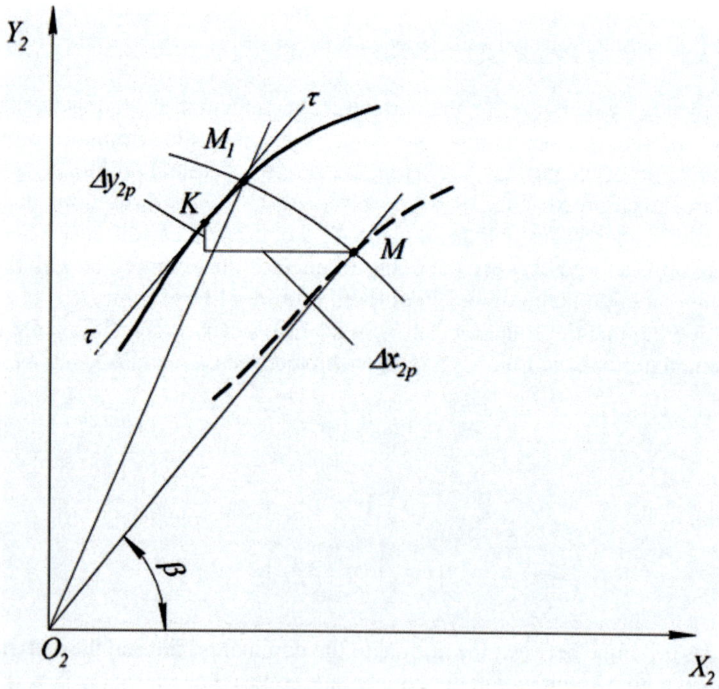

Fig. 10.11 Scheme for identifying the relationship of the arc deviation with the increment of the radius vector

Projecting both Δx_{2p} and Δy_{2p} on the direction MM_1 and calculating the actual length MM_1, taking into account the smallness of the errors, we write:

$$\Delta g = \Delta y_{2p} \cos \beta - \Delta x_{2p} \sin \beta + \left(\Delta y_{2p} \sin \beta + \Delta x_{2p} \cos \beta\right) \sin (\beta - \gamma), \quad (10.39)$$

where $\beta = \operatorname{arctg} \frac{y_{2H} + \Delta y_{2p}}{x_{2H} + \Delta x_{2p}} \cong \operatorname{arctg} \frac{y_{2H}}{x_{2H}}$; $\gamma = \operatorname{arctg} \frac{d\left(y_{2H} + \Delta y_{2p}\right)}{d\left(x_{2H} + \Delta x_{2p}\right)} \cong \operatorname{arctg} \frac{dy_{2H}}{dx_{2H}}$.

The obtained expressions (10.35), (10.36), (10.37), (10.38) and (10.39) make automating one of the stages of calculating (predicting) errors possible, since, if they are available, it is possible to calculate errors on a computer using the general equations of real tooth profiles as well as the information about the place of an error occurrence on a real profile.

All the information necessary for the machine calculation of normalized errors can be obtained by presenting the accuracy indicators discussed above in a form that would reflect not only the value of the error modulus but also the coordinates of its application.

One of these types can be quaternionic [14]:

$$V(\Delta) = \Delta + x_{2H}\bar{i} + y_{2H}\bar{j} + z_{2H}\bar{k}. \quad (10.40)$$

In (10.40), it is considered $\Delta = \Delta F \cup \Delta\rho_K \cup \Delta g$. In the quaternionic representation of errors, information about gear errors can be described by the matrices

$$\widetilde{M}(V)_i = \begin{vmatrix} \Delta_{1i} & x_{2H1i} & y_{2H1i} & z_{2H1i} \\ \Delta_{2i} & x_{2H2i} & y_{2H2i} & z_{2H2i} \\ \ldots & \ldots & \ldots & \ldots \\ \Delta_{ni} & x_{2Hni} & y_{2Hni} & z_{2Hni} \end{vmatrix}, \quad i = \{L, R\} \quad (10.41)$$

The degree of discreteness in the presentation of information is determined by the required accuracy of calculating the errors. Performing the necessary actions on the components of the matrices $\widetilde{M}(V)_i$, one can calculate any of the standardized accuracy indicators.

We will provide several examples so that one can identify the structure of the calculation formulas and, in terms of the structure, proceed to considering the laws of error distribution.

Let us calculate, using arc deviations, the values of the greatest kinematic error, profile error, tooth direction error, and tooth thickness deviation along a constant chord, thus covering all groups of accuracy indicators (norms). Based on the definition of a kinematic error and the previously given definition of an arc deviation, we find:

$$\Delta\phi = \Delta g \frac{\cos \alpha}{r_b}, \quad (10.42)$$

where $\Delta\phi$ is the kinematic error.

r_b is the radius of the base circle of the gear.
α is the pressure angle.

Consequently, the largest kinematic error of the gear is:

$$F'_{ir} = \Delta\phi_{\max} - \Delta\phi_{\max} = \frac{1}{r_b}[\max(\Delta g \cos\alpha) - \min(\Delta g \cos\alpha)], \quad 0 \leq \phi \leq 2\pi. \tag{10.43}$$

Similarly, we obtain other normalized accuracy indicators:

$$f_{fr} = \max(\Delta g \cos\alpha) - \min(\Delta g \cos\alpha), \quad \Delta\tau_n \leq \phi \leq \Delta\tau_{(n+1)}; \tag{10.44}$$

$$F_{\beta r} = \max(\Delta g) - \min(\Delta g), \quad \phi = const; \tag{10.45}$$

$$A_{\text{ce}} = \frac{\Delta g_L - \Delta g_R}{\cos\alpha}. \tag{10.46}$$

The above mentioned equations show the presence of an unambiguous relationship between Δg and standardized accuracy indicators. Consequently, there is a functional relationship and other reduced errors with the normalized indicators of accuracy [13].

Thus, the calculation of the numerical values of the actual errors at each point of the gear and the normalized errors according to dependencies of the form (10.42), (10.43), (10.44), (10.45), and (10.46) can be performed if the actual values of the radius vector $\Delta\bar{r}_{2p}$ are known in order to determine the values of the matrices $\widetilde{M}(V)_i$ and, from them, the required errors.

The actual values of the radius vector $\Delta\bar{r}_{2p}$ based on the general Eqs. (10.23), (10.24), (10.25), (10.26), (10.27), (10.28), and (10.29) can be calculated only if the methods for determining the reduced primary errors and other parameters included in the matrices $\widetilde{M}_H$, $\widetilde{M}_{\Delta H}$, $\widetilde{M}_{\Delta 1}$, $\widetilde{M}_{\Delta 2}$, and $\widetilde{M}_{\Delta K}$ and the radius vectors $\bar{r}_{1H}$, $\Delta\bar{r}_{1i}$, and $\Delta\bar{r}_{1K}$ are known.

Information support, when searching for elements of the specified matrices and radius vectors, is determined by the nature of the problem being solved as well as by the types of matrices and radius vectors.

Matrix $\widetilde{M}_H$ elements are determined by known methods [16] by assigning parameters γ, a_w, ψ_y, and ψ_z particular values corresponding to specific gear cutting schemes. For example, for the hobbing and broaching processes, the matrix $\widetilde{M}_H$ elements are as follows:

$$\widetilde{M}_{H(h)} = \begin{pmatrix} 1 & 0 & 0 & 0 \\ 0 & \cos\gamma & -\sin\gamma & 0 \\ 0 & \sin\gamma & \cos\gamma & k\phi_1 \\ 0 & 0 & 0 & 1 \end{pmatrix}; \quad \widetilde{M}_{H(b)} = \begin{pmatrix} 1 & 0 & 0 & 0 \\ 0 & 1 & 0 & 0 \\ 0 & 0 & 1 & \psi \\ 0 & 0 & 0 & 1 \end{pmatrix},$$

where $\gamma = \text{const}$, $a_w = \text{const}$, and $k = \text{const}$.

The values of the radius vector $\overline{r}_{1H}$ can be determined by the well-known methods of analytical geometry or mapping methods [21, 22] as the values of the radius vector of the nominal surface (line) of the tool – in the event that the main donation (i.e., movement with cutting speed) is carried out by changing one or more parameters from γ, a_w, ψ_y, ψ_z, ϕ_1, and ϕ_2 – and as the values of the radius vector of the envelope of the cutting edges with the parameter defining the main ghost otherwise.

The elements depending on the reduced primary errors can be found in several ways, depending on the methods for determining the numerical values of the reduced errors, considered as functions.

10.2.1.3 Methods for Determining the Numerical Values of the Functions of Reduced Primary Errors

The given errors included in the matrices $\widetilde{M}_{\Delta H}$, $\widetilde{M}_{\Delta 1}$, and $\widetilde{M}_{\Delta K}$ and the increments of the radius vectors $\Delta\overline{r}_{1i}$ and $\Delta\overline{r}_{2YT}$ are geometric objects caused by various factors. Therefore, the methods for their determination can be:

(a) Geometric objects that appear on the geometry of the gear – the method of direct measurements and the method of geometric diagnostics.
(b) Objects caused by various factors and having certain connections with these factors (in accordance with the model shown in Fig. 10.1) – based on the results of controlling the parameters of the factors with their subsequent calculation according to the dependencies between the parameters of the factors and the given primary errors – according to the results of the forecast of the parameters of the factors affecting the accuracy of processing, with the subsequent calculation of the reduced primary errors.

Let us take a look at these methods.

Direct measurement method
The direct measurement method includes determining the numerical values of the reduced primary errors using special or universal instruments for linear and angular measurements. Its advantage lies in its ability to obtain information with high accuracy. Serious disadvantages of the method are the following: the need to stop the processing process, the need to use a large number of special devices, and the complexity of measurement automation.

The most expedient area of application of this method is considered as the control of the given primary errors in the processes in which the shaping is carried out by a limited number of fixed points on the tool surface.

Diagnostic Method
The derivation of the general equations of real tooth profiles in Sect. 10.2.1.1 made formulating and solving, in general, the problem of obtaining information about the given primary errors by diagnostics based on the results of monitoring the processed gear possible.

The advantages of the diagnostic methods are the following: they do not require stopping the production process, have high productivity, and require the use of only one gear measuring device, which can be quite versatile.

It is known that the solution of any identification problem, to which the considered diagnostic problem also belongs, presupposes setting the structure of equations describing the phenomenon under consideration and then determining the parameters of these equations.

Since Eqs. (10.23), (10.24), (10.25), (10.26), (10.27), (10.28), and (10.29) are general and describe the process of occurrence of errors on the profiles of the teeth of a machined gear, depending on the values of the given primary errors, they are the necessary basis for solving the first and second stages of the identification problem.

The problem of diagnosing the abovementioned primary errors is always solved for a specific gear cutting machine and tool.

Therefore, the matrices $\widetilde{M}_H$, $\widetilde{M}_{2H}$, and $\widetilde{M}_{H1}$ and also the radius vectors $\overline{r}_{1H}$ and $\overline{r}_{2H}$ are known. As a result of monitoring the machined gear, arc deviations or increments of the radius of curvature are determined at different points of the gear, i.e., matrices of the form (10.41) are also known.

The unknown are the components of the matrices $\widetilde{M}_{\Delta H}$, $\widetilde{M}_{\Delta 1}$, and $\widetilde{M}_{\Delta 2}$ and the components of the increments of the radius vectors $\Delta\overline{r}_{1i}$ and $\Delta\overline{r}_{2YT}$ – a total of 24 functions of the envelope parameters. In some cases, when controlling a gear, it is removed from the machine, its angular position relative to the selected coordinate systems is not fixed, and one more angular parameter is added to these 24 functions – the initial phase of the gear rotation angle.

The aforementioned determines the formulation of a diagnostic problem. The methodology for its solution is based on the following:

Without limiting the generality of the statement of the problem, we can assume that information about Δg or $\Delta\rho_K$ can be obtained from the points that have undergone processing. In addition, measurements Δg or $\Delta\rho_K$ can be made at the points of actual tangency of the real polyline, obtained by a limited number of cutting edges of the tool, with the envelope of these cut marks. Therefore, operator (10.22) can be ignored when specifying the structure of the equations of the diagnostic process.

Taking this into account, the initial equations describing the structure of the model of the occurrence of errors can be written in the following forms:

$$\overline{r}_{2p} = \widetilde{M}_{\Delta 2}(\overline{r}_{2i} - \Delta\overline{r}_{2YT}); \tag{10.47}$$

$$\overline{r}_{2i} = \widetilde{M}_{\Sigma H}\overline{r}_{1i}; \tag{10.48}$$

$$\left(\frac{\partial\overline{r}_{1i}}{\partial u} \times \frac{\partial\overline{r}_{1i}}{\partial v}\right)\left(\widetilde{M}_{\Sigma H}^{-1}\frac{\partial\widetilde{M}_{\Sigma H}}{\partial\phi_1} \cdot \overline{r}_{1i}\right) = 0; \tag{10.49}$$

$$\left(\frac{\partial\overline{r}_{1i}}{\partial u}\times\frac{\partial\overline{r}_{1i}}{\partial v}\right)\left(\widetilde{M}_{\Sigma H}^{-1}\frac{\partial\widetilde{M}_{\Sigma H}}{\partial\psi}\cdot\overline{r}_{1i}\right)=0; \tag{10.50}$$

$$\widetilde{M}_{\Sigma H}=\widetilde{M}_{2H}\cdot\widetilde{M}_{\Delta H}\cdot\widetilde{M}_{H}\cdot\widetilde{M}_{H1}\cdot\widetilde{M}_{\Delta K}\cdot\widetilde{M}_{\Delta 1};\quad \widetilde{M}_{\Delta K}=\widetilde{E}; \tag{10.51}$$

$$\overline{r}_{1i}=\overline{r}_{1H}+\Delta\overline{r}_{1i}. \tag{10.52}$$

Let us presume:

$$\widetilde{M}_{\Delta i}=\widetilde{E}+\Delta\widetilde{M}_{\Delta i}, \tag{10.53}$$

where $\widetilde{E}$ is the unit matrix and $\Delta\widetilde{M}_{\Delta i}$ is a matrix, all elements of which are equal to the corresponding elements of the matrix $\widetilde{M}_{\Delta i}$, except for the diagonal ones, which are equal to zero.

Taking (10.53), with the possibility of neglecting the terms of the second order of smallness, into account, we transform matrix (10.51) to the form:

$$\begin{aligned}\widetilde{M}_{\Sigma H}&=\widetilde{M}_{2H}\left(\widetilde{E}+\Delta\widetilde{M}_{\Delta H}\right)\cdot\widetilde{M}_{H}\cdot\widetilde{M}_{H1}\left(\widetilde{E}+\Delta\widetilde{M}_{\Delta 1}\right)=\\&=\widetilde{M}_{2H}\cdot\widetilde{M}_{H}\cdot\widetilde{M}_{H1}+\widetilde{M}_{2H}\cdot\widetilde{M}_{H}\cdot\widetilde{M}_{H1}\cdot\Delta\widetilde{M}_{\Delta 1}+\widetilde{M}_{2H}\cdot\Delta\widetilde{M}_{\Delta H}\cdot\widetilde{M}_{H}\cdot\widetilde{M}_{H1}=\\&=\widetilde{M}_{\Sigma H0}+\widetilde{M}_{\Sigma H0}\cdot\Delta\widetilde{M}_{\Delta 1}+\widetilde{M}_{2H}\cdot\Delta\widetilde{M}_{\Delta H}\cdot\widetilde{M}_{H}\cdot\widetilde{M}_{H1}.\end{aligned} \tag{10.54}$$

Here, $\widetilde{M}_{\Sigma H0}$ is the nominal value of the matrix $\widetilde{M}_{\Sigma H}$.

Taking into account (10.52) and (10.54), the expression for the calculation of (10.48), also up to small second-order smallness, can be written as:

$$\begin{aligned}\overline{r}_{2i}&=\widetilde{M}_{\Sigma H0}\cdot\overline{r}_{1n}+\widetilde{M}_{\Sigma H0}\cdot\Delta\widetilde{M}_{\Delta 1}\cdot\overline{r}_{1H}+\widetilde{M}_{2H}\cdot\Delta\widetilde{M}_{\Delta H}\cdot\widetilde{M}_{H}\cdot\widetilde{M}_{H1}\cdot\overline{r}_{1H}+\widetilde{M}_{\Sigma H0}\cdot\Delta\overline{r}_{1i}=\\&=\overline{r}_{2H}+\widetilde{M}_{\Sigma H0}\cdot\Delta\widetilde{M}_{\Delta 1}\cdot\overline{r}_{1H}+\widetilde{M}_{2H}\cdot\Delta\widetilde{M}_{\Delta H}\cdot\widetilde{M}_{H}\cdot\widetilde{M}_{H1}\cdot\overline{r}_{1H}+\widetilde{M}_{\Sigma H0}\cdot\Delta\overline{r}_{1i}.\end{aligned} \tag{10.55}$$

As a result, we obtain (10.47) in the form

$$\begin{aligned}\overline{r}_{2p}=\ &\overline{r}_{2H}+\widetilde{M}_{\Sigma H0}\cdot\Delta\widetilde{M}_{\Delta 1}\cdot\overline{r}_{1H}+\widetilde{M}_{2H}\cdot\Delta\widetilde{M}_{\Delta H}\cdot\widetilde{M}_{H}\cdot\widetilde{M}_{H1}\cdot\overline{r}_{1p}\\&+\widetilde{M}_{\Sigma H0}\cdot\Delta\overline{r}_{1i}+\Delta\widetilde{M}_{\Delta 2}\cdot\overline{r}_{1n}-\Delta\overline{r}_{2YT}.\end{aligned} \tag{10.56}$$

To reveal the constraint Eqs. (10.49) and (10.50), at the first stage, we find intermediate expressions for calculating the partial derivatives using dependence (10.55).

Considering that the given primary errors, reflecting the violation of the shaping movement, in the general case, are functions of the envelope parameters, and the given primary errors of the instrumental surface also depend on the coordinates of this surface, after differentiation, we get the following:

$$\begin{aligned}\frac{\partial \overline{r}_{2i}}{\partial \phi_1} = \frac{\partial \overline{r}_{2H}}{\partial \phi_1} + \widetilde{M}_{\Sigma H0} \cdot \frac{\partial \Delta \widetilde{M}_{\Delta 1}}{\partial \phi_1} \cdot \overline{r}_{1H} + \frac{\partial \widetilde{M}_{\Sigma H0}}{\partial \phi_1} \cdot \Delta \widetilde{M}_{\Delta 1} \cdot \overline{r}_{1H} + \\ + \widetilde{M}_{2H} \cdot \Delta \widetilde{M}_{\Delta H} \cdot \frac{\partial \left(\widetilde{M}_H \cdot \widetilde{M}_{H1} \cdot \overline{r}_{1H}\right)}{\partial \phi_1} + \widetilde{M}_{2H} \cdot \frac{\partial \Delta \widetilde{M}_{\Delta H}}{\partial \phi_1} \cdot \widetilde{M}_H \cdot \widetilde{M}_{H1} \cdot \overline{r}_{1H} + \\ + \frac{\partial \widetilde{M}_{2H}}{\partial \phi_1} \cdot \Delta \widetilde{M}_{\Delta 1} \cdot \widetilde{M}_H \cdot \widetilde{M}_{H1} \cdot \overline{r}_{1H} + \frac{\partial \widetilde{M}_{\Sigma H0}}{\partial \phi_1} \cdot \Delta \overline{r}_{1i} = \frac{\partial \overline{r}_{2H}}{\partial \phi_1} + \overline{K};\end{aligned} \tag{10.57}$$

$$\begin{aligned}\frac{\partial \overline{r}_{2i}}{\partial \psi} = \frac{\partial \overline{r}_{2H}}{\partial \phi_1} + \widetilde{M}_{\Sigma H0} \cdot \frac{\partial \Delta \widetilde{M}_{\Delta 1}}{\partial \psi} \cdot \overline{r}_{1H} + \frac{\partial \widetilde{M}_{\Sigma H0}}{\partial \psi} \cdot \Delta \widetilde{M}_{\Delta 1} \cdot \overline{r}_{1H} + \\ + \widetilde{M}_{2H} \cdot \Delta \widetilde{M}_{\Delta H} \cdot \frac{\partial \left(\widetilde{M}_H \cdot \widetilde{M}_{H1} \cdot \overline{r}_{1H}\right)}{\partial \psi} + \widetilde{M}_{2H} \cdot \frac{\partial \Delta \widetilde{M}_{\Delta H}}{\partial \psi} \cdot \widetilde{M}_H \cdot \widetilde{M}_{H1} \cdot \overline{r}_{1H} + \\ + \frac{\partial \widetilde{M}_{2H}}{\partial \psi} \cdot \Delta \widetilde{M}_{\Delta 1} \cdot \widetilde{M}_H \cdot \widetilde{M}_{H1} \cdot \overline{r}_{1H} + \frac{\partial \widetilde{M}_{\Sigma H0}}{\partial \psi} \cdot \Delta \overline{r}_{1i} = \frac{\partial \overline{r}_{2H}}{\partial \psi} + \overline{L};\end{aligned} \tag{10.58}$$

$$\begin{aligned}\frac{\partial \overline{r}_{2i}}{\partial u} = {} & \frac{\partial \overline{r}_{2H}}{\partial u} + \widetilde{M}_{\Sigma H0} \cdot \Delta \widetilde{M}_{\Delta 1} \cdot \frac{\partial \overline{r}_{1H}}{\partial u} + \widetilde{M}_{2H} \cdot \Delta \widetilde{M}_{\Delta H} \cdot \widetilde{M}_H \cdot \widetilde{M}_{H1} \cdot \frac{\partial \overline{r}_{1H}}{\partial u} \\ & + \widetilde{M}_{\Sigma H0} \cdot \frac{\partial \Delta \overline{r}_{1i}}{\partial u} = \frac{\partial \overline{r}_{2H}}{\partial u} + \overline{M};\end{aligned} \tag{10.59}$$

$$\begin{aligned}\frac{\partial \overline{r}_{2i}}{\partial v} = {} & \frac{\partial \overline{r}_{2H}}{\partial v} + \widetilde{M}_{\Sigma H0} \cdot \Delta \widetilde{M}_{\Delta 1} \cdot \frac{\partial \overline{r}_{1H}}{\partial v} + \widetilde{M}_{2H} \cdot \Delta \widetilde{M}_{\Delta H} \cdot \widetilde{M}_H \cdot \widetilde{M}_{H1} \cdot \frac{\partial \overline{r}_{1H}}{\partial v} \\ & + \widetilde{M}_{\Sigma H0} \cdot \frac{\partial \Delta \overline{r}_{1i}}{\partial v} = \frac{\partial \overline{r}_{2H}}{\partial v} + \overline{N}.\end{aligned} \tag{10.60}$$

The constraint equations in the coordinate system $S_2\{X_2, Y_2, Z_2\}$ are as follows:

$$\left(\frac{\partial \overline{r}_2}{\partial u} \times \frac{\partial \overline{r}_2}{\partial v}\right) \frac{\partial \overline{r}_2}{\partial \phi_1} = 0; \tag{10.61}$$

$$\left(\frac{\partial \overline{r}_2}{\partial u} \times \frac{\partial \overline{r}_2}{\partial v}\right) \frac{\partial \overline{r}_2}{\partial \psi} = 0. \tag{10.62}$$

Substituting the expressions for the partial derivatives from (10.57) to (10.62) in these equations, we obtain:

$$\left[\left(\frac{\partial \overline{r}_{2H}}{\partial u} + \overline{M}\right) \times \left(\frac{\partial \overline{r}_{2H}}{\partial v} + \overline{N}\right)\right] \left(\frac{\partial \overline{r}_{2H}}{\partial \phi_1} + \overline{K}\right) = 0; \tag{10.63}$$

$$\left[\left(\frac{\partial \overline{r}_{2H}}{\partial u} + \overline{M}\right) \times \left(\frac{\partial \overline{r}_{2H}}{\partial v} + \overline{N}\right)\right] \left(\frac{\partial \overline{r}_{2H}}{\partial \psi} + \overline{L}\right) = 0. \tag{10.64}$$

Making transformations of (10.63) and (10.64) and taking into account that

$$\left(\frac{\partial \overline{r}_{2H}}{\partial u} \times \frac{\partial \overline{r}_{2H}}{\partial v}\right) \frac{\partial \overline{r}_{2H}}{\partial \phi_1} = 0;$$

$$\left(\frac{\partial \overline{r}_{2H}}{\partial u} \times \frac{\partial \overline{r}_{2H}}{\partial v}\right) \frac{\partial \overline{r}_{2H}}{\partial \psi} = 0,$$

we obtain

$$\begin{aligned} &\frac{\partial \overline{r}_{2H}}{\partial u} \times \overline{N} \cdot \frac{\partial \overline{r}_{2H}}{\partial \phi_1} + \overline{M} \times \frac{\partial \overline{r}_{2H}}{\partial v} \cdot \frac{\partial \overline{r}_{2H}}{\partial \phi_1} + \overline{M} \times \overline{N} \cdot \frac{\partial \overline{r}_{2H}}{\partial \phi_1} + \frac{\partial \overline{r}_{2H}}{\partial u} \times \frac{\partial \overline{r}_{2H}}{\partial v} \cdot \overline{K} + \\ &\quad + \frac{\partial \overline{r}_{2H}}{\partial u} \times \overline{N} \cdot \overline{K} + \overline{M} \times \frac{\partial \overline{r}_{2H}}{\partial v} \cdot \overline{K} + \overline{M} \times \overline{N} \cdot \overline{K} = 0; \end{aligned} \tag{10.65}$$

$$\begin{aligned} &\frac{\partial \overline{r}_{2H}}{\partial u} \times \overline{N} \cdot \frac{\partial \overline{r}_{2H}}{\partial \psi} + \overline{M} \times \frac{\partial \overline{r}_{2H}}{\partial v} \cdot \frac{\partial \overline{r}_{2H}}{\partial \psi} + \overline{M} \times \overline{N} \cdot \frac{\partial \overline{r}_{2H}}{\partial \psi} + \frac{\partial \overline{r}_{2H}}{\partial u} \times \frac{\partial \overline{r}_{2H}}{\partial v} \cdot \overline{L} + \\ &\quad + \frac{\partial \overline{r}_{2H}}{\partial u} \times \overline{N} \cdot \overline{L} + \overline{M} \times \frac{\partial \overline{r}_{2H}}{\partial v} \cdot \overline{L} + \overline{M} \times \overline{N} \cdot \overline{L} = 0. \end{aligned} \tag{10.66}$$

In the above equations,

$$\begin{aligned} \overline{K} = {} & \widetilde{M}_{\Sigma H0} \cdot \frac{\partial \Delta \widetilde{M}_{\Delta 1}}{\partial \phi_1} \cdot \overline{r}_{1H} + \frac{\partial \widetilde{M}_{\Sigma H0}}{\partial \phi_1} \cdot \Delta \widetilde{M}_{\Delta 1} \cdot \overline{r}_{1H} + \\ & + \widetilde{M}_{2H} \cdot \Delta \widetilde{M}_{\Delta H} \cdot \frac{\partial \left(\widetilde{M}_H \cdot \widetilde{M}_{H1} \cdot \overline{r}_{1H}\right)}{\partial \phi_1} + \widetilde{M}_{2H} \cdot \frac{\partial \Delta \widetilde{M}_{\Delta H}}{\partial \phi_1} \cdot \widetilde{M}_H \cdot \widetilde{M}_{H1} \cdot \overline{r}_{1H} + \\ & + \frac{\partial \widetilde{M}_{2H}}{\partial \phi_1} \cdot \Delta \widetilde{M}_{\Delta 1} \cdot \widetilde{M}_H \cdot \widetilde{M}_{H1} \cdot \overline{r}_{1H} + \frac{\partial \widetilde{M}_{\Sigma H0}}{\partial \phi_1} \cdot \Delta \overline{r}_{1i}; \end{aligned} \tag{10.67}$$

$$\begin{aligned} \overline{L} = {} & \widetilde{M}_{\Sigma H0} \cdot \frac{\partial \Delta \widetilde{M}_{\Delta 1}}{\partial \psi} \cdot \overline{r}_{1H} + \frac{\partial \widetilde{M}_{\Sigma H0}}{\partial \psi} \cdot \Delta \widetilde{M}_{\Delta 1} \cdot \overline{r}_{1H} + \\ & + \widetilde{M}_{2H} \cdot \Delta \widetilde{M}_{\Delta H} \cdot \frac{\partial \left(\widetilde{M}_H \cdot \widetilde{M}_{H1} \cdot \overline{r}_{1H}\right)}{\partial \psi} + \widetilde{M}_{2H} \cdot \frac{\partial \Delta \widetilde{M}_{\Delta H}}{\partial \psi} \cdot \widetilde{M}_H \cdot \widetilde{M}_{H1} \cdot \overline{r}_{1H} + \\ & + \frac{\partial \widetilde{M}_{2H}}{\partial \psi} \cdot \Delta \widetilde{M}_{\Delta 1} \cdot \widetilde{M}_H \cdot \widetilde{M}_{H1} \cdot \overline{r}_{1H} + \frac{\partial \widetilde{M}_{\Sigma H0}}{\partial \psi} \cdot \Delta \overline{r}_{1i}; \end{aligned} \tag{10.68}$$

$$\overline{M} = \widetilde{M}_{\Sigma H0} \cdot \Delta \widetilde{M}_{\Delta 1} \cdot \frac{\partial \overline{r}_{1H}}{\partial u} + \widetilde{M}_{2H} \cdot \Delta \widetilde{M}_{\Delta H} \cdot \widetilde{M}_H \cdot \widetilde{M}_{H1} \cdot \frac{\partial \overline{r}_{1H}}{\partial u} + \widetilde{M}_{\Sigma H0} \cdot \frac{\partial \Delta \overline{r}_{1i}}{\partial u}; \tag{10.69}$$

$$\overline{N} = \widetilde{M}_{\Sigma H0} \cdot \Delta \widetilde{M}_{\Delta 1} \cdot \frac{\partial \overline{r}_{1H}}{\partial v} + \widetilde{M}_{2H} \cdot \Delta \widetilde{M}_{\Delta H} \cdot \widetilde{M}_H \cdot \widetilde{M}_{H1} \cdot \frac{\partial \overline{r}_{1H}}{\partial v} + \widetilde{M}_{\Sigma H0} \cdot \frac{\partial \Delta \overline{r}_{1i}}{\partial v}. \tag{10.70}$$

The Eqs. (10.65), (10.66), (10.67), (10.68), (10.69), and (10.70) are nonlinear and do not allow finding a solution to the diagnostic problem in an explicit form. Therefore, the method of approximate solution can be applied, which consists of the following:

Earlier it was indicated that, due to the dependence of forces, kinematic parameters, temperatures, wear, etc. on the values of the envelope parameters, the given primary errors are considered as their functions. Therefore, the following dependencies are accepted:

$$\begin{aligned}&\Delta\phi_{xi}=\Delta\phi_{xi}(\phi_1,\psi);\quad \Delta\phi_{yi}=\Delta\phi_{yi}(\phi_1,\psi);\quad \Delta\phi_{zi}=\Delta\phi_{zi}(\phi_1,\psi);\\ &\Delta X_i=\Delta X_i(\phi_1,\psi);\quad \Delta Y_i=\Delta Y_i(\phi_1,\psi);\quad \Delta Z_i=\Delta Z_i(\phi_1,\psi).\end{aligned} \tag{10.71}$$

Considering the relationship between the parameters, ϕ_1, ψ and u, v, similarly, one can write:

$$\begin{aligned}&\Delta r_{x1}=\Delta r_{x1}(u,v);\quad \Delta r_{y1}=\Delta r_{y1}(u,v);\quad \Delta r_{z1}=\Delta r_{z1}(u,v);\\ &\Delta r_{x2YT}=\Delta r_{x2YT}(\phi_1,\psi);\quad \Delta r_{y2YT}=\Delta r_{y2YT}(\phi_1,\psi);\quad \Delta r_{z2YT}=\Delta r_{z2YT}(\phi_1,\psi).\end{aligned} \tag{10.72}$$

The expressions for the functions (10.71) and (10.72) are unknown. Therefore, they must be approximated either by polynomials or the Fourier series. Experience shows that it is advisable to approximate the expressions for calculating the increments of the angles of rotation by the Fourier series and the increments of linear displacements by power polynomials.

The degree of the polynomial and the number of Fourier series are set proceeding from the fact that the total number of functions of the reduced primary errors to be identified is 24. Therefore, taking into account the phase uncertainty of the measured errors in matrix (10.41) as well as the fact that the degree of the polynomial or the number of terms of the Fourier series cannot be more than the number N of points over which the approximation is performed, one can get the upper bounds:

$$n\leq\frac{N}{24}-1. \tag{10.73}$$

The lower limit is 1, so

$$n\geq 1. \tag{10.74}$$

It should be noted that the modern theory of the accuracy of gears considers the parameters of the primary errors as constants, i.e., as a special case of (10.73) and (10.74), which corresponds to $n = 1$ for polynomials and to $n = 3$ for the Fourier series.

Thus, choosing n on the basis of the indicated constraints and expanding Eqs. (10.56), (10.65), and (10.66) into a coordinate form, after finding the partial derivatives of the approximating functions and substituting them in (10.65) and (10.66), we obtain a system of scalar equations with unknown coefficients of polynomials and Fourier series.

Based on this system, for example, by the least squares method [14, 25], it is possible to establish the values of the unknown coefficients of the approximating functions.

In future, each function of the primary error can be decomposed into components: geometric, kinematic, power, temperature, and wear. This is explained by the fact that four out of five components of the total error can be predicted (forecast methods will be discussed below). The fifth term is defined as the difference between the total error and the four terms.

Based on the aforementioned, the general algorithm for solving the problem of diagnosing the accuracy of the elements of an MFTP system based on the results of monitoring a processed gear is shown in Fig. 10.12.

The algorithm is presented in an iterative form, which provides an increase in the accuracy of determining the functions of the given primary errors and can be implemented on a computer.

Devices that provide initial data for the implementation of the algorithm can operate both by the rolling method and by other methods arising from possible technological impact schemes. They are constructed proceeding from the provision of obtaining the information described by matrix (10.45).

A diagram of a device that provides the necessary information for solving the diagnostic problem (working by the run-in method) is shown in Fig. 10.13.

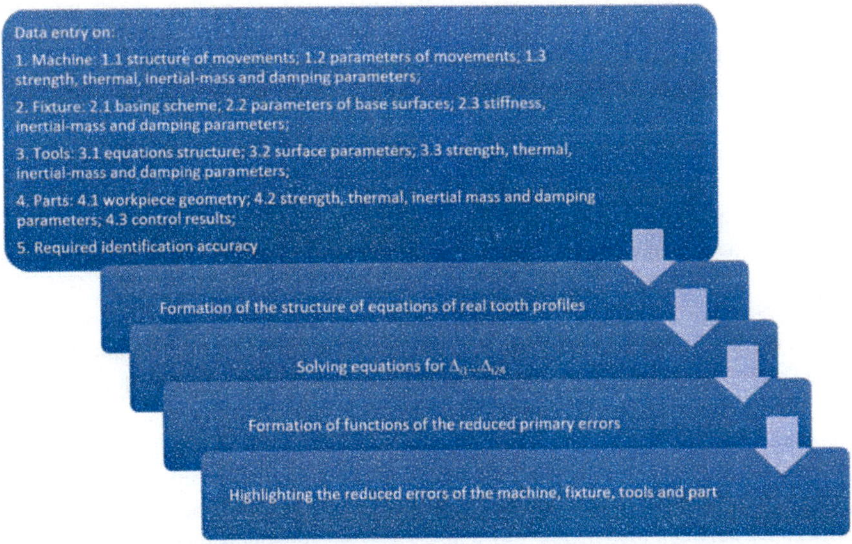

Fig. 10.12 Algorithm for solving the diagnosis of the accuracy of the elements of an MFTP system

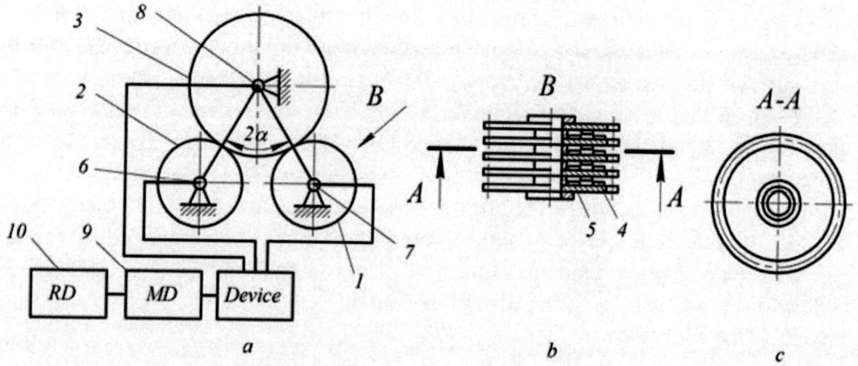

Fig. 10.13 Diagram of a diagnostic device that works by the run-in method

The device contains a frame on which adjustable axles are manteled for installing two measuring gears 1 and 2 as well as for installing controlled gear 3.

Measuring gears 1 and 2 are integral (Fig. 10.13b). They consist of coaxially mounted blocks with the possibility of angular displacement of thin toothed discs, the number of which is more than two.

In the process of testing, each disc must contact controlled gear 3. Therefore, each disc of the measuring gears, except the one on each gear, is spring-loaded in the direction of rotation. The non-spring-loaded discs are rigidly connected to the hubs.

During control, one of the measuring gears is the driving one and the other is the driven one. This ensures the engagement of the measuring gears with the opposite profiles of the teeth of the controlled gear. The possibility of angular displacement of the gear disks of the measuring gears relative to each other ensures the control of the error functions in different axial sections, which corresponds to different rows of matrix (10.41).

By attaching one of the toothed discs to the hub on each measuring gear, a rotation transmission and a single measurement base are provided for each disc. The toothed discs are equipped with sensors 4 and 5 with mutual angular displacement. In addition, the device provides sensors 6 and 7 for measuring the angles of rotation of the driving toothed disk of gear 1 and the driven toothed disk of gear 2 as well as sensor 8 for measuring the angle of rotation of the controlled gear. For receiving, converting, and analyzing signals from sensors, there is multichannel electronic device 9 and recording device 10. By comparing the signals of sensors 6 and 8, the single-profile kinematic error of the left tooth profiles and by comparing the signals of sensors 7 and 8, the single-profile kinematic error of the right tooth profiles are estimated.

The one-profile kinematic error is divided into two components – radial and tangential. For this, the tangential component obtained by comparing the signals from sensors 6 and 7 is vectorially subtracted from the total kinematic error. The difference between the signals from sensors 6 and 7 is proportional to the tangential component of the total kinematic error if the angle between the center transmission

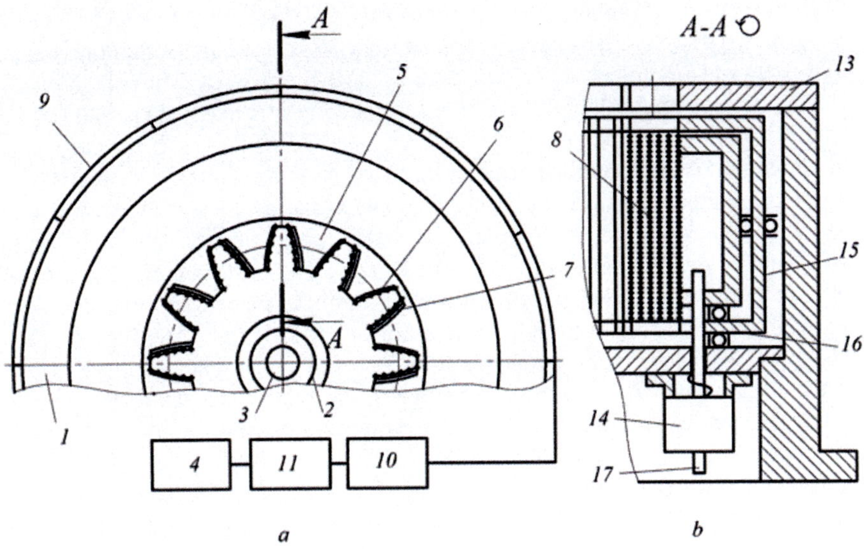

Fig. 10.14 Diagram of a device for diagnosing errors in an MFTP system, working in a discrete-contact way

lines formed by the measuring gears with the controlled gear is equal to the doubled angle of engagement in these gears. This was proved earlier in the studies by Ternyuk [38] and Ternyuk and Khlus [43].

The advantages of a device made according to this scheme include the following: the ability to obtain all the necessary information in one cycle of the gear control; the possibility of readjustment (versatility); the possibility of building on the basis of commercially available kinematometers.

Another example of a device for diagnosing errors in an MFTP system is the one shown in Fig. 10.14.

The device implements a scheme of technological influence with a class $K = 6 \cdot 2 \cdot 3 \cdot 1$, which is the most productive when controlling a gear. The point measurement method allows to directly determine the values of the errors included in matrix (10.41).

The device contains annular housing 1, mounting and clamping device 2 with drive 3, made, for example, in the form of a power cylinder with a rod, recording device 4, and composite measuring gear 5. This gear is equipped with gear blocks 6 and 7 with plates of the right and left profiles teeth. The blocks are installed with the possibility of angular displacement around the axis of adjusting clamping device 2. The shape of the plates is congruent with the profiles of the teeth of the controlled gear. However, this requirement is not mandatory since the placement of the sensors on plates of other shapes is also possible. Sensor 8 is evenly spaced on the active surface of the tooth. It has linear characteristics and can be capacitive, inductive, strain gauge, etc. The outputs of the sensors are connected by conductor 9 to switch

10 connected to computing device 11 containing a program input unit and connected to recording device 4. If necessary, between the switch and the computing device, converters and amplifiers can be turned on to match the characteristics of the system.

The mechanism for orientation of the controlled gear is fixed on the body, which can be made, for example, in the form of two plates 13 with grooves for the passage and direction of the controlled gear.

For the angular displacement of the gear blocks, in order to adjust the width of the cavity of the composite measuring gear and, thereby, install the blocks in the working and loading (nonworking) positions, drive 14 for the angular displacement of these blocks is provided, which is made, for example, in the form of a simple cam mechanism containing wedge-shaped plates – cams 15 and 16 (Fig. 10.14) – interacting by means of lever 17 with drive 3 of setting clamping device 2.

On the body, there are stops 18 for fixing the toothed blocks in the nominal (reference, working) position.

To automate the removal of a controlled gear, the device can be equipped with ejectors. Switch 10 can be configured as single-row or multi-row (Fig. 10.14). The switch contains contacts located along the annular perimeter, connected by means of conductors with linear displacement transducer 8 and a slider installed coaxially with the ring with contact elements, equipped with a rotation drive. The slider is connected to computing device 11 by conductors pointed in the figure by arrows.

A single-row commutator allows entering information about errors at various points of the controlled gear sequentially, which is necessary if a single-input computing device is used, and a multi-row one is parallel-sequential. This provides the use of a multi-input computing device and an increase in the speed of the device.

The highest productivity of the device is achieved when each sensor 5 is connected to its own input of the computing device. In this case, switch 10 has the structure of a conventional interface device.

The device works as follows:

All sensors are preliminarily adjusted to a zero signal either by the reference gear or by a special template. Then, the gear to be inspected is set and fixed on the setting jig. After that, the drive is switched on and the angular displacement of the toothed discs is made until they stop in special projections (basic position). This ensures contact of the sensors with the surfaces of the teeth and the pickup of signals from them through the commutator to the computing device. According to the program entered from block 12, the computing device calculates the necessary coefficients of the approximating functions of the primary reduced errors and takes further actions to identify errors in accordance with the algorithm shown in Fig. 10.12. The results obtained are recorded by device 4.

The advantages of the device include its high performance. The disadvantage is its low versatility.

It is obvious that it is possible to use other diagnostic devices to implement other classes of technological impact schemes.

The main advantages of the diagnostic method are the abovementioned high productivity, the possibility of automation, and the ability to control without stopping the production process of the next gear. However, it cannot be directly carried out at the time of processing, which makes it difficult to use it in adaptive precision devices.

The factorial method is devoid of this disadvantage.

Factorial Method

In accordance with the developed model of the occurrence and accounting of gear errors (Fig. 10.1), the values of the primary errors are the result of the influence of strictly defined factors. This serves as the basis for the development of a factorial method for assessing the values of the reduced primary errors. To develop a method, it is necessary to determine the relationship between the parameters characterizing the factors and the values of the reduced primary errors and then to establish possible methods for assessing these parameters for various processes.

Determination of the dependencies between the parameters of the factors and the given primary errors, like any other dependencies of the functional type, can be carried out using two approaches:

- By conducting experiments
- By modeling (analytically)

The establishment of dependencies by conducting experiments, including methods based on the rational planning of experiments, is described in detail in the literature [1, 9]. Experimental methods can be used to establish all the necessary dependencies. However, the disadvantage of these methods is the need for experimentation. In addition, the results obtained can only be applied under conditions in which the experiments were performed, i.e., they have a limited scope.

An analytical approach is free from these drawbacks. Appropriate models are required for its application. The fundamental approach to the construction of models was developed by Prof. B.M. Bazrov [3].

When applied to the problem under consideration, the use of analytical models can be provided as follows:

Let us represent the *i*th given primary error as an additive function:

$$\Delta_i = \sum_{j=1}^{n} f_{ij}\left(\overline{\Pi}_j\right), \tag{10.75}$$

where f_{ij} is the function of the influence of the *j*th factor on the –th reduced primary error; $\overline{\Pi}_j$ is the vector of parameters of the *j*th factor; and *n* is the number of factors.

In accordance with the general model of the occurrence and accounting of errors (Fig. 10.1) when processing gears, the disturbing factors can be geometric inaccuracies, the parameters of which are the vector $\overline{\Pi}_g$ of geometric deviations of the MFTP system's elements; kinematic deviations, characterized by deviations of speeds $\Delta\overline{V}_k$ and trajectories from their nominal values; temperature errors, the parameters of which are temperatures T° at various points of the MFTP system;

elastic deformations $\Delta\overline{Y}$,as a result of the impact of forces $\overline{P}$; wear, characterized by the parameters of its distribution on the elements of the MFTP system; and internal stresses, characterized by the tensors of their components.

Considering that internal stresses, as a destabilizing factor, can be removed by 98% or more by appropriate heat treatment, which is always carried out in the processing of high-quality gears [9], the component of the reduced primary errors caused by residual internal stresses, due to its small absolute value, is excluded from consideration.

Taking this into account, we obtain $n = 5$, and dependence (10.75) is transformed to the form:

$$\Delta_i = f_g(\overline{\Pi}_g) + f_k(\Delta\overline{V}_k) + f_t(\overline{T}_t) + f_f(\overline{P}_f) + f_w(\overline{W}_w), \tag{10.76}$$

The indices "g," "k," "t," "f," and "w"in (10.76) denote the functions of the influence of geometric inaccuracies, kinematic errors, temperature factor, forces, and wear, respectively.

Considering the MFTP system as a common mechanism, the influence of the vector $\overline{\Pi}_g$ on Δ_i can be determined in a general matrix way [24]:

$$\{\Delta i\} = \widetilde{M}_{ij} \cdot \overline{\Pi}, \tag{10.77}$$

where $\widetilde{M}_{ij}$ is the matrix of the influence of the parameters of geometric inaccuracies on the primary errors of the MFTP system, the elements of which reflect the geometric interaction of mechanisms, aggregates, and system parts. Methods for compiling the matrix $\widetilde{M}_{ij}$ are described in the study by Portman [24].

The type and parameters of the function $f_k(\Delta\overline{V}_k)$ can be found by considering the MFTP system as a mechanical system with many masses, moving under the influence of time-varying forces. The main issues of the influence of kinematic factors on the arising errors in the processing of gears under conditions of free rolling, including those with attached masses, are discussed in detail in the works of Prof. Yu. N. Sukhorukov [33, 34]. Based on these works, it is possible to write, in a general form, a matrix differential equation of the second order, which makes it possible to determine the influence of kinematic factors ($\Delta\overline{V}_k$) on each of the given primary errors:

$$\widetilde{m}\dot{V} + \widetilde{\eta}\Delta V + \widetilde{c}s = \widetilde{F}, \tag{10.78}$$

where

$\widetilde{m}$ is the mass matrix of the MFTP system.
$\dot{V}$ is the acceleration matrix.
ΔV is the velocity matrix.
s is the matrix of displacements.
$\widetilde{\eta}$ is the matrix of damping coefficients.

$\widetilde{c}$ is the stiffness matrix.
$\widetilde{F}$ is the matrix of acting forces.

System (10.78) can be solved by determining it for a specific MFTP system, as a rule, only by numerical methods.

There are many heat sources operating in an MFTP system: the cutting process, friction units, etc. Therefore, its temperature field is not uniform, but it is characterized by different temperatures at different points.

Having chosen the characteristic points within the dimensional chain, including the *j*th primary error, the influence of the temperature factor on the given primary errors can be determined based on the model of thermal expansion of bodies:

$$\Delta_{jt} = \sum_{i=1}^{m} \alpha_{ti} l_i (T_i - T_0) - \sum_{j=1}^{p} \alpha_{tj} l_j (T_j - T_0), \tag{10.79}$$

where

l_i is the length of the *i*th section of the MFTP system's dimensional chain, which has a temperature T_i.
α_{ti} is the coefficient of linear expansion.
T_0 is the temperature baseline at which temperature errors are assumed to be zero (usually $T_0 = 20\,°\,\mathrm{C}$).
m is the number of increasing links of the dimensional chain.
p is the number of reducing links of the dimensional chain.

The influence of forces on the reduced primary errors can be determined on the basis of matrix Eq. (10.78) by solving it and establishing dependencies of the form:

$$\Delta_{if} = P_i \sum_{i=1}^{n} \frac{1}{c_i}. \tag{10.80}$$

In other words, Eq. (10.80) from the general solution of Eq. (10.78) includes the terms containing the compliance coefficients of the elements of the MFTP system.

The impact of wear on these primary factors can be assessed in two ways: direct and indirect.

In direct assessment, the wear at the *i*th point of the *j*th element of the MFTP system can be considered as a common geometric inaccuracy and its effect on the reduced primary error in this case is described by Eq. (10.77).

The indirect evaluation system is based on the established fact that the amount of wear is closely correlated with forces, temperatures, vibrations [18], and the duration of their exposure.

With this approach, the effect of wear on the given primary errors can be reflected as follows:

$$\frac{dw_i}{dt} = aP_i^q + bT_i^l + cM_i^s; \quad w_i = \int_0^t \left(aP_i^q + bT_i^l + cM_i^s\right)dt \tag{10.81}$$

where

a, b, and c are the coefficients.
q, l, and s are the indicators of the degree of influence of forces, temperatures, and vibration level on the amount of wear at the ith point of the system, respectively.
M_i is the level of vibrations at the ith point of the system.

Thus, knowing the parameters of the factors, according to Eqs. (10.79), (10.80), and (10.81), detailed for each specific MFTP system, it is possible to calculate the given primary errors and, from them, the directly normalized errors of gears. Factors, in turn, can be either measured or calculated (predicted). Measurement of factor parameters can be carried out by well-established traditional methods:

- Geometric inaccuracies – devices for linear and angular measurements
- Kinematic inaccuracies – kinematometers
- Temperature parameters – temperature sensors
- Parameters of forces – dynamometers
- Wear – devices for linear and angular measurements for direct methods of control and devices for measurement of the level of vibrations, thermometers, and dynamometers (strain gauges, inductive, etc.) for indirect control methods

The advantage of the factor parameter measuring method is the possibility of implementation in the processing, which is an important prerequisite for creation of adaptive accuracy control systems. The disadvantage is the difficulty of controlling geometric inaccuracies due to their large number. However, this shortcoming can be overcome by taking into account the important property of individual components of errors – systematic – to retain their values for a sufficiently long time (i.e., to maintain stability over time).

The obtained results allow one to proceed to the consideration of methods of analytical error forecasting.

10.2.2 *Methods of Analytical Forecasting of Normalized Errors of Gears*

Methods of analytical forecasting of normalized errors of gears are used for the calculation of a range of errors of a certain kind (10.1) or for the calculation of values of actual errors with a set degree of reliability.

Therefore, the main prerequisite for their implementation is to obtain dependencies of form (10.1) and dependencies that allow calculating the values of actual errors with the required degree of reliability.

Of all the possible formulations of analytical error forecasting, we single out and solve the two most general formulations, which consist of developing error forecasting techniques for the case of taking into account all possible sources of errors in forecasting error scattering ranges (calculation-probability method and adaptive errors).

10.2.2.1 Calculation-Probability Method

Based on the analysis of the dependencies of types (10.43), (10.44), (10.45), and (10.46) [13], we can establish that the whole set of indicators of the accuracy of gears is divided into two groups. The first one includes indicators that take into account the direction of deviation from the face value with its sign, for example, indicators of lateral clearance, known as step deviation. The second one includes the indicators considered without their sign and defined as the module of a difference of extreme values.

Accordingly, one can write the following general dependencies to determine the indicators:

(a) The first group:

- Current value

$$\text{Н}\Pi_l = \left(U_i \Delta K_i + U_j \Delta K_j\right); \tag{10.82}$$

- Limit (normalized) value

$$[\text{Н}\Pi_l]_r = \left(U_i \Delta K_i + U_j \Delta K_j\right)\text{extremum}; \tag{10.83}$$

(b) The second group:

- Current value

$$\text{Н}\Pi_m = \left(U_i \Delta K_i - U_j \Delta K_j\right); \tag{10.84}$$

- Limit (normalized) value

$$[\text{Н}\Pi_m]_r = \left(U_i \Delta K_i - U_j \Delta K_j\right)\text{extremum}; \tag{10.85}$$

where

$$\Delta K_{i(j)} = V(\Delta_1) \bigotimes V(\Delta_2) \bigotimes \ldots \bigotimes V(\Delta_n) \tag{10.86}$$

$U_{i(j)}$ is the transfer coefficient of influence $\Delta K_{i(j)}$ on the normalized accuracy index, determined in a similar manner as (10.43), (10.44), (10.45), and (10.46).
$\bigotimes$ is the sign of multiplication of quaternions [14].

Based on the properties of the compositions of the laws of error distribution [42] and taking into account the obtained expressions (10.83) and (10.85) and the fact that $\Delta K_{i(j)}$ depends on the 24 primary errors, each of which is a function of 5 groups of parameters, we can say, on the basis of the limit theorems of probability theory [25], that the values НП_l and НП_m will be distributed according to Gauss's laws (shifted) and the modulus of difference (also shifted due to the presence of time constant – systematic – components).

Thus, we can write:

$$q(\text{НП}_l) = \frac{1}{\sigma_1 \sqrt{2\pi}} \exp\left[-\frac{1}{2}\left(\frac{\text{НП}_l - \xi_1}{\sigma_1}\right)^2\right]; \tag{10.87}$$

$$q(\text{НП}_m) = \frac{\text{НП}_m}{\sigma_2^2} \exp\left[-\frac{1}{2}\left(\frac{\text{НП}_m - \xi_2}{\sigma_2}\right)^2\right], \tag{10.88}$$

where

q is the density of distribution.
ξ_1 and ξ_2 are the mathematical expectations.
σ_1 and σ_2 are the variances of the considered quantities.

Expressions for calculating mathematical expectations and variances, established by taking into account the correlation of error modules, have the form:

$$\xi_1 = M(U_i \Delta K_i) + M\big(U_j \Delta K_j\big); \tag{10.89}$$

$$\xi_2 = M(U_i \Delta K_i) - M\big(U_j \Delta K_j\big); \tag{10.90}$$

$$\sigma_1 = \sqrt{\sigma^2_{U_i\Delta K_i} + \sigma^2_{U_j\Delta K_j} + 2\rho_{U_i\Delta K_i U_j\Delta K_j}\sigma_{U_i\Delta K_i}\sigma_{U_j\Delta K_j}}; \tag{10.91}$$

$$\sigma_2 = \sqrt{\sigma^2_{U_i\Delta K_i} + \sigma^2_{U_j\Delta K_j} - 2\rho_{U_i\Delta K_i U_j\Delta K_j}\sigma_{U_i\Delta K_i}\sigma_{U_j\Delta K_j}}, \tag{10.92}$$

where $\rho_{U_i\Delta K_i U_j\Delta K_j}$ is the correlation coefficient between $U_i\Delta K_i$ and $U_j\Delta K_j$.

The obtained expressions allow solving the problems of predicting normalized errors with the required reliability. Based on (10.87) and (10.88) and taking into account (10.89), (10.90), (10.91), and (10.92), we can write:

$$P\left([\text{НП}_l]^- < \text{НП}_l < [\text{НП}_l]^+\right) = \Phi\left(\frac{[\text{НП}_l]^+ - \xi_1}{\sigma_1}\right) - \Phi\left(\frac{[\text{НП}_l]^- - \xi_1}{\sigma_1}\right); \quad (10.93)$$

$$P\left(\text{НП}_m < [\text{НП}_m]^+\right) = \frac{1}{\sigma_2^2}\exp\left[-\frac{1}{2}\left(\frac{[\text{НП}_m]^+ - \xi_2}{\sigma_2}\right)^2\right], \quad (10.94)$$

where

$\Phi(x)$ is the Laplace integral.
$P(x)$ is the probability of exceeding the value x of the calculated value.

The use of dependencies (10.93) and (10.94) makes it possible to implement a computational-probabilistic analytical method for predicting errors. As a result of calculations, a range of distribution of errors are defined.

Example Let us consider the prediction of the radial beating of a ring gear for spur gears milled by a single-pass worm cutter. To obtain a solution in an analytical form, we assume that all factors, except base errors, are constant. Then,

$$\Delta F_{\text{R}} = \Delta Y_2 \sin(\phi_2 + \alpha) + C_{\text{R}}; \quad \Delta F_{\text{L}} = -\Delta Y_2 \sin(\phi_2 - \alpha) + C_{\text{L}},$$

where

$C_{\text{R}} = \text{const}$ and $C_{\text{L}} = \text{const}$;
$F_{rr} = 2\Delta Y_2$ и $P(F_{rr} \leq F)_r = P\left(\Delta Y_2 \leq \frac{F_{rr}}{2}\right)$, т.к. $M(\Delta Y_2) = 0$, $\xi_2 = 0$.

$$\sigma_{(F_{rr})} = \sqrt{\sigma^2_{(\Delta Y_2)} + \sigma^2_{(\Delta Y_2)} + 2\sigma_{(\Delta Y_2)}\sigma_{(\Delta Y_2)}} = 2\sigma_{(\Delta Y_2)}.$$

Information on the range of error distributions is sufficient to solve the main optimization problems using the traditional approach to product accuracy, based on the establishment of restrictions on the elements of the process and the MFTP system from condition (10.1).

However, with adaptive accuracy methods, being aware of only the error scattering ranges is not enough. It is necessary to know the actual value of the error at each point of the tooth of the machined gear.

The latter is provided using the computational-adaptive method of error prediction.

10.2.2.2 Computational-Adaptive Method

The error prediction computational-adaptive method is based on the following:

All errors are classified into two types: changeable over time (discrete or functional) and unchangeable over time (within the period T_c; under review). Thus, errors with graphs of change over time, similar to those shown in Fig. 10.15, are highlighted.

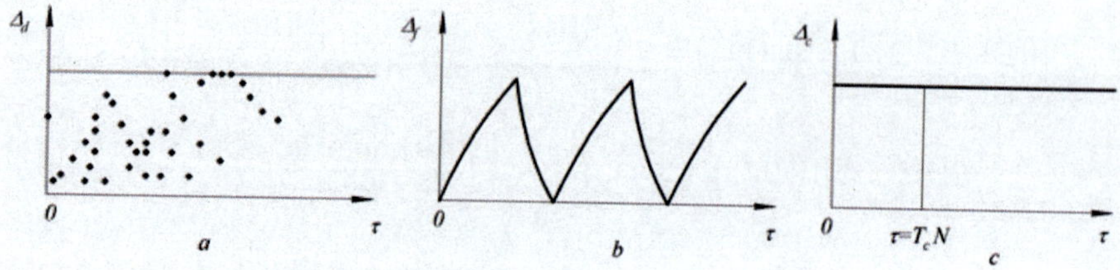

Fig. 10.15 Graphs of changes in errors over time

Examples are errors that change discretely over time including errors in the basing of workpieces and errors in adjusting a machine to size when milling. Such errors have a graph of change as shown in Fig. 10.15a. Errors that change functionally over time (Fig. 10.15b) include temperature errors, wear errors of the blade tool made of high-speed steel, and others. The constant changes within the considered period (systematic) errors include the actual collisions of the machine and usually appear idle (geometric and part of the kinematics). These errors are characterized by the graph shown in Fig. 10.15c.

The outcome of this classification of errors can be determined by

$$\Delta_n = \Delta_d + \Delta_f + \Delta_c, \tag{10.95}$$

The terms "Δ_d, Δ_f, Δ_c" in (10.95) are errors that change discretely and functionally and those that do not change, respectively.

Values Δ_c can be determined once before processing the first gear and are considered constant over time T_c.

The types of functional dependencies can also be determined a priori. To calculate functional dependencies, one needs to know the parameters that act as arguments. These parameters (forces, temperatures, vibration levels, etc.) can, in turn, be either predicted or monitored. The prediction of parameters is performed according to the correlation and regression equations commonly used in the theory of material cutting or on the basis of these equations using the methods of similarity theory [30].

Control can be performed in the same manner as with the factor method of determining the given primary errors. Similarly, information can be obtained about the components that change discretely. Given that the parameters characterizing the functional and discrete change of errors are relatively few, the sensors for the rapid acquisition of information about their values can be constantly on the machine.

As a result of entering a priori information, functions (10.82), (10.83), (10.84), (10.85), and (10.86) change from probabilistic to deterministic. This means that the reliability of the forecast on them will be equal to that of the supplied source information. For example, in the case of entering a priori information about the value ΔY_2 (from the previous example), the value will be determined as $F_{rr} = 2\Delta Y_2$ with a reliability equal to that of the information about ΔY_2. As this information can be obtained as a result of direct control, the reliability of the prediction of coins should be as high as desired and determined by the accuracy of measurements ΔY_2.

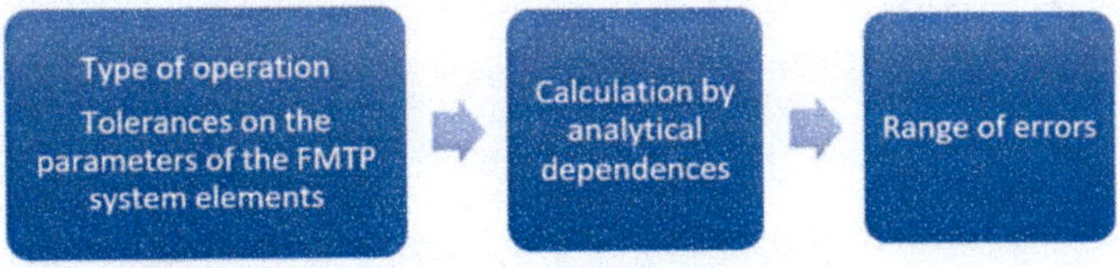

Fig. 10.16 Scheme for calculation-probability forecasting of errors

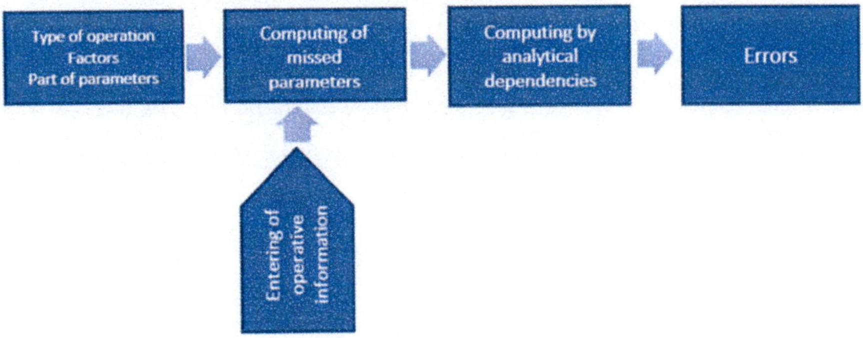

Fig. 10.17 Scheme for computational-adaptive method of error prediction

Based on the above, the general scheme for implementation of computational-probabilistic and computational-adaptive methods of error prediction can be depicted as shown in Figs. 10.16 and 10.17, respectively.

The main result of the research conducted in Sect. 10.2 was the development and study of a system model of the process of real tooth formation, reflecting the general laws of this process. As a result, the general equations of the real surfaces of the teeth of machined gears are obtained. It is established that the whole set of primary errors of any MFTP system can be reduced to the 24 listed primary errors, linearly independent but correlated in modulus. To determine such errors, methods of direct measurements, diagnostics, and factorial are proposed. The problem of diagnostics (identification) of errors of an MFTP system is set in the general form and solved. New gear measuring tools have been proposed for its implementation. It is proved that the nonlinear nature of the equations of real tooth surfaces with respect to the given primary errors necessitates the use of special methods for summation of errors and excludes the possibility of applying the principle of independence of errors. Methods for forecasting errors with a given forecast reliability have been developed. It is proved that when entering operational information, it is possible to predict not only the ranges of distribution of normalized errors but also the values of the actual deviations of the real points of the tooth profiles from the nominal ones.

For development of forecasting methods in the case of multifactorial processes, one can also use the relatively recent methodology of the system information approach [19, 20].

The obtained results are the basis for the following studies of the methods of control and management of the accuracy of gears.

10.3 Methods and Systems for Controlling the Accuracy of Gears

10.3.1 A Full Range of Possible Ways to Control the Accuracy of Gears

Traditional gear control systems [2, 9, 35, 36] are based on two methods:

- Direct method, providing information on the accuracy of the part by measuring the actual errors of a machined gear
- Indirect method, based on periodic verification of the accuracy of the elements of an MFTP system

The scheme for obtaining information using the first method of control is shown in Fig. 10.18.

The advantages of control systems based on the first method are high accuracy and reliability of the information obtained. However, there are significant disadvantages, which are as follows:

- Difficulties of control directly at the moment of processing of a detail
- Impossibility of application for the decision of problems of optimum operational control of accuracy "on indignation"
- The need to introduce special operations (transitions) of control, requiring additional complexity and the use of many complex gear tools, which do not always have sufficient performance and the required level of automation

The scheme for obtaining information using the second method of control is shown in Fig. 10.19.

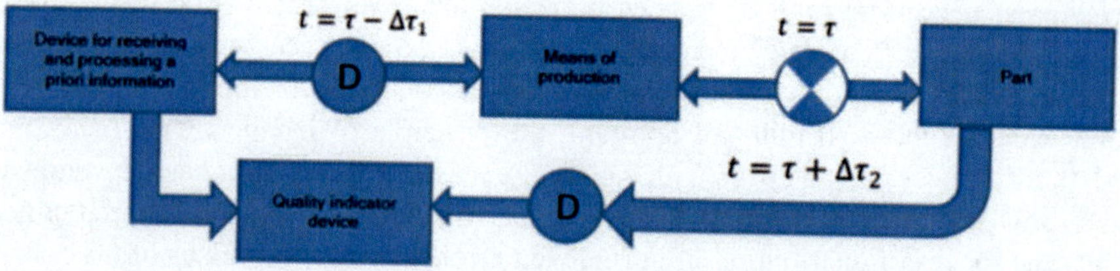

Fig. 10.18 Scheme for obtaining information using the direct method of control

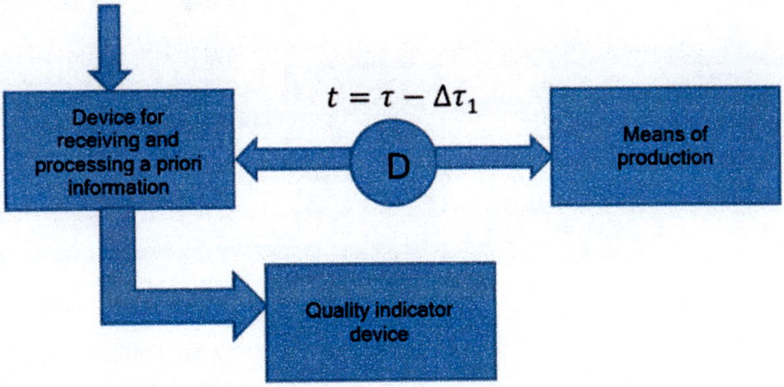

Fig. 10.19 Scheme for obtaining information using the indirect method of control

Systems based on the second method of obtaining information, although less time-consuming than the first one, cannot be considered optimal, as they often do not allow both ensuring the permitted degree of risk of missing a defective part and obtaining information on the current error values, and, therefore, they cannot be used in adaptive accuracy control systems.

Analysis of connections in the general model of occurrence of errors (Fig. 10.1) shows that the possibilities of control of gears are not limited to two considered ways of receiving information. The vector of normalized accuracy indicators $\overline{\text{НП}}$ is related to the vector $\overline{\Delta n}$ of the given primary errors:

$$\overline{\text{НП}}_i = f_i\left(\overline{\Delta n}_j\right), \quad i \in n, \quad j \in \{1, 2, \ldots, 24\}. \tag{10.96}$$

Therefore, control systems can be based on obtaining information about primary errors (10.96). Given that in this case, the information about the normalized accuracy consists of separate information about the accuracy of the elements of the MFTP system, control systems based on this method of control can be called an element-by-element one (unlike the previously considered system, which provides direct measurement). The scheme for obtaining information using the element-by-element method of control is shown in Fig. 10.20.

According to Fig. 10.1, each of these primary errors is the result of the influence of strictly defined factors, the parameters of which can be measured. Therefore, one can establish the relation:

$$\overline{\Delta n}_j = \phi_j\left(\overline{\Pi}_k\right), \quad \overline{\text{НП}}_i = f_i\left[\phi_j\left(\overline{\Pi}_k\right)\right], \quad k \in N. \tag{10.97}$$

It follows from (10.97) that, knowing the types of functions f_i and ϕ_j and controlling the parameters of the factors $\overline{\Pi}_k$, one can univocally assert the accuracy of the gear. Control systems based on this method of obtaining information can be called factorial (Fig. 10.21).

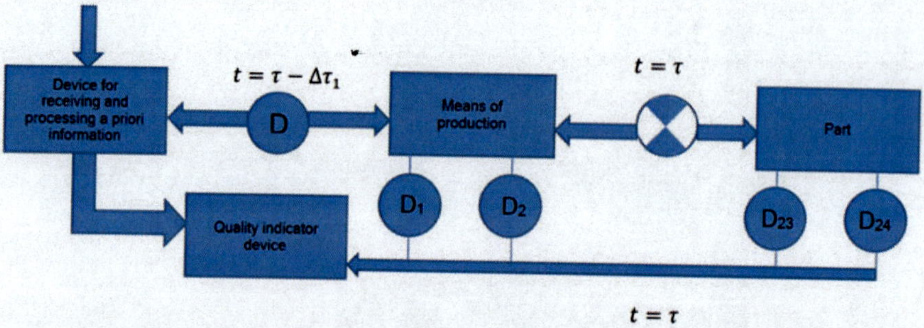

Fig. 10.20 Scheme for obtaining information using the element-by-element method of control

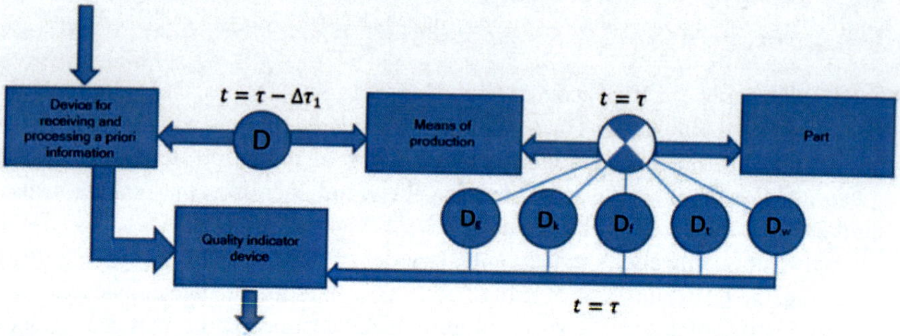

Fig. 10.21 Scheme for obtaining information using the factorial method of control

These control methods correspond to three different levels of relationships in the model shown in Fig. 10.1. It is not possible to select other levels corresponding to these methods. Therefore, given that the method of indirect control (Fig. 10.19) is a special case of the element-by-element method of control, based on research, three fundamentally different methods of control that are possible to implement can be asserted: direct, element-by-element, and factorial.

The last two methods of control can be directly carried out at the time of tooth formation, and, therefore, they can be used as the basis for adaptive accuracy control systems that implement the principles of control of both "perturbation" and "deviation" [17].

Based on the first method of control, it is possible to build only such accuracy control systems that implement the principle of control of "deviation" (i.e., using feedback) and, therefore, require additional workflows to correct the detected deviations in accuracy.

The three listed methods of control, which can be called simple, correspond to the four combined ones.

A list of all possible control methods classified by the method of obtaining information is provided in Table 10.1. It also provides information on the possible principles of accuracy control that can be implemented using the applied control methods.

Table 10.1 Ways to control the accuracy of gears

№. (class)	Information-obtaining method	Control mean type	Control mean, possible for realization
1	Measurement of the normalized parameters of the accuracy of a ring gear	Simple (direct)	By feedback
2	Measurement of the primary errors	Simple (element-by-element)	Outraged
3	Measurement of the parameters of the factors causing errors	Simple (factorial)	Outraged
4	Measurement of the normalized indicators of accuracy of a ring gear + measurement of the resulting primary errors	Combined	By feedback or combined
5	Measurement of the normalized indicators of accuracy of a ring gear + measurement of the parameters of the factors causing errors	Combined	By feedback or combined
6	Measurement of the given primary errors + measurement of the parameters of the factors causing errors	Combined	Outraged, by feedback or combined
7	Measurement of the normalized indicators of accuracy of a ring gear + measurement of the resulting primary errors + measurement of the parameters of the factors causing errors	Combined	By feedback or combined

Table 10.2 List of classes of methods of control of gears

Degree of completeness of control over time (volume of gears produced)	Degree of completeness of control on parameters	Control method class
Full	Full	Deterministic
	Partial	Probabilistic
Partial	Full	Probabilistic
	Partial	Probabilistic

Using one or another method of gear control, one can measure either all or some of the parameters that ultimately determine the accuracy of a gear and control either in full, at the time of production (or its volume), or not in full.

If we put the degree of completeness of control by factors and volume in a batch of gears on the basis of classification, the list of the possible classes of methods of control will be as shown in Table 10.2.

The classes of control methods given in Table 10.2 cover all possible control systems. Those classes that do not provide complete information on time or parameters are classified as probabilistic. One control class that provides complete information is classified as deterministic.

Let us move on to consider the degree of risk of missing a defective part for the specified set of control methods, taking into account their organization into different classes: deterministic or probabilistic.

10.3.2 Analysis of Measurement Errors and the Degree of Risk of Missing a Defective Part in Various Control Systems

Let us first consider the general methodology and then specify it for each method of control. The degree of risk will be assessed by the probability of missing a defective part:

$$P = P[(\text{Н}\Pi_{\Phi} > [\text{Н}\Pi_K]) \cap (\text{Н}\Pi_{\Phi} > \text{Н}\Pi_K)] \tag{10.98}$$

where $\text{Н}\Pi_{\Phi}$ and $\text{Н}\Pi_K$ are the actual and measured values of the monitored parameters, respectively.

By choosing the parameters to be inspected and the frequency of inspections, as well as the accuracy of measuring instruments, one can control the degree of risk of missing a defective part (10.98). In order to study the possibilities of this management, we will develop models of real control processes.

The most common method of control is the factor method. Having matched each level of connections of the model shown in Fig. 10.1 to its corresponding operator, taking into account that the primary information for each method is obtained by measurement, we find a general operator model of the factor control process

$$J[\Delta_{\Pi i}] = R_{\Pi i}^{(1)} \left\{ \bigcap_{j=1}^{24} \left[\bigcap_{k=1}^{5} R_{\Pi jk}^{(2)} \left(R_{U3}^{(3)} \Pi_k \right) \right] \right\}, \tag{10.99}$$

where

J is the sign of information about the object specified in parentheses.
$R_{U3}^{(3)}$ is operator of measurement of the ith parameter.
R_{Π} is the recalculation operator.

The upper index of the operator indicates the level of communication in the model shown in Fig. 10.1. The number of operators is determined by the number of the types of factors and the primary errors given, respectively.

From formula (10.99), by limit transition, it is possible to obtain models corresponding to the element-by-element and direct control methods:

$$J[\Delta_{\Pi i}] = R_{\Pi i}^{(1)} \left[\bigcap_{j=1}^{24} \left(R_{U3j}^{(2)} \Delta_{\Pi j} \right) \right] \tag{10.100}$$

$$J[\Delta_{\Pi i}] = R_{U3i}^{(1)} \text{Н}\Pi_{Ti}, \tag{10.101}$$

Equations (10.99), (10.100), and (10.101) are models of ideal control processes. Models of real control processes are built on their basis, and structures of control subsystems are also synthesized. The synthesis of structures is performed by taking into account the fact that:

- To each operator in dependencies (10.99), (10.100), and (10.101), the subsystem should be put in accordance.
- Subsystems can be combined.
- Equations (10.99), (10.100), and (10.101) reflect only the target actions implemented at the stage P.

Therefore, in order to obtain full-fledged control subsystems (their structures), it is necessary to include in their composition those technical means and performers that can perform auxiliary and preparatory functions.

Since Eqs. (10.99), (10.100), and (10.101) describe all possible methods of control, on the basis of such a technique, the full range of the possible structures of control systems is revealed, including the synthesis of new systems. Construction of models of real control processes is performed based on the following:

Each measurement or recalculation operator makes an error in the error value to be determined. These errors are different in magnitude for different operators and for different parameters within the same operator. The error of measurements $\boldsymbol{\Omega}_u$ is determined by the accuracy of measuring instruments and the error of recalculation $\boldsymbol{\Omega}_n$ – the quality of the applied mathematical model – by the degree of its adequacy to the process.

The dependencies provided in Sect. 10.2 on the relationship between the increments of coordinates (as a function of the above primary errors) and the normalized accuracy of the gears allow calculations be performed with arbitrarily high accuracy, as they are based on strict geometric relationships. Therefore, the operator error $R_{ni}^{(1)}$ can often be neglected. The task is to evaluate the impact of measurement errors and errors in the conversion operator $R_{ni}^{(2)}$.

Let us denote the actual value of the operator as

$$R^{(j)}_{\text{П}(U3)i} = R^{(j)}_{\text{НП}(U3)i} + \Delta R^{(j)}_{\text{П}(U3)i}, \tag{10.102}$$

where $R^{(j)}_{\text{НП}(U3)i}$ and $\Delta R^{(j)}_{\text{П}(U3)i}$ are the nominal (exact) values of the operator and its increment (error), respectively.

Then, taking into account that dependence (10.102) reflects the most general type of control, from which others follow as special cases, we can write:

$$\text{НП}_{Ki} = \left(R^{(1)}_{\text{НП}i} + \Delta R^{(1)}_{\text{П}i}\right)\left\{\bigcap_{j=1}^{24}\left[\bigcap_{k=1}^{5}\left(R^{(2)}_{\text{НП}K} + \Delta R^{(2)}_{\text{П}K}\right)\times\left(R^{(3)}_{\text{Н}U3K} + \Delta R^{(3)}_{U3K}\right)\text{П}_K\right]\right\}. \tag{10.103}$$

Up to small second-order smallness, Eq. (10.103) can be transformed as follows:

$$\text{НП}_{Ki} = \text{НП}_{\Phi} + \Delta R^{(1)}_{\Pi i}\left\{\bigcap_{j=1}^{24}\left[R^{(2)}_{\text{НП}K}\cdot\left(R^{(3)}_{\text{Н}U\text{З}K}\cdot\Pi_K\right)\right]\right\} + R^{(1)}_{\text{НП}i}\left\{\bigcap_{j=1}^{24}\left[\bigcap_{k=1}^{5}\Delta R^{(2)}_{\Pi K}\cdot\left(R^{(3)}_{\text{Н}U\text{З}K}\cdot\Pi_K\right)\right]\right\} + \\ + R^{(1)}_{\text{НП}i}\left\{\bigcap_{j=1}^{24}\left[\bigcap_{k=1}^{5}R^{(2)}_{\text{НП}K}\cdot\left(\Delta R^{(3)}_{U\text{З}K}\cdot\Pi_K\right)\right]\right\}, \tag{10.104}$$

where

$$\text{НП}_{\Phi} = R^{(1)}_{\text{НП}i}\left\{\bigcap_{j=1}^{24}\left[\bigcap_{k=1}^{5}R^{(2)}_{\text{НП}K}\cdot\left(R^{(3)}_{\text{Н}U\text{З}K}\cdot\Pi_K\right)\right]\right\}. \tag{10.105}$$

Based on (10.104) and taking into account that expression (10.105) is the exact value of the parameters, we can find the total control error:

$$\Omega_{Ki} = \Delta R^{(1)}_{\Pi i}\left\{\bigcap_{j=1}^{24}\left[R^{(2)}_{\text{НП}K}\cdot\left(R^{(3)}_{\text{Н}U\text{З}K}\cdot\Pi_K\right)\right]\right\} + R^{(1)}_{\text{НП}i}\left\{\bigcap_{j=1}^{24}\left[\bigcap_{k=1}^{5}\Delta R^{(2)}_{\Pi K}\cdot\left(R^{(3)}_{\text{Н}U\text{З}K}\cdot\Pi_K\right)\right]\right\} + \\ + R^{(1)}_{\text{НП}i}\left\{\bigcap_{j=1}^{24}\left[\bigcap_{k=1}^{5}R^{(2)}_{\text{НП}K}\cdot\left(\Delta R^{(3)}_{U\text{З}K}\cdot\Pi_K\right)\right]\right\}, \tag{10.106}$$

This error corresponds to the factor control method. In the case of the corresponding element-by-element control, it is necessary to put

$$R^{(3)}_{\text{Н}U\text{З}K} = 1,\quad \Delta R^{(3)}_{U\text{З}K} = 0.$$

Under direct control

$$R^{(1)}_{\text{НП}i} = R^{(2)}_{\text{Н}Ki} = 1,\quad \Delta R^{(j)}_{\Pi i} = \Delta R^{(j)}_{\Pi K} = 0.$$

Given the large number of components included in (10.106), it can be argued that the total control error will be distributed according to the normal law with parameters

$$M(\Omega_{Ki}) = \sum_{j=1}^{24} M\left(\Delta R^{(1)}_{\Pi j}\right)\left[R^{(2)}_{\text{НП}K}\left(R^{(3)}_{\text{Н}U\text{З}K}\cdot\Pi_K\right)\right] + \sum_{j=1}^{24} R^{(1)}_{\text{НП}j}\left[\sum_{k=1}^{5} M\left(\Delta R^{(1)}_{\Pi K}\right)R^{(3)}_{\text{Н}U\text{З}K}\cdot\Pi_K\right] + \\ + \sum_{j=1}^{24} R^{(1)}_{\text{НП}j}\left[\sum_{k=1}^{5} R^{(2)}_{\text{НП}K} M\left(\Delta R^{(3)}_{U\text{З}K}\right)\Pi_K\right]; \tag{10.107}$$

$$\sigma(\Omega_{Ki}) = \sqrt{\begin{array}{l}\left\{\sum_{j=1}^{24}\sigma\left(\Delta R^{(1)}\right)\left[R_{\text{HП}K}^{(2)}\left(R_{HU3K}^{(3)}\cdot\Pi_K\right)\right]\right\}^2 + \left\{\sum_{j=1}^{24}R_{\text{HП}j}^{(1)}\left[\sum_{k=1}^{5}\sigma\left(\Delta R^{(2)}\right)R_{HU3K}^{(3)}\cdot\Pi_K\right]\right\}^2 + \\ +\left\{\sum_{j=1}^{24}R_{\text{HП}j}^{(1)}\left[\sum_{k=1}^{5}R_{\text{HП}K}^{(2)}\sigma\left(\Delta R^{(3)}\right)\Pi_K\right]\right\}^2\end{array}};$$

(10.108)

where $M(x)$ and $\sigma(x)$ are the signs of mathematical expectation and variance, respectively.

Assuming independence of control errors from gear errors, the degree of risk of missing a defective part can be calculated depending on

$$\widetilde{\text{P}} = P(\text{H}\Pi_{\Phi} > [\text{H}\Pi_K]) \cdot P(\text{H}\Pi_{\Phi} > \text{H}\Pi_K) \tag{10.109}$$

From dependence (10.109), it follows that the degree of risk depends not only on the accuracy of control but also on the stability of the technological process that determines the value of P. To obtain a degree of risk not greater than the specified one, in the case of an unstable process, one should use more sophisticated measuring instruments.

The obtained dependencies are the basis for solving direct and inverse problems related to the organization of optimal gear control systems. On the basis of general equations, we will consider the features of various control systems. In this case, due to the deep study of direct control systems in the works by Taits and Markov [35] and Taits [35], we exclude them from consideration.

10.3.3 Element-by-Element Control Systems

In accordance with Eq. (10.100), systems implementing the element-by-element method must perform the function (operator) of measuring the given primary errors. These systems can be built on the basis of the sensors of linear and angular measurements. The task is to use such sensors to measure all 24 (in deterministic systems) or some (in probabilistic systems) of these primary errors. In accordance with the functions associated with obtaining, converting, and transmitting information on Eq. (10.100), the required technical means (subsystems), and combining them, we obtain a variant of the structural scheme of technical means implementing the element-by-element control system in the most general form, i.e., determined (Fig. 10.22).

The figure shows:

S_i – sensors of the linear and angular measurements
P_i – parameters
A – converter amplifier

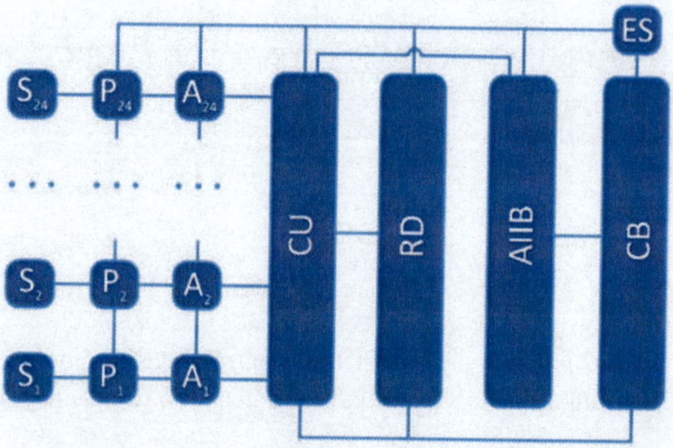

Fig. 10.22 A variant of the structure of the technical means that implements the element-by-element method of control

CU – calculating unit
RD – recording device
AIIB – a priori information input block
CB – control block
ES – energy source

Based on dependencies (10.106) and (10.109), it is possible to find a measurement error for this type of system and the degree of risk of missing a defective part:

$$\Omega_{Ki} = \Delta R^{(1)}_{\Pi i}\left\{\bigcap_{j=1}^{24}\left[R^{(2)}_{\text{Н}\Pi K}\cdot\Pi_K\right] + R^{(1)}_{\text{Н}\Pi i}\bigcap_{j=1}^{24}\left[\Delta R^{(2)}_{\Pi K}\cdot\Pi_K\right]\right\}; \tag{10.110}$$

$$\widetilde{\text{P}} = P(\text{Н}\Pi_{\Phi} > [\text{Н}\Pi_K])\cdot 2\Phi\left[\frac{\text{Н}\Pi_K - M(\Omega_K)}{\sigma(\Omega_K)}\right] \tag{10.111}$$

The dependencies are provided for a deterministic class of systems; in the case of probabilistic systems, the following should be taken into account. Depending on (10.110), the measured values of the parameters should be replaced by the predicted values of those parameters that are not measured, and, instead of the measurement error for the same parameters, the forecast error should be substituted.

The probability

$$P(\text{Н}\Pi_{\Phi} > [\text{Н}\Pi_K]) \tag{10.112}$$

in (10.111) should be determined on the basis of the fact that the process at the previous stage of control is in working order and therefore (10.112) is:

$$P(\text{Н}\Pi_{\Phi} > [\text{Н}\Pi_K]) = P_y(\text{Н}\Pi_{\Phi} > [\text{Н}\Pi_K]) \tag{10.113}$$

where P_y is the conditional probability of the occurrence of a defective part, calculated in view that at the time $t - \Delta t$, the process was operational.

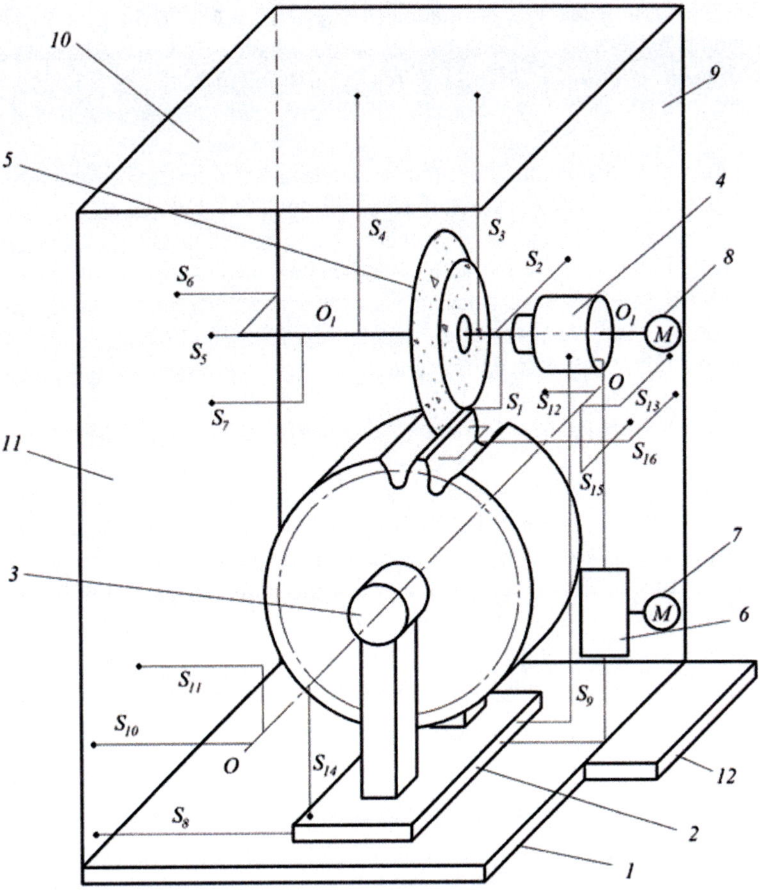

Fig. 10.23 An example of the technical means that implement the element-by-element control system

Dependence (10.113) is general and suitable for all control systems. An example of the element-by-element control system obtained by specifying the scheme in Fig. 10.22 in relation to the tooth grinder is shown in Fig. 10.23.

The system is intended for an operative assessment of accuracy and is directly located on the machine. The machine is composed of frame 1 on which Table 10.2 is mounted with clamping device 3 having a geometric axis $O-O$. Tool headstock 4 is also attached to the frame, carrying tool 5, the axis of which is marked as $O_1 - O_1$. Synchronization of the movements of the part and the tool is provided by kinematic circuit 6 equipped with drive 7. The tool rotates from independent drive 8.

The error in determining the increment of the radius vector $\Delta\bar{r}_{1i}$ is monitored by the sensor S_1. This is possible due to the point method of forming with a constant position of the point of contact on the tool surface. The same sensor also determines the components of the matrix $\widetilde{M}_{\Delta 1}$ because the base of rotation is the axis of the tool.

The position of the axis of rotation of the tool relative to the axis of the part is determined by sensors S_2, S_3, S_4, S_5, S_6, and S_7, which are paired with each other and determine the coordinates of the axes relative to three mutually perpendicular planes (positions 9, 10, and 11 in Fig. 10.23). Jointly with the sensors S_8 and S_9 that determine the degree of consistency of tool and workpiece movements, the sensors S_2–S_7 provide information about the components of the matrix. To determine the components of the matrix $\widetilde{M}_{\Delta 2}$, there are sensors S_{10}–S_{15} that measure the position of the part $\widetilde{M}_{\Delta 2}$ relative to the axis of rotation. The increment components of the radius vector $\Delta \overline{r}_{2YT}$ can be measured with a multi-input sensor S_{16}.

Thus, the listed sensors determine all the received primary errors, which are shown on the processed gear using the considered method of forming. The results of their readings, after appropriate conversion and amplification, are processed in device 12.

An essential advantage of a system of this type is its ability to obtain the necessary information about the accuracy of the machined gear directly at the time of processing using only two types of sensors (linear and angular measurements). However, such systems are difficult to implement in the case of methods of forming with linear or surface contact in the pair "tool surface–workpiece" due to the complexity of the operational assessment of the components of the increment of the radius vector $\Delta \overline{r}_{1i}$ at each point of contact. Therefore, in such cases, it is necessary to build either an incomplete system, in which the assessment of the accuracy of the tool is given a priori, or combined systems.

10.3.4 Factor Control Systems

According to the previous definition, factor systems are based on the control of the parameters of the factors that cause errors. In this case, either all parameters or their individual groups can be controlled. Given that each group of parameters corresponds to its own instrumentation, it is advisable to identify a complete set of different types of factor systems with different groups of controlled parameters.

Given that a purely predictive control system based on a priori error estimation is a system with a number of controlled parameter groups equal to zero, and if we take into account that there can be only five different groups of parameters, we can establish that factor systems can be of six types: 0-factor (prognostic), 1-factor, ..., 5-factor (full-factor).

The variants of structural schemes of control systems obtained on the basis of dependence (10.101) are displayed in Fig. 10.24 – prognostic – and in Fig. 10.25 – n-factor, $n = 1, 2, \ldots, 5$.

The figures show:

S_i – sensors of groups of factor parameters
t_c – control time; t – current time corresponding to the moment of processing
AIIB – a priori information input block

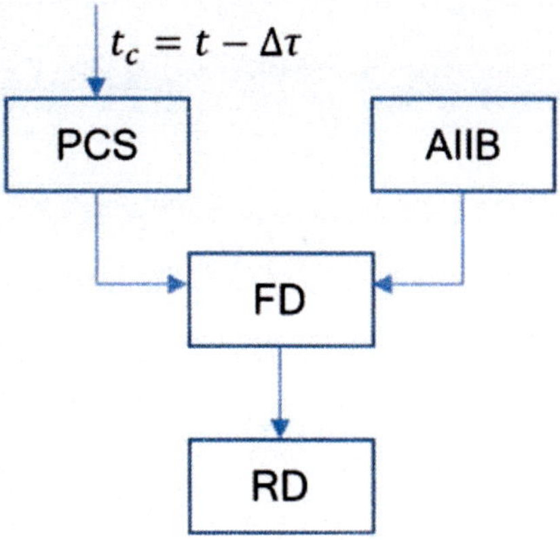

Fig. 10.24 Block diagram of a 0-factor control system

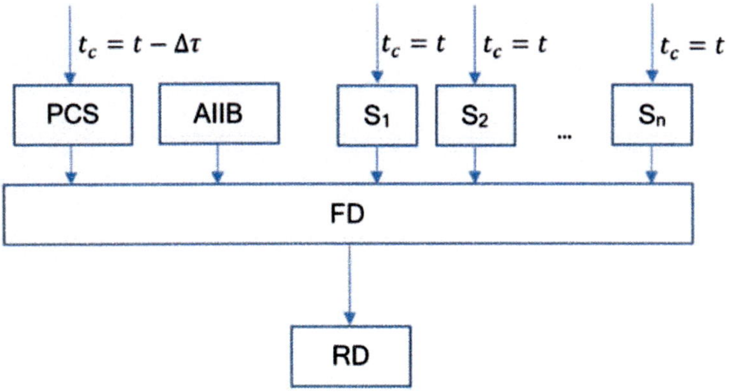

Fig. 10.25 Block diagram of the *n*-factor control system

FD – forecasting device
RD – recording device
PCS – system of preliminary control (forecasting)

The free arrow shows the input of the source information.

Obviously, there can be only one prognostic and full-factor system. The number of types of factor systems is equal to the number of combinations without repetitions from 5, i.e., one-factor – 5, two-factor – 10, three-factor – 10, four-factor – 5.

If we mark the action of control by the sign C, and the groups of parameters related to factors:

G – geometric
K – kinematic
F – force
T – temperature
W – wear

Table 10.3 Types of factor control systems

№.	Structural model	Name (type)
1	C	0-factor
2	C(G)	1-factor
3	C(K)	1-factor
4	C(F)	1-factor
5	C(T)	1-factor
6	C(W)	1-factor
7	C(G) + C(K)	2-factor
8	C(G) + C(F)	2-factor
9	C(G) + C(T)	2-factor
10	C(G) + C(W)	2-factor
11	C(K) + C(F)	2-factor
12	C(K) + C(T)	2-factor
13	C(K) + C(W)	2-factor
14	C(F) + C(T)	2-factor
15	C(F) + C(W)	2-factor
16	C(T) + C(W)	2-factor
17	C(G) + C(K) + C(F)	3-factor
18	C(G) + C(K) + C(T)	3-factor
19	C(G) + C(K) + C(W)	3-factor
20	C(G) + C(F) + C(T)	3-factor
21	C(G) + C(F) + C(W)	3-factor
22	C(K) + C(F) + C(T)	3-factor
23	C(K) + C(F) + C(W)	3-factor
24	C(F) + C(T) + C(W)	3-factor
25	C(T) + C(W) + C(K)	3-factor
26	C(G) + C(T) + C(W)	3-factor
27	C(G) + C(F) + C(T) + C(W)	4-factor
28	C(G) + C(F) + C(T) + C(K)	4-factor
29	C(G) + C(F) + C(W) + C(K)	4-factor
30	C(G) + C(T) + C(W) + C(K)	4-factor
31	C(F) + C(T) + C(W) + C(K)	4-factor
32	C(G) + C(K) + C(F) + C(T) + C(W)	5-factor

Then we can identify the types of factor control systems as shown in Table 10.3.

Thus, Table 10.3 indicates that there may be 32 types of factor control systems that differ in the completeness of the information received. One of them, i.e., 0-factor is completely probabilistic (prognostic), and the other one – full-factor – is determined. Other systems are partially probabilistic.

The choice of one or another type of factor control system is determined by the characteristics of the machining process: the nature of contact between the tool surface and the workpiece, the heat of the process, the magnitude of forces, the complexity of kinematics, requirements for gear accuracy, and others.

It follows from the analysis of dependence (10.106) that with an increasing number of groups of controlled parameters, the reliability of the accuracy estimates increases.

In the case of a 0-factor system, the probability $P = P(\Omega_K < \xi)$ of obtaining a prediction error with a value smaller than that can be determined on the basis of (10.107) and (10.108)

$$P(\Omega_K < \xi) = 2\Phi\left[\frac{\xi - M(\Omega_K)}{\sigma(\Omega_K)}\right], \tag{10.114}$$

considering

$$q(\Omega_K) = \frac{1}{\sigma(\Omega_K)\sqrt{2\pi}} \exp\left[-\frac{1}{2}\left(\frac{\Omega_K - M(\Omega_K)}{\sigma(\Omega_K)}\right)^2\right]. \tag{10.115}$$

Since, in (10.114), $P \rightarrow \min$ while sn (10.115) $\sigma(\Omega_K) \rightarrow \max$, use of the 0-factor system is appropriate only for coarse gears in sustainable production.

The system of type C(G) is characterized by the presence of sensors of geometric inaccuracies of the elements of an MFTP system. Such a system is effective in all processes because geometric inaccuracies always comprise a significant proportion of the total error. However, due to the complexity of controlling geometric inaccuracies at the time of processing, the most appropriate area of its application are processes with relatively simple kinematics of movements during formation, especially in the simplified version of the system, when only unstable deviations are controlled, i.e., those that change discretely or functionally during the processing of each gear, and stable deviations (systematic errors) are taken into account a priori.

The system type C(K) contains sensors of kinematic errors occurring in the kinematic circuits of an MFTP system. It is effective in processes with complex kinematics of movements or in processes carried out by free-running methods.

The type C(F) system uses force sensors. The actual elastic deformations in an MFTP system are determined by the magnitudes of the forces. In addition, the amount of wear is correlated with the amount of wear. Such a system is recommended for use in processes in which high cutting forces occur or in the case of a nonrigid MFTP system.

The type C(T) system requires the inclusion of temperature sensors at various points in an MFTP system's dimensional circuit. At a known temperature, temperature deformations are determined and the amount of wear is more accurately predicted. This system is recommended for processes with a large proportion of temperature components in the total error.

Finally, the system type C(W) provides control of tool wear parameters. This control can be direct or indirect (see Sect. 10.2). The system is recommended for processes with unstable wear of tools, for example, hard alloy, which are brittle.

Two-, three-, and four-factor control systems increase the amount of operational information about gear errors. Therefore, they increase the reliability of the assessment of accuracy.

Each of these systems has the advantages and disadvantages of combined single-factor systems. Therefore, it has a field of application corresponding to those of combined single-factor systems, slightly expanded in the range of the degrees of accuracy due to the increase in operational information.

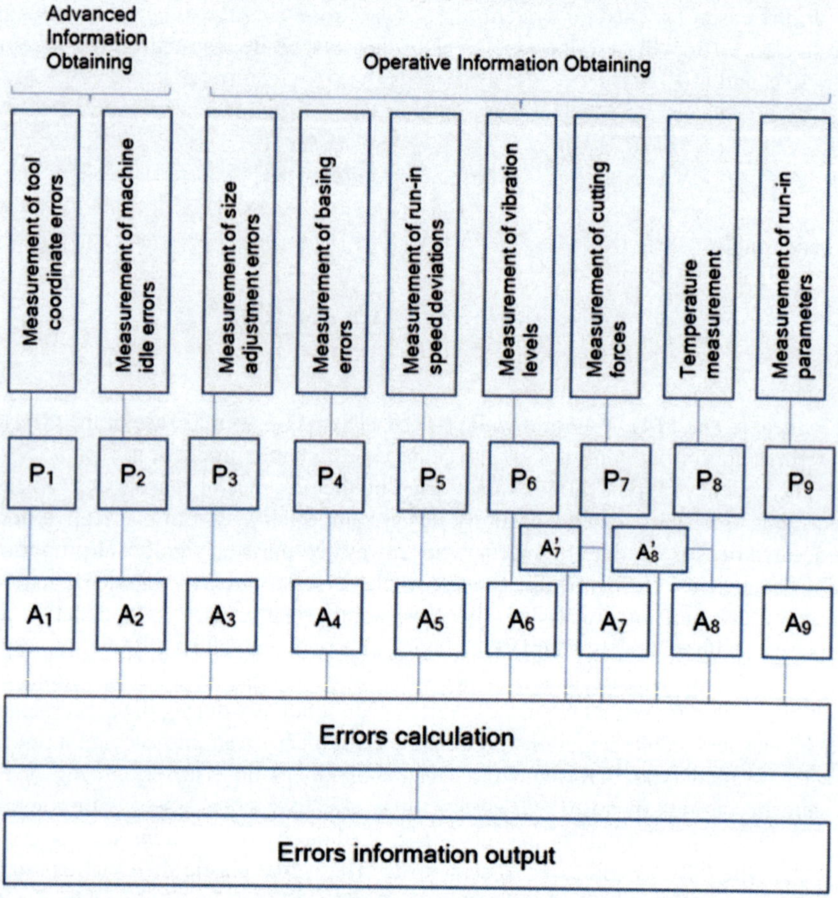

Fig. 10.26 General scheme for factor control

In a full-factor system having the form C(G) + C(K) + C(F) + C(T) + C(W), the measurement error is minimal. In accordance with dependence (10.106), it is determined by the accuracy of factor measurements and that of the models used to calculate the components of these errors. Theoretically, following from dependence (10.106), the probability of a measurement error can be reduced to any arbitrarily small value by increasing the accuracy of the measuring instruments used and reducing $\Delta R_{\Pi}^{(i)}$.

Both full-factor and n-factor $n \in \{1, 2, 3, 4, 5\}$ systems can have different designs. The following is an example of a five-factor system. Since systems with a smaller number of controlled parameter groups can be obtained on the basis provided by excluding the corresponding groups of sensors from consideration, due to their particularity, examples of such systems are not considered in this chapter. The general scheme for the factor control process is shown in Fig. 10.26.

A variant of the scheme for the structure of technical means implementing the factor control method, obtained on the basis of the general methodology, is shown in Fig. 10.27.

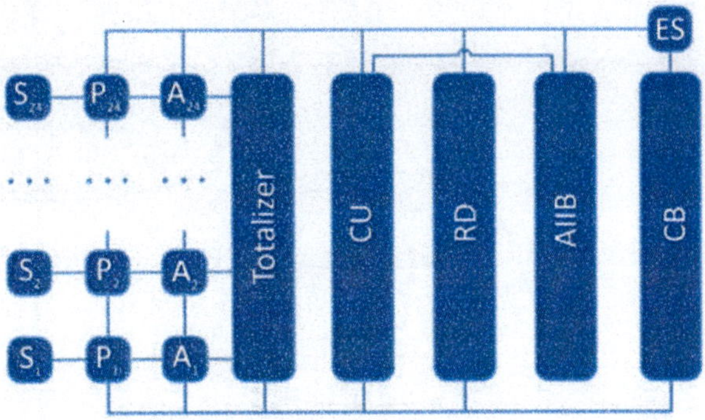

Fig. 10.27 A variant of the structure of technical means that implement the factor control method

The symbols in Figs. 10.26 and 10.27 correspond to those previously adopted. According to Fig. 10.26, when organizing factor control processes, some information can be introduced a priori. This is appropriate due to the stability of certain groups of geometric and kinematic errors and the difficulty of organizing control in the processing of all components. However, most of the information is directly obtained at the time of processing.

Thus, the process of factor control can be rationally organized as follows: persistent errors must be identified before processing. These are usually machine idling errors and static primary tool errors. The information obtained can be converted into electrical signals and amplified by the required amount in proportion to the coefficients of influence on the determined errors of the ring gear. These signals must then be transferred to a multi-channel computing device.

Furthermore, directly in processing, the measurement of variable components is performed: envelope parameters, to determine the actual coordinates of the points of occurrence of temperature deviations, cutting force, vibration level (in the case of an indirect assessment of tool wear); deviations of the running speed (if rolling processing is used); and variables on each gear of geometric errors (setting of details and adjustment of a strike on the size).

Information about these parameters must be also translated into electrical signals, which are amplified and entered into a multi-channel computing device. In a multichannel computing device, the signals are summed up and the desired values of the controlled errors are calculated, for example, in accordance with the dependencies previously provided in Sect. 10.2.

The values of the calculated accuracy indicators are entered into either the recording device or the executive body that controls the accuracy of the device. Since this method provides current information about errors at any point of the ring gear, the computing device by conversion receives actual information about any indicator of the accuracy of the ring gear, such as radial runout, radial error, profile error, tooth direction error, etc.

As an example, Fig. 10.28 shows a diagram explaining the factor control of errors F_{rr}, f'_{fr}, $F_{\beta r}$, and A_{Hr} at the time of gear forming.

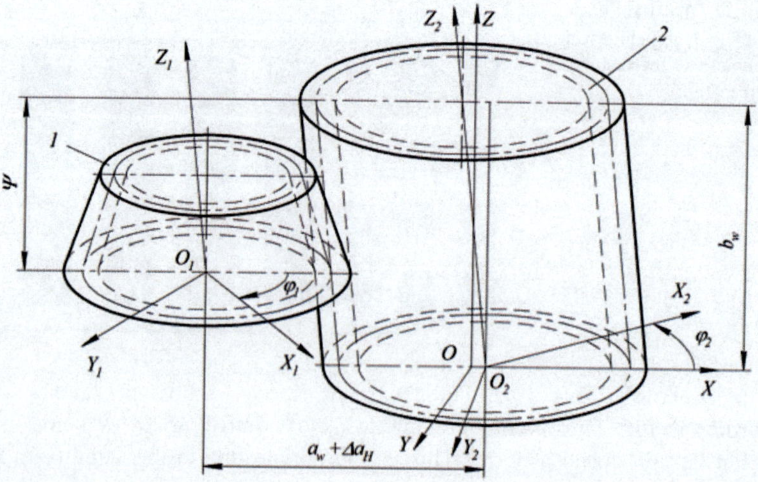

Fig. 10.28 Scheme for factor control of errors in the gear shaping process

In the diagram and in the following calculations, individual, insignificant details are skipped for simplification. In Fig. 10.28, the tool is marked by position 1 and the machined gear is position 2. In the considered processing method (Fig. 10.28), the envelope parameters are ψ and ϕ_2. The width of the gear is indicated by b_w and the center distance by a_w.

To implement the method before shaping, the idle errors of the machine (stable components) are determined: non-parallelism of the tool stroke relative to the axis of the device in the radial and tangential planes $\Delta\Phi_x(\psi)$ and $\Delta\Phi_y(\psi)$, respectively. The static primary errors of the tool on the right and left lines of action – $\Delta F_{\mathrm{R}}(\phi_2)$ and $\Delta F_{\mathrm{L}}(\phi_2)$, respectively – are measured directly relative to the axis of rotation.

The received information is converted into electrical signals and amplified for the subsequent calculation of errors on the values:

- For F_{rr} : $\Delta F_{\mathrm{R}}(\phi_2) \rightarrow K_1 \Delta F_{\mathrm{R}}(\phi_2) = \dfrac{1}{2\sin\alpha} \Delta F_{\mathrm{R}}(\phi_2)$;
  $\Delta F_{\mathrm{L}}(\phi_2) \rightarrow K_2 \Delta F_{\mathrm{L}}(\phi_2) = \dfrac{1}{2\sin\alpha} \Delta F_{\mathrm{L}}(\phi_2)$;
- For f'_{fr} : $\Delta F_{\mathrm{R}}(\phi_2) \rightarrow K_3 \Delta F_{\mathrm{R}}(\phi_2) = \dfrac{1}{2\sin\alpha} \Delta F_{\mathrm{R}}(\phi_2)$;
  $\Delta F_{\mathrm{L}}(\phi_2) \rightarrow K_4 \Delta F_{\mathrm{L}}(\phi_2) = \dfrac{1}{2\sin\alpha} \Delta F_{\mathrm{L}}(\phi_2)$;
- For $F_{\beta r}$: $\Delta\Phi_x(\psi) \rightarrow K_5 \Delta\Phi_x(\psi) = \sin\alpha \Delta\Phi_x(\psi)$;
  $\Delta\Phi_y(\psi) \rightarrow K_6 \Delta\Phi_y(\psi) = \cos\alpha \Delta\Phi_y(\psi)$;
- For A_{Hr} : $\Delta F_{\mathrm{R}}(\phi_2) \rightarrow K_7 \Delta F_{\mathrm{R}}(\phi_2) = \dfrac{\Delta F_{\mathrm{R}}(\phi_2)}{2\sin\alpha}$;
  $\Delta F_{\mathrm{L}}(\phi_2) \rightarrow K_4 \Delta F_{\mathrm{L}}(\phi_2) = \dfrac{\Delta F_{\mathrm{L}}(\phi_2)}{2\sin\alpha}$;
  $\Delta\Phi_x(\psi) \rightarrow K_9 \Delta\Phi_x(\psi) = 1 \cdot \Delta\Phi_x(\psi)$.

where K_1–K_9 are the gain coefficients equal to the coefficients of the influence of primary errors on the controlled indicators of the accuracy of the gear. These coefficients, as well as the following in the example, are taken from the work by Ternyuk [38].

Additionally, directly at the time of formation, the values of the envelope parameters ψ and ϕ_2 are controlled, for example, by the photoelectric method. These parameters are the values of the arguments in the functions of primary errors, including the given ones. The following are also monitored:

- Part T_p and tool T_t temperature, e.g., by artificial or natural thermocouple methods
- Radial component of cutting force $P_z = P_z(\phi_2)$ as a function ϕ_2
- Vibration level $\mu = \mu(\phi_2, \psi)$ as a function of ϕ_2 and ψ
- Deviation of running speed $\omega = \omega(\phi_2)$ as a function ϕ_2
- Errors in the base of the workpiece: turns at angles $\Delta\psi_x$ and $\Delta\psi_y$ around the axes, mutually perpendicular to each other and perpendicular to the axis of the device, as well as displacements ΔX and ΔY along these axes
- Machine setting error $\Delta a_H = \Delta a_H(\phi_2)$

Being a priori, the received operational information is converted into electric signals, for example, by sensors, and, then signals are amplified to values as follows:

(a) For calculation of F_{rr}: $\Delta X_2 \to K_{10}\Delta X_2 = \left(\frac{\Delta X_2 \cos\phi_2}{\sqrt{\Delta X_2^2 + \Delta Y_2^2}}\right)\Delta X_2$; $\Delta Y_2 \to K_{11}\Delta Y_2 = \left(\frac{\Delta Y_2 \cos\phi_2}{\sqrt{\Delta X_2^2 + \Delta Y_2^2}}\right)\Delta Y_2$;

$$P_z(\phi_2) \to K_{13}P_z(\phi_2) = j_p \cdot P_z(\phi_2);$$

$$T_t \to K_{14}T_t = \left(\frac{1}{T_t} f(\phi_2) \cdot b \int_0^t T_t^l dt\right) T_t;$$

$$P_z(\phi_2) \to K_{15}P_z(\phi_2) = \left(\frac{1}{P_z} f(\phi_2) \cdot a \int_0^t P_z^q dt\right) P_z(\phi_2);$$

$$\Delta a_H(\phi_2) \to K_{12}\Delta a_H(\phi_2) = 1 \cdot \Delta a_H(\phi_2);$$

$$\mu(\phi_2, \psi) \to K_{16}\mu(\phi_2, \psi) = \left(\frac{1}{\mu} f(\phi_2) \cdot c \int_0^t \mu^c dt\right) \mu(\phi_2, \psi);$$

(b) For calculation of f'_{fr}: the same as for F_{rr}

(c) For calculation of $F_{\beta r}$: $\Delta\psi_x \to K_{17}\Delta\psi_x = \left(\frac{\Delta X_2 \psi}{\sqrt{\Delta\psi_x^2 + \Delta\psi_y^2}}\right)\Delta\psi_x$; $\Delta\psi_y \to K_{18}\Delta\psi_y = \left(\frac{\Delta\psi_y \psi}{\sqrt{\Delta\psi_x^2 + \Delta\psi_y^2}}\right)\Delta\psi_y$;

$$P_z(\phi_2) \rightarrow K_{19} P_z(\phi_2) = c\psi^3 P_z(\phi_2);$$

(d) For calculation of A_{Hr}: the same as for F_{rr} and additionally:

$$T_p \rightarrow K_{20} T_p = -\frac{mz_p}{2} \alpha_{Tp} T_p; \quad T_t \rightarrow K_{21} T_t = -\frac{mz_t}{2} \alpha_{Tt} T_t;$$

The received signals are transferred into a multi-channel computing device, in which of the relevant components and the issuance of information about the current errors or the definition of normalized errors is summed up. In the latter case, the calculation is based on:

$$\begin{aligned} F_{rr} = & \max_{\phi_2} [K_1 \Delta F_{\mathrm{R}}(\phi_2) + K_2 \Delta F_{\mathrm{L}}(\phi_2) + K_{10} \Delta X_2 + K_{11} \Delta Y_2 + K_{12} \Delta a_H(\phi_2) + K_{13} P_z(\phi_2)] - \\ & - \min_{\phi_2} [K_1 \Delta F_{\mathrm{R}}(\phi_2) + K_2 \Delta F_{\mathrm{L}}(\phi_2) + K_{10} \Delta X_2 + K_{11} \Delta Y_2 + K_{12} \Delta a_H(\phi_2) + K_{13} P_z(\phi_2)]; \\ & 0 \le \phi_2 \le 2\pi; \ \psi = 0{,}5 b_w; \end{aligned}$$

$$\begin{aligned} f'_{fr} = & \max_{\phi_2} [K_1 \Delta F_{\mathrm{R}}(\phi_2) + K_2 \Delta F_{\mathrm{L}}(\phi_2) + K_{12} \Delta a_H(\phi_2) + K_{14} T_t + K_{15} P_z(\phi_2) + K_{16} \mu(\phi_2)] - \\ & - \min_{\phi_2} [K_1 \Delta F_{\mathrm{R}}(\phi_2) + K_2 \Delta F_{\mathrm{L}}(\phi_2) + K_{12} \Delta a_H(\phi_2) + K_{14} T_t + K_{15} P_z(\phi_2) + K_{16} \mu(\phi_2)]; \\ & (i-1)\tau \le \phi_2 \le i\tau; \ i = 1, \ldots, z; \ \psi = 0{,}5 b_w; \end{aligned}$$

$$\begin{aligned} F_{\beta r} = & \max_{\psi} \left[K_5 \Delta \Phi_x(\psi) + K_6 \Delta \Phi_y(\psi) + K_{17} \Delta \psi_x + K_{18} \Delta \psi_y + K_{19} P_z(\phi_2)\right] - \\ & - \min_{\psi} \left[K_5 \Delta \Phi_x(\psi) + K_6 \Delta \Phi_y(\psi) + K_{17} \Delta \psi_x + K_{18} \Delta \psi_y + K_{19} P_z(\phi_2)\right]; \\ & \phi_2 = i\tau; \ i = 1, \ldots, z; \ \psi = b_w; \end{aligned}$$

$$\begin{aligned} A_{Hr} = & \min_{\phi_2} \left[\begin{array}{l} K_7 \Delta F_{\mathrm{R}}(\phi_2) + K_8 \Delta F_{\mathrm{L}}(\phi_2) + K_9 \Delta \Phi_x(\psi) + K_{10} \Delta X_2 \\ \quad + K_{11} \Delta Y_2 + K_{12} \Delta a_H(\phi_2) + K_{13} P_z(\phi_2) + K_{20} T_p + K_{21} T_t \end{array} \right]; \\ & 0 \le \phi_2 \le 2\pi; \ \psi = 0{,}5 b_w. \end{aligned}$$

This example does not use rolling speed deviation information. However, it is required when calculating fluctuations in the length of the general normal, run-in errors, and other errors.

Thus, according to the proposed method of factor control, any error of a gear can be controlled. At the same time, the need to control the gear outside the machine is eliminated, which saves time and reduces the number of used measuring instruments. In addition, it is possible to organize adaptive accuracy control.

These benefits should ensure that the factor control systems are more widespread in future. For this purpose, it is expedient to organize the release of gear processing machines equipped with the corresponding measuring, converting, amplifying, and computing tools.

10.3.5 Combined Control Systems

With the correct combination, combined control systems can realize the main benefits of all simple control systems. In addition, this type of system is the only one that allows obtaining redundant information about errors and thus provides an opportunity to build highly effective control tools that allow control "themselves." Block diagrams of combined systems are a combination of schemes of simple systems, and they are not given separately.

Each type of control system produces its own measurement error $\boldsymbol{\Omega}_K$. This error appears with the probability $P(\boldsymbol{\Omega}_K \leq \xi)$. Therefore, the probability $q(\xi)$ of obtaining objective information about errors with an error not exceeding ξ is

$$q(\xi) = 1 - P(\boldsymbol{\Omega}_K \leq \xi); \tag{10.116}$$

The measure of information corresponds to the probability obtained from (10.116)

$$J(\xi) = \log_2 \frac{1}{q(\xi)}. \tag{10.117}$$

Let $J^{(1)}_{(\xi)}$ correspond to the measure of information obtained in the process in which the first type of control system is used; $J^{(2)}_{(\xi)}$ corresponds to the second and $J^{(3)}_{(\xi)}$ corresponds to the third. Then, if all three types of control systems are simultaneously implemented, then the total information instead of (10.117) will be:

$$J^{(\Sigma)}_{(\xi)} = J^{(1)}_{(\xi)} + J^{(2)}_{(\xi)} + J^{(3)}_{(\xi)}. \tag{10.118}$$

This type of system, in which three methods of control are used, is called a three-species system. In the particular case, one species may be missing, which will correspond to two species. Commonly used systems (simple) are one-species. Since $J^{(1)}_{(\xi)}$, $J^{(2)}_{(\xi)}$, and $J^{(3)}_{(\xi)}$ are a measure of information obtained in different ways, but with the same value, it becomes obvious that these measures can be judged by comparing them on the basis of the serviceability of control devices.

Taking into account the different accuracy of control, the condition of correctness of realization of three-type control can be written in the form:

$$\left(\left|J^{(1)}_{(\xi 1)} - J^{(2)}_{(\xi 2)}\right| \leq \xi\right) \cap \left(\left|J^{(2)}_{(\xi 2)} - J^{(3)}_{(\xi 3)}\right| \leq \xi\right) \cap \left(\left|J^{(1)}_{(\xi 1)} - J^{(3)}_{(\xi 3)}\right| \leq \xi\right). \tag{10.119}$$

The same can be done for the two-species:

$$\left|J^{(1)}_{(\xi 1)} - J^{(2)}_{(\xi 2)}\right| \leq \xi. \tag{10.120}$$

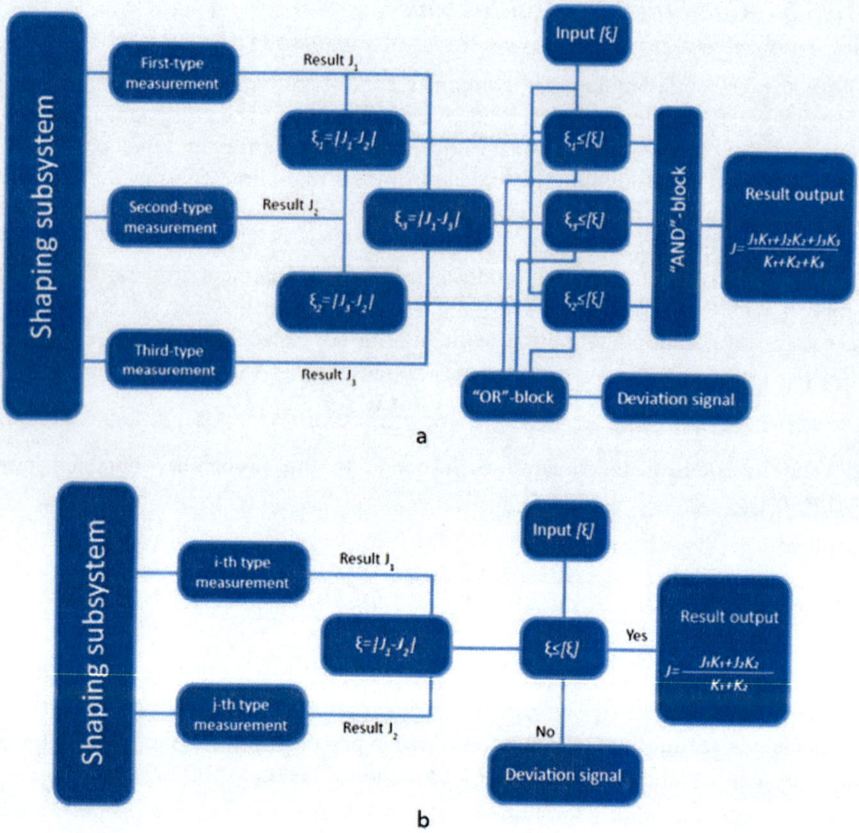

Fig. 10.29 Schemes for information-redundant three- and two-type processes of gear control

Since complete information about errors can be provided by each of the simple systems, according to (10.118), three- and two-type control systems can be redundant in the information sense. This allows using inequalities (10.119) and (10.120) to build highly reliable and high-precision complex systems. Such systems implement the processes shown in Fig. 10.29.

Figure 10.29a displays a two-species system, and Fig. 10.29b shows a three-species system. When using redundant information systems, an increase in reliability is achieved due to the possibility of failure indication using inequalities (10.119) and (10.120). Improving accuracy is carried out by averaging errors. Thus, the value can be taken as a measure of information in combined systems

$$J_{(\xi)}^{(K)} = \frac{\sum_{i=1}^{n} K_i J(\xi)}{\sum_{i=1}^{n} K_i}, \tag{10.121}$$

where K_i, $i \in \{1, 2, 3\}$ is the reduction factor (weights).

The values of these coefficients can be selected inversely proportional to the error variances of the respective control systems. In view of the need for rationing K_i, one can write:

$$K_i = \frac{\sqrt[n]{\sum_{i=1}^{n} \sigma_i^2}}{\sigma_i}.$$

In this approach, taking into account the obtained equation from (10.121), it follows that

$$J_{(\xi)}^{(K)} > J_{(\xi)}^{(i)}, \quad i \in \{1, 2, 3\}.$$

This is proof of the possibility of improving the accuracy of measurements (obtaining more reliable information) about the errors when using two- or three-type gear control systems.

By studying the general system model of the process of errors and by applying the general method of the synthesis of structures, a full range of possible methods and control systems can be identified, their capabilities can be evaluated for recommendations for applications, and a method of synthesis of the new control systems can be developed. The complete set of control methods includes seven elements, three of which are simple and four are combined. In addition to methods based on direct control of normalized accuracy, the measurement of gear errors can be performed by element-by-element, factor, simple, and combined methods. Element-by-element, factor, and combined methods (based on them) of control can provide information about errors directly at the time of their occurrence, which provides an opportunity to build, on their basis, adaptive accuracy control systems implementing the principle of "perturbation" control. Systems of direct control, which are delayed in relation to the process of formation, do not allow building such systems, but can only ensure the implementation of the principle of management "feedback." This requires extra effort to correct mistakes. To calculate the actual measurement errors, as well as the degree of risk of missing a defective part, a general model is proposed, from which the dependencies for the calculation of these values using certain methods and control systems follow as a special case. The possibility of building information-redundant gear control systems has been proven, which allows creating highly reliable and high-precision control systems that provide the required performance.

The obtained results are the basis for the synthesis of highly efficient gear control systems.

10.4 Technological Methods of Controlling the Accuracy of Machining Gears

10.4.1 The General Characteristics of the Methods of Controlling the Accuracy of Machining Gears

A general description of the methods of controlling the accuracy of machining gears can be obtained by highlighting the objectives and levels of control, control variables and the nature of their change over time, and the stages of implementation. It is known [4] that in the engineering practice of creating and operating systems, it is possible to set two types of optimization problems:

- Obtaining the desired effect at a minimum cost
- Obtaining the maximum effect when using the specified limited resources

Therefore, in the problems of optimal control of the accuracy of machining of gears, we can identify two possible goals:

- Ensuring the specified accuracy at a minimum of costs (first of all, labor intensity)
- Achieving maximum accuracy while limiting costs (including labor intensity)

Many possible levels of accuracy control can be identified by considering the technological process as a hierarchical formation, and it can be divided into operations, transitions, and passes [9]. Based on this, the general symbolic model of the structure of the technological process can be represented as:

$$|T_n| = \bigcap_{i=1}^{n} O_i \left[\bigcap_{j=1}^{m} \Pi_j^{(i)} \left(\bigcap_{\xi=1}^{k} P_\xi^{(ij)} \right) \right],$$

where

T_n is the technological process.
O_i is the ith operation.
$\Pi_j^{(i)}$ is the jth transition of the ith operation.
$P_\xi^{(ij)}$ is the ξth passage transition of the jth operation.
n, m, and k are the number of operations, transitions, and passes, respectively.

According to this model, the power of many possible levels of control of the accuracy of machining of gears equals four. Obviously, the general process control scheme (actions) shown in Fig. 10.30 can be implemented at each level.

Considering accuracy as one of the outputs of the system, in accordance with dependence (2.3) and Fig. 10.30, it can be claimed that accuracy control at each level, in general, can be provided by changing at time t_1 its system input $U(t_1)$ and system state $C_{t1} = (|X_S|_{t1}, R_{St1})$, which is characterized by its $|X_S|_{t1}$ structure and many connections between its parameters R_{St1} and processing time t_2. At any given

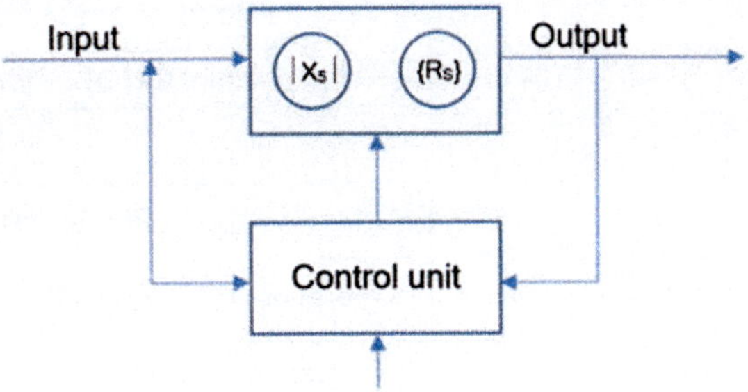

Fig. 10.30 General control scheme

time, management may cover all or only some of the factors that allow variation, i.e., it can be complete or partial. Depending on the nature of the change in control effects over time, we can differentiate between static control based on a priori process information and dynamic control based on operational process information. In addition, there can be combined control, when a part of the parameters or structure changes statically and the other part changes dynamically.

Depending on the implemented principle [7], it is possible to identify the control of accuracy "by deviation" and "by perturbation." The implementation of these principles is ensured, in general, by the actions of technical means.

It should be noted that Fig. 10.31 covers the characteristics of both of the possible methods of ensuring the required accuracy by satisfying conditions (10.1): restrictive, consisting of a restriction, i.e., static change of all factors effecting errors, and adaptive (or compensatory), consisting of operational, i.e., dynamic change of these factors.

The obtained results allow moving to a systematic study of accuracy management at all possible levels. Equations (10.23), (10.24), (10.25), (10.26), (10.27), (10.28), and (10.29) can be used as a general model of the achieved results.

10.4.2 Accuracy Management at the Process Level

In accordance with the model presented in Sect. 10.4.1 related to the process level, it can be written

$$\text{Н}\Pi_i = \eta_i[C_{t1},\ U_i(t_1),\ t_2] = \eta_i\left\{\left[\bigcap_{i=1}^{n} O_i(t_1),\ R_i(t_1)\right],\ U_3(t_1),\ t_2\right\}. \tag{10.122}$$

Thus, at the process level, accuracy can be controlled by changing the type, number, and order of operations as well as input parameters related to the workpiece and time. The introduction of an operation in the process is provided by the need to change the parameters of accuracy НП or roughness R_a, R_z of the teeth. Therefore, the decisive rule for forming the structure of the process can be written in the form

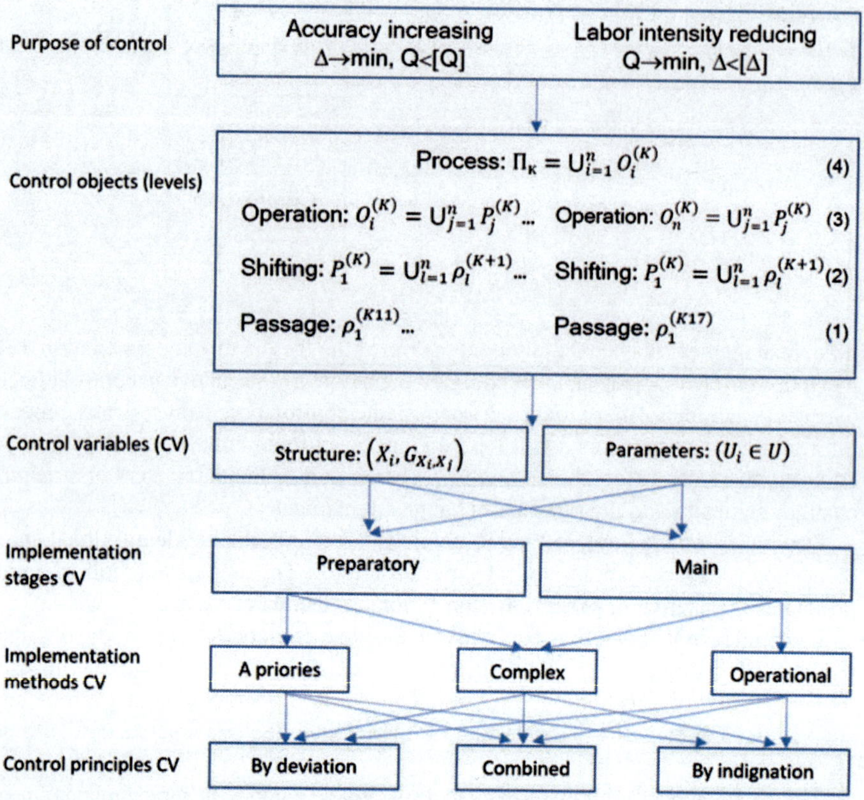

Fig. 10.31 Signs of accuracy control methods

$$\exists \varepsilon_{ij}^{(\Delta)} \left(\varepsilon_{ij}^{(\Delta)} < 1 \right) \cup \exists \varepsilon_{ij}^{(z)} \left(\varepsilon_{ij}^{(z)} < 1 \right). \tag{10.123}$$

where

$$\varepsilon_{ij}^{(\Delta)} = \frac{\text{Н}\Pi_{i-1}}{\text{Н}\Pi_i}; \quad \varepsilon_{ij}^{(z)} = \frac{R_{zi-1}}{R_{zi}} \cup \frac{R_{ai-1}}{R_{ai}}. \tag{10.124}$$

Conditions (10.123) and (10.124) mean that it is not necessary that all indicators of accuracy or roughness are improved at the same time at each operation. The possibility of compensating for individual components of errors in different operations, as well as the unequal efficiency of these operations, creates the preconditions for solving the problem of choosing the optimal values $\varepsilon_{ij}^{(\Delta)}$ and $\varepsilon_{ij}^{(z)}$. Since $\text{Н}\Pi_i = f(\Delta \overline{r}_2)$, in accordance with (10.23), (10.24), (10.25), (10.26), (10.27), (10.28), and (10.29), is affected not only by the actual deviations of the shape and size of the real surface compared to the nominal but also by the position of the

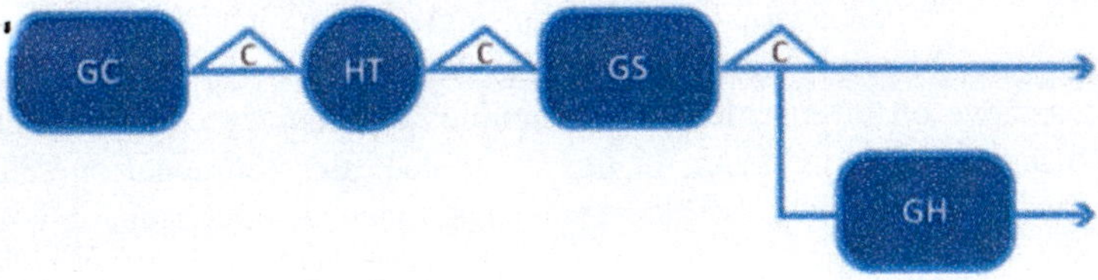

Fig. 10.32 Example of a process scheme with a variable structure

measuring (operational) base reflected by the elements of the matrix $\widetilde{M}_{\Delta 2}$, we can identify two fundamentally different types of operation machining, which affects the final accuracy:

- Operations on processing of the teeth influencing $\Delta \bar{r}_{2n}$ or on $\Delta \bar{r}_{2n}$ and $\widetilde{M}_{\Delta 2}$ at the same time
- Base surface treatment operations affecting $\widetilde{M}_{\Delta 2}$

The process can generally include any kind and any type of operation. Therefore, it can be claimed that at the process level, within the conditions of the physical feasibility of a system, there are no restrictions on the accuracy achieved. However, in the synthesis of real systems, when time constraints, applicability constraints, and cost constraints take effect, such constraints on the achieved accuracy are available. The accuracy that is actually realized is always provided by the capabilities of the operations, which are parts of the process.

The dependence analysis (10.122) shows that, in order to control accuracy, the process can have both a constant ($\bigcap_{i=1}^{n} O_i = \text{const}$) and a variable ($\bigcap_{i=1}^{n} O_i = \text{var}$) structure. Examples of processes with a constant structure indicating the achieved efficiency are described in many works, including the studies by Jeff and Hamada [9] and Ponomarev [23]. However, they are not considered here. The possibilities of the control of accuracy at the process level, by changing the structure the diagram of the technological process with a variable structure, are illustrated in Fig. 10.32.

The process includes gear cutting (GC), heat treatment (HT), and gear sharping (GS) and, if necessary, a second finishing operation – gear honing (GH). The control is marked with the letter "C."

According to the scheme, some parts that meet the requirements of accuracy and quality of the lateral surfaces of the teeth may not undergo gear honing. According to the practical implementation of the process, it is established that this part can reach 80% of the total volume of processed gears. Thus, in this example, the management of the process structure is provided in order to reduce labor intensity. The latter reaches 80% of the total labor intensity for gear honing, designed for the entire program of gear production.

Minimum composition processes – single-operation – have the least ability to control accuracy. In addition, their use is limited to the fact that modern gears, as a rule, require intermediate chemical–thermal treatment. However, their use is desirable, as such processes require the simplest technological systems in general. The possibility and expediency of application of single-operational processes are defined at the stage of synthesis in a technological system.

Let us move on to consider the possibilities of accuracy control by choosing the kind and type of operation. In this case, since the traditional operations are well-studied [36], we will consider multi-instrumental and database processing operations.

10.4.2.1 Possibilities of Accuracy Control Using Two- and Three-Tool Processing Methods

In multi-tool machining, each tool forms the tooth surfaces described by Eqs. (10.23), (10.24), (10.25), (10.26), (10.27), (10.28), and (10.29). Moreover, operator (10.24) selects from all possible cuts those that actually remain on the gear in the form of a real surface. In this regard, there is no fundamental difference between single and multi-tool processing (in the context of the laws of error). However, from the point of view of the possibilities of error management, it is expedient to consider finishing methods with the use of several tools, with the aim of revealing the patterns of different types of connections to the transformation of the original workpiece errors.

In practice, two- and three-tool processing methods are often implemented using, usually, disk multi-toothed tools such as rollers, shavers, and hons [9, 34]. There are many designs of machines that implement these methods, which are fundamentally different from each other with regard to the kinematic connections between tools as well as between tools and gears.

Based on practice [9], each variant has its own patterns of error conversion in the processing. Therefore, the choice of optimal options for newly designed machines or the establishment of rational areas of operation of the existing equipment should be made with due regard to the achieved accuracy of processing.

The most common and some new schemes of two- and three-tool processing are shown in Fig. 10.33.

In Fig. 10.33, the circles indicate the machined gears and tools. The heavy solid lines show rigid kinematic connections. The thin short arrows show the possible movements of the elements of the MFTP system during processing. The absence of a symbol of support in the center of the gear means that the processing is performed without fixing the position of the axle in space.

The values of the connection parameters characterizing the scheme are provided in Table 10.4.

In this case, the scheme number in the table corresponds to the figure number in Fig. 10.33. The following marks are accepted:

Φ – angular coordinate parameter (Fig. 10.33)
e, β – eccentricity parameters
Δa – offset parameter
1–2, 2–3, ..., 1–4 – connections, the presence of which is reflected by one
O, T – one- and two-profile processing

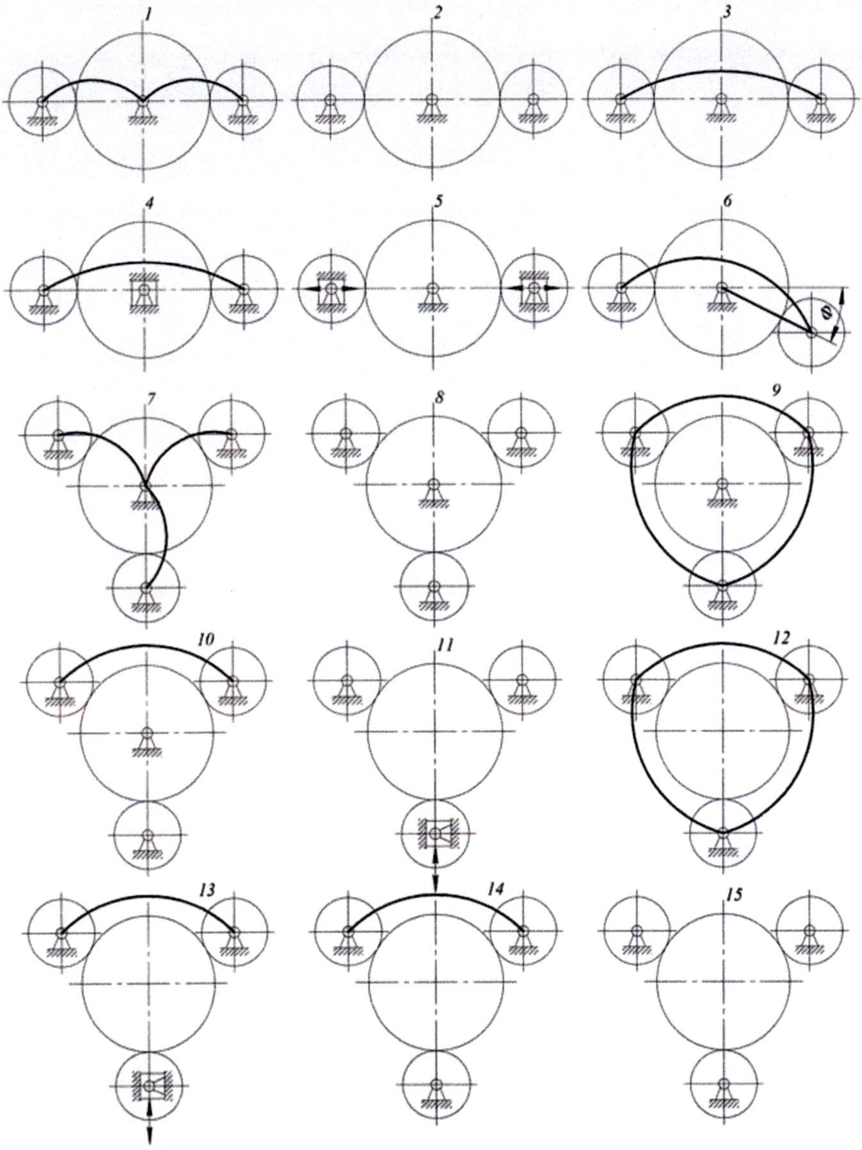

Fig. 10.33 Schemes for two- and three-tool processing

Considering the features introduced in Table 10.4, it can be determined that other processing schemes are possible, which, however, have not yet become widespread due to the complexity of their implementation or lack of significant advantages over the above.

Table 10.4 The characteristics of connection parameters of multi-tool processing circuits

	Parameters							Solid relations						Type of machine gearing					
														1–4		2–4		3–4	
№.	β_2	β_3	e	β	Δa_1	Δa_2	Δa_3	1–4	2–4	3–4	1–2	2–3	3–1	O	T	O	T	O	T
1	0	0	0	0	0	0	0	1	1	–	0	–	–	1	–	1	–	–	–
														–	1	–	1	–	–
2	0	0	0	0	0	0	0	0	0	–	0	–	–	1	–	1	–	–	–
														–	1	–	1	–	–
3	0	0	0	0	0	0	0	0	0	–	1	–	–	1	–	1	–	–	–
														–	1	–	1	–	–
4	0	0	e	0	0	0	0	0	0	–	1	–	–	1	–	1	–	–	–
														–	1	–	1	–	–
5	0	0	0	0	Δa_1	Δa_2	0	0	0	–	0	–	–	–	1	–	1	–	–
														–	1	–	1	–	–
6	0	Φ	0	0	0	0	0	0	0	–	1	–	–	1	–	1	–	–	–
														–	1	–	1	–	–
7	30°	30°	0	0	0	0	0	1	1	1	0	0	0	1	–	1	–	1	–
														–	1	–	1	–	1
8	30°	30°	0	0	0	0	0	0	0	0	0	0	0	1	–	1	–	1	–
														–	1	–	1	–	1
9	30°	30°	0	0	0	0	0	0	0	0	1	1	1	1	–	1	–	1	–
														–	1	–	1	–	1
10	30°	30°	0	0	0	0	0	0	0	0	0	1	0	1	–	1	–	1	–
														–	1	–	1	–	1
11	30°	30°	e	β	0	0	Δa_3	0	0	0	0	0	0	–	1	–	1	–	1
														–	1	–	1	–	1
12	30°	30°	e	β	0	0	0	0	0	0	1	1	1	1	–	1	–	1	–
														–	1	–	1	–	1

13	30°	30°	e	β	0	0	Δa_3	0	0	0	0	1	0	–	1	–	1	–	1
														–	1	–	1	–	1
14	30°	30°	e	β	0	0	0	0	0	0	0	1	0	1	–	1	–	1	–
														–	1	–	1	–	1
15	30°	30°	e	β	0	0	0	0	0	0	0	0	0	–	1	–	1	–	1
														–	1	–	1	–	1

Let us analyze the processing schemes listed in Table 10.4, based on the following assumptions: In finishing methods of processing teeth, the cutting forces directed tangentially to the profiles (tangential) are small compared to the normal one [29]. Therefore, their influence, in order to simplify the calculations, will be ignored. The flexibility of an MFTP system affects the conversion of errors in gear processing [40]. However, it does not change the nature of these transformations but only causes quantitative differences between the calculated values of the amplitudes [41]. Therefore, we will ignore the elastic deformations of the MFTP system. Provided that only the low-frequency components of the total error will be considered in the future – as its main part, determined by the scheme of multi-tool processing, forces of inertia and damping, due to their small influence [38], we will also ignore.

When studying the regularities of error transformation, we take into account the following conditions that are met with rationally organized processing [41]:

1. A closed circuit of the dimensional circuit of an MFTP system. This closure is ensured by rigid kinematic connections and the presence of sufficient allowance, by continuous pressing of the tools to the workpiece, or by stabilizing the center distance and providing sufficient allowance
2. Power (dynamic) balance of an MFTP system, due to the presence of force interactions between the elements of the MFTP system
3. Continuity of the gear, provided by its relatively high mechanical properties that do not allow significant deformation of the teeth and rim

Let us write the expressions that reflect these conditions in general. The relationship between the increments of the lines of action is derived from the closed condition of the MFTP system [36, 41]:

$$\Delta T^{(i)} = \frac{\Delta F_{\mathrm{R}}^{(i)} - \Delta F_{\mathrm{L}}^{(i)}}{2\cos\alpha}; \tag{10.125}$$

$$\Delta H^{(i)} = \frac{\Delta F_{\mathrm{R}}^{(i)} - \Delta F_{\mathrm{L}}^{(i)}}{2\sin\alpha}; \tag{10.126}$$

$$\Delta F_{\mathrm{R1}}^{(i)} = \Delta F_{\mathrm{R0}}^{(i)} - \Delta t_{\mathrm{R}}^{(i)} \tag{10.127}$$

$$\Delta F_{\mathrm{L1}}^{(i)} = \Delta F_{\mathrm{L0}}^{(i)} - \Delta t_{\mathrm{L}}^{(i)} \tag{10.128}$$

where

$\Delta T^{(i)}$ is the tangential increment of the lines of action of the processed gear.

$\Delta t_{\mathrm{R,L}}^{(i)}$ is the allowance; the lower index indicates the pass number (stroke) and profile (R – right, L – left) and the upper index indicates the tool number.

The relationship between allowance and cutting force (N) based on the study by Samoilov and Syromyatnikov [29] is assumed to be linear:

$$N_{\mathrm{R}} = C_p \Delta t_{\mathrm{R}}; \quad N_{\mathrm{L}} = C_p \Delta t_{\mathrm{L}}, \tag{10.129}$$

where $C_p = \text{const}$.

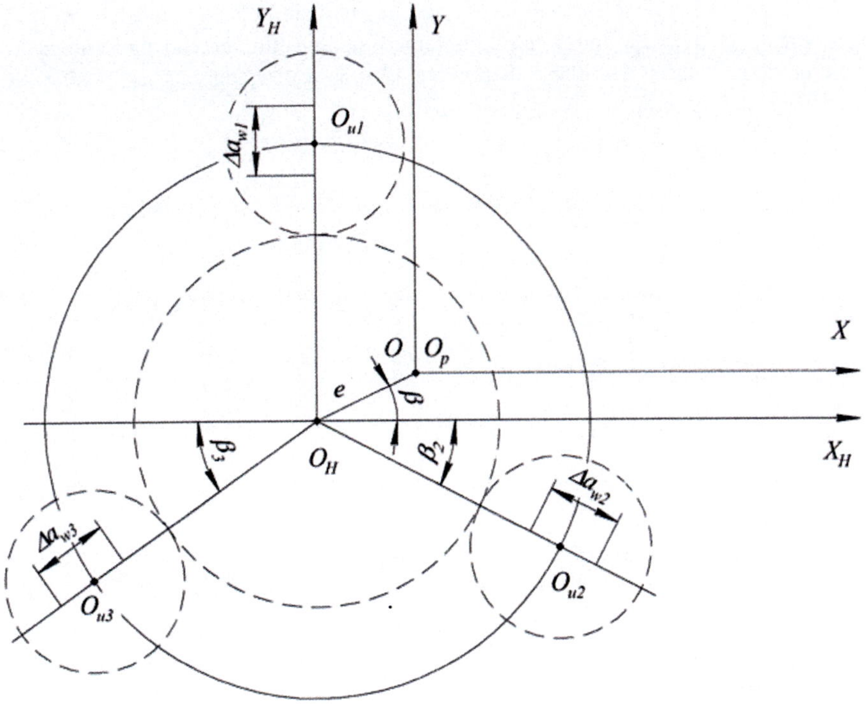

Fig. 10.34 Scheme for applied coordinate systems

On the basis of (10.125), (10.126), (10.127), (10.128), and (10.129), we obtain for two-profile processing:

$$P_r = \frac{N_R + N_L}{2\sin\alpha} = C_p \frac{\Delta t_R + \Delta t_L}{2\sin\alpha}; \tag{10.130}$$

$$P_t = \frac{N_R - N_L}{2\cos\alpha} = C_p \frac{\Delta t_R - \Delta t_L}{2\cos\alpha}; \tag{10.131}$$

The index "r"in (10.130) denotes the radial cutting force "t"in (10.131), which is tangential.

Equation (10.131) confirms the legitimacy of the assumption that the tangential components of the cutting forces are small ($P_t = 0$ at $\Delta t_R = \Delta t_L$).

Let us consider the coordinate system: $S_H\{X_H, O_H, Y_H\}$ – fixed (Fig. 10.34) and movable coordinate system $S\{X, O, Y\}$, which does not rotate, but has a center O that coincides with the geometric center O_p of the workpiece.

The axes of the coordinate systems are parallel, and the centers are shifted by e and phase β. The angular positions of the axes on which the centers O_{u2} and O_{u3} are located are determined by the parameters β_2 and β_3, respectively. These parameters, in general, can be variable. The offsets of the workpiece centers are determined by the parameters a_{w1}, a_{w2}, and a_{w3}. The directions of possible offsets are displayed in

Fig. 10.34. Dependencies for the calculation of increments of lines of action in coordinate systems S_H and S can be obtained, taking into account the influence of geometric eccentricity on these increments [36] as well as the angular positions of the gears, set by the parameters β_2 and β_3:

$$\Delta F_{LH}^{(1)} = \Delta F_{\mathrm{L}}^{(1)} + e\sin(\beta + \alpha); \tag{10.132}$$

$$\Delta F_{RH}^{(1)} = \Delta F_{\mathrm{R}}^{(1)} - e\sin(\beta - \alpha); \tag{10.133}$$

$$\Delta F_{LH}^{(2)} = \Delta F_{\mathrm{L}}^{(2)} + e\sin(\beta + \alpha + 90^\circ + \beta_2); \tag{10.134}$$

$$\Delta F_{RH}^{(2)} = \Delta F_{\mathrm{R}}^{(2)} - e\sin(\beta - \alpha + 90^\circ + \beta_2); \tag{10.135}$$

$$\Delta F_{LH}^{(3)} = \Delta F_{\mathrm{L}}^{(3)} + e\sin(\beta + \alpha + 270^\circ - \beta_3); \tag{10.136}$$

$$\Delta F_{RH}^{(3)} = \Delta F_{\mathrm{R}}^{(3)} - e\sin(\beta - \alpha + 270^\circ - \beta_3). \tag{10.137}$$

Conditions that reflect the synchronicity of rotation of the tool and the workpiece can be presented in the form

$$\Delta T_i^{(i)} = \mathrm{const.} \tag{10.138}$$

The constancy of the center distance is expressed

$$\Delta H_i^{(i)} = \mathrm{const.} \tag{10.139}$$

The condition reflecting changes in center distance can be written as:

$$\Delta H_{j+1}^{(i)} = \Delta H_j^{(i)} - \Delta a_{w(j+1)}^{(i)} = \Delta H_j^{(i)} - \frac{\Delta t_{\mathrm{R}(j+1)}^{(i)} + \Delta t_{\mathrm{L}(j+1)}^{(i)}}{2\sin\alpha}. \tag{10.140}$$

The power equilibrium conditions of the elements of an MFTP system, like any other solids, should reflect the equality of the zero sums of the moments of cutting forces and reactions of the supports relative to the workpiece axis and the sums of the projections of forces and reactions on two mutually perpendicular axes OX and OY.

A scheme for forces and reactions operating in general is provided in Fig. 10.35.

Based on this scheme, the condition of the zero sum of moments can be written as:

$$C_p \sum_{j=1}^{n} \left(\Delta F_{\mathrm{L0}}^{(j)} - \Delta F_{\mathrm{L1}}^{(j)} \right) - C_p \sum_{j=1}^{n} \left(\Delta F_{\mathrm{R0}}^{(j)} - \Delta F_{\mathrm{R1}}^{(j)} \right) - \sum_{j=1}^{n} R_j; \quad n = \{2, 3\}, \tag{10.141}$$

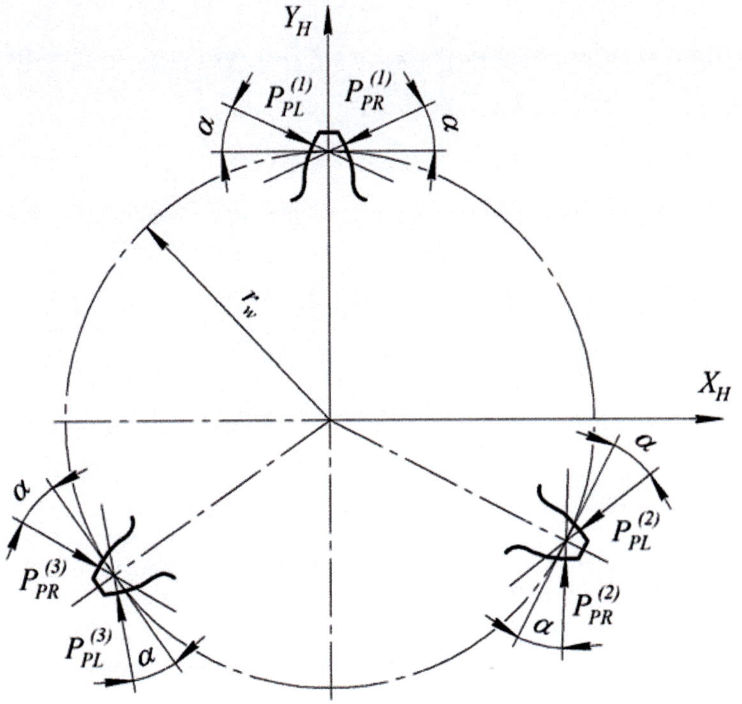

Fig. 10.35 Scheme for acting forces and reactions

where $\sum_{j=1}^{n} R_j$ is the sum of the reactions of the rigid bonds. In the absence of connections, $-\sum_{j=1}^{n} R_j = 0$.

Similarly, we can write the condition of the zero sum of moments relative to the centers of each instrument as:

$$C_p\left(\Delta F_{\mathrm{L0}}^{(i)} - \Delta F_{\mathrm{L1}}^{(i)}\right) - C_p\left(\Delta F_{\mathrm{R0}}^{(i)} - \Delta F_{\mathrm{R1}}^{(i)}\right) - R_i = 0. \tag{10.142}$$

In the absence of connections or drive, $R_i = 0$. The condition of equality of zero land projections of forces and reactions on the axes OY and OX can be written in the following form (see Fig. 10.35):

$$\begin{aligned} C_p\Big\{&\Big[\left(\Delta F_{\mathrm{L0}}^{(1)} - \Delta F_{\mathrm{L1}}^{(1)}\right) - \left(\Delta F_{\mathrm{R0}}^{(1)} - \Delta F_{\mathrm{R1}}^{(1)}\right)\Big]\sin\alpha - \Big[\left(\Delta F_{\mathrm{R0}}^{(2)} - \Delta F_{\mathrm{R1}}^{(2)}\right) - \left(\Delta F_{\mathrm{L0}}^{(2)} - \Delta F_{\mathrm{L1}}^{(2)}\right)\Big]\cos\alpha\cos\beta_2 - \\ &- \Big[\left(\Delta F_{\mathrm{R0}}^{(2)} - \Delta F_{\mathrm{R1}}^{(2)}\right) + \left(\Delta F_{\mathrm{L0}}^{(2)} - \Delta F_{\mathrm{L1}}^{(2)}\right)\Big]\sin\alpha\sin\beta_2 - \Big[\left(\Delta F_{\mathrm{L0}}^{(3)} - \Delta F_{\mathrm{L1}}^{(3)}\right) - \left(\Delta F_{\mathrm{R0}}^{(3)} - \Delta F_{\mathrm{R1}}^{(3)}\right)\Big]\cos\alpha\cos\beta_3 - \\ &- \Big[\left(\Delta F_{\mathrm{L0}}^{(3)} - \Delta F_{\mathrm{L1}}^{(3)}\right) + \left(\Delta F_{\mathrm{R0}}^{(3)} - \Delta F_{\mathrm{R1}}^{(3)}\right)\Big]\sin\alpha\sin\beta_3\Big\} + R_y = 0; \end{aligned} \tag{10.143}$$

$$
\begin{aligned}
&C_p\Big\{\Big[\Big(\Delta F_{\mathrm{L0}}^{(1)}-\Delta F_{\mathrm{L1}}^{(1)}\Big)-\Big(\Delta F_{\mathrm{R0}}^{(1)}-\Delta F_{\mathrm{R1}}^{(1)}\Big)\Big]\cos\alpha-\Big[\Big(\Delta F_{\mathrm{R0}}^{(2)}-\Delta F_{\mathrm{R1}}^{(2)}\Big)-\Big(\Delta F_{\mathrm{L0}}^{(2)}-\Delta F_{\mathrm{L1}}^{(2)}\Big)\Big]\cos\alpha\sin\beta_2-\\
&-\Big[\Big(\Delta F_{\mathrm{R0}}^{(2)}-\Delta F_{\mathrm{R1}}^{(2)}\Big)+\Big(\Delta F_{\mathrm{L0}}^{(2)}-\Delta F_{\mathrm{L1}}^{(2)}\Big)\Big]\sin\alpha\cos\beta_2+\Big[\Big(\Delta F_{\mathrm{L0}}^{(3)}-\Delta F_{\mathrm{L1}}^{(3)}\Big)-\Big(\Delta F_{\mathrm{R0}}^{(3)}-\Delta F_{\mathrm{R1}}^{(3)}\Big)\Big]\cos\alpha\sin\beta_3+\\
&+\Big[\Big(\Delta F_{\mathrm{L0}}^{(3)}-\Delta F_{\mathrm{L1}}^{(3)}\Big)+\Big(\Delta F_{\mathrm{R0}}^{(3)}-\Delta F_{\mathrm{R1}}^{(3)}\Big)\Big]\sin\alpha\cos\beta_3\Big\}+R_x=0;
\end{aligned}
\tag{10.144}
$$

Finally, the condition of gear continuity, fulfilled in the conditions of application of rigid communication between tools, can be represented by the following dependencies:

$$\Delta F_{\mathrm{L}i}^{(1)}=\Delta F_{\mathrm{L}i}^{(2)}=\Delta F_{\mathrm{L}i}^{(3)}; \tag{10.145}$$

$$\Delta F_{\mathrm{R}i}^{(1)}=\Delta F_{\mathrm{R}i}^{(2)}=\Delta F_{\mathrm{R}i}^{(3)}; \tag{10.146}$$

The obtained equations are enough to analyze the basic regularities of error transformations for any of the multi-tool processing schemes shown in Fig. 10.33. Graphs of change of initial errors depending on the processing time for the considered schemes are displayed in Fig. 10.36 (solid lines indicate graphs of initial errors and dotted lines show emerging errors).

In two-tool machining of gears with two (Fig. 10.33, diagram 1) or three (Fig. 10.33, diagram 7) tools with a rigid kinematic connection between each, the tool and the gear at a fixed position of the gear axis in space during machining tools and workpiece rotate synchronously; therefore, $\Delta F_{\mathbf{R}i}=$ const and $\Delta F_{\mathbf{L}i}=$ const. This is why in the considered processing schemes, the initial errors are always corrected regardless of their type (Fig. 10.36a).

When a free to rotate gear, is rotated by several gear-finishing tools with motionless axis of rotation (Fig. 10.33, Schemes 2 and 8) for each tool the same conditions are fulfilled, as at one-tool processing when the gear is leading that for the first time prof. B.A. Thaits [36]. Therefore, the nature of the error transformation in this case will be the same as in single-tool processing: the geometric eccentricity will be corrected by transforming it into kinematic eccentricity. The original kinematic eccentricity will not change (Fig. 10.36b).

When machining two (Fig. 10.33, diagram 3) or three (Fig. 10.33, diagram 9) tools with a rigid kinematic connection of tools with each other, free rotation of the gear when fixing the position of its axis in space is performed:

- The condition of constancy of the center distance between each of the tools and the gear
- The condition of equality, zero moments of forces acting on the right and left profiles of the teeth
- The condition of continuity of a gear

Taking into account the previously obtained equations that reflect these conditions and solving them together, we will get

- For two-tool processing:

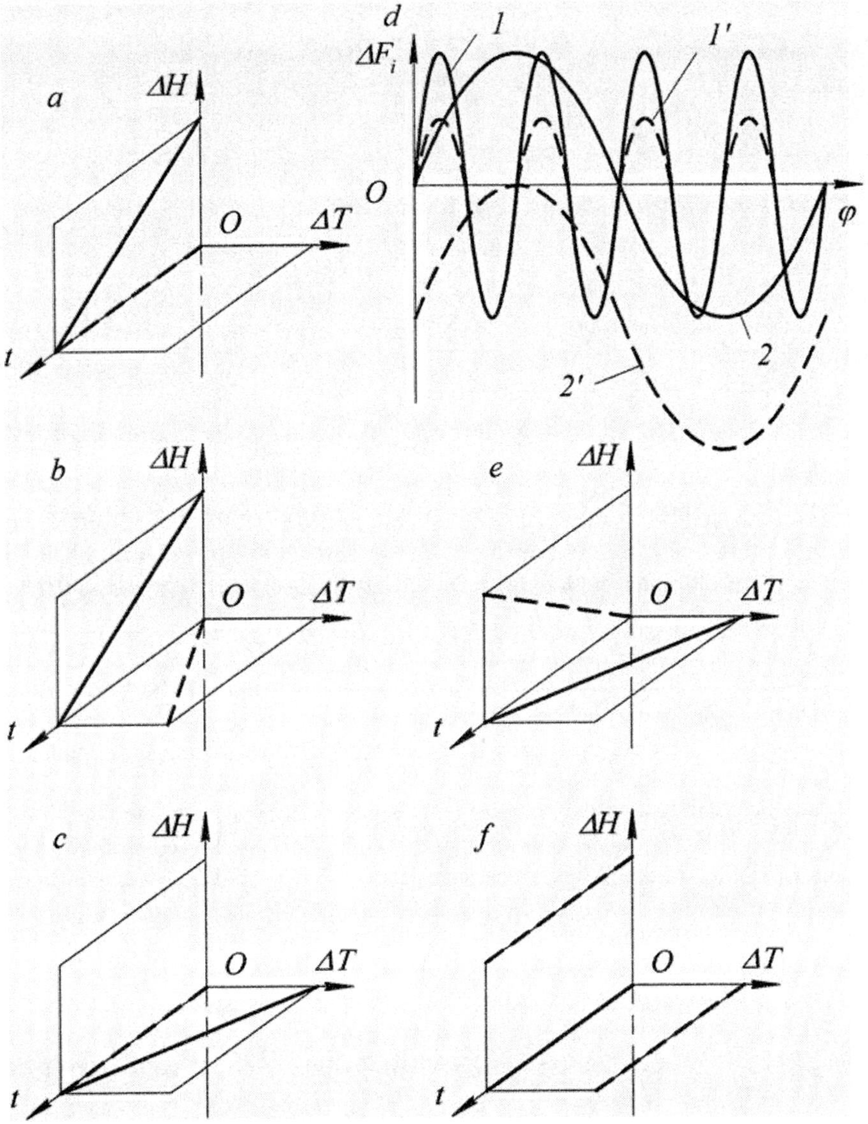

Fig. 10.36 Graph of error change

$$\Delta F_{\mathrm{L}i}^{(1)} = \frac{1}{4}\sum_{i=1}^{2}(\Delta F_{\mathrm{L}i} - \Delta F_{\mathrm{R}i}); \tag{10.147}$$

$$\Delta F_{\mathrm{R}i}^{(1)} = -\frac{1}{4}\sum_{i=1}^{2}(\Delta F_{\mathrm{L}i} - \Delta F_{\mathrm{R}i}); \tag{10.148}$$

$$\Delta T^{(1)} = \frac{1}{2}\sum_{i=1}^{2}\Delta T_i; \tag{10.149}$$

- For three-tool processing:

$$\Delta F_{\mathrm{L}i}^{(1)} = \frac{1}{6}\sum_{i=1}^{3}(\Delta F_{\mathrm{L}i} - \Delta F_{\mathrm{R}i}); \tag{10.150}$$

$$\Delta F_{\mathrm{R}i}^{(1)} = -\frac{1}{6}\sum_{i=1}^{3}(\Delta F_{\mathrm{L}i} - \Delta F_{\mathrm{R}i}); \tag{10.151}$$

$$\Delta T^{(1)} = \frac{1}{3}\sum_{i=1}^{3}\Delta T_i; \tag{10.152}$$

The analysis [41] of Eqs. (10.147), (10.148), (10.149), (10.150), (10.151), and (10.152) taking into account (10.138) shows that when processed according to the considered scheme, the initial errors with the period 2π are completely corrected regardless of their type, since, in this case, $\sum_{i=1}^{3}\Delta T_i = 0$ for $n = 2$ and $n = 3$, so $\Delta F_{\mathrm{L}}^{(1)} = \Delta F_{\mathrm{R}}^{(1)} = 0$. The graph of error changes is shown in Fig. 10.36c. Errors with a period other than 2π may not be corrected at all (for example, in the case of two-tool machining with a period $(2k + 1)\pi$, where k, an arbitrary positive integer, is required) or may be partially corrected.

To completely correct the errors of any frequency, it is necessary to change the phases of the location of the tools during processing. This is achieved using Scheme 6 shown in Fig. 10.33. When processing gears with two tools, by changing their angular position to the gear (Fig. 10.33, Scheme 6), the demonstrations for calculations – increments of line of action (10.147) and (10.148) – remain valid because they are derived for the general case.

The initial errors are as follows:

$$\Delta F_{\mathrm{L}0} = l_\eta \sin\left[k(\phi + \alpha)\right]; \tag{10.153}$$

$$\Delta F_{\mathrm{R}0} = -l_\eta \sin\left[k(\phi - \alpha + \beta_i)\right], \tag{10.154}$$

where l_η is the amplitude of error of the kth frequency.

Substituting (10.153) and (10.154) in (10.147) and (10.148), we obtain:

$$\Delta F_{\mathrm{L}1} = \frac{l_\eta}{4}\{\sin\left[k(\phi + \alpha)\right] + \sin\left[k(\phi - \alpha + \beta_i)\right]\}; \tag{10.155}$$

$$\Delta F_{\mathrm{R}1} = -\frac{l_\eta}{4}\{\sin\left[k(\phi + \alpha)\right] + \sin\left[k(\phi - \alpha + \beta_i)\right]\}. \tag{10.156}$$

Assuming $\Delta F_{\mathrm{L}1} = 0$ and $\Delta F_{\mathrm{R}1} = 0$ in (10.155) and (10.156), we can find:

$$k(\beta_i + 2\alpha) = 180^\circ \tag{10.157}$$

It follows from (10.157) that the choice of β_i, for example, by its continuous change, can compensate for any initial errors. This is a condition for the synthesis of a new method of processing gears.

Free running of a gear with three tools and an unfixed position of its axis in space (Fig. 10.33, diagram 11), in addition to the condition of force equilibrium in accordance with Eqs. (10.141), (10.143), and (10.144), performed at $\sum_{i=1}^{n} R_i = 0$, $R_y = P$, $R_x = 0$, provides conditions of continuous two-profile engagements for each tool given by Eq. (10.140) as well as the equilibrium conditions of the tools given by Eq. (10.142).

Solving these equations together, we obtain:

$$\Delta H_i = \Delta H_{01} - \Delta a_w; \tag{10.158}$$

$$\Delta T_1^{(1)} = \Delta T_1^{(0)}; \tag{10.159}$$

$$\Delta F_{\mathrm{L}i}^{(1)} = \Delta F_{\mathrm{L}}^{(0)} - \frac{P}{3C_p}; \tag{10.160}$$

$$\Delta F_{\mathrm{R}i}^{(1)} = \Delta F_{\mathrm{R}}^{(0)} - \frac{P}{3C_p}. \tag{10.161}$$

Analysis of Eqs. (10.158), (10.159), (10.160), and (10.161) shows that the initial kinematic errors do not change with this processing scheme. There is only the removal of the allowance on both lines of action. The graph of change in initial errors is illustrated in Fig. 10.36d.

When machining gears with three tools with a rigid kinematic connection of the tools with each other, having free movement of the gear with an axis not fixed in space (Fig. 10.33, diagram 12), the following conditions are met:

- Power balance of the workpiece (10.141), (10.143), and (10.144)
- Continuity of the gear (10.145) and (10.146)
- Constancy of center distances between tools (10.139)

Solving these equations, in view of the conditions (10.132), (10.133), (10.134), (10.135), (10.136), and (10.137), we obtain the following equations for calculating the increments of the lines of action of the processed gear:

$$\Delta T_1^{(1)} = \frac{1}{3}\left[(2\Delta T_1 - \Delta T_2 - \Delta T_3)\cos^2\alpha - \sqrt{3}(\Delta H_2 - \Delta H_3)\sin^2\alpha + \Delta T_1 + \Delta T_2 + \Delta T_3\right]; \tag{10.162}$$

$$\Delta H_1^{(1)} = \frac{2\Delta H_1 \sin\alpha - (\Delta H_2 + \Delta H_3)\sin^2\alpha + \sqrt{3}(\Delta T_2 - \Delta T_3)\cos^2\alpha}{2\sin\alpha + 3\cos^2\alpha + \sin^2\alpha}; \tag{10.163}$$

$$\Delta F_{\mathrm{R}}^{(1)} = \left[\frac{-2\Delta H_1 \sin\alpha - (\Delta H_2 + \Delta H_3)\sin^2\alpha + \sqrt{3}(\Delta T_2 - \Delta T_3)\cos^2\alpha}{2\sin\alpha + 3\cos^2\alpha + \sin^2\alpha}\right]\sin\alpha - \\ -\frac{\cos\alpha}{3}\left[(2\Delta T_1 - \Delta T_2 - \Delta T_3)\cos^2\alpha - \sqrt{3}(\Delta H_2 - \Delta H_3)\sin^2\alpha + \Delta T_1 + \Delta T_2 + \Delta T_3\right]; \tag{10.164}$$

$$\Delta F_{\mathrm{L}}^{(1)} = \left[\frac{2\Delta H_1 \sin\alpha + (\Delta H_2 + \Delta H_3)\sin^2\alpha + \sqrt{3}(\Delta T_2 - \Delta T_3)\cos^2\alpha}{2\sin\alpha + 3\cos^2\alpha + \sin^2\alpha}\right]\sin\alpha - \\ -\frac{\cos\alpha}{3}\left[(2\Delta T_1 - \Delta T_2 - \Delta T_3)\cos^2\alpha - \sqrt{3}(\Delta H_2 - \Delta H_3)\sin^2\alpha + \Delta T_1 + \Delta T_2 + \Delta T_3\right]. \tag{10.165}$$

Analyzing Eqs. (10.162), (10.163), (10.164), and (10.165), we can show [41] that when processing gears according to the considered scheme, the geometric eccentricity is not completely corrected ($\Delta H_i^{(i)} \neq 0$). These practices confirm this [9]. The initial radial error can be increased by converting the kinematic eccentricity into the geometric one. This means that the considered processing scheme exerts a reverse effect on the observed one when machining with one, two, or three tools in free-running conditions and a fixed gear axis. The graph of change in the initial errors is shown in Fig. 10.36d.

The analysis shows that the schemes of multi-tool processing differ significantly in terms of the nature of the transformation of the original errors. The most complete error correction can be achieved when machining gears with a rigid connection between the tools and the gear when fixing the gear axis during machining. However, tools for machining with a rigid kinematic connection between tools and the gear are usually complex, rigid, and inaccurate, given that in their dividing circuits, as a rule, there is a spatial transmission or a complex electronic device such as an "electric shaft." Therefore, in many cases, it is more expedient to use methods of processing by two or three tools with a rigid kinematic connection, with free running in of a gear and a fixed position of its axis in space. Such methods, as the tool axes can be parallel, are implemented by simple devices that contain high-precision gears in their dividing circuits. In some cases, in order to control the relationship between tangential and radial errors, it is advisable to use functionally redundant devices that would implement some or all of the described processing schemes.

10.4.2.2 Precision Control by Changing the Positions of the Base Surfaces of Gears

The accuracy of gears controlled by changing the position of the base surfaces is usually performed after chemical heat treatment operations [23]. The task is to restore the positions of the base surfaces (central hole and end face, necks, center holes) so that the errors of the ring gear measured relative to these (restored) surfaces are minimal. If the bases are incorrectly restored, then the center of the ring gear may shift from its optimal position along three mutually perpendicular axes, namely, e_x,

e_y, and e_z, as well as rotations around the same axes by the angles ψ_x, ψ_y, and ψ_z. Offsets along the axes OX and OY will cause increments of the lines of action

$$\Delta F_{\mathrm{R}} = \sqrt{e_x^2 + e_y^2}\sin(\phi + \xi + \alpha); \tag{10.166}$$

$$\Delta F_{\mathrm{L}} = -\sqrt{e_x^2 + e_y^2}\sin(\phi + \xi - \alpha); \tag{10.167}$$

These increments, in turn, will increase the magnitude of the kinematic error $2\sqrt{e_x^2 + e_y^2}$.

The offset e_z along the axis OZ will not increase the kinematic error, as measured in the axial plane of the gear; however, it can change the size, shape, and location of the contact spot in spatial transmissions that prevent movement "by themselves" along the central axis, such as conical hypoid, spherical, and others.

Rotation around the axis OZ also does not cause any additional errors in round gears due to the fact that their dividing surface is the body of rotation. However, rotations around the axes OX and OY will cause changes in the contact spot in all cases and kinematic errors in the case of an asymmetric location of the crown on the hub. These changes will be as follows: let $X_{2T} = f_1(u_2, v_2)$, $Y_{2T} = f_2(u_2, v_2)$, and $Z_{2T} = f_3(u_2, v_2)$ be the equations of the surfaces of the teeth received for the operation of restoring the bases, for example, after chemical–thermal treatment. Then, for round gears, we can write the equation of the radius vector of the coordinates of the teeth relative to the new base

$$\overline{r}_{2H} = \widetilde{M}_{\Delta\psi}\overline{r}_{2T}, \tag{10.168}$$

where

$$\widetilde{M}_{\Delta\psi} = \begin{vmatrix} \cos\Delta\psi & 0 & \sin\Delta\psi \\ 0 & 1 & 0 \\ \sin\Delta\psi & 0 & \cos\Delta\psi \end{vmatrix}; \tag{10.169}$$

$$\Delta\psi = \sqrt{\Delta\psi_x^2 + \Delta\psi_y^2}. \tag{10.170}$$

From Eq. (10.168), taking into account (10.169) and (10.170), up to small first-order smallness, we can find [38]:

$$\Delta F_{\mathrm{L}} = -\Delta F_{\mathrm{R}} = f_3(u_2, v_2)\Delta\psi\sin(\tau_n + \beta_c); \tag{10.171}$$

where

τ is the angular pitch of the teeth.
n is the number of the tooth on which the error is measured.
$\beta_c = \text{const}$ is the error phase.

Increments (10.171) cause changes in the error of the direction of the tooth by the size

$$F_{\beta 2(\mathrm{L,\ R})} = -b_w \Delta\psi, \tag{10.172}$$

On the other hand, increments of the lines of action will increase the kinematic error by [38]:

$$F'_{i2} = e_n \Delta\psi, \tag{10.173}$$

In Eqs. (10.172) and (10.173), b_w is the gear width and e_n is the distance from the center of the skew to the center of the gear.

The shape and relative position, as well as the size of the teeth when changing the position of the base surfaces, do not change, so the smoothness and backlash in the transmission also do not change (although the gap itself, as seen from (10.166) and (10.167) changes), measured independently of the positions of the bases (for example, on tooth thickness, the size on rollers, length of the general normal). Based on the above, we can draw a conclusion that the optimal performance of operations to change the position of the base surfaces will happen if the errors characterizing the kinematic accuracy and contact spot in the transmission are minimized.

The reference surface of a real gear can be significantly different from a round one. Its contour can be described with sufficient accuracy by the Fourier equation [23]:

$$r_{\Phi} = r_H + a_0 + \sum_{i=1}^{n} a_i \sin(i\phi_2 + \beta_i), \tag{10.174}$$

where r_H is the nominal value of the radius of the dividing surface in the calculated cross section of the end plane and a_i, $i \in \{0, 1, \ldots, n\}$ is Fourier series coefficient.

The contour of the longitudinal line of each tooth can be described similarly

$$\Delta F_{\beta r T0} = a_0 + \sum_{k=1}^{n} a_k \sin(k\phi_2 + \beta_k). \tag{10.175}$$

The equation of a circle of radius r_{0H} shifted relative to the center by a quantity e_T with phase β_T has the form:

$$r_{0H} = r_H - \Delta + e_T \sin(\phi_2 + \beta_T). \tag{10.176}$$

Taking into account Eqs. (10.174), (10.175), and (10.176), as well as dependencies (10.166), (10.167), (10.169), and (10.170), the problem of optimal control of the accuracy of tooth processing in the considered operations (passes) can be written to find e_T, β_T, $\Delta\psi$, and β_ψ based on the conditions:

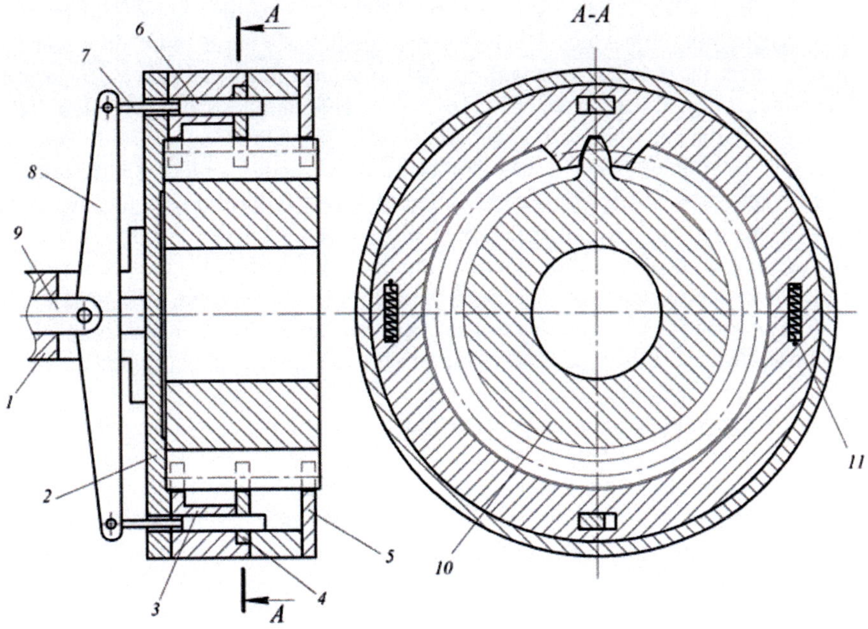

Fig. 10.37 Device for centering and clamping of gears

$$\max_{\phi}\,(r_{\Phi} - r_H) \rightarrow \min\,; \tag{10.177}$$

$$\max_{\phi}\left[\left(\widetilde{M}_{\Delta\psi}\cdot\overline{r}_{2T}\right) - r_H\right] \rightarrow \min\,. \tag{10.178}$$

The problem has a numerical solution. Various devices can be used for the technical implementation of this task [23]. In accordance with Eqs. (10.177) and (10.178), devices must provide a choice of extreme points and base them on the workpiece. If a gear element (subsystem) is proposed at each point to make a potentially possible contact and these elements can be combined with a single drive, then using 7.1, an automatic device can be synthesized, the general view of which is shown in Fig. 10.37 in two mutually perpendicular projections.

The device comprises shank 1, which is fastened to housing 2. Three gear disks 3, 4, and 5 are coaxially mounted on the housing. A wedge mechanism is provided for the angular displacement of toothed disk 4 around its axis. Cam 6 of this mechanism has the shape of a trapezoid in the tangential plane, i.e., a plane passing through the cam perpendicular to the radius. Cam 6 is mounted with axial movement onto the grooves of the housing and is connected by rod 7 to rocker arm 8. Rocker arm 8 is hinged to rod 9 of the drive, which can be, for example, a pneumatic cylinder. To unclip installed gear 10, spring mechanism 11 is provided, which returns toothed disk 4 to its original position at the upper position of cam 6.

Centering and clamping of the gear of the chuck is performed as follows: gear 10 is installed in the cavities of the gears with emphasis on the end surface of the device. Then, the drive is switched on. The drive receives axial movement of rod 9 with rocker arm 8. Rocker arm 8 moves cam 6. This cam, resting on one side of the housing with the other interacting with toothed disk 4, makes it rotate around its axis. Due to the rotation of disk 4, a choice of the gap between the teeth of gear 10 and the teeth of gears 3, 4, and 5 is provided. As a result, gear 10 is centered and clamped by the teeth of the gears. Because clamping disk 4 rotates relative to its axis and the pressure angles at all possible points of contact, determined by the tooth groove profile of discs 3, 4, and 5, exceed the friction angles, self-centering is ensured at any position of gear 10. The grooves of the gears can have a trapezoidal shape, which minimizes the radial beating of the ring gear or a shape congruent with the tooth profiles of the machined gear, which minimizes the oscillation of the measuring center distance for gear rotation. After clamping gear 10, the bases are processed and changed relative to their original position.

In some cases, when there are no devices as described above, the optimization of base recovery operations is performed by taking into account the use of traditional multi-cam cartridges [23]. The task of minimizing errors in these cases is to solve two problems: reducing the radial errors of the chuck surface and eliminating or minimizing the influence of the initial errors of the workpiece on the centering process. Methods for solving this problem are described in detail in the studies by Ternyuk et al. [37] and Ternyuk [38]. It was experimentally proved that, by optimizing the processes of restoration of base surfaces, it is possible to increase the accuracy by 2–3 degrees (for carburized gears) [23]. Studies show the possibility of significant changes in errors in operations for the treatment of base surfaces.

10.4.2.3 Influence of Process Input Parameters on the Final Accuracy of Gears

In relation to the process, the input parameters are those that characterize the workpiece as well as the technological environment. The influence of the environment is taken into account in each pass of the tool. Therefore, at the process level, this factor can be studied by summing up the effects of each pass (stroke), which is discussed below. The parameters characterizing the workpiece have direct and indirect effects on accuracy. The direct effect can be detected by the errors of the base surfaces directly determine the values of the errors of basing (matrix components $\widetilde{M}_{\Delta 2}$), and that in case of insufficient allowance, according to Eq. (10.24), the tooth profiles will be "black". Indirect influence is expressed through the influence of force, temperature, wear, and kinematic errors.

Based on the study of the direct influence of the parameters of the workpieces on the errors of basing [36], one should consider the influence of these parameters on the possibility of processing without "blacks." Based on the method in the study by Ternyuk [38], we consider two offset diagrams of changes in the mean $\Delta_{H\Phi}$ radial

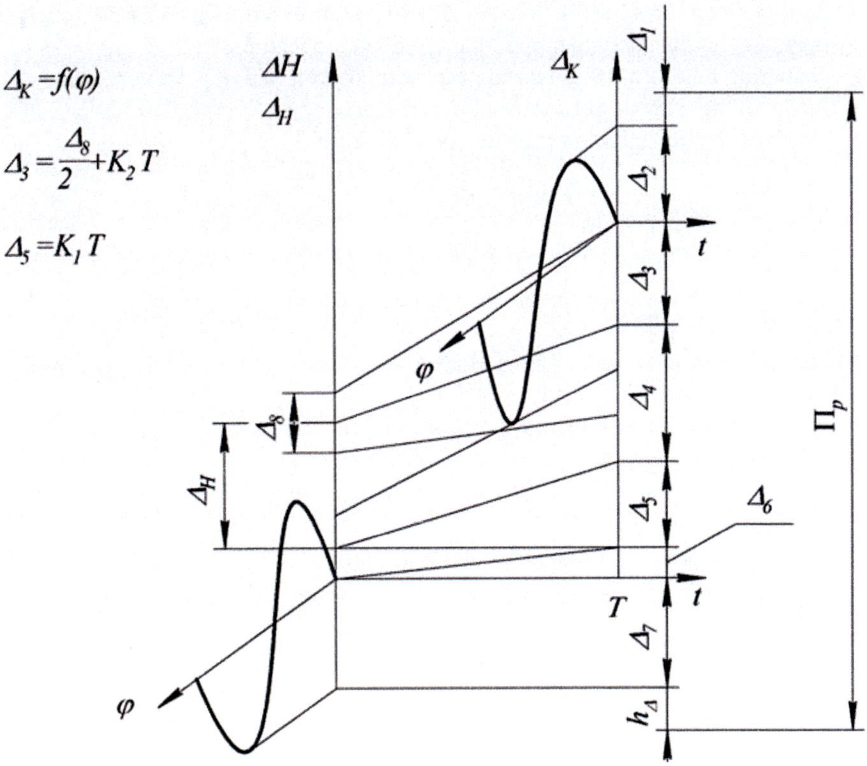

Fig. 10.38 Determination of the estimated allowance

increments within the gear speed Δ_H, with the lines of action of the gears entering the operation from $(i-1)$th during the sub-adjustment period T_M. Here, $\Delta_{H\Phi}$ is the error of the size of the machine.

These diagrams are shown in Fig. 10.38, which, in turn, shows the coordinate systems $\Delta HO\phi$ and graphs of changes in the angle of rotation of the gear of the periodic components of the radial increments of the lines of action of the gear.

The location of the coordinate systems $\Delta HO\phi$ corresponds to the boundary points of the chart at the time of processing $(t=0)$at time $t=T_M$.

From the upper point of the graph $\Delta H=f(\phi)$, the possible error of adjustment Δ_H of the machine realizing i-th operation, on the size is postponed, and from the bottom – depth of defective layer of surfaces of teeth h_g. The closing link of the constructed dimensional chain corresponding to the general-type MFTP system is the ultimate calculated (not actual) allowance h_p, which must be adjusted to remove the machine to process the entire batch of gears without traces of blackness. It follows from Fig. 10.38 that the value of the calculated allowance is affected by the following errors:

- Periodic increments of the lines of action caused by errors of both the previous operation and the installation on the considered operation
- Scattering field of medium-sized gears coming from the $(i - 1)$th operation
- Errors of adjustment of the size of the machine
- Depth of the defective layer of the tooth surfaces

By detailing the structure of the MFTP system and setting dimensional prices between elements, each of these errors can be presented through others. Thus, the periodic increments of the lines of action of the treated gear are affected by:

- The initial periodic increments of the lines of action, arising on the $(i - 1)$th operation $\Delta F_{n(i-1)}$
- The radial error of basing the workpiece e_{pi} directly on the machine performing the ith operation
- The radial beating of the adjusting and clamping device – e_{ni}
- The part $F_{\beta z(i-1)}$ of the error of the direction of the tooth of the gear (taper and other components), having a period of manifestation for the rotation of the gear less than 2π
- The duration of the inter-adjustment period T_M

The scattering field of medium-sized workpieces depends on:

- The error $\Delta H_{(i-1)}$ of adjustment to the size of the machine performing the $(i - 1)$ th operation
- The error $\Delta b_{(i-1)}$ of execution of the size by the preconfigured machine
- The duration of the inter-adjustment period T_M

The error of setting up the machine to the size can be determined if the following are known:

- Error of setting the center distance Δa_w
- Non-parallelism of the machine table in the radial plane $\Delta_{\Phi p}$

The depth of the defective layer h_g is determined by the requirements for the quality of the tooth surfaces and is usually determined by the condition:

$$h_g = \max\left\{\Delta h_p, \Delta h_O\right\} + \Delta h_c, \tag{10.179}$$

In Eq. (10.179),Δh_p, Δh_O are the values of "ripples" [36] and cut profiles, respectively, and Δh_c is the depth of a continuous layer of material with inadmissible structural transformations.

These components are summarized in Table 10.5, which shows their transfer coefficients and distribution laws.

The table shows u_{i1}, u_{i2} – the transfer coefficients of the influence of the primary errors on the radial and tangential planes, respectively, which are the calculated allowances. The first is necessary for the organization of processes with a two-profile machining scheme, including the free-run method, and the second for processes with a single-profile machining scheme by methods that provide a rigid connection between the tool and the gear.

Table 10.5 The characteristics of the primary errors affecting the estimated allowance

№.	Name of the primary error	Designation	Gear ratios		Law of distribution
			u_{i1}	u_{i2}	
Vector errors					
1	Fluctuation of the measuring center distance	$F''_{1r(i-1)}$	1	–	Rayleigh's law
2	Single-profile kinematic error	$F'_{1r(i-1)}$	–	1	Rayleigh's law
3	Radial base error	e_{pi}	1	1	Rayleigh's law
4	Radial beating of the device	e_{ni}	1	1	Rayleigh's law
Scalar errors					
1	Size execution error on $(i-1)$th operation	$\Delta_{H(i-1)}$	1	$2tg\alpha$	Gaussian law
2	Size execution error on $(i-1)$th operation	$\Delta_{B(i-1)}$	1	$2tg\alpha$	Gaussian law
3	Effective part of the tooth direction error	$F''_{\beta r(i-1)}$	$\frac{1}{2\sin\alpha}$	1	Gaussian law
4	Error of machine settings on the *i*th operation	Δ_{Hi}	1	$2tg\alpha$	Gaussian law
5	Nonparallel movement of the machine table In the radial plane In the tangential plane ·	$\Delta_{\Phi p}$ $\Delta_{\Phi T}$	1 –	– 1	Gaussian law Gaussian law
Systematic errors					
1	Depth of the defective layer	h_g	$\frac{1}{\sin\alpha}$		–

As known, when summing up the vector errors, the total errors will be distributed according to Rayleigh's law, and, when summing up the scalar errors,the total errors will be distributed according to Gauss's law. Therefore, based on the study by Ternyuk [38], the law of distribution of the allowance can be represented by a composition of the laws of Rayleigh and Gauss, and, then, the probability density of the distribution of the permitted allowance Π_p is:

$$q(\Pi_p) = \frac{\sigma_p\left(\Pi_p - \Delta\Sigma - \frac{h_g}{2\sin\alpha}\right)}{\left(\sigma_p^2 + \sigma_r^2\right)\sqrt{\sigma_p^2 + \sigma_r^2}} \exp\left[-\frac{\left(\Pi_p - \Delta\Sigma - \frac{h_g}{2\sin\alpha}\right)^2}{2\left(\sigma_p^2 + \sigma_r^2\right)}\right], \tag{10.180}$$

where $\sigma_p = \frac{1}{36}\sqrt{\sum_{i=1}^{n}\left[6,48h_{gi}^2 + \left(1 - h_{gi}\right)^2\right]^2 \Delta_i^2}$, here, Δ_i are the error values that are numerically equal:

- When calculating the allowance for processing gears by free-running methods according to the two-profile scheme:

$$\left(F''_{ir} - \frac{b_\Phi T_M}{2}\right); \quad e_{pi} = \max\left\{\Delta\delta_i; \Delta\psi_i e_H\right\}; \quad e_{ni};$$

- When calculating the allowance for machining gears on machines with a rigid kinematic connection between the part and the tool:

$$\left(F''_{ir} - \frac{b_\Phi T_M}{2}\, tg\,\alpha\right); \quad e_{pi} = \max\left\{\Delta\delta_i\, tg\,\alpha; \Delta\psi_i e_H\, tg\,\alpha\right\}; \quad e_{ni}\, tg\,\alpha.$$

Here,

b_Φ is the coefficient of the influence of time T_M on F''_{ir} in the regression equation;
$\Delta\delta_i$, $\Delta\psi_i$ are the algebraic differences of radial errors of basing and angles of the skew of preparation on $(i-1)$th and ith operations, respectively;
e_H is the greatest distance from the center of the ring gear to the point of intersection of the axes of the mandrel and the gear

$$\sigma_r = \frac{1}{6}\sqrt{\Delta^2_{B(i-1)} + \left(\Delta_{H(i-1)} + b_H T_M\right)^2 + \Delta^2_{Hi} + \left(\frac{\Delta\Phi_p b}{L_{bH}}\right) + \left(\frac{F^*_{\Delta r(i-1)}}{2\sin\alpha}\right)^2};$$

$$\Delta\Sigma = \frac{1}{2}\left[\Delta_{B(i-1)} + \Delta_{H(i-1)} + b_H T_M + \Delta_{Hi} + \frac{\Delta\Phi_p b}{L_{bH}} + \frac{F^*_{\Delta r(i-1)}}{2\sin\alpha}\right];$$

Here, L_{bH} is the base length determined by $\Delta\Phi_p$ and
b_H is the coefficient of influence T_M on Δ_{Hi} in the regression equation.

Dependencies for the calculation of σ_p, σ_r, and $\Delta\Sigma$ are made taking into account the linear change of primary errors over time, which was experimentally proved in the study by Krivchenko [15].

Integrating expression (10.180) and solving it relatively, we obtain the value of Π_p as follows:

$$\Pi_p = \sqrt{2\left(\sigma_p^2 + \sigma_r^2\right)\ln\frac{100\sigma_p}{P\sqrt{\sigma_p^2 + \sigma_r^2}}} + \Delta\Sigma + \frac{h_g}{2\sin\alpha}, \tag{10.181}$$

where P is the permitted probability of the actual value of the allowance beyond the calculated value.

Dependence (10.181) is the basis for calculating the optimal allowance for finishing gears. On the basis of this dependence, it is possible to analyze the possible schemes of arrangement of allowances and tolerances on them. Figure 10.39 shows the traditional layout of allowances for multi-operational processes.

A feature of such a scheme is the presence of a guaranteed minimum allowance, which must be removed. However, it can be established that such a layout is not optimal for either the achieved accuracy or the complexity of machined gears. Given condition (10.24), as well as the linear relationship between the allowance $[h_g]$ and

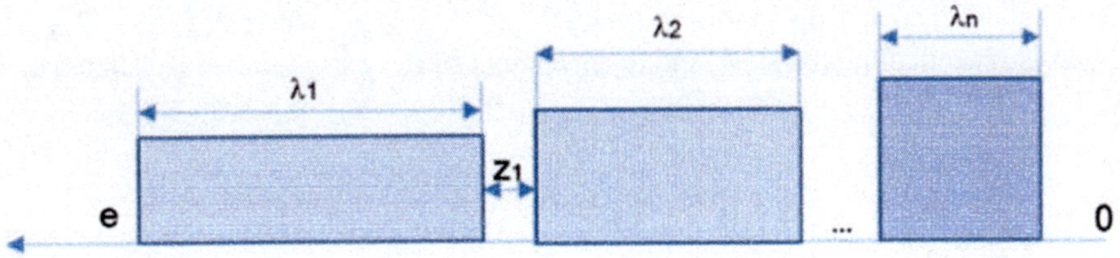

Fig. 10.39 Traditional layout of allowances

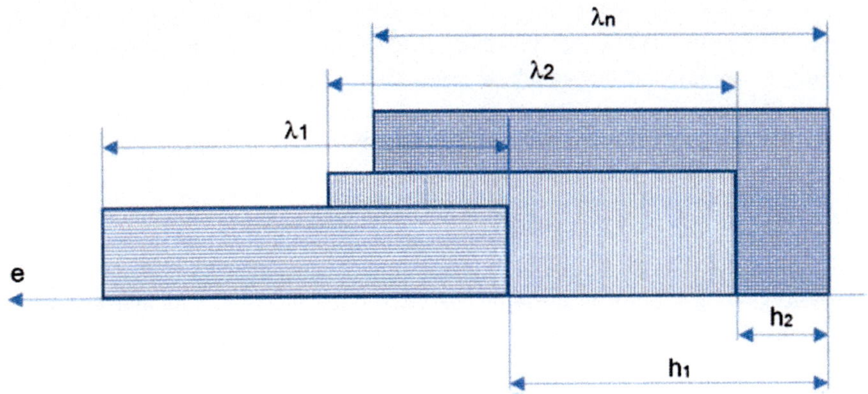

Fig. 10.40 Optimal layout of allowances

labor intensity and the need to obtain a defective layer on the surfaces of the teeth that is no more than permissible, the most rational layout of allowances can be depicted as shown in Fig. 10.40.

Ways to ensure the implementation of the scheme in Fig. 10.40 can be identified by analyzing Eq. (10.181) to calculate the allowance. It is obvious that the allowance can be reduced by minimizing all errors included in Eq. (10.181). In order to develop measures to reduce these errors, it is necessary to know the patterns of their occurrence. The main factors causing the harmonic increments of the lines of action of the processed gear, especially radial, are the errors of basing. Therefore, an important measure to reduce the allowance is to minimize the harmful effects of basing errors.

When machining different parts, it is often advisable to use the principle that all operations involve the use of the same surfaces as a base. When machining gears, this principle is often irrational or impossible due to the geometric features of the crown, the presence of operations to restore the base surfaces, and the conditions of formation [9].

It follows from Eq. (10.144) that if the conditions $\widetilde{M}_{\Delta 2i} = \widetilde{M}_{\Delta 2(i+1)}$, $i \in \{1, \ldots, n\}$ are observed in multi-operational (multi-transition, multi-pass) processes and in base recovery operations to ensure $\widetilde{M}_{\Delta rn} = -\widetilde{M}_{\Delta 2\delta}$, where $\widetilde{M}_{\Delta rn} = -\widetilde{M}_{\Delta 2\delta}$ is the matrix reflecting the skew and displacement of the part in these operations, then

$$\Delta\overline{r}_{2p} \rightarrow \min\,, \quad h_p \rightarrow \min\,. \tag{10.182}$$

To ensure $\widetilde{M}_{\Delta 2i} = \widetilde{M}_{\Delta 2(i+1)}$, i.e., Eq. (10.182), it is necessary to have the same installation errors (basing and fixing) at each tooth machining operation. Moreover, they do not have to be minimal. This makes it possible to significantly expand the tolerances on the base surfaces of the workpieces during their intermediate processing. For example, if this principle is not met, then the tolerances on the workpieces are limited by the kinematic errors and the errors of the contact spot. For the most common medium-modular gears of 6–8 degrees of accuracy, they are 0.02–0.05 mm.

At the same time, in compliance with this principle, the limiting factor is not the accuracy of the gear but the consistency of the depth of the carburized (nitrocarburized) layer on the base surfaces. For medium-module gears, the permissible fluctuation in the depth of the carburized layer is 0.2–0.5 mm. This means that the implementation of the principle of optimal basing allows one to expand the tolerances on the base surfaces of the gears by up to 10 times. Due to the expansion of these tolerances, it is possible to significantly build up the procurement subsystem.

The principle of optimal basing can be made possible in two ways: active and passive. The active path involves measuring installation errors in the previous operation and then reproducing them in the subsequent one. Its implementation requires appropriate measuring tools and mechanisms for small movements, which can provide an accurate restoration of the position of the technological axis. The passive path involves the processing of gears in intermediate operations relative to the same technological axis. With great accuracy, the passive path is realized through the use of satellite mandrels as well as combined instruments. An example of the effective use of satellite mandrels is the processing of gears on the ROTA-FZ-200 system [30]. A general view of a high-rigidity satellite mandrel is shown in Fig. 10.41.

The figure shows: 1 – base surface and 2 – installation cone.

An example of the use of combined tools is a circular broach with internal teeth, equipped with profile spring-loaded rollers.

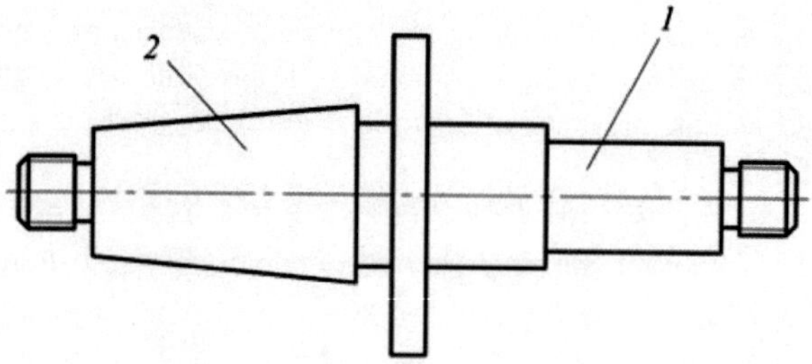

Fig. 10.41 Satellite mandrel with increased rigidity

The scattering field of a medium-sized gear can be reduced by [38]:

- The optimal setting of machines performing the $(i - 1)$th operation
- Reduction of inter-adjustment periods
- Use of one setting to perform the $(i - 1)$ th and i th operations
- Adaptive control settings

The optimal size adjustment of gear cutting machines can be performed according to universal methods [32]. Algorithms for calculating the optimal levels of tuning of certain types of gear processing machines are described in the study by Ternyuk [38].

Most gear machines allow many limb parameters to be adjusted. The methods for reducing sizing errors, when setting limb distances, are the same for all machines. Therefore, to address this issue, applying the recommendations provided in the study by Solomentsev et al. [32] is required.

Shortening of the inter-tuning periods reduces the length of the scatter plot. Optimality is provided at the automatic sub-adjustment of the machine after processing each detail. The calculation of the optimal parameters of the sub-adjustment can be performed by known methods, for example, those described in the study by Solomentsev et al. [32].

Using the same setting to perform the $(i - 1)$th and ith operations is possible when using combined tools, for example, in the aforementioned circular broach.

Adaptive control of the setting of gearing machines can be provided by the use of special devices, for example, such as those in the study by Solomentsev et al. [32]. The method of controlling the adjustment of gear processing machines is not fundamentally different from the method of controlling the adjustment of other machines. Therefore, the results presented in the study by Solomentsev et al. [32] and other works can be used.

The depth of the defective layer h_g can be adjusted by changing the processing modes, tool geometry, coolant, and a fundamental change in the type of operation performed before the considered. This determines the ability to control the value of the allowance in order to minimize it. The minimum value of the allowance $(\Pi_{p\min})$ is provided by the integrated implementation of these measures. Moreover, the value $\Pi_{p\min}$ required for machining parts without traces of blackness is not the same for different processing methods. It can be as small as desired with a corresponding reduction in the parameters included in the dimensional chain containing Π_p. Therefore, it must be determined by the general parametric optimization of the technological system.

Thus, the studies presented in Sect. 10.4.2 show that the possibilities of controlling the accuracy of gear processing at the level of the technological process are determined by the capabilities of the operations included in the process and the parameters of the original workpiece.

Now, we proceed to considering the operation level of accuracy control.

10.4.3 Operational Accuracy Control

Both at the process level and the operation level, accuracy control can be achieved by changing the structure and parameters. The elements of the structure of the machining operation include the following:

- The type of transition used (elements X_i)
- The number of transitions used
- The connections between transitions, determined by the order of succession of transitions as well as the interaction of their parameters

The parameters of the operation are as follows: parameters of structure, modes, input and output parameters of the workpiece, and others.

The types of transitions used in gear machining operations are strictly limited. As a rule, these are the transitions of installation (Π_y), removal (Π_c), machining (Π_M), and control (Π_K). A reinstall transition (Π_n) is also possible.

The general structure of the operation can be described by a symbolic model:

$$O_n = \bigcap_i \Pi_i, \quad i \in \{y, c, M, K, n\}. \tag{10.183}$$

From this model, one can obtain any particular kind of transition sequence.

Machining transitions have the greatest influence on the formation of accuracy. The nature and content of machining transitions determine the nature and the name of the operation. The minimum composition of the operation follows from (10.183):

$$O_n = \Pi_y \cap \Pi_M \cap \Pi_c. \tag{10.184}$$

Processes with gear reinstallation are practically not used in the practice of gear processing. However, they are possible, and, in some cases, they are appropriate.

In view of Eq. (10.184), it can be noted that the main possible way to control the accuracy of machining gears at the level of operation is the choice of machining transitions.

10.4.4 Precision Control at the Machining Transition Level

The structure of a transition is determined by the type, number, and sequence of passes as well as the relationship between them and their parameters. The transition parameters are structure, mode, and input and output parameters of the workpiece. The type of passage (working stroke) is determined by the level of concentration of the technological impact. A large number of different types of passes allow controlling the accuracy within a wide range. The number of passes affects the accuracy of processing due to two factors arising from the analysis of the general model of occurrence and accounting of errors (Fig. 10.1) as well as Eqs. (10.23), (10.24), (10.25), (10.26), (10.27), (10.28), and (10.29):

- Changes in cutting forces and temperatures, which lead to changes in the recoverable deformations and wear of the tool within the processing of all teeth of the gear
- Changes in the magnitudes and phases of tool errors, which lead to a change in the position of points with extreme values ΔF and cutting sections of the profiles of the teeth of the gear's most protruding parts of the tool when moving the contact point on the tool surface

With regard to the processing of gears, one can write:

$$\Delta_{\mathrm{res}} = \frac{f\left(t_p + \Delta_{\mathrm{uc}}\right)}{C_{\mathrm{np}}}, \tag{10.185}$$

In (10.185):

$\Delta_{\mathbf{res}}$ is the residual error.
$\Delta_{\mathbf{uc}}$ is the initial error.
$C_{\mathbf{np}}$ is the reduced rigidity of the MFTP system.
$f(x)$ is a function of dependence of force P_p on the depth t_p of cutting.

Figure 10.42 illustrates the change in residual deformation from the number of tool passes.

The graph of change of the recovered thermal deformations described by dependence has a similar character:

$$\Delta_{\mathrm{res}} = \sum_{i=1}^{n} 2\pi e_i\left[\phi\left(t_p + \Delta_{\mathrm{uc}}\right) - T_0\right], \tag{10.186}$$

In Eq. (10.186), $\phi(x)$ is a function of the dependence of temperature on the depth of cutting.

The only difference is the relatively smaller effect of temperature deformations compared to that of force deformations.

Fig. 10.42 Graph of change in residual deformation

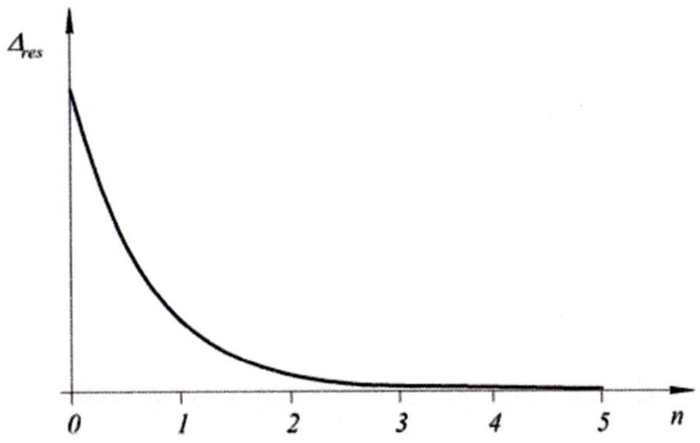

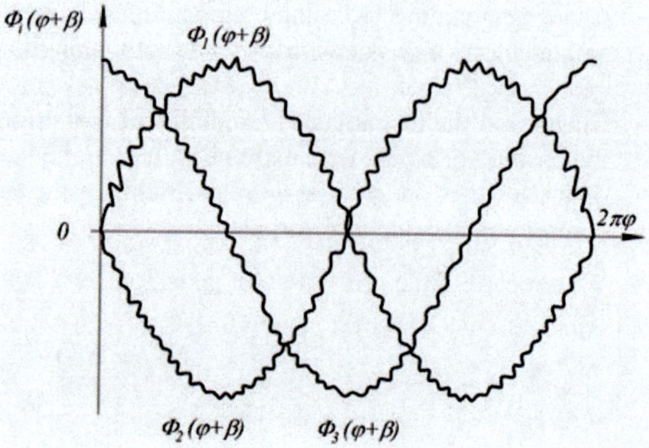

Fig. 10.43 Graph of functions on two passes of the tool

The mechanism of change of accuracy of gears due to a change of a phase and/or amplitude of errors of an MFTP system at repeated passes works differently. It is shown at a rather small giving on incision or their absence. It is explained as follows: each pass of the tool corresponds to its function of excessive increments of the line of action:

$$\Delta F_{in3\delta} = \Phi_i(\phi + \beta_i). \tag{10.187}$$

In general, function (10.187) changes both in shape and in phase when performing repeated passes, i.e., $\Phi_i(\phi + \beta_i) \neq \Phi_{(i+1)}(\phi + \beta_{(i+1)})$, where i, $(i + 1)$ are pass numbers.

Figure 10.43 shows two functions corresponding to two tool passes for a particular case in which the type of function does not change.

The kinematic error of the processed gear (F'_{ir}) is characterized by the scope of the function $\Phi_i(\phi + \beta_i)$. Given the periodicity of this function by the angle of rotation, one can write:

$$\begin{aligned}
F'_{ir1} &= \max \Phi_1(\phi + \beta_1) - \min \Phi_1(\phi + \beta_1);\\
F'_{ir2} &= \max\,[\Phi_1(\phi + \beta_1) \cap \Phi_2(\phi + \beta_2)] - \min\,[\Phi_1(\phi + \beta_1) \cap \Phi_2(\phi + \beta_2)];\\
&\ldots\\
F'_{irn} &= \max \left[\bigcap_{i=1}^{n} \Phi_i(\phi + \beta_i)\right] - \min\left[\cap_{i=1}^{n} \Phi_i(\phi + \beta_i)\right],
\end{aligned} \tag{10.188}$$

where n is the passage number.

Cyclic errors are characterized by the same dependencies but within the range of angle ϕ change by one angular step. If we take into account that the protruding sections of the profiles are cut during repeated passes, as the other sections are already cut (Fig. 10.43), it is not difficult to see that by changing the phase β_i of

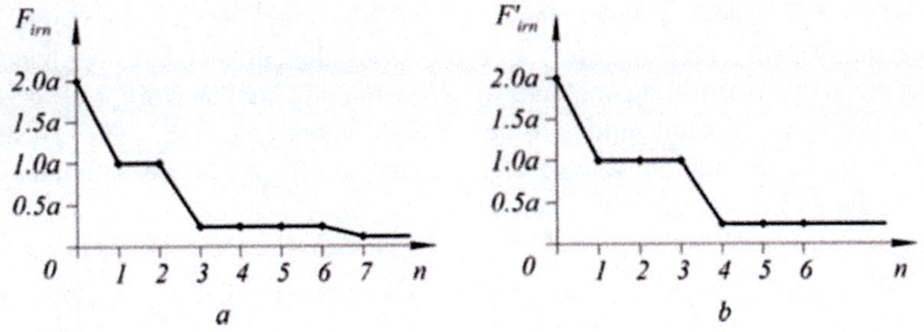

Fig. 10.44 Examples of changes in a single-profile kinematic error depending on the number of passes of the tool

function relative to the function $\Phi_i(\phi + \beta_i)$ phase $\Phi_{i-1}(\phi + \beta_{i-1})$ and not significantly changing the function F'_{irn} values in accordance with Eq. (10.188). Figure 10.44 shows two examples of changes in a single-profile kinematic error F'_{irn} for cases in which the functions $\Phi_i(\phi + \beta_i)$ are described by sinusoids of the form $a_i \sin(\phi + \beta_i)$.

Figure 10.44a corresponds to the case in which the period of the function corresponds to 2π, whereas Fig. 10.44b corresponds to $\frac{3}{4}\pi$. The first case can be implemented on any gear machine by reinstalling the workpiece, sequentially rotating it by 90° on each pass. The second one can be implemented when gear shaping with the numbers of teeth of the tool z_1 and the processed gear z_2 determined from the ratio $\frac{z_1}{z_2} = \frac{1}{3}$.

These examples show that the practical implementation of the effect of improving the accuracy of repeated passes can be performed in many operations. In some cases, optimization is achieved by measuring or predicting the function $\Phi_i(\phi + \beta_i)$ on each pass and then by finding and implementing the optimal phase β_i of displacement of the tool or gear according to the condition $F'_{irn} \rightarrow \min$. In other cases, accuracy can be automatically increased by carrying out nursing work moves without interrupting the processing.

Let us find the conditions under which the specified effect of an increase of accuracy on repeated working moves of the tool is realized. Based on the analysis of Fig. 10.43, we can establish that to implement the effect, it is necessary to ensure:

- Changes in the shape or phase of the function $\Phi_i(\phi + \beta_i)$
- No cutting at points corresponding to $\min\Phi_i(\phi + \beta_i)$ to a depth greater than $\max\Phi_i(\phi + \beta_i) - \max[\Phi_i(\phi + \beta_i) \cap \Phi_{(i+1)}(\phi + \beta_{(i+1)})]$

Changing the shape of the function $\Phi_i(\phi + \beta_i)$ is characteristic of abrasive machining processes in the best manner possible, when each stroke significantly changes the shape of the tool due to its re-editing. Changing phase β_i of the function $\Phi_i(\phi + \beta_i)$ can be carried out in almost all gear processes. The absence of cutting at the points corresponding to $\min\Phi_i(\phi + \beta_i)$ to the specified depth can be carried out only in processes in which technical means for stabilizing synchronous rotation or mutual angular position of the tool and preparation (rigid kinematic communication, connected masses, etc.) are used. For processes that do not have such devices (carried out by free-rolling methods or with braking of the workpiece or tool), cutting occurs at all points of the profiles, at least on one side of the teeth. Therefore, for such processes, the considered effect cannot be fully realized. Obviously, the smaller the difference, the less accurate the process will be

$$\Delta^* = \left\{ \max \Phi_i(\phi + \beta_i) - \max \left[\Phi_i(\phi + \beta_i) \cap \Phi_{(i+1)}\left(\phi + \beta_{(i+1)}\right)\right]\right\} - t_{\min}, \tag{10.189}$$

where $t_{\min}$ is the depth of cut at the corresponding points $\min\Phi_i(\phi + \beta_i)$.

When $\Delta^* < 0$, the process of formation will be unstable, i.e., diverging in accuracy. Instability is inherent in the processing of the involute and other profiles by methods of free running or braking of the workpiece. This is due to the fact that with such processing methods, the cutting conditions that determine the cutting forces are not the same at different points of the tooth profiles [34, 41]. The greatest removal of metal with a constant force of pressing the tool to the workpiece occurs at the foot of the tooth. Therefore, in such cases, even when processing absolutely accurate workpieces, the errors of the tooth profile may increase along with increasing the number of passes. Figure 10.45a illustrates what has been said.

According to previous studies [41], it can be stated that the absolute value of the error asymptotically tends to its limit, which provides dynamic stability (not corresponding to the minimum processing error).

Improving the accuracy of the gear profile during machining by free-running and braking of the workpiece can be achieved by eliminating the factors causing condition (10.189) or by changing the beginning of the development phase of the unstable process. Elimination of the factors causing condition (10.189) is possible at the expense of management of the sizes entering into dependence.

The value t is the easiest element to control. A change in t can be achieved by forcible rotation, creating uneven braking of the workpiece or by changing the

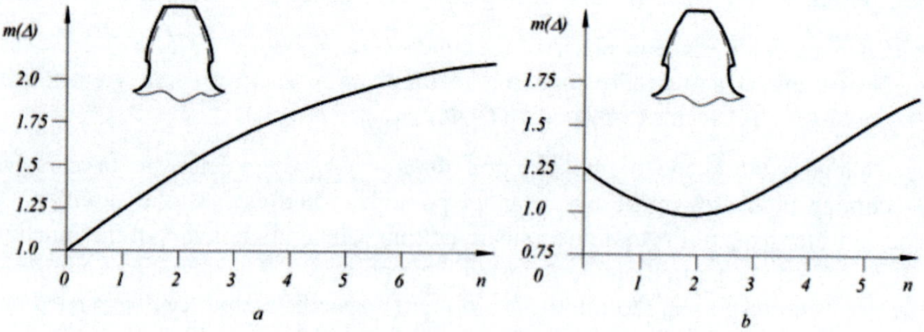

Fig. 10.45 Error change diagrams

design of the tool so that at a constant pressing force, the tool at different points of contact with the workpiece profile provides constant removal of material.

It is possible to change the beginning of the phase of unstable development of the molding process, for example, by pre-modifying the workpiece allowance [41]. The graph of change in the error of the tooth profile during repeated passes of the tool in this case has the form shown in Fig. 10.45b. The graph is drawn for the case of processing the workpiece with an increased allowance at the foot of the tooth.

Various measures to increase the stability of the molding process when machining gears by free rolling or braking the workpiece (tool) are described in detail in the studies by Sukhorukov [34], Ternyuk [41], and other works. The method of processing that provides the highest accuracy in the implementation of a partially stable process is described in the study by Ternyuk [41].

The obtained laws are the basis for choosing the kind and type of machining pass as well as the purpose of its number.

10.4.5 Control of Accuracy at the Level of Passage during Mechanical Processing of Teeth

10.4.5.1 Analysis of Factors that Allow Variation

It is possible to identify the factors that allow variation at the level of the machining passage of teeth on the basis of the model of occurrence and accounting of errors (Fig. 10.1), general equations of real tooth profiles (10.23), (10.24), (10.25), (10.26), (10.27), (10.28), and (10.29), and Sect. 10.4.1. It follows from the latter that a passage is an elementary structural formation of the process and is not structured itself. At its level, accuracy can be controlled only by changing the parameters. Based on Eqs. (10.23), (10.24), (10.25), (10.26), (10.27), (10.28), and (10.29), it can be established that at the level of passage, the control parameters are those that determine:

- Movement of the end parts of the machine
- The shape of the producing surface
- The recoverable deformations of the teeth of the processed gear
- The blank

These parameter groups can be controlled in many ways implemented by different devices. The first three groups of parameters allow variation to directly counteract the abovementioned primary errors. Therefore, many methods of adaptive accuracy control based on the principle of compensation can be built on their basis. The fourth factor – the initial parameters of the workpiece – determines the magnitude of the primary errors, the possibility of processing the gear without traces of “blackness” (Eq. (10.24)), and the position of the measuring axis.

10.4.5.2 Compensatory Methods of Adaptive Accuracy Control

Using the general equations of real tooth profiles, we obtain the basic equations that reflect the laws of error compensation.

It is obvious that the system of Eqs. (10.23), (10.24), (10.25), (10.26), (10.27), (10.28), and (10.29) is valid for any values of the given primary errors and parameters of compensatory effects, including equal to zero. This meets the conditions of nominal shaping and can be written as:

$$\overline{r}_{2p} = \overline{r}_{2\Phi}; \tag{10.190}$$

$$\overline{r}_{2\Phi} = \overline{r}_{2i}; \tag{10.191}$$

$$\overline{r}_{2i} = \widetilde{M}_{\Sigma HH} \overline{r}_{1i}, \tag{10.192}$$

$$\left(\frac{\partial \overline{r}_{1H}}{\partial u} \times \frac{\partial \overline{r}_{1H}}{\partial v}\right)\left(\widetilde{M}_{\Sigma HH}^{-1} \frac{\partial \widetilde{M}_{\Sigma HH}}{\partial \phi_1} \cdot \overline{r}_{1H}\right) = 0; \tag{10.193}$$

$$\left(\frac{\partial \overline{r}_{1H}}{\partial u} \times \frac{\partial \overline{r}_{1H}}{\partial v}\right)\left(\widetilde{M}_{\Sigma HH}^{-1} \frac{\partial \widetilde{M}_{\Sigma HH}}{\partial \psi} \cdot \overline{r}_{1H}\right) = 0; \tag{10.194}$$

$$\widetilde{M}_{\Sigma HH} = \widetilde{M}_{2H} \cdot \widetilde{M}_{H} \cdot \widetilde{M}_{H1}. \tag{10.195}$$

Taking into account Eqs. (10.190), (10.191), (10.192), (10.193), (10.194), and (10.195), based on the account of the components of the first order of smallness, we will look for the required equations as follows: convert the matrix $\widetilde{M}_{\Sigma}$ to the form:

$$\begin{aligned}\widetilde{M}_{\Sigma} = \; & \widetilde{M}_{\Sigma H0} + \widetilde{M}_{\Sigma H0} \cdot \Delta\widetilde{M}_{\Delta 1} + \widetilde{M}_{2H} \cdot \Delta\widetilde{M}_{\Delta H} \cdot \widetilde{M}_{H} \cdot \widetilde{M}_{H1} + \Delta\widetilde{M}_{K1} \\ & + \widetilde{M}_{\Sigma H0} \cdot \Delta\widetilde{M}_{K2},\end{aligned} \tag{10.196}$$

In Eq. (10.196) $\Delta\widetilde{M}_{\Delta i} = \widetilde{M}_{0i} - \widetilde{E}$, where $\widetilde{E}$ is a single matrix.

Then, the expression for $\overline{r}_{2\Phi}$ will look like

$$\begin{aligned}\overline{r}_{2\Phi} = \; & \overline{r}_{2H} + \widetilde{M}_{\Sigma H0} \cdot \Delta\widetilde{M}_{\Delta 1} \cdot \overline{r}_{1H} + \widetilde{M}_{2H} \cdot \Delta\widetilde{M}_{\Delta H} \cdot \widetilde{M}_{H} \cdot \widetilde{M}_{H1} \cdot \overline{r}_{1H} + \widetilde{M}_{\Sigma H0}(\Delta\overline{r}_{1i} + \Delta\overline{r}_{1K}) + \\ & + \Delta\widetilde{M}_{K1} \cdot \overline{r}_{1H} + \widetilde{M}_{\Sigma H0} \cdot \Delta\widetilde{M}_{K2} \cdot \overline{r}_{1H} + \Delta\widetilde{M}_{\Delta 2} \cdot \overline{r}_{2H} - \Delta\overline{r}_{2YT} + \Delta\overline{r}_{2K}.\end{aligned} \tag{10.197}$$

The equations of relations after transformations can be reduced to the form:

$$\begin{aligned}& \frac{\partial \overline{r}_{2H}}{\partial u} \times \overline{N}_1 \cdot \frac{\partial \overline{r}_{2H}}{\partial \phi_1} + \overline{M}_1 \times \frac{\partial \overline{r}_{2H}}{\partial v} \cdot \frac{\partial \overline{r}_{2H}}{\partial \phi_1} + \overline{M}_1 \times \overline{N}_1 \cdot \frac{\partial \overline{r}_{2H}}{\partial \phi_1} + \frac{\partial \overline{r}_{2H}}{\partial u} \times \frac{\partial \overline{r}_{2H}}{\partial v} \cdot \overline{K}_1 + \\ & + \frac{\partial \overline{r}_{2H}}{\partial u} \times \overline{N}_1 \cdot \overline{K}_1 + \overline{M}_1 \times \frac{\partial \overline{r}_{2H}}{\partial v} \cdot \overline{K}_1 + \overline{M}_1 \times \overline{N}_1 \cdot \overline{K}_1 = 0;\end{aligned} \tag{10.198}$$

$$\frac{\partial \overline{r}_{2H}}{\partial u} \times \overline{N}_1 \cdot \frac{\partial \overline{r}_{2H}}{\partial \psi} + \overline{M}_1 \times \frac{\partial \overline{r}_{2H}}{\partial v} \cdot \frac{\partial \overline{r}_{2H}}{\partial \psi} + \overline{M}_1 \times \overline{N}_1 \cdot \frac{\partial \overline{r}_{2H}}{\partial \psi} + \frac{\partial \overline{r}_{2H}}{\partial u} \times \frac{\partial \overline{r}_{2H}}{\partial v} \cdot \overline{L}_1 +$$
$$+ \frac{\partial \overline{r}_{2H}}{\partial u} \times \overline{N}_1 \cdot \overline{L}_1 + \overline{M}_1 \times \frac{\partial \overline{r}_{2H}}{\partial v} \cdot \overline{L}_1 + \overline{M}_1 \times \overline{N}_1 \cdot \overline{L}_1 = 0. \tag{10.199}$$

where the dependencies for calculating $\overline{K}_1, \overline{L}_1, \overline{M}_1$, и $\overline{N}_1$ are in the following forms:

$$\overline{K}_1 = \widetilde{M}_{\Sigma H0} \cdot \frac{\partial \Delta \widetilde{M}_{\Delta 1}}{\partial \phi_1} \cdot \overline{r}_{1H} + \frac{\partial \widetilde{M}_{\Sigma H0}}{\partial \phi_1} \cdot \Delta \widetilde{M}_{\Delta 1} \cdot \overline{r}_{1H} +$$
$$+ \widetilde{M}_{2H} \cdot \Delta \widetilde{M}_{\Delta H} \cdot \frac{\partial \left(\widetilde{M}_H \cdot \widetilde{M}_{H1} \cdot \overline{r}_{1H}\right)}{\partial \phi_1} + \widetilde{M}_{2H} \cdot \frac{\partial \Delta \widetilde{M}_{\Delta H}}{\partial \phi_1} \cdot \widetilde{M}_H \cdot \widetilde{M}_{H1} \cdot \overline{r}_{1H} +$$
$$+ \frac{\partial \widetilde{M}_{2H}}{\partial \phi_1} \cdot \Delta \widetilde{M}_{\Delta 1} \cdot \widetilde{M}_H \cdot \widetilde{M}_{H1} \cdot \overline{r}_{1H} + \frac{\partial \widetilde{M}_{\Sigma H0}}{\partial \phi_1} \cdot \Delta \overline{r}_{1i} + \frac{\partial \widetilde{M}_{\Sigma H0}}{\partial \phi_1} \cdot \Delta \overline{r}_{1K} + \widetilde{M}_{\Sigma H0} \cdot \frac{\partial \Delta \widetilde{M}_{\Delta K2}}{\partial \phi_1} \cdot \overline{r}_{1H} +$$
$$+ \frac{\partial \widetilde{M}_{\Sigma H0}}{\partial \phi_1} \cdot \Delta \widetilde{M}_{\Delta K2} \cdot \overline{r}_{1H} + \frac{\partial \widetilde{M}_{2H}}{\partial \phi_1} \cdot \Delta \widetilde{M}_{\Delta K2} \cdot \widetilde{M}_H \cdot \widetilde{M}_{H1} \cdot \overline{r}_{1H} + \frac{\partial \Delta \widetilde{M}_{\Delta K1}}{\partial \phi_1} \cdot \overline{r}_{1H}; \tag{10.200}$$

$$\overline{L}_1 = \widetilde{M}_{\Sigma H0} \cdot \frac{\partial \Delta \widetilde{M}_{\Delta 1}}{\partial \psi} \cdot \overline{r}_{1H} + \frac{\partial \widetilde{M}_{\Sigma H0}}{\partial \psi} \cdot \Delta \widetilde{M}_{\Delta 1} \cdot \overline{r}_{1H} +$$
$$+ \widetilde{M}_{2H} \cdot \Delta \widetilde{M}_{\Delta H} \cdot \frac{\partial \left(\widetilde{M}_H \cdot \widetilde{M}_{H1} \cdot \overline{r}_{1H}\right)}{\partial \psi} + \widetilde{M}_{2H} \cdot \frac{\partial \Delta \widetilde{M}_{\Delta H}}{\partial \psi} \cdot \widetilde{M}_H \cdot \widetilde{M}_{H1} \cdot \overline{r}_{1H} +$$
$$+ \frac{\partial \widetilde{M}_{2H}}{\partial \psi} \cdot \Delta \widetilde{M}_{\Delta 1} \cdot \widetilde{M}_H \cdot \widetilde{M}_{H1} \cdot \overline{r}_{1H} + \frac{\partial \widetilde{M}_{\Sigma H0}}{\partial \psi} \cdot \Delta \overline{r}_{1i} + \frac{\partial \widetilde{M}_{\Sigma H0}}{\partial \psi} \cdot \Delta \overline{r}_{1K} + \widetilde{M}_{\Sigma H0} \cdot \frac{\partial \Delta \widetilde{M}_{\Delta K2}}{\partial \psi} \cdot \overline{r}_{1H} +$$
$$+ \frac{\partial \widetilde{M}_{\Sigma H0}}{\partial \psi} \cdot \Delta \widetilde{M}_{\Delta K2} \cdot \overline{r}_{1H} + \frac{\partial \widetilde{M}_{2H}}{\partial \psi} \cdot \Delta \widetilde{M}_{\Delta K2} \cdot \widetilde{M}_H \cdot \widetilde{M}_{H1} \cdot \overline{r}_{1H} + \frac{\partial \Delta \widetilde{M}_{\Delta K1}}{\partial \psi} \cdot \overline{r}_{1H}; \tag{10.201}$$

$$\overline{M}_1 = \widetilde{M}_{\Sigma H0} \cdot \Delta \widetilde{M}_{\Delta 1} \cdot \frac{\partial \overline{r}_{1H}}{\partial u} + \widetilde{M}_{2H} \cdot \Delta \widetilde{M}_{\Delta H} \cdot \widetilde{M}_H \cdot \widetilde{M}_{H1} \cdot \frac{\partial \overline{r}_{1H}}{\partial u} + \widetilde{M}_{\Sigma H0} \cdot \frac{\partial \Delta \overline{r}_{1i}}{\partial u} + \widetilde{M}_{\Sigma H0} \cdot \frac{\partial \Delta \overline{r}_{1K}}{\partial u}; \tag{10.202}$$

$$\overline{N}_1 = \widetilde{M}_{\Sigma H0} \cdot \Delta \widetilde{M}_{\Delta 1} \cdot \frac{\partial \overline{r}_{1H}}{\partial v} + \widetilde{M}_{2H} \cdot \Delta \widetilde{M}_{\Delta H} \cdot \widetilde{M}_H \cdot \widetilde{M}_{H1} \cdot \frac{\partial \overline{r}_{1H}}{\partial v} + \widetilde{M}_{\Sigma H0} \cdot \frac{\partial \Delta \overline{r}_{1i}}{\partial v} + \widetilde{M}_{\Sigma H0} \cdot \frac{\partial \Delta \overline{r}_{1K}}{\partial v}. \tag{10.203}$$

In special cases, the equations of connection can be simplified by equating to zero the terms containing the products of derivatives or errors and derivatives. Such cases correspond to the consideration of the conditions for compensation of low-frequency errors, for which small errors and their first derivatives are provided. At high frequencies ($h \geq 3$ and more), these terms can no longer be neglected. Equations (10.197), (10.198), (10.199), (10.200), (10.201), (10.202), and (10.203) provide answers to the main questions of the error compensation theory considered below. In this theory, which underlies adaptive methods of accuracy control, the central task is to determine the structures and elements (parameters) of the matrices $\Delta\widetilde{M}_{K1}$ and $\Delta\widetilde{M}_{K2}$, as well as the increments of the radius vectors $\overline{r}_{1K}$ and $\overline{r}_{2K}$, the known values of other parameters included in Eqs. (10.23), (10.24), (10.25), (10.26), (10.27), (10.28), and (10.29), based on the condition of obtaining the highest accuracy of processing.

Since the highest accuracy corresponds to $\Delta\overline{r}_{1p}=0$, the validity of the equations of relations, the general equations of the theory of compensation of errors can be written in the form

$$\begin{aligned}&\widetilde{M}_{\Sigma H0}\cdot\Delta\widetilde{M}_{\Delta 1}\cdot\overline{r}_{1H}+\widetilde{M}_{2H}\cdot\Delta\widetilde{M}_{\Delta H}\cdot\widetilde{M}_{H}\cdot\widetilde{M}_{H1}\cdot\overline{r}_{1H}+\widetilde{M}_{\Sigma H0}(\Delta\overline{r}_{1i}+\Delta\overline{r}_{1K})+\\&+\Delta\widetilde{M}_{K1}\cdot\overline{r}_{1H}+\widetilde{M}_{\Sigma H0}\cdot\Delta\widetilde{M}_{K2}\cdot\overline{r}_{1H}+\Delta\widetilde{M}_{\Delta 2}\cdot\overline{r}_{2H}-\Delta\overline{r}_{2YT}+\Delta\overline{r}_{2K}\rightarrow 0;\end{aligned} \tag{10.204}$$

$$\begin{aligned}&\frac{\partial\overline{r}_{2H}}{\partial u}\times\overline{N}_1\cdot\frac{\partial\overline{r}_{2H}}{\partial\phi_1}+\overline{M}_1\times\frac{\partial\overline{r}_{2H}}{\partial v}\cdot\frac{\partial\overline{r}_{2H}}{\partial\phi_1}+\overline{M}_1\times\overline{N}_1\cdot\frac{\partial\overline{r}_{2H}}{\partial\phi_1}+\frac{\partial\overline{r}_{2H}}{\partial u}\times\frac{\partial\overline{r}_{2H}}{\partial v}\cdot\overline{K}_1+\\&+\frac{\partial\overline{r}_{2H}}{\partial u}\times\overline{N}_1\cdot\overline{K}_1+\overline{M}_1\times\frac{\partial\overline{r}_{2H}}{\partial v}\cdot\overline{K}_1+\overline{M}_1\times\overline{N}_1\cdot\overline{K}_1=0;\end{aligned} \tag{10.205}$$

$$\begin{aligned}&\frac{\partial\overline{r}_{2H}}{\partial u}\times\overline{N}_1\cdot\frac{\partial\overline{r}_{2H}}{\partial\psi}+\overline{M}_1\times\frac{\partial\overline{r}_{2H}}{\partial v}\cdot\frac{\partial\overline{r}_{2H}}{\partial\psi}+\overline{M}_1\times\overline{N}_1\cdot\frac{\partial\overline{r}_{2H}}{\partial\psi}+\frac{\partial\overline{r}_{2H}}{\partial u}\times\frac{\partial\overline{r}_{2H}}{\partial v}\cdot\overline{L}_1+\\&+\frac{\partial\overline{r}_{2H}}{\partial u}\times\overline{N}_1\cdot\overline{L}_1+\overline{M}_1\times\frac{\partial\overline{r}_{2H}}{\partial v}\cdot\overline{L}_1+\overline{M}_1\times\overline{N}_1\cdot\overline{L}_1=0.\end{aligned} \tag{10.206}$$

On the basis of the received equations, it is possible to implement simple ways of management (compensation) of the errors carried out at the expense of change:

- Movements of the end links of the machine (change of elements of the matrices $\Delta\widetilde{M}_{K1}$ and $\Delta\widetilde{M}_{K2}$)
- Shape and size of the producing surface (change $\Delta\overline{r}_{1K}$)
- Recoverable deformations of a detail (change $\Delta\overline{r}_{2K}$)

With three simple ones, four combined control methods are possible:

- Additional movements of the end parts of the machine + the shape and size of the producing surface
- Additional movements of the end parts of the machine + the restored deformations of a detail
- Shape and dimensions of the producing surface + the recoverable deformations of the part
- Additional movements of the end parts of the machine + the shape and size of recoverable deformations of the part of the producing surface r

In the general case, the elements of the matrices $\Delta\widetilde{M}_{K1}$ and $\Delta\widetilde{M}_{K2}$ and the increments of the radius vectors $\Delta\overline{r}_{1K}$ and $\Delta\overline{r}_{2K}$ can change over time in accordance with changes in the envelope parameters. This case can be called dynamic. In the particular case, the named elements, once installed before processing, may not change. This case determines the static type of compensation method. In general, the concept of type (dynamic or static) covers a method that is multi-circuit due to the need to control the indicators of kinematic accuracy, smoothness, contact spot, and lateral clearance in the transmission. It is obvious that within the method, if it is covered in parts, it is possible that some indicators are controlled on the basis of a dynamic method and others on the basis of a static one. The method, in which dynamic and static controls are applied together, can be called mixed.

Based on this and taking into account the possibility of covering only a part of the parameters, it is possible to distinguish classes, kinds, types, groups, and subgroups of the methods of error compensation, as shown in Table 10.6.

The table indicates the following:

(a) The types of methods: D – dynamic, S – static, M – mixed (combined)
(b) The groups of methods: C – complete, P – partial

Some of the methods listed in Table 10.6 are new. In particular, the methods based on dynamic control of tool size and shape, as well as some combined methods,

Table 10.6 Classification of ways to compensate for errors

Class	Kind	Type	Group
Simple	Control of formation movements	D	P
		S	P/C
		M	P/C
	Gear tooth deformation control	D	P
		S	P/C
		M	P/C
	Dimension control to tool shape	D	P
		S	P/C
		M	P/C
Combine	Control of formation movements + control of deformations of the teeth of a gear	D	P
		S	P/C
		M	P/C
	Control of formation movements + control of the sizes and the form of the tool	D	P
		S	P/C
		M	P/C
	Gear tooth deformation control + tool size and shape control	D	P
		S	P/C
		M	P/C
	Control of formation movements + control of deformations of the teeth of a gear + control of the sizes and the shape of the tool	D	P
		S	P/C
		M	P/C

have not been used before. Let us consider the possibilities of the selected methods of compensation. The condition of optimality of the compensation method based on the introduction of additional movements of the end parts of the machine can be written in the form:

$$\Delta K = \left(\Delta\overline{r}_{1p} - \Delta\overline{r}_2^{(g)}\right) \rightarrow \min\,, \tag{10.207}$$

where $\Delta\overline{r}_2^{(g)}$ is the required amount of profile modification.

It is obvious that the global extremum corresponds to $\Delta K = 0$. However, as will be shown below, such a condition is not always feasible. Based on Eqs. (10.204), (10.205), and (10.206), in general, we can write:

$$\Delta K = f\left(\Delta\widetilde{M}_{\Delta 1}, \Delta\widetilde{M}_{\Delta H}, \Delta\widetilde{M}_{\Delta 2}, \Delta\overline{r}_{1i}, \Delta\overline{r}_{2YT}, \Delta\overline{r}_2^{(g)} + \Delta\widetilde{M}_{K1}\cdot\overline{r}_{1H} + \widetilde{M}_{\Sigma H0}\cdot\Delta\widetilde{M}_{K2}\cdot\overline{r}_{1H}\right) \rightarrow 0; \tag{10.208}$$

$$F_1\left(\Delta\widetilde{M}_{\Delta 1}, \Delta\widetilde{M}_{\Delta H}, \Delta\widetilde{M}_{\Delta 2}, \Delta\overline{r}_{1i}, \Delta\overline{r}_{2YT}, \Delta\overline{r}_2^{(g)}, \Delta\widetilde{M}_{K1}, \Delta\widetilde{M}_{K2}, \frac{\partial\Delta\widetilde{M}_{K1}}{\partial\phi_1}, \frac{\partial\Delta\widetilde{M}_{K2}}{\partial\phi_1}\right) = 0; \tag{10.209}$$

$$F_2\left(\Delta\widetilde{M}_{\Delta 1}, \Delta\widetilde{M}_{\Delta H}, \Delta\widetilde{M}_{\Delta 2}, \Delta\overline{r}_{1i}, \Delta\overline{r}_{2YT}, \Delta\overline{r}_2^{(g)}, \Delta\widetilde{M}_{K1}, \Delta\widetilde{M}_{K2}, \frac{\partial\Delta\widetilde{M}_{K1}}{\partial\psi}, \frac{\partial\Delta\widetilde{M}_{K2}}{\partial\psi}\right) = 0. \tag{10.210}$$

Each of the matrices $\Delta\widetilde{M}_{K1}$,$\Delta\widetilde{M}_{K2}$ can generally contain six unknown elements. At the same time, Eq. (10.208) gives three scalar equations and Eqs. (10.209) and (10.210) give one. Therefore, the problem does not have a single solution.

If we consider that the elements of the matrices $\Delta\widetilde{M}_{K1}$ and (or) $\Delta\widetilde{M}_{K2}$ are functions of the two envelop parameters ϕ_1 and ψ, then we can establish that, if Eqs. (10.208), (10.209), and (10.210) are valid, which corresponds to the nature of the point of contact between the tool surface and the workpiece, three linearly independent components of the movements of compensation mechanisms are necessary for an unambiguous solution of the compensation problem. Such motions can be, for example, motions along any three mutually perpendicular coordinate axes. This is the case with the simplest compensation mechanism with three degrees of freedom (generalized controlled coordinates). An example of such a case is shown in Fig. 10.46, which illustrates a diagram of grinding cylindrical gear 1 with barrel-shaped teeth by disc wheel 2, for example, on a machine equipped with mechanism 3 of error compensation.

The figure shows: S – sensor and C – converter. The mechanism (Fig. 10.46) provides small controlled movements along three mutually perpendicular axes: one vertical and two horizontal, one of which is directed along the axis of the gear. Obviously, using additional movements from this mechanism, it is possible to

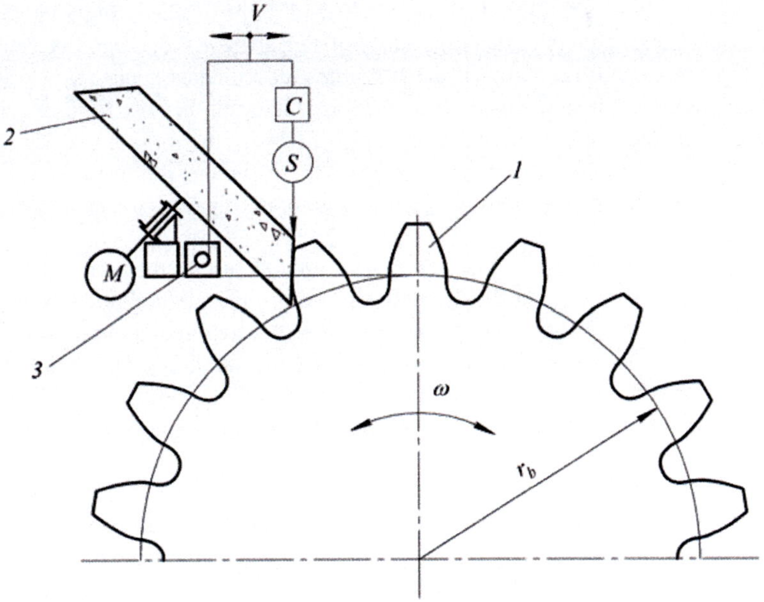

Fig. 10.46 Error control scheme for grinding gears with longitudinal crowning

provide any tooth profile (within the feasibility of an MFTP system, in general, and the conditions of formation [28], in particular).

For the special case corresponding to the machining of gears with straight teeth and the setting of the machine for machining with zero machine engagement, due to the possibility of moving the tooth surfaces to a flat producing surface parallel to themselves, the compensation mechanism can be simplified. In this particular case, only one controlled coordinate is required for full compensation – movement along the general normal of the tool and the gear being machined.

In the case of shaping with contact along a line corresponding to the transformation into the identity of one of the equations, for example, (10.210), we have only four equations in total to determine the compensatory motions. The fifth equation connecting the two curvilinear coordinates of the surface is missing. Therefore, taking into account that the components of the motions included in the matrices $\Delta\widetilde{M}_{K1}$ and (or) $\Delta\widetilde{M}_{K2}$ are functions of the envelope parameters ϕ_1 and ψ, and do not depend on u and v, it can be established that a full compensation of all component errors is impossible. Uncompensated errors are those of the producing surface, which are functions of u and v. In this case, the compensation problem is solved as an optimization.

When forming contact on the surface, both coupling Eqs. (10.209) and (10.210) become identities. Therefore, we have only three equations (scalar), on the basis of which we can determine the required parameters of compensatory motions. In such shaping, the envelope parameters are absent and, therefore, there are no equations that would link the equations of the generating surface to the envelope parameters.

Therefore, only a partial compensation of errors is possible during shaping with contact on the surface. Thus, on the basis of Eq. (10.208), one can determine the setting parameters as functions of the curvilinear coordinates u and v.

It is obvious that both linear and surface contact of the tool surface with the workpiece increasing the number of free parameters in the motion matrices $\Delta\widetilde{M}_{K1}$ and $\Delta\widetilde{M}_{K2}$ up to the maximum number of six increases the compensation possibilities, as it expands the possibilities of the best approximation of function (10.208) to zero.

Similarly, it can be shown that, regardless of the nature of the contact of the tool surface with the workpiece, in the case of two-profile machining, only a partial compensation of errors is possible. Tool surface errors are uncompensated. That is, it can be argued that with an increasing level of concentration of the technological impact on each tooth of the gear, the possibility of an error compensation decreases.

This refers to the methods of dynamic full error compensation. These methods require the use of special adaptive monitoring and control systems.

Most gear machines are not currently equipped with such systems. Therefore, from a practical point of view, it is interesting to identify ways to improve accuracy by compensating for movements or other techniques without the use of special adaptive systems. These methods are of the static type and provide only a partial compensation of the individual components of errors. Let us consider examples of such methods.

Example 1 Improving the accuracy of gears by eliminating the influence of the shape of the tool on the shape of the treated surface. As it follows from the general equations of the real profiles of the teeth (10.23), (10.24), (10.25), (10.26), (10.27), (10.28), and (10.29), $\Delta\overline{r}_2 = f(\Delta\overline{r}_{1i}) = \text{const}$ at $\overline{r}_{1H} = \overline{r}_{1j} = \text{const}$. Therefore, under this condition, the shape of the resulting envelope does not depend on the shape of the tool surface and, while maintaining this shape within the machining cycle, the errors of the tool will not affect the errors of the machined gears. The condition of maintaining the shape of the tool surface within the machining cycle can be ensured with high accuracy using high-strength tool materials, such as superhard synthetics, or using tool surfaces that allow high-speed travel parallel to themselves (e.g., disc wheels). In this case, the condition $\overline{r}_{1H} = \overline{r}_{1j} = \text{const}$ means that the entire surface of each tooth of the gear must be machined by the same point of the tool surface. The specified conditions at the time of processing of involute gears are realized on the machines working using a disk tool with zero angle of a profile. At the same time, an increase in accuracy in comparison with processing by a circle $\alpha_u \neq 0$ at 0.5–1 degree according to the norms of smoothness is provided.

Example 2 Improving the accuracy of gears by introducing time constant radial displacements of the tool (control of the movements of the end parts of the machine). The scheme of the method is shown in Fig. 10.47.

Gear 2 is machined by hob 1 mounted eccentrically with eccentricity e_{Φ} and phase β_{Φ}. The method is based on the creation due to the eccentric installation of the cutter such increments of lines of action, which compensate for the individual

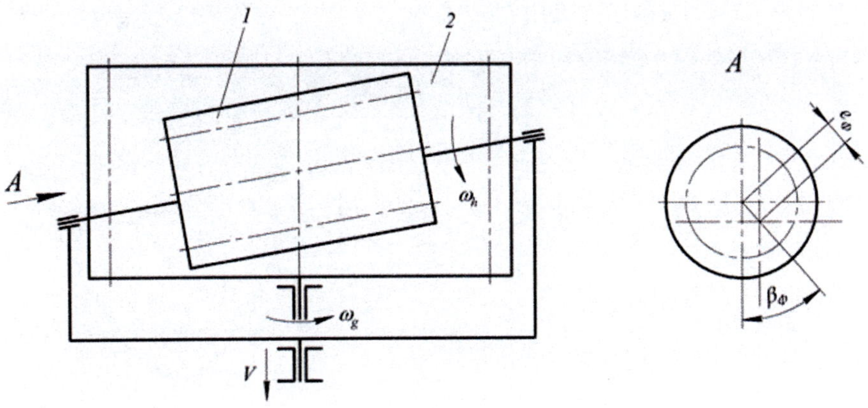

Fig. 10.47 The scheme of a milling gear's eccentrically installed hob

components of the total error. The possibilities of the method are derived from the following: the function of the total error of the gear on the increment of the line of action during machining without its displacement can be expressed by the Fourier equation:

$$\Delta F_{0\delta} = a_0 + \sum_{i=1}^{n} a_i \sin(i\phi_i + \beta_i). \tag{10.211}$$

Improving accuracy is possible if at least one harmonic component of the *i*th frequency is provided due to the tool offset:

$$\Delta F_u = e_\Phi \sin(i\phi_1 + \beta_\Phi). \tag{10.212}$$

Functions of form (10.212) can be obtained for many rotating tools with a nonzero profile angle by providing the tools with appropriate eccentricities. When gear milling with a worm cutter, the two-profile error can be decomposed into a Fourier series of type (10.211).

From the general error (Fig. 10.48), we can distinguish the component of the tooth frequency (curve "a"), which, in almost all cases, when using machines that meet the standards of kinematic accuracy, is dominant due to errors in the manufacture and installation of the cutter.

The component of the tooth frequency error is described by the expression:

$$\Delta H_z = a_z \sin\left(z\phi_1 + \beta_z\right).$$

If the cutter is given a radial offset with eccentricity phase β_i (Fig. 10.47), then this offset will cause radial increments of the lines of action of the machined gear equal to

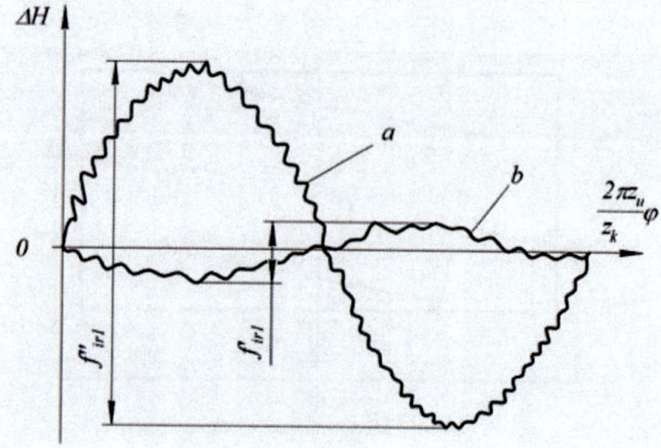

Fig. 10.48 Error diagram

$$\Delta H_{\Phi} = e \sin (z\phi_1 + \beta_i).$$

Under the condition $e = a_z$ and $\beta_i = \pi + \beta_z$, the error of the tooth frequency will be compensated, as in this case

$$\Delta H_{\Sigma} = \Delta H_z + \Delta H_{\Phi} = a_z \sin \left(z\phi_1 + \beta_z\right) + e \sin (z\phi_1 + \beta_i).$$

Similarly, when gear shaping, by displacing the tool in the radial direction, it is possible to change the increments of the lines of action of the machined gear according to the law:

$$\Delta F_{\mathrm{L(R)}} = \pm e_u \sin \left(\frac{z_k}{z_u}\phi_1 \mp \alpha + \beta_u\right),$$

where

e_u is the eccentricity (displacement) of the gear shaper cutter.
z_u is the number of teeth of the tool.
z_k is the number of teeth of the processed gear.
β_u is the eccentricity phase.

By varying the parameters e_u, z_u, and β_u within the allowable values, it is possible to compensate for the various components of the total error of the gear. For example, for $z_u = z_k$, we get the ability to compensate for the geometric eccentricity of the gear. With single-profile processing in this case, a compensation of kinematic eccentricity is also possible, as it is known [11, 36] that kinematic eccentricity causes increments of the lines of action with the period 2π.

Thus, the general method of compensating for errors by introducing time constant displacements of the tool is as follows: it is necessary to identify the frequency of errors that can be caused by tool misalignment. Then, from the total error, the component of the selected frequency is selected by setting its amplitude and phase. Thereafter, the instrument is provided with an eccentricity equal in magnitude to the amplitude of the selected component and opposite in phase. For the practical implementation of the above, the tools must be installed on the mandrels (devices) that provide precise continuous control of the magnitude of the eccentricity and its phase. The achieved increase in accuracy when milling gears of accuracy class "B" is equal to the ratio $\frac{f''_{ir1}}{f'''_{ir1}} = 1{,}5...2{,}0$, which corresponds to an increase in accuracy by 2–3 degrees in terms of smoothness of the transmission.

Example 3 Improving the accuracy of gears by compensating for the high-frequency components of the oscillations of the running speed. This method is feasible when processing gears by methods of incomplete (partial) [22] shaping, for example, with a blade tool. For the first time, it was proposed for gear milling in the study by Khlus [11]. It can be applied to operations of gearing, gear planing of spur gears, and others carried out by running in with interruption of the process. The essence of the method is that the profiling cuts are carried out with a frequency equal to or the multiple of the frequency of high-frequency fluctuations of the running speed. Frequency control is performed by selecting the number of profiling edges (chip grooves) of the cutter [11], machining modes (number of double strokes – when gnashing and planing), and other factors.

Figure 10.49 [11] illustrates a graph of the tool rotation angle ϕ_u depending on the angle of rotation of the part ϕ_2 in the presence of high-frequency oscillations of the run-in speed (curve "a").

In the case $f_p = f_v$ of frequency cuts (in the figure, the cuts are marked by straight vertical segments), they are located on the curve "a" and at the same time on the line "b," parallel to the line "c," representing the nominal dependence of rotation angles $\phi_u = i\phi_2 + c_1$, where $c_1 = \text{const}$. It is obvious that due to the roundness of the gear, the shift of the straight line along the axis $O\phi_u$ does not affect the cyclic error.

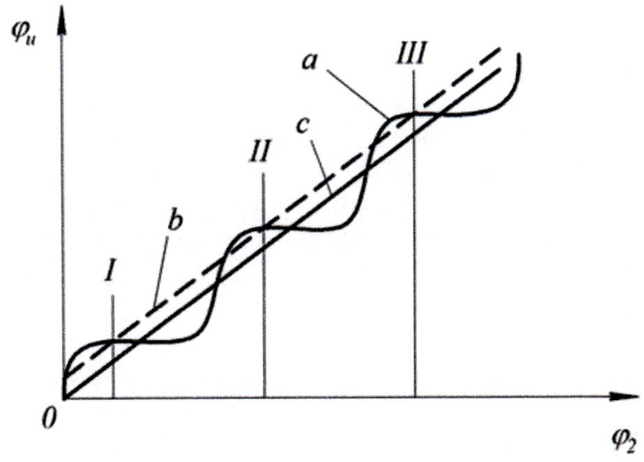

Fig. 10.49 Dependency chart $\phi_u = f(\phi_2)$ with indication of cuts

For the scheme of technological influence with a class $K = 1 \cdot 2 \cdot 1 \cdot 1 \cdot 1$ (realized, for example, by gear milling by a worm mill), in this case, a full compensation of the high-frequency components of the fluctuations of speed of run-in of one frequency is provided. For other schemes, such as $K = 2 \cdot 2 \cdot 1 \cdot 1 \cdot 1 \cdot 1$ of a gearing class, only a partial compensation will be provided, as the profiling cuts have a length along the angle of rotation of the gear (Sections 1-1, 2-2, ..., II-II in Fig. 10.49).

The optimal condition of the compensation method based on the control of the shape and size of the producing surface is described, as in the first case, by dependence (10.207). However, in this case, due to the fact that changes in the size and shape of the generating surface can be a function of the curvilinear coordinates u and v, based on Eqs. (10.204), (10.205), and (10.206), it can be argued that a full compensation of any type of initial error is possible, regardless of the level of concentration of the technological impact.

In this case, the equations that allow determining the required changes in the shape and size of the producing surface have the forms:

$$\widetilde{M}_{\Sigma H0} \cdot \Delta\widetilde{M}_{\Delta 1} \cdot \overline{r}_{1H} + \widetilde{M}_{2H} \cdot \Delta\widetilde{M}_{\Delta H} \cdot \widetilde{M}_{H} \cdot \widetilde{M}_{H1} \cdot \overline{r}_{1H} + \widetilde{M}_{\Sigma H0}(\Delta\overline{r}_{1i} + \Delta\overline{r}_{1K})$$
$$+ \Delta\widetilde{M}_{\Delta 2} \cdot \overline{r}_{2H} - \Delta\overline{r}_{2YT} \to 0.$$

$$\frac{\partial\overline{r}_{2H}}{\partial u} \times \overline{N}_1^* \cdot \frac{\partial\overline{r}_{2H}}{\partial\phi_1} + \overline{M}_1^* \times \frac{\partial\overline{r}_{2H}}{\partial v} \cdot \frac{\partial\overline{r}_{2H}}{\partial\phi_1} + \overline{M}_1^* \times \overline{N}_1^* \cdot \frac{\partial\overline{r}_{2H}}{\partial\phi_1} + \frac{\partial\overline{r}_{2H}}{\partial u} \times \frac{\partial\overline{r}_{2H}}{\partial v} \cdot \overline{K}_1^* +$$
$$+ \frac{\partial\overline{r}_{2H}}{\partial u} \times \overline{N}_1^* \cdot \overline{K}_1^* + \overline{M}_1^* \times \frac{\partial\overline{r}_{2H}}{\partial v} \cdot \overline{K}_1^* + \overline{M}_1^* \times \overline{N}_1^* \cdot \overline{K}_1^* = 0;$$
$$\frac{\partial\overline{r}_{2H}}{\partial u} \times \overline{N}_1^* \cdot \frac{\partial\overline{r}_{2H}}{\partial\psi} + \overline{M}_1^* \times \frac{\partial\overline{r}_{2H}}{\partial v} \cdot \frac{\partial\overline{r}_{2H}}{\partial\psi} + \overline{M}_1^* \times \overline{N}_1^* \cdot \frac{\partial\overline{r}_{2H}}{\partial\psi} + \frac{\partial\overline{r}_{2H}}{\partial u} \times \frac{\partial\overline{r}_{2H}}{\partial v} \cdot \overline{L}_1^* +$$
$$+ \frac{\partial\overline{r}_{2H}}{\partial u} \times \overline{N}_1^* \cdot \overline{L}_1^* + \overline{M}_1^* \times \frac{\partial\overline{r}_{2H}}{\partial v} \cdot \overline{L}_1^* + \overline{M}_1^* \times \overline{N}_1^* \cdot \overline{L}_1^* = 0.$$

where $\overline{K}_1^*$, $\overline{L}_1^*$, $\overline{M}_1^*$, и $\overline{N}_1^*$ are values of $\overline{K}_1$, $\overline{L}_1$, $\overline{M}_1$, и $\overline{N}_1$, calculated considering $\Delta\widetilde{M}_{K1} = \Delta\widetilde{M}_{K2} = 0$ и $\overline{r}_{2K} = 0$.

A theoretically predictable possibility of a full compensation of any type of error due to control of the size and the form of a producing surface till now does not have sufficiently developed technical maintenance. Additional work in this direction is required. Methods are known to compensate for gear errors by controlling the shape and size of the tool surface, for example, using corrected sharpening of cutters, changing the profile angle of gear tools, and other parameters. However, devices for a full dynamic error compensation have not been described in the literature.

The creation of methods that would provide a full compensation for errors has recently been facilitated by the new research works on changes in the geometry of various materials based on the phenomena of the "memory" of metals and alloys, "giant magnetostriction," and others. The third of the considered simple ways consists of giving to teeth of a gear of the restored deformations equal in size and return on a sign to the arising errors.

Since the real surface of each tooth of a machined gear is two-parameter, due to the fact that the radius vector $\overline{r}_{2K}$ receives the necessary increments, it is obvious that it is possible to provide full compensation for any initial errors.

The system of equations obtained from (10.204) to (10.206) in this case has the forms:

$$\widetilde{M}_{\Sigma H0} \cdot \Delta\widetilde{M}_{\Delta 1} \cdot \overline{r}_{1H} + \widetilde{M}_{2H} \cdot \Delta\widetilde{M}_{\Delta H} \cdot \widetilde{M}_{H} \cdot \widetilde{M}_{H1} \cdot \overline{r}_{1H} + \widetilde{M}_{\Sigma H0} \Delta\overline{r}_{1i} + \Delta\widetilde{M}_{\Delta 2} \cdot \overline{r}_{2H}$$
$$- \Delta\overline{r}_{2YT} + \overline{r}_{2H} = 0.$$

$$\frac{\partial\overline{r}_{2H}}{\partial u} \times \overline{N}_1^{**} \cdot \frac{\partial\overline{r}_{2H}}{\partial\phi_1} + \overline{M}_1^{**} \times \frac{\partial\overline{r}_{2H}}{\partial v} \cdot \frac{\partial\overline{r}_{2H}}{\partial\phi_1} + \overline{M}_1^{**} \times \overline{N}_1^{**} \cdot \frac{\partial\overline{r}_{2H}}{\partial\phi_1} + \frac{\partial\overline{r}_{2H}}{\partial u} \times \frac{\partial\overline{r}_{2H}}{\partial v} \cdot \overline{K}_1^{**} +$$
$$+ \frac{\partial\overline{r}_{2H}}{\partial u} \times \overline{N}_1^{**} \cdot \overline{K}_1^{**} + \overline{M}_1^{**} \times \frac{\partial\overline{r}_{2H}}{\partial v} \cdot \overline{K}_1^{**} + \overline{M}_1^{**} \times \overline{N}_1^{**} \cdot \overline{K}_1^{**} = 0;$$

$$\frac{\partial\overline{r}_{2H}}{\partial u} \times \overline{N}_1^{**} \cdot \frac{\partial\overline{r}_{2H}}{\partial\psi} + \overline{M}_1^{**} \times \frac{\partial\overline{r}_{2H}}{\partial v} \cdot \frac{\partial\overline{r}_{2H}}{\partial\psi} + \overline{M}_1^{**} \times \overline{N}_1^{**} \cdot \frac{\partial\overline{r}_{2H}}{\partial\psi} + \frac{\partial\overline{r}_{2H}}{\partial u} \times \frac{\partial\overline{r}_{2H}}{\partial v} \cdot \overline{L}_1^{**} +$$
$$+ \frac{\partial\overline{r}_{2H}}{\partial u} \times \overline{N}_1^{**} \cdot \overline{L}_1^{**} + \overline{M}_1^{**} \times \frac{\partial\overline{r}_{2H}}{\partial v} \cdot \overline{L}_1^{**} + \overline{M}_1^{**} \times \overline{N}_1^{**} \cdot \overline{L}_1^{**} = 0.$$

The technical implementation of this method by the existing technical means in full is still impossible due to the peculiarities of the physical and mechanical properties of the materials from which the gears are made: steel, plastics, and others. However, a partial error compensation based on the control of the recoverable deformations of the part is possible, for example, by modifying the teeth of the blanks for shaving, in order to equalize the elastic deformations. This determines the need for further research into the field of technical means that would allow the full implementation of the last two ways of controlling the accuracy of gears at the level of the aisle.

Obviously, each combined method of accuracy control has the capabilities of the simple methods included in it. Therefore, there is no need to consider each of them separately. Taking into account the fact that a number of the considered methods can provide a theoretically complete compensation of errors, the maximum realized reserves of an increase in accuracy of processing of gears are established. According to (10.204), these reserves can no longer be limited by the whole set of sources of MFTP errors but by errors in testing control actions. The latter are performed by special monitoring and regulating devices, for example, similar to those described in the study by Ehrlenspiel et al. [4]. Therefore, the maximum achievable accuracy of the gears will be equal to the accuracy of these devices. The accuracy of tracking and adjusting devices is determined by errors of measuring subsystems and errors of the mechanisms of small movements. It is at the level of tenths of a micrometer.

The results obtained above on the study of multilevel accuracy control allow proceeding to the development of the principles of optimizing the accuracy of gears.

10.4.6 The Principles of Optimization of the Accuracy of Gears

1. The principles of optimization of the specified accuracy of gears come from the following. In accordance with the presented results, the fulfillment of conditions (10.1) entails the limitation of the structures, parameters, and laws of change of the producing subsystem. Therefore, the problem of optimization of accuracy can be solved only within the solution of the general problem of complex optimization of technological systems.
2. Since accuracy management is purposeful, it is rational to identify and rank the factors that have a significant impact on the achievement of the goal under conditions (10.1).
3. In order to formalize the solution process and implement it on a computer, it is necessary to discretize the levels of the change of parameters.
4. Restrictive methods of ensuring accuracy, providing static control of the structures and parameters, as a rule, do not require the inclusion of additional subsystems in an MFTP system. Therefore, due to the simplicity and availability of implementation, static control methods are preferable to dynamic ones. However, in view of ensuring high accuracy, dynamic methods are more effective.
5. Since the area of the base surfaces of gears, in most cases, is less than the total area of the teeth and their configuration is simpler in multi-operational processes, it is advisable to introduce operations to restore the base surfaces that affect $\Delta\widetilde{M}_{\Delta 2}$.

The control of gear errors differs in terms of purpose, level of implementation, control variables, stage and nature of implementation, degree of completeness, the principle of generating control action, and the law of its change over time. The total management is multilevel. Technological methods of controlling the accuracy of machining gears can be implemented at the level of the technological process as a whole, operation, transition, and stroke (passage). Many ways to control accuracy at each level are closed. The possibilities and areas of rational application of methods are various. In general, accuracy can be controlled by changing the structure and parameters (including input) at all levels. Adaptive-compensatory control by changing the parameters can be carried out at the level of the working stroke. There are seven ways to adaptively compensate for accuracy control, including three simple ones that involve changing the movements of the MFTP end points, tool sizes and shapes, and recoverable gear deformations. The required parameters of compensatory effects can be determined on the basis of the solution of the general equations provided in this section. When using the method based on changing the movements of the end links of an MFTP system, a full error compensation is possible only in the case of point contact in the "tool–workpiece" pair and single-profile processing. When using other methods, a full compensation of errors is possible, regardless of the level of concentration of the technological impact. However, devices for the technical implementation of these methods in full have not yet been developed. An extremely high accuracy can be achieved by methods that theoretically provide a complete compensation of errors and is determined by the accuracy of measuring instruments and the accuracy of the mechanisms of small movements. Theoretical principles have been developed to solve the problems of optimization accuracy. Their implementation is possible within the framework of solving the general problem of complex optimization of a technological system.

10.5 Conclusion

The general laws of the processes of real shaping can be reflected by the common-for-all methods of machining equations of the real profiles of teeth, containing a closed set of linearly independent, but modularly correlated, given primary errors.

Based on these equations, new forecasting methods with the required reliability of error distribution ranges and necessary for adaptive accuracy of the actual values of deviations of real points of profiles from nominal ones are considered, the problem of an MFTP system's error identification is considered and solved, and ways of systematization of research in the field of accuracy of gears of various, including new, kinds are planned.

Many ways to control the accuracy of gears, which are distinguishable by the methods of obtaining information, include three simple and four combined methods. The simple ones include direct, element-by-element, and factorial methods. Combined methods are combinations of simple ones. Element-by-element, factorial, and combined methods can provide information directly at the time of shaping, which provides an opportunity to build highly efficient adaptive control systems for the accuracy of dental processing. On the basis of the combined methods of control, it is possible to build information-redundant systems that allow not only an increase in the accuracy of control but also in the control themselves, to ensure a given reliability of the measurement results.

The control of gear errors differs in terms of purpose, level of implementation, control variables, stage and nature of implementation, degree of completeness, the principle of generating control effects, and the law of its change over time. Total management is multilevel. Many ways to control accuracy at each level are closed. The possibilities and areas of rational application of methods are various. A number of methods can theoretically provide a complete compensation for errors. Such methods include control of the size and shape of the tool and control of the recoverable deformations of the part. In special cases, when a single-profile machining with point contact of the tool and the workpiece is used, a method based on the control of the movements of the final links of an MFTP system can also provide a full compensation.

An extremely high accuracy can be achieved by methods that theoretically provide a complete compensation of errors and is determined by the accuracy of measuring instruments and the accuracy of small displacement mechanisms (at the level of tenths of a micrometer).

References

1. Adler, Y.P., Markova, E.V., Granovsky, Y.V.: Planning of Experiment in Finding the Optimal Conditions. Nauka Publ, Moscow (1976) 284 p. [In Russian]
2. Alikulov, D.E.: Modern Methods of Gear Error Control. NIIMASH Publ, Moscow (1971) 60 p. [In Russian]
3. Bazrov, B.M.: Determination of the total error in part's machining. Bullet. Mech. Eng. **8**, 8–11 (1978) [In Russian]
4. Ehrlenspiel, K., Kiewert, A., Lindemann, U.: Cost-Efficient Design. Springer (2007) 559 p
5. Erikhov, M.L.: On the issue of the synthesis of meshing with point contact. In: Theory of Gears in Machines, pp. 78–91. Mashinostroenie Publ, Moscow (1966) [In Russian]
6. Erikhov, M.L.: Principles of taxonomy, methods of analysis and issues of synthesis of gearing schemes. Doctor. tech. Sciences Thesis. Khabarovsk (1972). [In Russian]
7. Feldbaum, A.A., Butkovsky, A.G.: Methods of the theory of automatic control. – M.: Nauka, 744 p (1971) [In Russian]
8. Gulida, E.N.: Technology of Finishing Operations for Cylindrical Gears Processing. Vysshaya Shkola Publ, Lviv (1977) 167 p. [In Russian]
9. Jeff, C., Hamada, M.: Experiments: Planning, Analysis, and Optimization, 2nd edn. Wiley (2011)
10. Kalashnikov, S.N., Kogan, G.I., Kopf, I.A., et. al.: Production of gears. Handbook / Ed. B.-A. Thaits. – M.: Mashinostroenie, 708 p (1975) [In Russian]
11. Khlus, A.A. Influence of generating speed deviations on the tooth profile of an involute spur gear. – The theory of mechanisms and machines, no. 9, 1970, pp. 129–132. [In Russian]
12. Khlus, A.A., Dudin, V.V., Ternyuk, N.E.: Influence of high-frequency oscillations of generating speed on the accuracy of machined gears. Theory of mechanisms and machines, no. 22, 107–112 (1977) [In Russian]
13. Khlus, A.A., Ternyuk, N.E.: Geometric method for gear errors calculation using arc deviations. – Theory of mechanisms and machines, no. 28 (1980) [In Russian]
14. Korn, G.A., Korn, T.M.: Mathematical Handbook for Scientists and Engineers. Definitions, Theorems, and Formulas for Reference and Review. Dover Publications, Inc, New York (2000) 1152 p
15. Krivchenko, G.I.: Study of the cutting process of cylindrical gears with spiral teeth by the radial feed method (on the example of a food separator drive). Candidate of Technical Sciences Thesis. Krasnodar (1973). [In Russian]
16. Litvin, F.L., Fuentes, A.: Gear Geometry and Applied Theory. Cambridge University Press, New York (2004)
17. Lee, T.H., Adams, G.E., Gaines, W.M.: Computer Process Control: Modeling and Optimization. Wiley (1968)
18. Lishinsky, L.Y.: Optimization of Cutting Mode of Machines and Automatic Lines Based on the Use of Computers. NIIMASH Publ, Moscow (1974) 135 p. [In Russian]
19. Lutsky, S.: Processes and Systems in Technosphere: Discrete-Probabilistic Information Systems. LAP LAMBERT Academic Publishing (2017) 108 p
20. Lutskyy, S.: System-information approach to uncertainty of process and system parameters. Innov. Technol. Sci. Solut. Indus. **3**(17), 91–105 (2021)
21. Perepelitsa, B.A.: Mappings of Affine Space in the Theory of Surface Shaping by Cutting. Vysshaya Shkola Publ, Kharkov (1981) 152 p. [In Russian]
22. Perepelitsa, B.A.: Development of the theory of shaping and design of cutting tools based on multi-parameter mappings. Doc. Tech. Sciences Thesis. Kharkov, 387 p (1982) [In Russian]
23. Ponomarev, V.P.: Optimization of Machining Processes for Carburized Gears. South Ural Book Publishing House, Chelyabinsk (1974) 226 p. [In Russian]
24. Portman, V.T.: Universal method for calculating the accuracy of mechanical devices. Bullet. Mech. Eng. **7**, 12–16 (1981) [In Russian]

25. Pugachev, V.S.: Probability Theory and Mathematical Statistics for Engineers. Pergamon Press (1984) 450 p
26. Radzevich, S.P.: Gear Cutting Tools: Science and Engineering, 2nd edn, p. 637. Taylor & Francis/CRC Press (2018)
27. Radzevich, S.P.: Theory of Gearing: Kinematics, Geometry, and Synthesis, 2nd edn, p. 935. Taylor & Francis/CRC Press (2018)
28. Rodin, P.R.: Fundamentals of Surfaces Shaping by Cutting. Vishcha Shkola Publ, Kyiv (1977) 192 p. [In Russian]
29. Samoilov, S.I., Syromyatnikov, I.S.: Cutting forces in shaving processing of large-module gears. Machines and tools, No. 3, 20–22 (1965) [In Russian]
30. Shaumyan, G.A.: Integrated Automation of Production Processes. M.: Mashinostroenie Publ (1973) 639 p. [In Russian]
31. Silin, S.S.: Similarity method for cutting of materials. – M.: Mashinostroenie, 19710. – 152 p. [In Russian]
32. Solomentsev, Y.M., Mitrofanov, V.G., Protopopov, S.P., et.al.: Adaptive controlling of technological processes (on metal-cutting machines). M.: Mashinostroenie, 536 p (1980) [In Russian]
33. Sukhorukov, Y.N.: Influence of dynamic factors on the accuracy and productivity of abrasive methods for gears finishing by free generating methods. In: Improving the Quality of Production and Manufacturing Technology for General Purpose Gearboxes. – Kyiv: Knowledge, pp. 23–24 (1977) [In Russian]
34. Sukhorukov, Y.N.: Technological bases for finishing of involute spur gears hardened to high hardness using free generating methods. Doct. Tech. Sciences Thesis. Omsk., 432 p (1972) [In Russian]
35. Taits, B.A., Markov, N.N.: Precision and Control of Gears. Mashinostroenie Publ (1978) 136 p. [In Russian]
36. Taits, B.A.: Precision and Control of Gear Wheels. Mashinostroenie Publ (1972) 367 p. [In Russian]
37. Ternyuk, N.E., Dudin, V.V., Khlus, A.A.: Reducing of Gear Errors at Grinding of Base Holes. UkrNIINTI, KhTSNTI, Kharkov (1974) 4 p. [In Russian]
38. Ternyuk, N.E.: Investigation of the main laws of accuracy management in spur gears production. Cand. Tech. Sciences Thesis. Kharkov. 184 p (1975) [In Russian]
39. Ternyuk, N.E., Kotlyarov, B.S., Digtenko, V.G.: Rotary schemes for gears processing. In: Improving the Reliability and Reducing the Metal Consumption of Gears and Gearboxes for General Engineering Applications: Conf. Proceedings, p. 1010, Sevastopol (1983) [In Russian]
40. Ternyuk, N.E., Listopad, N.F., Khlus, A.A.: Influence of MFTP system parameters on the accuracy of gears during shaving process. Theory of mechanisms and machines, no. 19, 115–120 (1975) [In Russian]
41. Ternyuk, N.E.: Conversion of gear errors at two- and three-tool processing methods. Theory of mechanisms and machines, no. 210 (1980). [In Russian]
42. Ternyuk, N.E.: Structural-parametric synthesis of automated technological systems for gears production. In: Improving the Technical Level of Gears of Energy-Saturated Tractors, pp. 54–56, Kharkov (1983) [In Russian]
43. Ternyuk, N.E., Khlus, A.A.: Gear and linkage mechanisms for gears tangential errors detection. In: Theory of Mechanisms and Machines, p. 171. Alma-Ata, Science Publ (1977) [In Russian]
44. Ternyuk, N.E.: Technological processes of gears machining. In: Encyclopedic Handbook in 40 Volumes. Vol. III-3 Manufacturing Technology of Machine Elements, pp. 809–816, Moscow (2000)

Chapter 11
Elliptical Gear Drives

Boris M. Klebanov

11.1 Introduction

Among the noncircular gears intended for gear drives with a cyclically changing gear ratio, elliptical gears are the simplest in analytical interpretation and the most producible. The aim of this chapter is to improve the method of calculating the strength of elliptical gears, published by the author earlier [1].

11.2 Geometry: Basic Equations

In an elliptical gear, the centrode in mesh with the cutter (a hob, a rack, or a shaper) is elliptical. This means that the ellipse is the reference curve (pitch curve) of an elliptical gear, and the gear drive comprised of two such gears can be schematically represented by two identical ellipses. Ellipses 1 and 2 are located as shown in Fig. 11.1a and touch each other at point P, which is the instantaneous center of rotation of one ellipse relative to the other. The ellipses rotate around their fixed foci F_1 and F_2 while touching each other permanently, and, during rotation, the point of their tangency P is always located on the center line F_1F_2. There is no slippage between the ellipses at point P.

The ellipse is characterized by the following dimensions: major axis length $2a$, minor axis length $2b$, and the distance between the foci (points F and F^*) that equals $2c$, where

$$c = \sqrt{a^2 - b^2}.$$

B. M. Klebanov
Aerospace Industry, Israel, USA

S. P. Radzevich, *Novikov/Conformal Gearing*,
https://doi.org/10.1007/978-3-031-10019-2_11

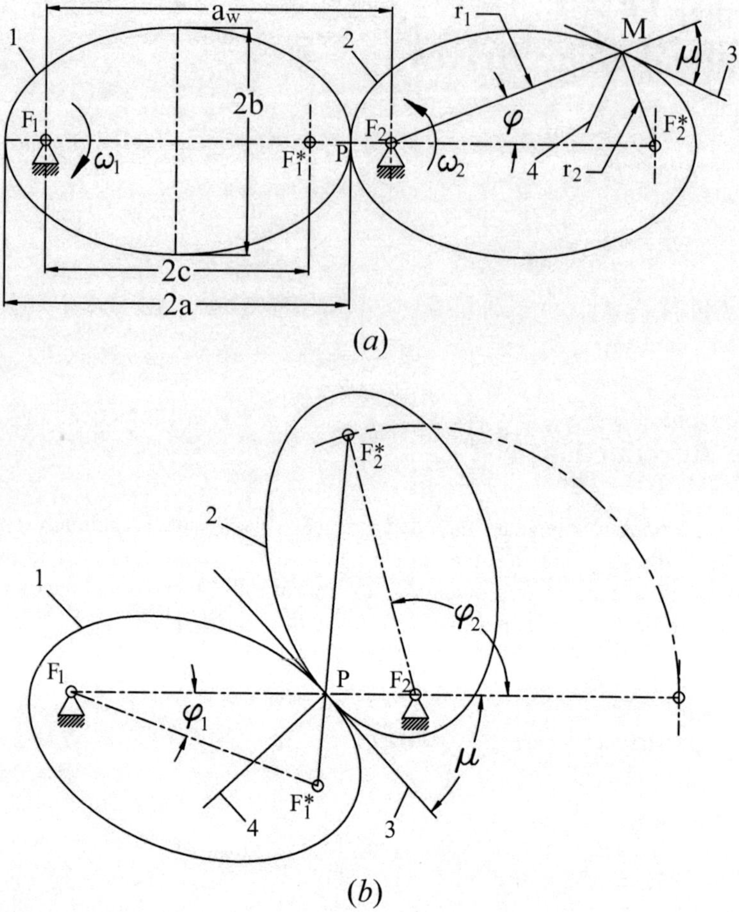

Fig. 11.1 Elliptical gear drive centrodes

Factor $e = {}^{c}\!/_{a}$ is called the "eccentricity of ellipse." The smaller the minor axis as compared with the major axis, the bigger is the focal distance $2c$ and the greater is the eccentricity e of the ellipse. With $c = 0$, $e = 0$, $a = b$, the ellipse turns into a circle.

In polar coordinates, with the center in the focus of the ellipse, the radius vector r_1 of any point M on the ellipse equals [2]:

$$r_1 = \frac{a(1 - e^2)}{1 - e\ \cos\varphi} \tag{11.1}$$

Line 3 that is tangent to the ellipse at point M makes with the extension of radius vector of this point angle μ that is given by

$$\tan\mu = \frac{1 - e\ \cos\varphi}{e\ \sin\varphi} \tag{11.2}$$

As stated earlier, the elliptical gears rotate around the fixed axes passing through the foci points F_1 and F_2 of their reference ellipses, and, so, the center distance $a_w = 2a$.

The instantaneous gear ratio at each angle of rotation equals the ratio of distances between point P and the centers of rotation F_1 and F_2 (see Fig. 11.1b):

$$u = \frac{\omega_2}{\omega_1} = \frac{F_1P}{F_2P}.$$

Since $F_1P = r_1$ can be obtained from Eq. (11.1) for each angle φ_1 and $F_2P = 2a - F_1P$, the instantaneous gear ratio can be calculated for any angular position of the gears.

In the position shown in Fig. 11.1a ($\varphi_1 = 0$), the gear ratio is maximal:

$$u_{\max} = \frac{\omega_{2\max}}{\omega_1} = \frac{a+c}{a-c} = \frac{1+e}{1-e}. \tag{11.3}$$

At $\varphi_1 = 180°$, the gear ratio is minimal:

$$u_{\min} = \frac{\omega_{2\min}}{\omega_1} = \frac{a-c}{a+c} = \frac{1-e}{1+e}. \tag{11.4}$$

The radius of curvature R of the ellipse at point M is determined as follows:

$$R = \frac{(r_1 r_2)^{1.5}}{ab}. \tag{11.5}$$

Here, the radii r_1 and r_2 are the distances from point M to the foci; see Fig. 11.1a: $r_1 = MF_2$ and $r_2 = MF_2^*$. The distance r_1 is obtained from Eq. (11.1) and $r_2 = 2a - r_1$.

Line 3 that is the tangent to the ellipse at point M is obtained as a perpendicular to line 4, which is the bisector of the angle between r_1 and r_2 $\left(\angle F_2MF_2^*\right)$. The angle μ between the extension of the radius vector r_1 of the point M and line 3 is used for determining the working pressure angle α_{wt} (see Sect. 11.3).

Figure 11.1b demonstrates the graphical layout of the elliptical centrodes at any angle of rotation. In this case, centrode 1 is rotated clockwise (CW) from its initial position by angle φ_1 about the fixed focus point F_1. Then, a line is drawn between focus F_1^* and point P, which is the intersection of ellipse 1 with the center line F_1F_2, and ellipse 2 is rotated counter-CW (CCW) about its fixed focus F_2 in such a way that focus F_2^* lies on the extension of line F_1^*P. After that, we draw bisector line 4 and tangent line 3.

From Fig. 11.1b, we obtain several useful parameters:

- Angle φ_2 at given φ_1
- Angle μ at given φ_1
- Instant gear ratio at given φ_1:

$$u_{\varphi 1} = \frac{\omega_2}{\omega_1} = \frac{F_1P}{F_2P}.$$

From Fig. 11.1b, we can also see that the distances from point P to the foci of both ellipses are pairwise equal: $PF_1 = PF_2^*$ and $PF_1^* = PF_2$. Hence, the radii of curvature of both ellipses at point P are equal. In the generating process, the motion of the cutter is determined by the varying radius of curvature of the centrode, and, so, the right and the left profiles of each tooth are slightly different. To simplify the calculations, we will proceed from the assumption that the working profiles of two engaged teeth are identical and that their radii of curvature at point P correspond to the radius of curvature R of the ellipse at this point.

Note: The shape of the teeth flanks is not involute because the centrode is noncircular. The variation of the centrode curvature leads to the formation of a complex surface of the tooth flank that can hardly be expressed by a short mathematical formula. However, there is no fundamental difference with involute gears for which the radius of curvature of the involute profile also varies from the top to the root of the tooth, and only the curvature radii at the pitch point are usually considered in the design formula. We are going to apply the same approach when designing elliptical gear drives.

Based on the generating method of tooth cutting, the following relations can be stated (by analogy with involute gears):

- The nominal tooth thickness measured along the arc of the centrode equals the space width of the basic rack, that is, $0.5\pi m$.
- In an involute gear, the radius of curvature ρ of the tooth profile on the pitch radius $r_\mathrm{p} = 0.5mz$ is given by:

$$\rho = r_\mathrm{p} \sin \alpha_\mathrm{P},$$

where α_P is the profile angle of the basic rack. The same equation can also be used for elliptical gears, but the radius of curvature of the elliptical centrode R (see Eq. (11.5)) shall be substituted for the pitch radius of the round centrode r_p:

$$\rho = R \sin \alpha_\mathrm{P}. \tag{11.6}$$

The magnitude of radius R depends on the location of the tooth defined in the polar coordinates by angle φ. This angle cannot be exactly calculated until the gear data, such as the centrode dimensions a and b, module m, and number of teeth z, are known. Tentatively, for design calculations, the angle φ can be taken equal to, for example, the angle of the rotation of the gear, in which the maximum force is applied to the teeth.

The approach to calculating the profiles of the teeth of noncircular gears can be found in the study by Radzevich [3].

Obviously, the length L of the perimeter of the centrode must satisfy the following equation:

$$L = \pi\, m\, z, \tag{11.7}$$

where m = module and z = number of teeth.

The length L of an ellipse is given by a quickly convergent series:

$$L = \pi(a+b)\left[1 + \frac{\lambda^2}{4} + \frac{\lambda^4}{64} + \frac{\lambda^6}{256} + \frac{25\lambda^8}{16{,}384} + \frac{49\lambda^{10}}{65{,}536} + \frac{441\lambda^{12}}{1{,}048{,}576} + \ldots\right], \tag{11.8}$$

where $\lambda = (a - b)/(a + b)$.

Usually, the first three terms of the series (after 1) provide the deviation from the true value, which can be neglected.

11.3 Kinematics and Dynamics of Elliptical Gear Drives

The equations of motion are slightly different for the half-turn starting from position $u_{\mathbf{max}}$ shown in Figure 11.2a (let us call it "first half-turn") and the half-turn starting from position $u_{\mathbf{min}}$ shown in Figure 11.2b ("second half-turn"). In the calculations, the initial position of the gears in each half-turn is characterized by $\varphi_1 = 0$.

The relationship between φ_2 and φ_1, called the transmission function of an elliptic gear drive, has an analytical expression [4]:

$$\tan\frac{\varphi_2}{2} = C\tan\left(\frac{\varphi_1}{2}\right),$$

or, that is the same as

$$\varphi_2 = 2\ \mathrm{arc\ tan}\left[C\ \tan\left(\frac{\varphi_1}{2}\right)\right], \tag{11.9}$$

where factor $C = \frac{1+e}{1-e}$ for the first half-turn and $C = \frac{1-e}{1+e}$ for the second half-turn.

The angular speed of the driven gear ω_2, provided that the angular speed of the driving gear ω_1 is constant and hence $\varphi_1 = \omega_1 \cdot t$, equals:

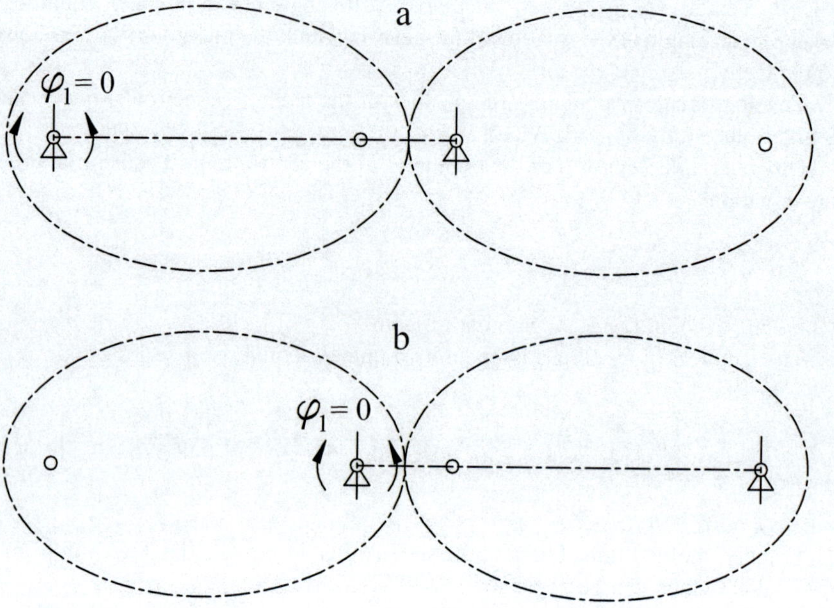

Fig. 11.2 The initial positions of the 1st (**a**) and the 2nd (**b**) half-turns

$$\omega_2 = \dot{\varphi}_2 = \frac{C\,\omega_1}{1 + \left(C^2 - 1\right)\,\sin^2(\varphi_1/2)}. \tag{11.10}$$

The rotational acceleration of the driven gear equals:

$$\varepsilon_2 = \ddot{\varphi}_2 = -\frac{C\,\left(C^2 - 1\right)}{2}\,\frac{\sin\varphi_1}{\left[1 + \left(C^2 - 1\right)\,\sin^2(\varphi_1/2)\right]^2}\,\omega_1^2. \tag{11.11}$$

In Eqs. (11.10) and (11.11), factor C shall be substituted like in Eq. (11.9): separately for the first half-turn and the second half-turn.

Diagrams of φ_2; $\varphi_2 - \varphi_1$; ω_2/ω_1; and ε_2/ω_1^2 as functions of angle φ_1 are presented in Figs. 11.3, 11.4, 11.5, and 11.6, respectively.

Note: In these and subsequent diagrams, two half-turns are combined into one full turn where the interval from 0° to 180° represents the first half-turn and that from 180° to 360° the second half-turn.

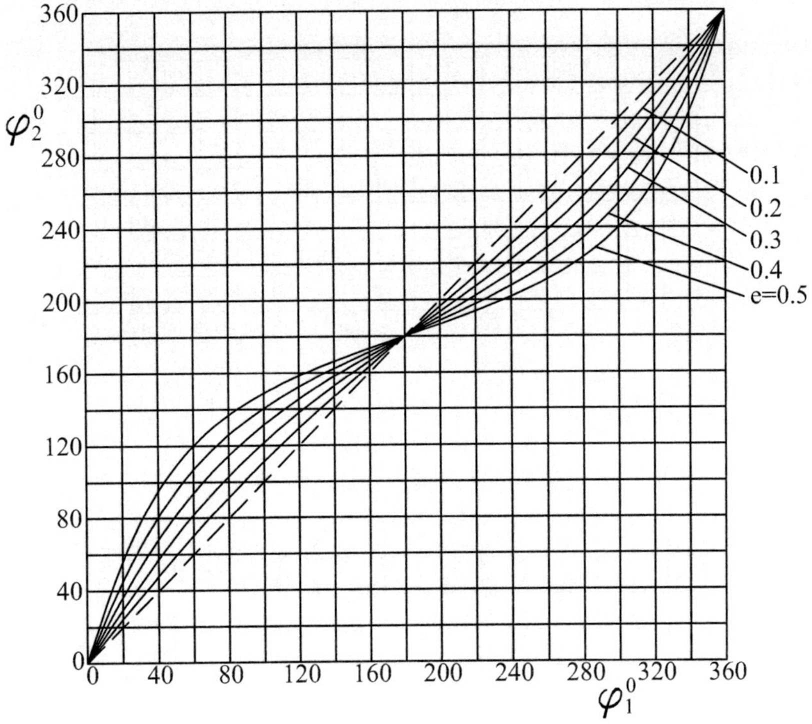

Fig. 11.3 Diagrams of $\varphi_2 = f(\varphi_1)$

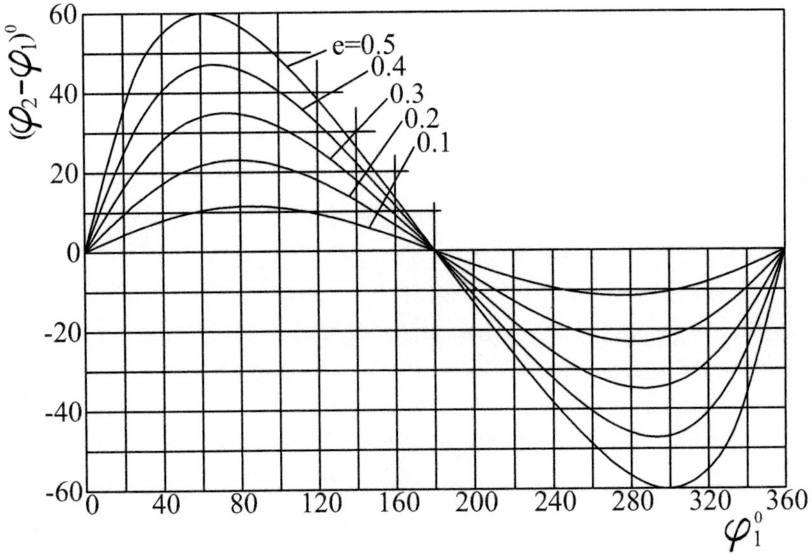

Fig. 11.4 Diagrams of $(\varphi_2 - \varphi_1) = f(\varphi_1)$

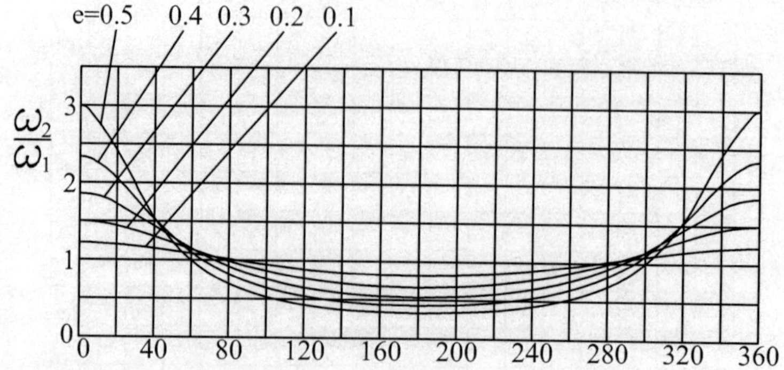

Fig. 11.5 Diagrams of $(\omega_2/\omega_1) = f(\varphi_1)$

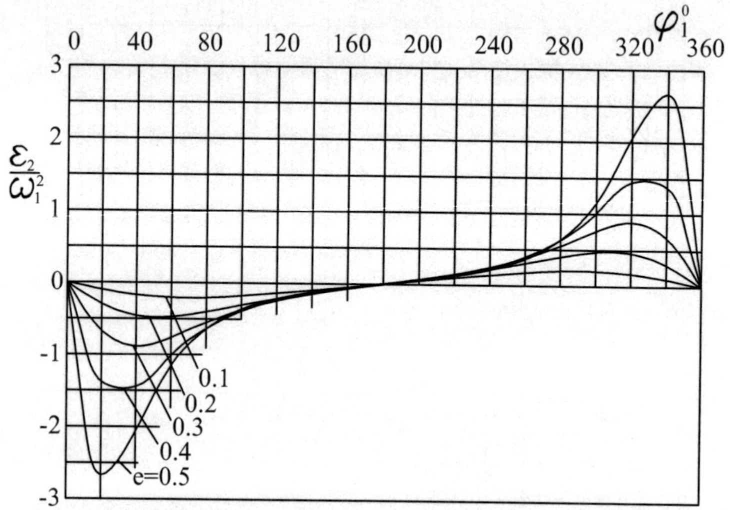

Fig. 11.6 Diagrams of $\left(\varepsilon_2/\omega_1^2\right) = f(\varphi_1)$

11.4 Pressure Angles in Elliptical Gear Drives

Pressure angle is the angle between the direction of motion of a driven body and the direction of the force applied to it by a driving body. In the best-case scenario, when these directions coincide (i.e., the pressure angle equals 0), the efficiency of mechanism is maximal. As the pressure angle grows, the forces in the mechanism and the energy losses grow, and, finally, if the pressure angle nears $(90\,^\circ - \alpha_f)$, where α_f is the angle of friction, the translation of motion between the bodies becomes impossible due to friction forces.

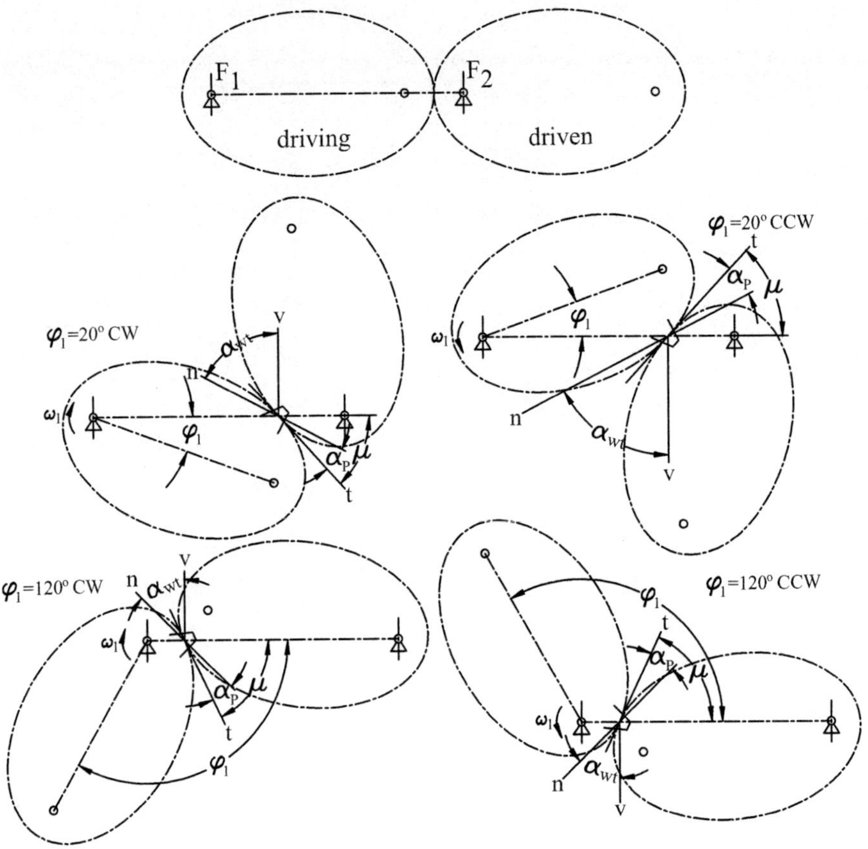

Fig. 11.7 Examples of angular positions within the first half-turn

Since point P is located on the center line at any angular position of the elliptical gears, the direction of motion at this point is always perpendicular to the center line. The direction of the tooth force that is perpendicular to the tooth profile is determined by two components. The first, obviously, is the angle of the tooth profile, and, at point P, it equals $\boldsymbol{\alpha}_{\mathbf{P}}$. The second component, which determines the angular position of the tooth with respect to the center line, is the angle of inclination of the common tangent to the center line (angle μ; see Fig. 11.1). Examples of the combinations of angles $\boldsymbol{\alpha}_{\mathbf{P}}$ and μ adding up to the pressure angle $\boldsymbol{\alpha}_{\mathbf{wt}}$ are given in Fig. 11.7 (for the first half-turn) and in Fig. 11.8 (for the second half-turn). Here, the direction of motion is denoted by "*v*," the common tangent by "*t*," and the normal to the working profile of the driving tooth by "*n*."

Angle μ can be calculated as follows [2]:

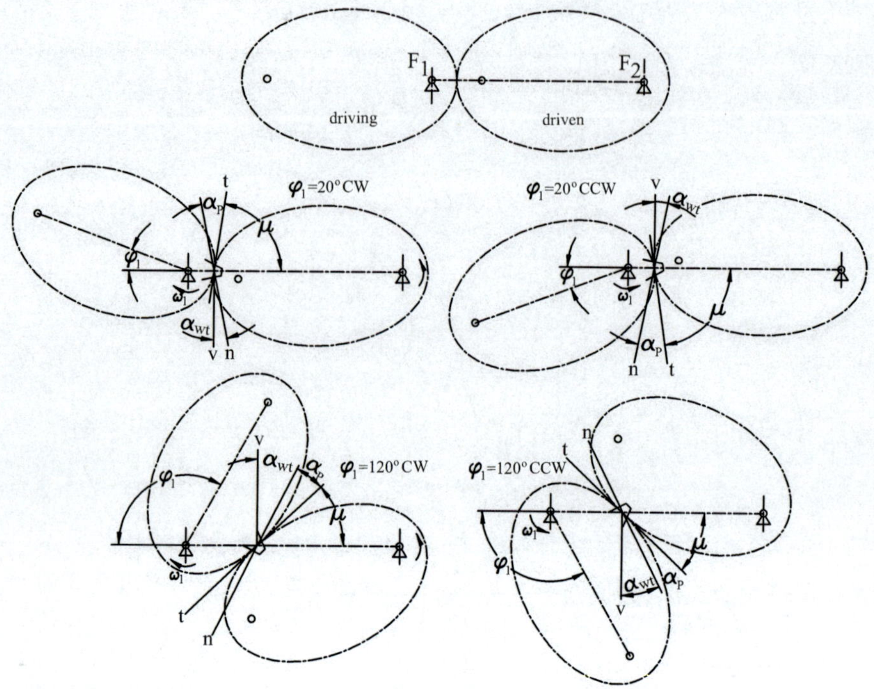

Fig. 11.8 Examples of angular positions within the second half-turn

- For the first half-turn:

$$\mu_1 = \arctan\frac{1 - e\ \cos\varphi_1}{e\sin\varphi_1} \tag{11.12}$$

where φ_1 is the angle of rotation from position u_{max} shown in Figure 11.2a;

- For the second half-turn:

$$\mu_2 = \arctan\frac{1 + e\ \cos\varphi_1}{e\sin\varphi_1} \tag{11.13}$$

where φ_1 is the angle of rotation from position u_{min} shown in Figure 11.2b.

From Figs. 11.7 and 11.8, one can see that the pressure angle equals

- For the first half-turn:

$$\alpha_{\text{wt1}} = 90\,° - (\mu - \alpha_{\text{P}}). \tag{11.14}$$

- For the second half-turn:

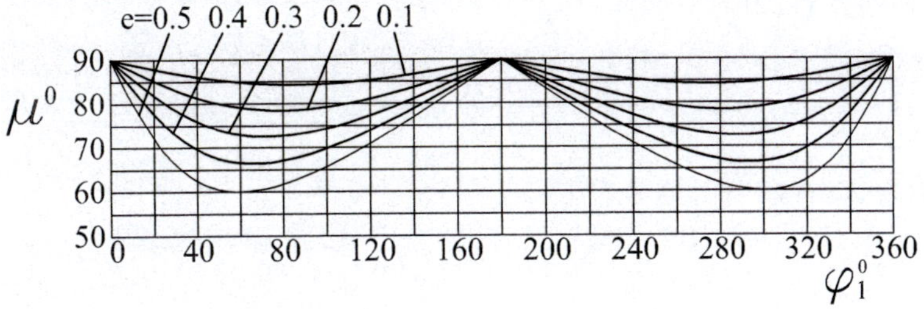

Fig. 11.9 Diagrams of $\mu = f(\varphi_1)$

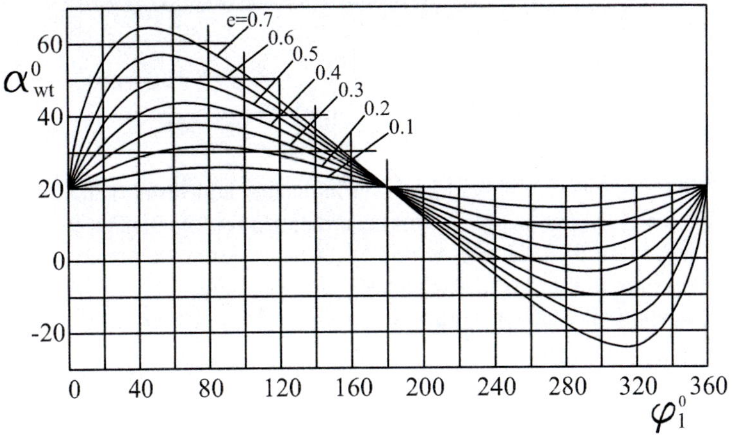

Fig. 11.10 Diagrams of $\alpha_{wt} = f(\varphi_1)$

$$\alpha_{wt2} = 90\,^\circ - (\mu + \alpha_P). \tag{11.15}$$

Here, α_P = angle of the basic rack profile.

Diagrams of angles μ and α_{wt} (at $\alpha_P = 20°$) are presented in Figs. 11.9 and 11.10, respectively. It is recommended [2] not to exceed $\alpha_{wt} = |50°|$.

Note: Is there any difference in essence between the positive and negative values of the pressure angle α_{wt}? There is no difference in this case. The "plus" or "minus" is obtained when the angle is calculated using Eqs. 11.14 and 11.15. In a two-dimensional problem, if $\alpha_{wt} = 0$, then the vectors of motion and of applied force are collinear. If $\alpha_{wt} \neq 0$, then the force vector deviates from collinearity by angle α_{wt} in the CW (we are free to sign it as "positive") or CCW direction (then it will be "negative"). However, only the modulus $|\alpha_{wt}|$ matters in our case.

Example 11.1 The eccentricity of the ellipse $e = 0.5$ and $\alpha_P = 20°$. What is the pressure angle at $\varphi_1 = 160°$ in the second half-turn?

From Eq. (11.13):

$$\mu_2 = \arctan \frac{1 + 0.5 \cos 160°}{0.5 \sin 160°} = \arctan 3.1 = 72.12°.$$

From Eq. (11.15):

$$\alpha_{wt2} = 90° - (72.12° + 20°) = -2.12°.$$

As stated above, the sign "minus" does not matter.

From Figs. 11.9 and 11.10, we can see that, in order to not exceed the pressure angle of 50°, the ellipse eccentricity should not exceed $e = 0.5$.

11.5 Tooth Load in Elliptical Gear Drives

Figure 11.11 presents an elliptical gear pair in the first half-turn (a) and in the second half-turn (b). Here, 1 and 2 are the driving and the driven gears, respectively, 3 is the common tangent to the centrodes at point P, 4 is the normal to the center line F_1F_2, and 5 is the normal to the working profiles of the driving teeth at point P.

Force F_n applied to the engaged teeth originates from the resistance torque on the driven shaft and is found as follows:

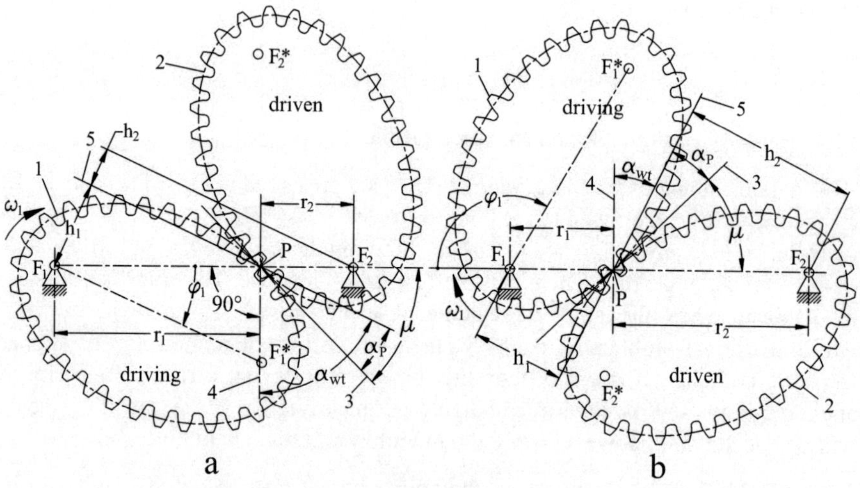

Fig. 11.11 Pressure angle α_{wt} and the arm of the tooth force h_2

$$F_{\mathrm{n}} = \frac{T_2}{h_2}\ \mathrm{N}. \tag{11.16}$$

Here, T_2 = torque on the driven shaft, N·mm:

$$T_2 = T_{\mathrm{S}} + T_{\mathrm{J}}.$$

T_{S} is the part of T_2 caused by the useful service torque on the output member of the driven mechanism. Torque T_{J} is caused by inertia forces of the driven parts, including the driven elliptical gear, which may increase or reduce the torque applied to the meshing teeth.

Torque T_{J} is given by

$$T_{\mathrm{J}} = J_{\mathrm{eq.}} \frac{\varepsilon_2}{\omega_1^2} \omega_1^2\ \mathrm{N \cdot mm}. \tag{11.17}$$

Here,

- ε_2/ω_1^2 values can be calculated using Eq. (11.11) or taken from Fig. 11.6 and
- $J_{\mathrm{eq.}}$ is the equivalent mass moment of inertia determined as follows:

$$J_{\mathrm{eq.}} = \sum_{i=1}^{k} J_i \left(\frac{\omega_i}{\omega_1}\right)^2\ \mathrm{N \cdot mm \cdot s^2}, \tag{11.18}$$

where J_i = mass moment of inertia of the rotating member i of the driven mechanism.

ω_i = average angular speed of member i.
ω_1 = angular speed of the driving gear.**Example 11.2** Here are the formulas to determine the mass moments of inertia J for two typical rotating elements:

- For a solid cylinder with diameter d and face width b:

$$J_{\mathrm{cyl.}} = \frac{\gamma}{g} \frac{\pi d^4}{32} b\ \mathrm{N \cdot mm \cdot s^2},$$

where γ = specific weight of the material (for steel, $\gamma = 7.7 \cdot 10^{-5}$ N/mm^3).

g = acceleration of gravity ($g = 9.81 \cdot 10^3$ mm/s^2).
d and b are in mm.

- For a solid elliptical body, the mass moment of inertia about the axis passing through the focus point equals:

$$J_{\text{ell.}} = \frac{\gamma}{g}\frac{\pi a^4}{4} b_g \left(2 + 3\, e^2\right)\sqrt{1 - e^2}\, \text{N} \cdot \text{mm} \cdot \text{s}^2, \tag{11.19}$$

where a = semi-major axis of the ellipse (mm);

e = eccentricity of the ellipse.
b_g = face width of the gear (mm).
γ and g are the same as above.

The mass moments of inertia of more complicated shapes can be calculated using well-known methods.

Member h_2 in Eq. (11.16) is the arm of force F_n about the center of rotation of the driven gear (focus point F_2). From Fig. 11.11, one can see that for both the first and the second half-turns,

$$h_2 = r_2 \cos \alpha_{wt}.$$

Since $r_2 = 2a - r_1$ and the r_1 value is determined for the first half-turn from Eq. (11.1),

$$r_2 = 2a - \frac{a\,(1 - e^2)}{1 - e \cos \varphi_1}.$$

Hence, for the first half-turn,

$$\frac{h_2}{a} = \frac{1 + e^2 - 2e \cos \varphi_1}{1 - e \cos \varphi_1} \cos \alpha_{wt}.$$

For the second half-turn,

$$r_1 = \frac{a\,(1 - e^2)}{1 + e \cos \varphi_1},$$

and, hence,

$$\frac{h_2}{a} = \frac{1 + e^2 + 2\,e \cos \varphi_1}{1 + e \cos \varphi_1} \cos \alpha_{wt}.$$

The functions $h_2/a = f(\varphi_1)$ are diagrammed for the full turn in Fig. 11.12.

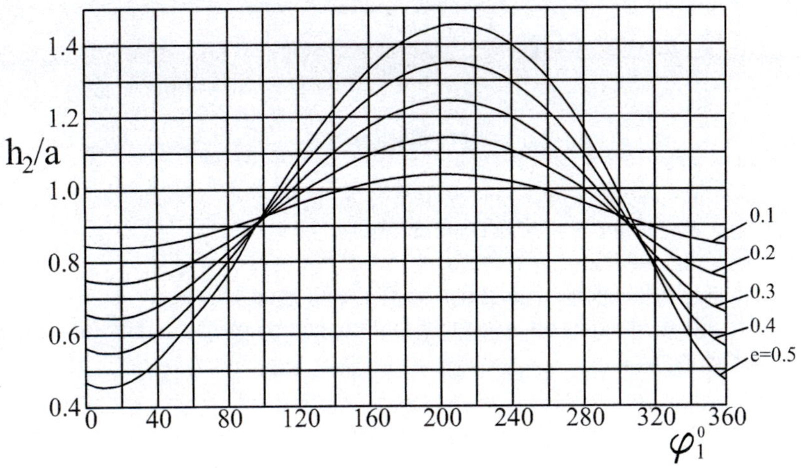

Fig. 11.12 Diagrams of $(h_2/a) = f(\varphi_1)$

11.6 Calculation of the Tooth Strength of Elliptical Gears Using the Colloquial Calculations of Involute Spur Gears

In practice, the strength of moderately loaded gear drives is estimated using simplified formulas. For the assessment of resistance to pitting, factor K is often used for involute gears:

$$K = \frac{2\,T_1(u+1)}{b_g\,d_1^2 u}\,\frac{\text{N}}{\text{mm}^2}. \tag{11.20}$$

Here, T_1 = torque applied to the pinion (N·mm).
d_1 = reference diameter of the pinion (mm).
b_g = gear face width (mm).
u = gear ratio.

The value of factor K obtained by this formula should not exceed a certain admissible limit that depends on the strength of the gear material. For spur gears of tempered steel, $K_{\mathbf{adm}} \approx 0.008\text{HB} - 0.87\ \text{N/mm}^2$; for spur gears with teeth induction hardened to HRC ≥ 50, $K_{\mathbf{adm}} \approx 3.5\ \text{N/mm}^2$ [1].

To understand how to adopt factor K to strengthen the calculation of elliptical gears, let us analyze from where Eq. (11.20) comes.

Pitting is the fatigue failure of the working surfaces of the teeth flanks caused by cyclic loading with a high contact stress σ_H. For a linear contact, the contact stress is determined by the Hertz formula:

$$\sigma_{\mathrm{H}} = 0.418\sqrt{\frac{F_{\mathrm{n}}\,E}{\rho_{\mathrm{eq.}}\,b_{\mathrm{g}}}\,\frac{\mathrm{N}}{\mathrm{mm}^2}}. \tag{11.21}$$

Here, $F_{\mathrm{n}} = 2\,T_1/(d_1 \cdot \cos\alpha_{\mathrm{P}})$ = normal force transmitted by the engaged pair of teeth (N);

T_1 = torque applied to the pinion shaft (N·mm).

d_1 = reference diameter of the pinion (mm).

E = modulus of elasticity (N/mm^2).

b_{g} = length of the contact line, for straight teeth equals the gear face width (mm).

ρ_{eq} = equivalent radius of curvature of the contacting surfaces (mm) that equals in external engagement:

$$\rho_{\mathrm{eq.}} = \frac{\rho_1 \rho_2}{\rho_1 + \rho_2}.$$

The curvature radius of the involute obtained in generating roll depends on the curvature radius of the centrode. In involute spur gear pairs with the sum of profile shifts $x_1 + x_2 = 0$, the centrodes are the reference circles d_1 and d_2, and the curvature radii of the involute profiles ρ_1 and ρ_2 in point P equal:

$$\rho_1 = 0.5\,d_1 \sin\alpha_{\mathrm{P}}.$$

$$\rho_2 = 0.5 d_2 \sin\alpha_{\mathrm{P}} = 0.5 u d_1 \sin\alpha_{\mathrm{P}}.$$

hence,

$$\rho_{\mathrm{eq.}} = \frac{d_1\,u}{1+u}\,0.5\,\sin\alpha_{\mathrm{P}}.$$

Substituting expressions for F_{n} and $\rho_{\mathrm{eq.}}$ in Eq. (11.21), we obtain after rearrangement:

$$\frac{2\,T_1\,(u+1)}{d_1^2\,b_{\mathrm{g}}\,u} = \sigma_{\mathrm{H}}^2\,\frac{\sin(2\alpha_{\mathrm{P}})}{0.7\,E} = \sigma_{\mathrm{H}}^2 \cdot C = K.$$

Here, C is constant for most gears because $\alpha_{\mathrm{P}} = 20°$ and the modulus of elasticity E is approximately the same for steel used in gear production. Thus, the value of factor K reflects the admissible contact stress and mainly depends on the strength of the material of the teeth (see above).

It was noted above that the shape of the teeth flanks of the elliptical gears is not involute because the centrode is noncircular. The variation of curvature of the centrode leads to the formation of a complex surface of the tooth flank. It is necessary to take into consideration that the calculation of gears (involute and non-involute) for resistance to pitting in principle cannot be mathematically accurate due to the complexity of both the tooth geometry and the processes occurring in the contact zone of the teeth. Let us start from the contact stress that is determined by the Hertz formula, with the radius of curvature of the involute in the pitch point. First, the radius of curvature of the involute varies throughout its length. Second, roughness and tooth profile errors lead to significant deviations of the tooth surface curvature from the calculated value. Third, the combination of rolling and sliding in the contact of the teeth with high contact stresses causes wear and plastic deformation of the surface layers of the metal. This not only changes the geometry of the teeth flanks but also changes the microstructure of the thin surface layers. For these reasons, it makes no sense to enter all the complexities of the geometry of the tooth profile of the elliptical gears because the admissible value of factor K in any case should be determined experimentally, just like it is done for involute gear drives.

Note: Other factors, such as speed, required service life, operating regime (for example, heavy duty of light duty), etc. influence the $K_{\mathbf{adm}}$ value remarkably. Here, we use the recommendations for moderate working conditions.

Let us go back to the question of the adaptation of Eq. (11.20) to the strength calculation of elliptical gears:

- For elliptical gears, in Eq. (11.20), the expression $2T_1/d_1$ that determines the peripheral (tangential) force in the contact of teeth shall be replaced by $F_n \cdot \cos\alpha_P$ to get the physical equivalent.
- Member d_1 that determines the radius of curvature of the tooth profile shall be replaced by the doubled radius of curvature R of the gear centrode; see Eq. (11.5).
- $u = 1$ because in each angular position, the radii of curvature of the elliptical centrodes at point P are equal.

As a result, Eq. (11.20) and the pitting strength condition take the following form for elliptical gear drives:

$$K = \frac{F_n \cos\alpha_P}{R\, b_g} \leq K_{\mathrm{adm}} \frac{\mathrm{N}}{\mathrm{mm}^2}. \tag{11.22}$$

The radius of curvature of the centrode R at point P obtained from Eq. (11.5) can be transformed as follows:

$$\frac{R}{a} = \frac{1}{\sqrt{1-e^2}} \left[\frac{1 - e^4 - 2e(1-e^2)\cos\varphi_1}{(1 - e\,\cos\varphi_1)^2} \right]^{1.5}.$$

Function $R/a = f(\varphi_1)$ is diagrammed in Fig. 11.13.

The difference between Eqs. (11.20) and (11.22) is that in Eq. (11.20), torque T_1 is given independently of the gears dimensions, whereas in Eq. (11.22), force F_n (i.e., torque T_2; see Eq. (11.16)) depends on the dimensions of the driven gear whose mass moment of inertia influences torque T_2. This fact makes the process of dimensioning of elliptical gears iterative. For the first approximation, the position

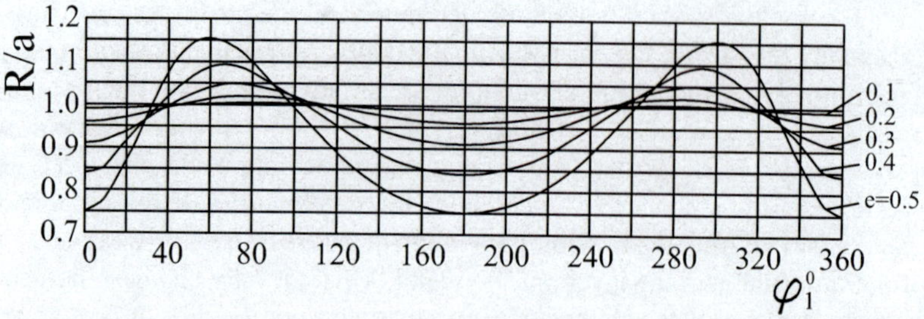

Fig. 11.13 Diagrams of $(R/a) = f(\varphi_1)$

shown in Fig. 11.2a can be chosen. In this angular position, there are no inertia forces and the arm h_2 of force F_n is close to minimum (see Fig. 11.12). The next iteration should be calculated for the position of maximum acceleration.

11.7 Example of the Design of an Elliptical Gear Drive

It is required to design an elliptical gear drive, which has the maximal angular speed of the driven shaft seven times higher than its minimal angular speed. The service torque applied to the driven shaft is $T_S = 100$ N·m and the rotational speed $n_1 = 600$ rpm ($\omega_1 = 62.83$ s^{-1}). The straight teeth with core hardness 280–320 HB are induction hardened to 50–52 HRC, so $K_{adm} = 3.5$ N/mm^2. The mass moment of inertia of the parts connected to the driven gear, such as the shaft, coupling, cam, and others is: $J_M = 5$ N·mm·s^2.

From Eqs. (11.3) and (11.4):

$$\frac{\omega_{2\max}}{\omega_{2\min}} = \frac{u_{\max}}{u_{\min}} = \left(\frac{1+e}{1-e}\right)^2.$$

From Fig. 11.14, we find for $\omega_{2\max}/\omega_{2\min} = 7$, $e \approx 0.45$. From Fig. 11.12, we find for $e = 0.45$ and $\varphi_1 = 360°$, $h_2/a \approx 0.52$.

Since $\varepsilon = 0$ in this angular position, the inertia torque is also zero and the tooth force equals:

$$F_n = \frac{T_S}{h_2} = \frac{10^5}{0.52\,a}\ \text{N}.$$

From Fig. 11.13, we find for $e = 0.45$, $R/a \approx 0.8$. At this stage, we are to choose the face width of the gear as part of its biggest diameter, which equals $2a$. Let us set $b_g = 0.5a$. In this case, from Eq. (11.22):

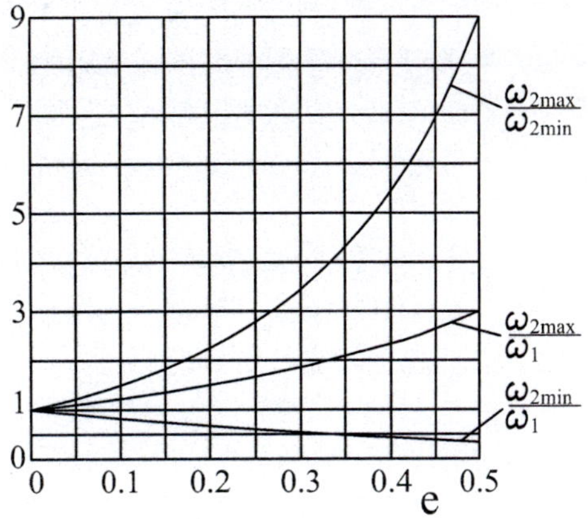

Fig. 11.14 The kinematic effect of elliptical gear drives depending on the ellipse eccentricity *e*

$$3.5 = \frac{10^5}{0.52a} \cdot \frac{\cos 20^\circ}{0.5a \cdot 0.8a} = \frac{4.518 \cdot 10^5}{a^3}.$$

Here, we obtain $a = 50.5$ mm and $b_g = 0.5a = 25.25$ mm. This is the result of the first approximation that provided us with the dimensions of the driven gear required for the calculation of inertia forces.

Now, we calculate "the first iteration" for the position with the highest inertial load. The mass moment of inertia of the driven gear can be calculated using Eq. (11.19):

$$J_{\text{ell.}} = \frac{7.7 \cdot 10^{-5}}{9.81 \cdot 10^3} \cdot \frac{\pi \cdot 50.5^4}{4} 25.25 \left(2 + 3 \cdot 0.45^2\right) \sqrt{1 - 0.45^2} = 2.357 \text{ N} \cdot \text{mm} \cdot \text{s}^2.$$

From Fig. 11.6, we can see that the maximal positive acceleration at $e = 0.45$ takes place between $\varphi_1 = 330°$ and $\varphi_1 = 340°$. Suppose $\varphi_1 = 335°$. The maximum acceleration rate can be taken from Fig. 11.15 based on Fig. 11.6:

$$\varepsilon_{2\,\max} \approx 2\,\omega_1^2 = 2 \cdot 62.83^2 = 7.9 \cdot 10^3 \text{ s}^{-2}.$$

Hence, the torque created by the acceleration of the driven gear equals:

$$T_J = (J_M + J_{\text{ell.}})\varepsilon_{2\,\max} = (5 + 2.357) \cdot 7.9 \cdot 10^3 = 5.8 \cdot 10^4 \text{ N} \cdot \text{mm}.$$

Hence, the total torque in the position of maximum acceleration, i.e., at $\varphi_1 = 335°$, equals:

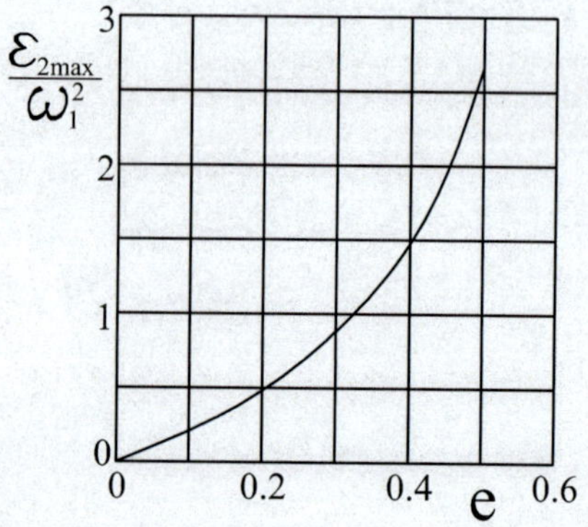

Fig. 11.15 Diagrams of $(\varepsilon_{2\,\max}/\omega_1^2)=f(e)$

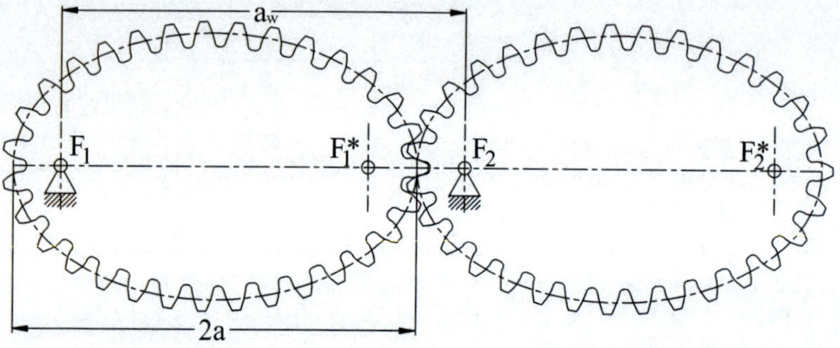

Fig. 11.16 An elliptical gear pair

$$T_2 = T_S + T_J = 10^5 + 5.8 \cdot 10^4 = 1.58 \cdot 10^5\ \text{N} \cdot \text{mm}.$$

From Figs. 11.12 and 11.13: at $\varphi_1 = 335°$, $h_2/a \approx 0.65$, and $R/a \approx 0.92$. From Eq. (11.16):

$$F_n = \frac{1.58 \cdot 10^5}{0.65a}.$$

From Eq. (11.22):

$$3.5 = \frac{1.58 \cdot 10^5}{0.65a} \cdot \frac{\cos 20\,°}{0.5a \cdot 0.92a} = \frac{4.966 \cdot 10^5}{a^3}.$$

Here, we obtain $a = 52.2$ mm and $b_g = 26.1$ mm.

"For the second iteration," we calculate the moment of inertia of the gear very quickly because, while maintaining geometric similarity, it is proportional to the fifth grade of the scale:

$$J_{\text{ell.}} = \left(\frac{52.2}{50.5}\right)^5 \cdot 2.357 = 2.78\ \text{N} \cdot \text{mm} \cdot \text{s}^2.$$

$$T_{\text{J}} = (5 + 2.78) \cdot 7.9 \cdot 10^3 = 6.15 \cdot 10^4\ \text{N} \cdot \text{mm}.$$

$$T_2 = 10^5 + 6.15 \cdot 10^4 = 1.615 \cdot 10^5\ \text{N} \cdot \text{mm}.$$

Because the relations h_2/a, R/a, and b_g/a are the same as in the first iteration,

$$3.5 = \frac{1.615 \cdot 10^5}{0.65a} \cdot \frac{\cos 20^\circ}{0.5a \cdot 0.92a} = \frac{5.076 \cdot 10^5}{a^3}.$$

Here, $a = 52.5$. We can see that the iterative process converges very quickly because the dimension obtained in the second iteration is only 0.5% bigger than that in the first iteration. The designer, however, has other ways to satisfy the strength requirement: for example, it is possible to reduce the $J_{\text{ell.}}$ value by removing the excessive metal from the driven gear body, to increase the b/a value, etc. So, for now, let us stop the iterations and set $a = 52.5$ mm. Other parameters of the elliptical centrode: $c = ea = 0.45{\cdot}52.5 = 23.625$ mm;

$$b = \sqrt{a^2 - c^2} = \sqrt{52.5^2 - 23.625^2} = 46.884\ \text{mm}.$$

Now, we continue designing the gear.

(a) Choice of the tooth module.

The maximal module is calculated from the condition of no undercutting of teeth. It is known that in gears produced with a standard cutting tool (profile angle $\alpha_{\mathbf{P}} = 20°$, addendum height $h_{\mathbf{aP}} = 1$ m) without profile shifting, the minimal number of teeth without undercutting $z_{\mathbf{min}} = 17$. In our case, the minimal radius of curvature of the centrode (see Fig. 11.13) $R_{\mathbf{min}} \approx 0.8a = 0.8{\cdot}52.5 = 42$ mm. The module for $z_{\mathbf{min}} = 17$ is found from the following relation:

$$2\,\pi\,R_{\min} = \pi\,m\,z_{\min};$$

$$m = \frac{2\,R_{\min}}{z_{\min}} = \frac{2 \cdot 42}{17} = 4.94 \rightarrow 5\ \text{mm}.$$

(an extremely small undercut is admissible.)

The minimal tooth module is determined based on the reasoning of the fracture resistance of the teeth. For preliminary estimation of the minimal tooth module of a spur gear, the following formula can be used [1]:

$$m \geq \frac{10\,T_1}{d_1 b_g \sigma_b}\ \text{mm}, \tag{11.23}$$

where T_1 = maximal torque on the pinion shaft (N·mm).
d_1 = reference diameter of the pinion (mm).
b_g = gear face width (mm).
σ_b = allowable bending stress (N/mm^2).
As applied to our case,

$$\frac{T_1}{d_1} = 0.5\,F_n \cos \alpha_P.$$

Force F_n shall be calculated for the second iteration (at $\varphi_1 = 335°$):

$$F_n = \frac{T_2}{h_2} = \frac{1.615 \cdot 10^5}{0.65 \cdot 52.5} = 4733\ \text{N}.$$

The minimum limiting bending stress σ_b for the induction hardened teeth with a core hardness of HB280 equals 230 N/mm^2. Substituting these data in Eq. (11.23), we obtain the minimum module:

$$m \geq \frac{10 \cdot 0.5 \cdot 4733 \cdot \cos 20°}{26.25 \cdot 230} = 3.68 \rightarrow 4\ \text{mm}.$$

Thus, the module shall be between 4 and 5 mm.

(b) Calculation of the tooth number.

From Eq. (11.7):

$$z = \frac{L}{\pi m} = \text{whole number}. \tag{11.24}$$

The length L of the centrode is determined from Eq. (11.8), where $a + b = 52.5 + 46.884 = 99.384$ mm and

$$\lambda = \frac{a-b}{a+b} = \frac{52.5 - 46.884}{52.5 + 46.884} = 0.056508;$$

$$\begin{aligned} L &= \pi \cdot 99.384 \left[1 + \frac{0.056508^2}{4} + \frac{0.056508^4}{64} + \frac{0.056508^6}{256} + \cdots\right] \\ &= 312.224\left[1 + 7.983 \cdot 10^{-4} + 1.59 \cdot 10^{-7} + 1.27 \cdot 10^{-10}\right] \\ &= 312.224 + 0.2492 + 0.0496 \cdot 10^{-3} + \cdots \\ &= 312.473\,\text{mm}. \end{aligned}$$

One can see that in this case, already the second term of the series (after 1) is negligible.

Returning to Eq. (11.24), we find the number of teeth at $m = 4$ mm:

$$z = \frac{312.473}{\pi \cdot 4} = 24.865811 \rightarrow 25.$$

Note: The number of teeth should be odd for ease of manufacture and assembly (see Fig. 11.16). If, according to Eq. (11.24), z turns out to be even, then it shall be increased or reduced by 1.

To fulfill Eq. (11.7), the centrode perimeter L must be brought in accordance with the real number of teeth. Since the length L is directly proportional to a (at the same ellipse eccentricity e), the final values of the semi-axes of the elliptical centrode are found as follows:

$$a = 52.5 \frac{25}{24.865811} = 52.7833 \text{ mm}.$$

$$c = 52.7833 \cdot 0.45 = 23.7525 \text{ mm}.$$

$$b = \sqrt{52.7833^2 - 23.7525^2} = 47.1370 \text{ mm}.$$

The center distance $a_w = 2a = 105.561$ mm.

(c) Design testing.

Let us control the strength of the teeth in the position of maximal acceleration. The mass moment of inertia of the driven gear is increased, and its mass moment of inertia finally equals:

$$J_{\text{ell.}} = \left(\frac{52.78}{50.5}\right)^5 \cdot 2.357 = 2.939 \text{ N} \cdot \text{mm} \cdot \text{s}^2.$$

$$T_J = (5 + 2.839) \cdot 7.9 \cdot 10^3 = 6.27 \cdot 10^4 \text{ N} \cdot \text{mm}.$$

$$T_2 = 10^5 + 6.27 \cdot 10^4 = 1.627 \cdot 10^5 \text{ N} \cdot \text{mm}.$$

$$F_n = \frac{T_2}{h_2} = \frac{1.627 \cdot 10^5}{0.65 \cdot 52.78} = 4742 \text{ N}.$$

From Eq. (11.22):

$$K = \frac{4742 \cos 20\,°}{0.92 \cdot 52.78 \cdot 26.4} = 3.48 < K_{\text{adm}} \frac{\text{N}}{\text{mm}^2}.$$

The bending strength of the teeth is checked using Eq. (11.23):

$$m \geq \frac{10 \cdot 0.5 \cdot 4742 \cdot \cos 20^\circ}{26.25 \cdot 230} = 3.67 \text{ mm}.$$

This requirement is fulfilled as well.

11.8 Conclusion

The method of the design and calculation of elliptical gears proposed here is based on the method commonly used in the practice of designing circular involute gears in which the load level of the gear teeth is determined by the value of factor K. The proposed method, with the help of diagrams, is quite simple. When the gear parameters are defined and the location of each tooth is known, a more accurate calculation of teeth for bending strength and resistance to pitting can be made.

However, the process of pitting of the tooth surfaces is highly complicated, and the simplified calculations are not reliable enough without experimental checks. The admissible values of factor K (K_{adm}) for involute gears are based on the results of testing gear drives in laboratories and in operation. The same, obviously, should be done for elliptical gear drives. The K_{adm} values recommended for involute gear drives can be used for elliptical gear drives as a first approximation.

References

1. Klebanov, B.M.: Groper, Morel. Power Mechanisms of Rotational and Cyclic Motions. CRC Press (2016)
2. Litvin, F.L., et al.: Noncircular Gears. Design and Generation. Cambridge University Press (2009)
3. Radzevich, S.P.: Theory of Gearing. Kinematics, Geometry and Synthesis, 2nd edn. CRC Press (2018)
4. Dwight, H.B.: Tables of Integrals and Other Mathematical Data. Prentice Hall PTR (1961)

Appendices

Appendix A: Elements of Vector Calculus

A "vector," the key to all the theories of part surface generation, is a triple real number (in most computer languages, these are usually called "floating point numbers") and is noted in a **bold** typeface, that is, **A** or **a**.

Care must be taken to differentiate between two types of vectors:

- ***Position vector***: A position vector runs from the origin of coordinate $(0, 0, 0)$ to a point (X, Y, Z), and its length gives the distance of the point from the origin. Its components are given by (X, Y, Z). The essential concept to understand about a position vector is that it is anchored to specific coordinates (points in space). A set of points that are used to describe the shape of all part surfaces can be considered as position vectors.
- ***Direction vector***: A direction vector differs from a position vector, in that it is "not" anchored to specific coordinates. Frequently, direction vectors are used in a form in which they have unit length; in this case, they are said to be "normalized." The most common application of a direction vector in the theory of part surface generation is to specify the orientation of a surface or ray direction. For this, we use a direction vector at right angles ("normal") and pointing away from the part surface. Such "normal" vectors are also the key in many calculations in the theory of part surface generation.

Vector calculus is a powerful tool for solving many geometrical and kinematical problems that pertain to the design and generation of part surfaces. In this book, "**vectors** are understood as quantities that have magnitude and direction and obey the law of addition."

S. P. Radzevich, *Novikov/Conformal Gearing*,
https://doi.org/10.1007/978-3-031-10019-2

A.1 Fundamental Properties of Vectors

The distance-and-direction interpretation suggests a powerful way to visualize a vector, that is, as a directed line segment or an arrow. The length of the arrow (at some predetermined scale) represents the magnitude of the vector, and the orientation of the segment and placement of the arrowhead (at one end of the segment or the other) represents its direction.

Vectors possess certain properties, the set of which is commonly interpreted as the set of the fundamental properties of vectors.

A.1.1 Addition

Given two vectors **a** and **b**, their sum $(\mathbf{a} + \mathbf{b})$ is graphically defined by joining the tail of **b** to the head of **a**. Then, the line from the tail of **a** to the head of **b** is the sum $\mathbf{c} = (\mathbf{a} + \mathbf{b})$.

A.1.2 Equality

Two vectors are equal when they have the same magnitude and direction. The position of the vectors is unimportant for equality.

A.1.3 Negation

The vector $-\mathbf{a}$ has the same magnitude as **a** but an opposite direction.

A.1.4 Subtraction

From the properties of "addition" and "negation," the following $\mathbf{a} - \mathbf{b} = \mathbf{a} + (-\mathbf{b})$ can be defined.

A.1.5 Scalar Multiplication

The vector $k\mathbf{a}$ has the same direction as **a**, with a magnitude k times that of **a**. k is called a scalar as it changes the scale of the vector **a**.

A.2 Mathematical Operations over Vectors

The following rules and mathematical operations can be determined from the above-listed fundamental properties of vectors.

A.2.1 Components of Vectors

Let us assume that a set of three vectors **a**, **b**, **c** and two scalars k and t are given. Then, vector addition and scalar multiplication have the following properties:

$$\mathbf{a}+\mathbf{b}=\mathbf{b}+\mathbf{a} \tag{A.1}$$

$$\mathbf{a}+(\mathbf{b}+\mathbf{c})=(\mathbf{a}+\mathbf{b})+\mathbf{c} \tag{A.2}$$

$$k(t\mathbf{a})=kt\mathbf{a} \tag{A.3}$$

$$(k+t)\mathbf{a}=k\mathbf{a}+t\mathbf{a} \tag{A.4}$$

$$k(\mathbf{a}+\mathbf{b})=k\mathbf{a}+k\mathbf{b} \tag{A.5}$$

The magnitude a of a vector **a** is:

$$a=|\mathbf{a}|=\sqrt{a_x^2+a_y^2+a_z^2} \tag{A.6}$$

where a_x, a_y, and a_z are the scalar components of **a**.

A unit vector $\bar{\mathbf{a}}$ in the direction of a vector **a** is:

$$\bar{\mathbf{a}}=\frac{\mathbf{a}}{|\mathbf{a}|}=\frac{\mathbf{a}}{a} \tag{A.7}$$

The components $\bar{a}_x$, $\bar{a}_y$, and $\bar{a}_z$ of a unit vector $\bar{\mathbf{a}}$ are also the direction cosines of the vector $\bar{\mathbf{a}}$:

$$\cos\alpha=\bar{a}_x \tag{A.8}$$

$$\cos\beta=\bar{a}_y \tag{A.9}$$

$$\cos\gamma=\bar{a}_z \tag{A.10}$$

It is common practice to denote the components $\bar{a}_x$, $\bar{a}_y$, and $\bar{a}_z$ by l, m, and n, respectively.

A.2.2 Scalar Product (or Dot Product) of Two Vectors

The formula:

$$\mathbf{a} \cdot \mathbf{b} = a_x b_x + a_y b_y + a_z b_z = |\mathbf{a}||\mathbf{b}| \cos \angle(\mathbf{a}, \mathbf{b}) \tag{A.11}$$

is commonly used for calculation of the scalar product of two vectors **a** and **b**.

Equation (A.11) can also be represented in the form:

$$\mathbf{a} \cdot \mathbf{b} = [\mathbf{a}]^T \cdot [\mathbf{b}] = \begin{bmatrix} a_x & a_y & a_z \end{bmatrix} \cdot \begin{bmatrix} b_x \\ b_y \\ b_z \end{bmatrix} \tag{A.12}$$

The angle $\angle(\mathbf{a}, \mathbf{b})$ between the two vectors **a** and **b** is calculated as:

$$\angle(\mathbf{a}, \mathbf{b}) = \cos^{-1}\left(\frac{\mathbf{a} \cdot \mathbf{b}}{|\mathbf{a}||\mathbf{b}|}\right) \tag{A.13}$$

The scalar product of the two vectors **a** and **b** features the following properties:

$$\mathbf{a} \cdot \mathbf{a} = |\mathbf{a}|^2 \tag{A.14}$$

$$\mathbf{a} \cdot \mathbf{b} = \mathbf{b} \cdot \mathbf{a} \tag{A.15}$$

$$\mathbf{a} \cdot (\mathbf{b} + \mathbf{c}) = \mathbf{b} \cdot \mathbf{a} + \mathbf{b} \cdot \mathbf{c} \tag{A.16}$$

$$(k\mathbf{a}) \cdot \mathbf{b} = \mathbf{a} \cdot (k\mathbf{b}) = k(\mathbf{a} \cdot \mathbf{b}) \tag{A.17}$$

If **a** is perpendicular to **b**, then:

$$\mathbf{a} \cdot \mathbf{b} = 0 \tag{A.18}$$

A.2.3 Vector Product (or Cross Product) of Two Vectors

The vector product of two vectors can be calculated from the formula:

$$\mathbf{a} \times \mathbf{b} = \left(a_y b_z - a_z b_y\right)\mathbf{i} + \left(a_z b_x - a_x b_z\right)\mathbf{j} + \left(a_x b_y - a_y b_x\right)\mathbf{k} \tag{A.19}$$

Here, in Eq. (A.19), **i**, **j**, and **k** are the unit vectors in the X, Y, and Z directions, respectively, of the reference system XYZ, in which the vectors **a** and **b** are specified.

The vector product possesses the following property: in case $\mathbf{a} \times \mathbf{b} = \mathbf{c}$, then the vector $\mathbf{c}$ is perpendicular to a plane through the vectors $\mathbf{a}$ and $\mathbf{b}$.

The vector product of two vectors $\mathbf{a}$ and $\mathbf{b}$ features the following properties:

$$\mathbf{a} \times \mathbf{b} = \begin{vmatrix} \mathbf{i} & \mathbf{j} & \mathbf{k} \\ a_x & a_y & a_z \\ b_x & b_y & b_z \end{vmatrix} \tag{A.20}$$

$$\mathbf{a} \times \mathbf{b} = |\mathbf{a}||\mathbf{b}|\mathbf{n} \sin \angle(\mathbf{a}, \mathbf{b}) \tag{A.21}$$

where the unit normal vector to the plane through the vectors $\mathbf{a}$ and $\mathbf{b}$ is denoted by $\mathbf{n}$:

$$|\mathbf{a} \times \mathbf{b}| = |\mathbf{a}||\mathbf{b}| \sin \angle(\mathbf{a}, \mathbf{b}) \tag{A.22}$$

Coordinates of the vector product $\mathbf{a} \times \mathbf{b}$ can also be expressed in the form:

$$|\mathbf{a} \times \mathbf{b}| = \begin{bmatrix} 0 & -a_z & a_y \\ a_z & 0 & -a_x \\ -a_y & a_x & 0 \end{bmatrix} \cdot \begin{bmatrix} b_x \\ b_y \\ b_z \end{bmatrix} = \begin{bmatrix} -a_z b_y + a_y b_z \\ -a_x b_z + a_z b_x \\ -a_y b_x + a_x b_y \end{bmatrix} \tag{A.23}$$

$$\mathbf{a} \times \mathbf{b} = -\mathbf{b} \times \mathbf{a} \tag{A.24}$$

$$\mathbf{a} \times (\mathbf{b} + \mathbf{c}) = \mathbf{a} \times \mathbf{b} + \mathbf{a} \times \mathbf{c} \tag{A.25}$$

$$(k\mathbf{a}) \times \mathbf{b} = \mathbf{a} \times (k\mathbf{b}) = k(\mathbf{a} \times \mathbf{b}) \tag{A.26}$$

$$\mathbf{i} \times \mathbf{j} = \mathbf{k},\ \mathbf{j} \times \mathbf{k} = \mathbf{i},\ \mathbf{k} \times \mathbf{i} = \mathbf{j} \tag{A.27}$$

If $\mathbf{a}$ is parallel to $\mathbf{b}$, then:

$$\mathbf{a} \times \mathbf{b} = 0 \tag{A.28}$$

A.2.4 Triple Scalar Product of Three Vectors

The product $(\mathbf{a} \times \mathbf{b}) \cdot \mathbf{c}$ is commonly referred to as the "triple scalar product" of the three vectors $\mathbf{a}$, $\mathbf{b}$, and $\mathbf{c}$.

The triple scalar product of the three vectors $\mathbf{a}$, $\mathbf{b}$, and $\mathbf{c}$ features the following properties:

$$(\mathbf{a}\times\mathbf{b})\cdot\mathbf{c}=(\mathbf{b}\times\mathbf{c})\cdot\mathbf{a}=(\mathbf{c}\times\mathbf{a})\cdot\mathbf{b} \tag{A.29}$$

$$(\mathbf{b}\times\mathbf{c})\cdot\mathbf{a}=\mathbf{a}\cdot(\mathbf{b}\times\mathbf{c}) \tag{A.30}$$

$$(\mathbf{a}\times\mathbf{b})\cdot\mathbf{c}=\mathbf{a}\cdot(\mathbf{b}\times\mathbf{c}) \tag{A.31}$$

$$\mathbf{a}\cdot(\mathbf{b}\times\mathbf{c})=\begin{vmatrix} a_x & a_y & a_z \\ b_x & b_y & b_z \\ c_x & c_y & c_z \end{vmatrix} \tag{A.32}$$

A.2.5 Triple Vector Product of Three Vectors

The product $(\mathbf{a}\times\mathbf{b})\times\mathbf{c}$ is commonly referred to as the "triple vector product" of the three vectors **a**, **b**, and **c**.

The product $(\mathbf{a}\times\mathbf{b})\times\mathbf{c}$ can be evaluated by two vector products. However, it can also be evaluated in a more simply manner by use of the identity:

$$(\mathbf{a}\times\mathbf{b})\times\mathbf{c}=(\mathbf{a}\cdot\mathbf{c})\mathbf{b}-(\mathbf{b}\cdot\mathbf{c})\mathbf{a} \tag{A.33}$$

It should be mentioned here that, in general, the triple vector products $(\mathbf{a}\times\mathbf{b})\times\mathbf{c}$ and $\mathbf{a}\times(\mathbf{b}\times\mathbf{c})$ are not equal:

$$(\mathbf{a}\times\mathbf{b})\times\mathbf{c}\neq\mathbf{a}\times(\mathbf{b}\times\mathbf{c}) \tag{A.34}$$

Analytical interpretations of many problems and results in the field of geometry of surfaces are much simpler when vector calculus is used.

A.2.6 Lagrange Equation for Vectors

For the purposes of calculation of the mixed product of vectors **a** and **b**, an equation:

$$(\mathbf{a}\times\mathbf{b})\cdot(\mathbf{a}\times\mathbf{b})=(\mathbf{a}\cdot\mathbf{a})(\mathbf{b}\cdot\mathbf{b})-(\mathbf{a}\cdot\mathbf{b})^2 \tag{A.35}$$

can be used.

Equation (A.35) is due to Lagrange.[1]

[1] Joseph-Louis Lagrange (January 25, 1736–April 10, 1813), a famous French mathematician, astronomer, and mechanician.

A.3 Similarity and Difference Between Vectors and Matrices

A vector **a** is commonly represented in the form:

$$\mathbf{a} = \mathbf{i}a + \mathbf{j}b + \mathbf{k}c \tag{A.36}$$

The same vector **a** allows for a matrix representation in one of the following forms:

$$\mathbf{a} = \begin{bmatrix} a \\ b \\ c \end{bmatrix} \tag{A.37}$$

$$\mathbf{a} = \begin{bmatrix} a \\ b \\ c \\ 1 \end{bmatrix} \tag{A.38}$$

$$\mathbf{a} = \begin{bmatrix} 1 & 0 & 0 & a \\ 0 & 1 & 0 & b \\ 0 & 0 & 1 & c \\ 0 & 0 & 0 & 1 \end{bmatrix} \tag{A.39}$$

Operations over vectors in a matrix representation (see the form Eq. (A.39)) is preferred as multiple coordinate system transformations are often required.

Vectors obey the commutative law, that is, the equalities $\mathbf{a} \times \mathbf{b} = \mathbf{b} \times \mathbf{a}$ and $\mathbf{a} \cdot \mathbf{b} = \mathbf{b} \cdot \mathbf{a}$ are valid for vectors. This is not always applicable to matrices.

Two kinds of products are valid for vectors, that is, the dot product $(\mathbf{a} \cdot \mathbf{b})$ and the cross product $(\mathbf{a} \times \mathbf{b})$ of vectors, which are not valid with respect to matrices.

Appendix B: Elements of the Differential Geometry of Surfaces

The discussion in this book is primarily focused on the elements of the theory of gear cutting tool design.

A gear and pinion tooth flank and their motion in space in relation to one another are analytically described in a reference system. An orthogonal Cartesian[2] reference

---

[2] ***René Descartes*** (March 31, 1596–February 11, 1650) (Latinized form: *Renatus* ***Cartesius***;), a French mathematician, philosopher, and writer.

system is a major kind of reference system that is commonly used for this purpose. The mutually perpendicular coordinate axes of a Cartesian coordinate system are conventionally labeled as *X*, *Y*, and *Z*.

In a Cartesian reference system, the axes can be oriented in either a left- or right-handed sense. A right-handed Cartesian reference system is preferred, and all algorithms and formulae used in this book assume a right-handed convention.

A coordinate system provides a numerical frame of reference for the three-dimensional space in which the theory is developed. Two coordinate systems are particularly useful to us: the ubiquitous Cartesian (*XYZ*) rectilinear system and the spherical polar (r, θ, φ) or angular system. Cartesian coordinate systems are the most commonly used, but angular coordinates are often helpful as well.

B.1 Specification of a Gear Tooth Flank

A gear tooth flank could be uniquely determined by two independent variables. Therefore, we give a gear tooth flank, $\mathcal{G}$ (see Fig. B.1), in most cases, by expressing

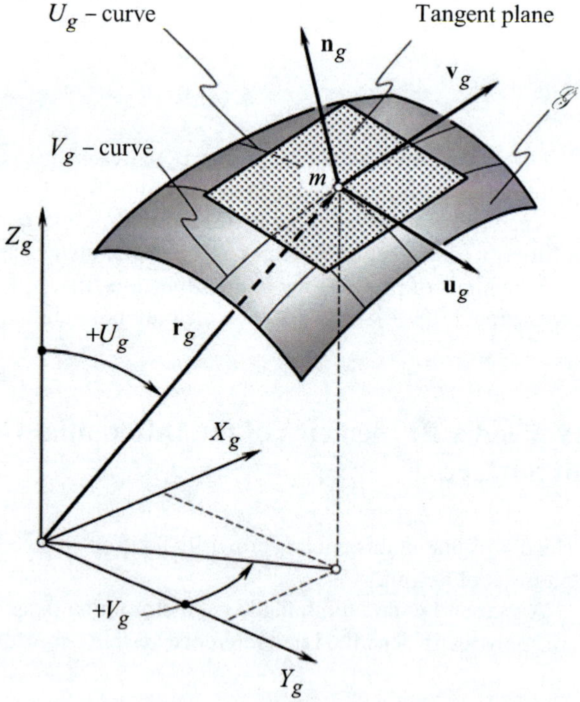

Fig. B.1 Principal parameters of the local topology of a gear tooth flank, $\mathcal{G}$

its rectangular coordinates X_g, Y_g, and Z_g as functions of two Gaussian[3] coordinates U_g and V_g in a certain closed interval:[4]

$$\mathcal{G} \quad \Rightarrow \quad \mathbf{r}_g = \mathbf{r}_g(U_g, V_g) = \begin{bmatrix} X_g(U_g, V_g) \\ Y_g(U_g, V_g) \\ Z_g(U_g, V_g) \\ 1 \end{bmatrix} \tag{B.1}$$

$$U_{1.g} \leq U_g \leq U_{2.g}; \quad V_{1.g} \leq V_g \leq V_{2.g}$$

where

$\mathbf{r}_g$ is the position vector of a point of the gear tooth flank $\mathcal{G}$.
U_g and V_g are the curvilinear coordinates (Gaussian coordinates) of the gear tooth flank $\mathcal{G}$.
X_g, Y_g, Z_g are the Cartesian coordinates of the point of the gear tooth flank $\mathcal{G}$.
$U_{1.\,g}$ and $U_{2.\,g}$ are the boundary values of the closed interval of the U_g– parameter
$V_{1.\,g}$ and $V_{2.\,g}$ are the boundary values of the closed interval of the V_g– parameter.

The parameters U_g and V_g must enter in Eq. (B.1) independently, which means that the matrix:

$$\mathbf{M} = \begin{bmatrix} \dfrac{\partial X_g}{\partial U_g} & \dfrac{\partial Y_g}{\partial U_g} & \dfrac{\partial Z_g}{\partial U_g} \\ \dfrac{\partial X_g}{\partial V_g} & \dfrac{\partial Y_g}{\partial V_g} & \dfrac{\partial Z_g}{\partial V_g} \end{bmatrix} \tag{B.2}$$

has a rank of 2.

Positions at which the rank is 1 or 0 are singular points; when the rank at all points is 1, then Eq. (B.1) represents a curve.

Other methods of surface specification are also known. Specification of a gear tooth flank by

- An equation in explicit form
- An equation in implicit form
- A set of parametric equations

are among the most frequently used methods of surface specification in practice.

It is assumed here and below that for any given kind of a gear tooth flank, $\mathcal{G}$, its specification can be converted either into a vector form or into a matrix form as it follows from Eq. (B.1).

---

[3] Johan Carl Friedrich Gauss (April 30, 1777–February 23, 1855), a famous German mathematician and physical scientist.

[4] All the equations that are valid for the gear tooth flank $\mathcal{G}$ are also valid for the pinion tooth flank $\mathcal{P}$.

B.2 Tangent Vectors and Tangent Plane; Unit Normal Vector

The following notation is proven to be convenient in the consideration below.

The first derivatives of $\mathbf{r}_g$ with respect to Gaussian coordinates U_g and V_g are designated as:

$$\frac{\partial \mathbf{r}_g}{\partial U_g} = \mathbf{U}_g \tag{B.3}$$

$$\frac{\partial \mathbf{r}_P}{\partial V_P} = \mathbf{V}_P \tag{B.4}$$

and for the unit tangent vectors:

$$\mathbf{u}_g = \frac{\mathbf{U}_g}{|\mathbf{U}_g|} \tag{B.5}$$

$$\mathbf{v}_g = \frac{\mathbf{V}_g}{|\mathbf{V}_g|} \tag{B.6}$$

correspondingly.[5]

The direction of the tangent line to the U_g- coordinate line through a given point m on the gear tooth flank $\mathcal{G}$ is specified by the unit tangent vector $\mathbf{u}_g$ (as well as by the tangent vector $\mathbf{U}_g$). Similarly, the direction of the tangent line to the V_g- coordinate line through the same point m on the gear tooth flank $\mathcal{G}$ is specified by the unit tangent vector $\mathbf{v}_g$ (as well as by the tangent vector $\mathbf{V}_g$).

The significance of the unit tangent vectors $\mathbf{u}_g$ and $\mathbf{v}_g$ becomes evident from the following considerations.

First, the unit tangent vectors $\mathbf{u}_g$ and $\mathbf{v}_g$ yield an equation of the tangent plane to the gear tooth flank $\mathcal{G}$ at a specified point m:

$$\text{Tangent plane} \quad \Rightarrow \quad \begin{bmatrix} \left[\mathbf{r}_{t.p} - \mathbf{r}_g^m\right] \\ \mathbf{u}_g \\ \mathbf{v}_g \\ 1 \end{bmatrix} = 0 \tag{B.7}$$

where

$\mathbf{r}_{t.\,p}$ is the position vector of a point of the tangent plane to the gear tooth flank $\mathcal{G}$ at a specified point m.

$\mathbf{r}_g^m$ is the position vector of the point m on the gear tooth flank $\mathcal{G}$.

[5] It is right to underline here that the unit tangent vectors $\mathbf{u}_P$ and $\mathbf{v}_P$ are dimensionless values as it is follows from Eqs. (B.5) and (B.6).

Second, the tangent vectors yield an equation of the perpendicular $\mathbf{N}_g$ and of the unit normal vector $\mathbf{n}_g$ to the gear tooth flank $\mathcal{G}$ at a given point m:

$$\mathbf{N}_g = \mathbf{U}_g \times \mathbf{V}_g \tag{B.8}$$

and

$$\mathbf{n}_g = \frac{\mathbf{N}_g}{|\mathbf{N}_g|} = \frac{\mathbf{U}_g \times \mathbf{V}_g}{|\mathbf{U}_g \times \mathbf{V}_g|} = \mathbf{u}_g \times \mathbf{v}_g \tag{B.9}$$

When the order of the multipliers in Eqs. (B.8) and (B.9) is properly chosen, then the unit normal vector $\mathbf{n}_g$ (as well as the normal vector $\mathbf{N}_g$) is pointed outward of the bodily side of the surface $\mathcal{G}$.

B.3 Local Frame

Two unit tangent vectors $\mathbf{u}_g$ and $\mathbf{v}_g$ along with the unit normal vector $\mathbf{n}_g$ comprise a local frame $\mathbf{u}_g$, $\mathbf{v}_g$, $\mathbf{n}_g$, having the origin at a current point m on the gear tooth flank $\mathcal{G}$. The unit tangent vector $\mathbf{u}_g$ is perpendicular to the unit normal vector $\mathbf{n}_g$ (that is, $\mathbf{u}_g \perp \mathbf{n}_g$) and the unit tangent vector $\mathbf{v}_g$ is also perpendicular to the unit normal vector $\mathbf{n}_g$ (that is, $\mathbf{v}_g \perp \mathbf{n}_g$). In general, the unit tangent vectors $\mathbf{u}_g$ and $\mathbf{v}_g$ are not perpendicular to one another, and they form a certain angle ω_g. In order to construct an orthogonal local frame, either the unit tangent vector $\mathbf{u}_g$ in the local frame ($\mathbf{u}_g$, $\mathbf{v}_g$, $\mathbf{n}_g$) must be substituted with a unit tangent vector $\mathbf{u}_g^*$ or the unit tangent vector $\mathbf{v}_g$ in the same local frame ($\mathbf{u}_g$, $\mathbf{v}_g$, $\mathbf{n}_g$) must be substituted with a unit tangent vector $\mathbf{v}_g^*$. For the calculation of the newly introduced unit tangent vectors $\mathbf{u}_g^*$ and $\mathbf{v}_g^*$, the following equations can be used:

$$\mathbf{u}_g^* = \mathbf{u}_g \times \mathbf{n}_g \tag{B.10}$$

$$\mathbf{v}_g^* = \mathbf{v}_g \times \mathbf{n}_g \tag{B.11}$$

It is convenient to choose that order of the multipliers in Eqs. (B.10) and (B.11), which preserves the orientation (the hand) of the original local frame ($\mathbf{u}_g$, $\mathbf{v}_g$, $\mathbf{n}_g$), namely, if the original local frame ($\mathbf{u}_g$, $\mathbf{v}_g$, $\mathbf{n}_g$) is right-hand-oriented, then the newly constructed local frame (either the local frame $\left(\mathbf{u}_g^*, \mathbf{v}_g, \mathbf{n}_g\right)$ or the local frame $\left(\mathbf{u}_g^*, \mathbf{v}_g, \mathbf{n}_g\right)$) should also be a right-hand-oriented local frame, and vice versa.

It should be pointed out here that another possibility to construct an orthogonal local frame is also available. A local frame of this kind is commonly referred to as a "Darboux[6] frame" and is briefly considered below in this section of this book.

The unit tangent vectors $\mathbf{u}_g$ and $\mathbf{v}_g$ to a surface $\mathcal{G}$ at a point m are of critical importance when solving practical problems in the field of gearing. This statement is proven by the numerous examples shown below.

B.4 Fundamental Forms of a Surface

Let us consider two other important issues concerning the gear tooth flank geometry – both are related to intrinsic geometry in the differential vicinity of a current surface point m.

The first fundamental form of a surface The first issue is the so-called "first fundamental form" $\mathbf{\Phi}_{1.\ g}$ of a gear tooth flank, $\mathcal{G}$. The metric properties of the gear tooth flank $\mathcal{G}$ are described by the first fundamental form $\mathbf{\Phi}_{1.\ g}$ of the surface. Usually, the first fundamental form $\mathbf{\Phi}_{1.\ g}$ is represented in a quadratic form as follows:

$$\mathbf{\Phi}_{1.g} \quad \Rightarrow \quad ds_g^2 = E_g\, dU_g^2 + 2F_g\, dU_g\, dV_g + G_g\, dV_g^2 \tag{B.12}$$

where

s_g is the linear element on the gear tooth flank $\mathcal{G}$ (s_g is equal to the length of a segment of a certain curve on the gear tooth flank $\mathcal{G}$).

E_g, F_g, G_g are the fundamental magnitudes of the first order at a surface point.

Equation (B.12) for the first fundamental form $\mathbf{\Phi}_{1.\ g}$ is known from many advanced sources. In the theory of gearing, another form of the analytical representation of the first fundamental form $\mathbf{\Phi}_{1.\ g}$ is proven to be useful:

$$\mathbf{\Phi}_{1.g} \quad \Rightarrow \quad ds_g^2 = [dU_g \quad dV_g \quad 0 \quad 0] \cdot \begin{bmatrix} E_g & F_g & 0 & 0 \\ F_g & G_g & 0 & 0 \\ 0 & 0 & 1 & 0 \\ 0 & 0 & 0 & 1 \end{bmatrix} \cdot \begin{bmatrix} dU_g \\ dV_g \\ 0 \\ 0 \end{bmatrix} \tag{B.13}$$

This kind of analytical representation of the first fundamental form $\mathbf{\Phi}_{1.\ P}$ was proposed by Prof. S.P. Radzevich (~2008).

[6]Jean Gaston Darboux (August 14, 1842–February 23, 1917), a French mathematician.

The practical advantage of Eq. (B.13) is that it can be easily incorporated into computer programs when multiple coordinate system transformations are used. The last is vital for the theory of gearing.

The fundamental magnitudes of the first order E_g, F_g, G_g can be calculated from the set of the following equations:

$$E_g = \mathbf{U}_g \cdot \mathbf{U}_g \tag{B.14}$$

$$F_g = \mathbf{U}_g \cdot \mathbf{V}_g \tag{B.15}$$

$$G_g = \mathbf{V}_g \cdot \mathbf{V}_g \tag{B.16}$$

Equations (B.14) through (B.16) can be represented in an expanded form as follows:

$$E_g = \frac{\partial \mathbf{r}_g}{\partial U_g} \cdot \frac{\partial \mathbf{r}_g}{\partial U_g} = \frac{\partial X_g}{\partial U_g} \cdot \frac{\partial X_g}{\partial U_g} + \frac{\partial Y_g}{\partial U_g} \cdot \frac{\partial Y_g}{\partial U_g} + \frac{\partial Z_g}{\partial U_g} \cdot \frac{\partial Z_g}{\partial U_g} \tag{B.17}$$

$$F_g = \frac{\partial \mathbf{r}_g}{\partial U_g} \cdot \frac{\partial \mathbf{r}_g}{\partial V_g} = \frac{\partial X_g}{\partial U_g} \cdot \frac{\partial X_g}{\partial V_g} + \frac{\partial Y_g}{\partial U_g} \cdot \frac{\partial Y_g}{\partial V_g} + \frac{\partial Z_g}{\partial U_g} \cdot \frac{\partial Z_g}{\partial V_g} \tag{B.18}$$

$$G_g = \frac{\partial \mathbf{r}_g}{\partial V_g} \cdot \frac{\partial \mathbf{r}_g}{\partial V_g} = \frac{\partial X_g}{\partial V_g} \cdot \frac{\partial X_g}{\partial V_g} + \frac{\partial Y_g}{\partial V_g} \cdot \frac{\partial Y_g}{\partial V_g} + \frac{\partial Z_g}{\partial V_g} \cdot \frac{\partial Z_g}{\partial V_g} \tag{B.19}$$

The fundamental magnitudes of the first order E_g, F_g, G_g are functions of the U_g- and V_g- coordinates of a point of the gear tooth flank $\mathcal{G}$. In their general form, these relationships can be represented in the form:

$$E_g = E_g(U_g, V_g) \tag{B.20}$$

$$F_g = F_g(U_g, V_g) \tag{B.21}$$

$$G_g = G_g(U_g, V_g) \tag{B.22}$$

It is important to point out here that the fundamental magnitudes E_g and G_g are always positive (that is, $E_g > 0$, $G_g > 0$) and that the fundamental magnitude F_g can be equal to zero ($F_g \geq 0$). This indicates that the first fundamental form $\mathbf{\Phi}_{1.\ g}$ at a point of the gear tooth flank $\mathcal{G}$ is always positively defined ($\mathbf{\Phi}_{1.\ g} \geq 0$) and cannot be of a negative value.

By use of the first fundamental form $\mathbf{\Phi}_{1.\ g}$, the following major parameters of the geometry of the gear tooth flank $\mathcal{G}$ can be calculated:

(a) The length of a curve-line segment on the gear tooth flank $\mathcal{G}$
(b) The square of the gear tooth flank $\mathcal{G}$ portion that is bounded by a closed curve on the surface
(c) The angle between any two directions on the gear tooth flank $\mathcal{G}$

The length s_g of a curve-line segment:

$$U_g = U_g(t) \tag{B.23}$$

$$V_g = V_g(t) \tag{B.24}$$

on the gear tooth flank $\mathcal{G}$ is given by the equation:

$$s_g = \int_{t_0}^{t} \sqrt{E_g\left(\frac{dU_g}{dt}\right)^2 + 2F_g\frac{dU_g}{dt}\frac{dV_g}{dt} + G_g\left(\frac{dV_g}{dt}\right)^2}\, dt \tag{B.25}$$

$$t_0 \leq t \leq t_1$$

For calculation of the square S_g of the gear tooth flank $\mathcal{G}$ patch Σ, which is bounded by a closed curve on the surface $\mathcal{G}$, the following equation can be used:

$$S_g = \iint_{\Sigma} \sqrt{E_g G_g - F_g^2}\, dU_g\, dV_g \tag{B.26}$$

Ultimately, the value of the angle ω_g between two given directions through a certain point m on the gear tooth flank $\mathcal{G}$ can be calculated from one of the equations below:

$$\cos \omega_g = \frac{F_g}{\sqrt{E_g G_g}} \tag{B.27}$$

$$\sin \omega_g = \frac{H_g}{\sqrt{E_g G_g}} \tag{B.28}$$

$$\tan \omega_g = \frac{H_g}{F_g} \tag{B.29}$$

For the calculation of the discriminant H_g of the first fundamental form $\Phi_{1.\,g}$, the following equation can be used:

$$H_g = \sqrt{E_g G_g - F_g^2} \tag{B.30}$$

It is assumed here that the discriminant H_g is always non-negative – that is, $H_g = +\sqrt{E_g G_g - F_g^2}$.

The first fundamental form $\Phi_{1.\,g}$ represents the length of a curve-line segment, and, thus, it is always non-negative – that is, the inequality $\Phi_{1.\,g} \geq 0$ is always valid.

The first fundamental form $\Phi_{1.\,g}$ remains the same when the surface is banding. This is another important feature of the first fundamental form $\Phi_{1.\,g}$.

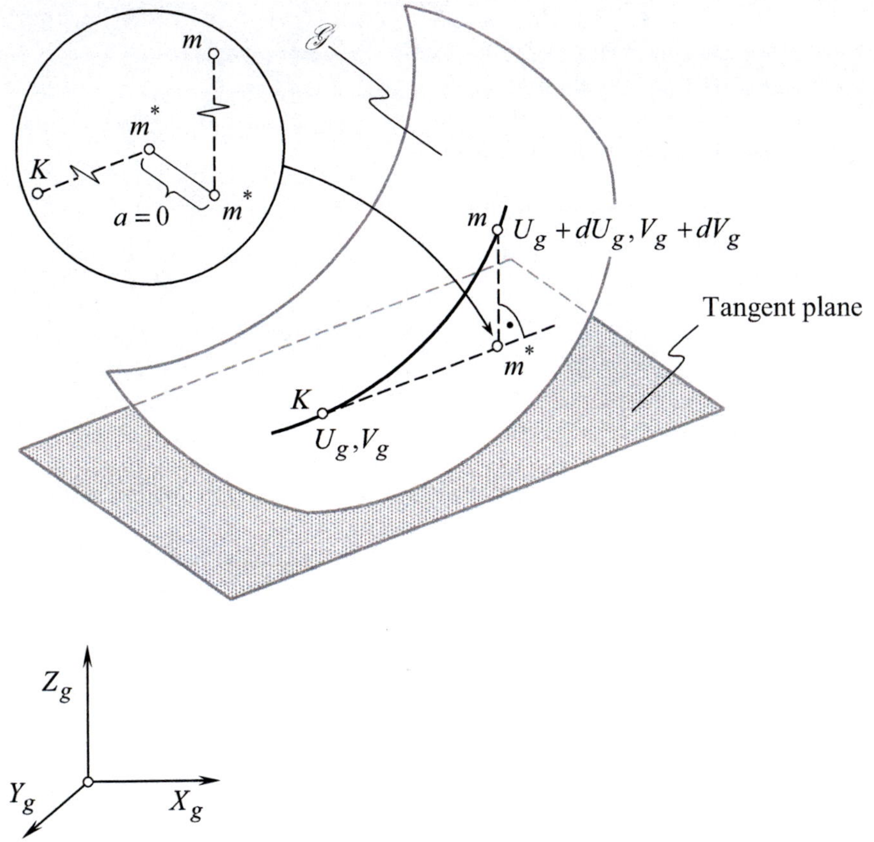

Fig. B.2 Definition of the second fundamental form $\mathbf{\Phi}_{2.\ g}$ at a point of a smooth gear tooth flank, $\mathcal{G}$

Second fundamental form of a surface The "second fundamental form" $\mathbf{\Phi}_{2.\ g}$ of the gear tooth flank $\mathcal{G}$ is another of the two abovementioned issues. The second fundamental form $\mathbf{\Phi}_{2.\ g}$ describes the curvature of a smooth regular surface, $\mathcal{G}$.

Let us consider a point K on the smooth regular part surface $\mathcal{G}$ (see Fig. B.2). The location of the point K is specified by two coordinates, U_g and V_g. A line through the point K is entirely located within the surface $\mathcal{G}$. A nearby point m is located within the line through the point K. The location of the point m is specified by the coordinates $U_g + dU_g$ and $V_g + dV_g$ as it is infinitesimally close to the point K. The closest distance of approach of the point m to the tangent plane through the point K is expressed by the second fundamental form $\mathbf{\Phi}_{2.\ g}$. Torsion of the curve Km is ignored. Therefore, the distance a is assumed to be equal to zero ($a = 0$).

The second fundamental form $\mathbf{\Phi}_{2.\ g}$ describes the curvature of the smooth, regular part surface $\mathcal{G}$. Usually, it is represented in a quadratic form (see Fig. B.2) as follows:

$$\Phi_{2.g} \quad \Rightarrow \quad -d\mathbf{r}_g \cdot d\mathbf{n}_g = L_g\, dU_g^2 + 2M_g\, dU_g\, dV_g + N_g\, dV_g^2 \tag{B.31}$$

Equation (B.31) is known from many advanced sources.

In the theory of gearing, another analytical representation of the second fundamental form $\mathbf{\Phi}_{2.\ g}$ is proven to be useful:

$$\Phi_{2.g} \quad \Rightarrow \quad [dU_g \quad dV_g \quad 0 \quad 0] \cdot \begin{bmatrix} L_g & M_g & 0 & 0 \\ M_g & N_g & 0 & 0 \\ 0 & 0 & 1 & 0 \\ 0 & 0 & 0 & 1 \end{bmatrix} \cdot \begin{bmatrix} dU_g \\ dV_g \\ 0 \\ 0 \end{bmatrix} \tag{B.32}$$

This analytical representation of the second fundamental form $\mathbf{\Phi}_{2.\ \mathcal{P}}$ was proposed by Prof. S.P. Radzevich (~2008).

Similar to Eq. (B.13), the practical advantage of Eq. (B.32) is that it can be easily incorporated into computer programs when multiple coordinate system transformations are used. The latter is vital for the theory of gearing.

In Eq. (B.32), the parameters L_g, M_g, and N_g designate the fundamental magnitudes of the second order.

By definition, the fundamental magnitudes of the second order are equal:

$$L_g = -\mathbf{U}_g \cdot \frac{\partial \mathbf{n}_g}{\partial U_g} = \mathbf{n}_g \cdot \frac{\partial \mathbf{U}_g}{\partial U_g} \tag{B.33}$$

$$M_g = -\frac{1}{2}\left(\mathbf{U}_g \cdot \frac{\partial \mathbf{n}_g}{\partial V_g} + \mathbf{V}_g \cdot \frac{\partial \mathbf{n}_g}{\partial U_g}\right) = \mathbf{n}_g \cdot \frac{\partial \mathbf{U}_g}{\partial V_g} = \mathbf{n}_g \cdot \frac{\partial \mathbf{V}_g}{\partial U_g} \tag{B.34}$$

$$N_g = -\mathbf{V}_g \cdot \frac{\partial \mathbf{n}_g}{\partial V_g} = \mathbf{n}_g \cdot \frac{\partial \mathbf{V}_g}{\partial V_g} \tag{B.35}$$

For the calculation of the fundamental magnitudes of the second order of the smooth regular gear tooth flank $\mathcal{G}$, the following equations can be used:

$$L_g = \frac{\frac{\partial \mathbf{U}_g}{\partial U_g} \times \mathbf{U}_g \cdot \mathbf{V}_g}{\sqrt{E_g G_g - F_g^2}} \tag{B.36}$$

$$M_g = \frac{\frac{\partial \mathbf{U}_g}{\partial V_g} \times \mathbf{U}_g \cdot \mathbf{V}_g}{\sqrt{E_g G_g - F_g^2}} = \frac{\frac{\partial \mathbf{V}_g}{\partial U_g} \times \mathbf{U}_g \cdot \mathbf{V}_g}{\sqrt{E_g G_g - F_g^2}} \tag{B.37}$$

$$N_g = \frac{\frac{\partial \mathbf{V}_g}{\partial V_g} \times \mathbf{U}_g \cdot \mathbf{V}_g}{\sqrt{E_g G_g - F_g^2}} \tag{B.38}$$

Equations (B.36) through (B.38) can be represented in an expanded form as follows:

$$L_g = \frac{\begin{vmatrix} \frac{\partial^2 X_g}{\partial U_g^2} & \frac{\partial^2 Y_g}{\partial U_g^2} & \frac{\partial^2 Z_g}{\partial U_g^2} \\ \frac{\partial X_g}{\partial U_g} & \frac{\partial Y_g}{\partial U_g} & \frac{\partial Z_g}{\partial U_g} \\ \frac{\partial X_g}{\partial V_g} & \frac{\partial Y_g}{\partial V_g} & \frac{\partial Z_g}{\partial V_g} \end{vmatrix}}{\sqrt{E_g G_g - F_g^2}} \tag{B.39}$$

$$M_g = \frac{\begin{vmatrix} \frac{\partial^2 X_g}{\partial U_g \partial V_g} & \frac{\partial^2 Y_g}{\partial U_g \partial V_g} & \frac{\partial^2 Z_g}{\partial U_g \partial V_g} \\ \frac{\partial X_g}{\partial U_g} & \frac{\partial Y_g}{\partial U_g} & \frac{\partial Z_g}{\partial U_g} \\ \frac{\partial X_g}{\partial V_g} & \frac{\partial Y_g}{\partial V_g} & \frac{\partial Z_g}{\partial V_g} \end{vmatrix}}{\sqrt{E_g G_g - F_g^2}} \tag{B.40}$$

$$N_P = \frac{\begin{vmatrix} \frac{\partial^2 X_g}{\partial V_g^2} & \frac{\partial^2 Y_g}{\partial V_g^2} & \frac{\partial^2 Z_g}{\partial V_g^2} \\ \frac{\partial X_g}{\partial U_g} & \frac{\partial Y_g}{\partial U_g} & \frac{\partial Z_g}{\partial U_g} \\ \frac{\partial X_g}{\partial V_g} & \frac{\partial Y_g}{\partial V_g} & \frac{\partial Z_g}{\partial V_g} \end{vmatrix}}{\sqrt{E_g G_g - F_g^2}} \tag{B.41}$$

The fundamental magnitudes of the second order L_g, M_g, and N_g are also functions of the U_g- and V_g- coordinates of a point of the gear tooth flank $\mathcal{G}$. In general, these relationships can be represented in the form:

$$L_g = L_g(U_g, V_g) \tag{B.42}$$

$$M_g = M_g(U_g, V_g) \tag{B.43}$$

$$N_g = N_g(U_g, V_g) \tag{B.44}$$

The discriminant T_g of the second fundamental form $\Phi_{2.\,g}$ can be calculated from the following equation:

$$T_g = \sqrt{L_g N_g - M_g^2} \tag{B.45}$$

We now come to a theorem, which is an essential justification for considering the differential geometry of surfaces in connection with the six fundamental magnitudes. It was first proven (1867) by Bonnet[7] and may be enunciated as follows:

Theorem B.1 When six fundamental magnitudes E_g, F_g, G_g and L_g, M_g, N_g are given, and when they fulfill the Gauss characteristic equation and the two Mainardi[8]–Codazzi[9] relations, they determine a gear tooth flank, $\mathcal{G}$, uniquely say as to its position and orientation in space.

This theorem is commonly referred to as the "main theorem in the theory of surface" or simply as the "Bonnet theorem." According to the main theorem, two surfaces that have identical first and second fundamental forms must be either congruent or symmetrical to one another.

By the use of the six fundamental magnitudes, all parameters of local geometry of a given part surface can be calculated.

B.5 Principal Directions on a Gear Tooth Flank

The direction of vectors of the principal directions $\mathbf{T}_{1.\,g}$ and $\mathbf{T}_{2.\,g}$ at a point of a gear tooth flank, $\mathcal{G}$, can be specified in terms of the ratio dU_g/dV_g. For the vectors $\mathbf{T}_{1.\,g}$ and $\mathbf{T}_{2.\,g}$ of the first and second principal directions, respectively, at a point m of a smooth, regular part surface $\mathcal{G}$, the corresponding values of the ratio dU_g/dV_g are calculated as the roots of the quadratic equation:

$$\begin{vmatrix} E_g\,dU_g + F_g\,dV_g & F_g\,dU_g + G_g\,dV_g \\ L_g\,dU_g + M_g\,dV_g & M_g\,dU_g + N_g\,dV_g \end{vmatrix} = 0 \tag{B.46}$$

The first principal plane section $C_{1.\,g}$ is perpendicular to the gear tooth flank $\mathcal{G}$ at a current surface point m and passes through the vector of the first principal direction $\mathbf{T}_{1.\,g}$. The second principal plane section $C_{2.\,g}$ is orthogonal to the gear tooth flank $\mathcal{G}$ at a current surface point m and passes through the vector of the second principal direction $\mathbf{T}_{2.\,g}$.

The principal directions $\mathbf{T}_{1.\,g}$ and $\mathbf{T}_{2.\,g}$ can be identified at any and all points of the smooth, regular gear tooth flank $\mathcal{G}$, except for umbilic points and flat points of the surface. At umbilic points of a surface, as well as at flat points, the principal directions cannot be identified.

---

[7] Pierre Ossian Bonnet (December 22, 1819–June 22, 1892), a French mathematician.

[8] Gaspare Mainardi (June 27, 1800–March 9, 1879), an Italian mathematician.

[9] Delfino Codazzi (March 7, 1824–July 21, 1873), an Italian mathematician.

In the theory of gearing, it is often preferred to not use the vectors $\mathbf{T}_{1.g}$ and $\mathbf{T}_{2.g}$ of the principal directions but, instead, to use the unit vectors $\mathbf{t}_{1.g}$ and $\mathbf{t}_{2.g}$ of the principal directions. The unit tangent vectors $\mathbf{t}_{1.g}$ and $\mathbf{t}_{2.g}$ are calculated from the equations:

$$\mathbf{t}_{1.g} = \frac{\mathbf{T}_{1.g}}{|\mathbf{T}_{1.g}|} \tag{B.47}$$

$$\mathbf{t}_{2.g} = \frac{\mathbf{T}_{2.g}}{|\mathbf{T}_{2.g}|} \tag{B.48}$$

respectively.

The unit tangent vectors $\mathbf{t}_{1.g}$ and $\mathbf{t}_{2.g}$ of the principal directions at a point m on the gear tooth flank $\mathcal{G}$ along with the unit normal vector $\mathbf{n}_g$ at the same point m comprise an orthogonal local frame ($\mathbf{t}_{1.g}, \mathbf{t}_{2.g}, \mathbf{n}_g$). All three unit vectors $\mathbf{t}_{1.g}$, $\mathbf{t}_{2.g}$, and $\mathbf{n}_g$ are mutually perpendicular to one another. The local frame ($\mathbf{t}_{1.g}, \mathbf{t}_{2.g}, \mathbf{n}_g$) is commonly referred to as a "Darboux frame."

B.6 Curvatures at a Point of a Part Surface

The first and second principal radii of curvature, i.e., $R_{1.g}$ and $R_{2.g}$, respectively, at a point of the gear tooth flank $\mathcal{G}$ are measured within the first and second principal plane sections $C_{1.g}$ and $C_{2.g}$, respectively. For the calculation of values of the principal radii of curvature, the following equation is commonly used:

$$R_g^2 - \frac{E_g N_g - 2F_g M_g + G_g L_g}{T_g} R_g + \frac{H_g}{T_g} = 0 \tag{B.49}$$

One should remember that algebraic values of the radii of principal curvature $R_{1.g}$ and $R_{2.g}$ relate to one another as $R_{2.g} > R_{1.g}$. In particular cases, at umbilic points on the gear tooth flank $\mathcal{G}$, no principal curvatures can be identified as all normal curvatures of the tooth surface $\mathcal{G}$ at an umbilic point are equal to one another.

Another two important parameters of the local topology of the gear tooth flank $\mathcal{G}$ are as follows:

- The mean curvature $\mathcal{M}_g$
- The intrinsic curvature (Gaussian curvature or full curvature) curvature $\mathcal{G}_g$

For the calculation of the curvatures $\mathcal{M}_g$ and $\mathcal{G}_g$, the following equations are commonly used:

$$\mathcal{M}_g = \frac{k_{1.g} + k_{2.g}}{2} = \frac{E_g N_g - 2F_g M_g + G_g L_g}{2 \cdot \left(E_g G_g - F_g^2\right)} \tag{B.50}$$

$$\mathcal{G}_g = k_{1.g} \cdot k_{2.g} = \frac{L_g N_g - M_g^2}{E_g G_g - F_g^2} \tag{B.51}$$

The expressions for the mean curvature $\mathcal{M}_g$ and for the Gaussian curvature $\mathcal{G}_g$:

$$\mathcal{M}_g = \frac{k_{1.g} + k_{2.g}}{2} \tag{B.52}$$

$$\mathcal{G}_g = k_{1.g} \cdot k_{2.g} \tag{B.53}$$

considered together yield a quadratic equation with respect to the principal curvatures $k_{1.\,g}$ and $k_{2.\,g}$:

$$k_g^2 - 2\,\mathcal{M}_g k_g + \mathcal{G}_g = 0 \tag{B.54}$$

The following formulae

$$k_{1.g} = \mathcal{M}_g + \sqrt{\mathcal{M}_g^2 - \mathcal{G}_g} \tag{B.55}$$

$$k_{2.g} = \mathcal{M}_g - \sqrt{\mathcal{M}_g^2 - \mathcal{G}_g} \tag{B.56}$$

are the solutions to Eq. (B.54).

Here, in Eqs. (B.55) and (B.56), the first principal curvature of the gear tooth flank $\mathcal{G}$ at a current point m is designated as $k_{1.\,g}$, and $k_{2.\,g}$ is designated as the second principal curvature of the gear tooth flank $\mathcal{G}$ at the same point m.

The principal curvatures $k_{1.\,g}$ and $k_{2.\,g}$ are the reciprocals of the corresponding principal radii of curvature $R_{1.\,g}$ and $R_{2.\,g}$:

$$k_{1.g} = \frac{1}{R_{1.g}} \tag{B.57}$$

$$k_{2.g} = \frac{1}{R_{2.g}} \tag{B.58}$$

The first principal curvature $k_{1.\,g}$ is always larger than the second principal curvature $k_{2.\,g}$ of the gear tooth flank $\mathcal{G}$ at a current point m – that is, the inequality:

$$k_{1.g} > k_{2.g} \tag{B.59}$$

is always valid.

This brief consideration of the major elements of part surface geometry makes it possible to introduce two definitions that are of critical importance for further discussion.

As already mentioned earlier in this section of the book, it was proven by Bonnet that the specification of the first and the second fundamental forms determines a unique surface if "Gauss' characteristic equation" and the "Mainardi–Codazzi relations of compatibility" are satisfied and those two surfaces that have identical first and second fundamental forms are congruent.[10] The six fundamental magnitudes determine a surface uniquely, except as to its position and orientation in space.

Specification of a surface in terms of the first and second fundamental forms is usually called the "natural kind" of surface representation. In general form, this kind of part surface representation can be expressed by a set of two equations:

$$\left.\begin{array}{r}\text{Natural form}\\ \text{of surface } \mathcal{G} \text{ representation}\end{array}\right| \Rightarrow \quad \mathcal{G} = \mathcal{G}\left(\boldsymbol{\Phi}_{1.g},\ \boldsymbol{\Phi}_{2.g}\right)$$
$$\times \begin{cases} \boldsymbol{\Phi}_{1.g} = \boldsymbol{\Phi}_{1.g}\left(E_g,\ F_g,\ G_g\right) \\ \boldsymbol{\Phi}_{2.g} = \boldsymbol{\Phi}_{2.g}\left(E_g,\ F_g,\ G_g,\ L_g,\ M_g,\ N_g\right) \end{cases} \tag{B.60}$$

Equation (B.60) can be derived from Eq. (B.1). A given gear tooth flank $\mathcal{G}$ can be expressed in both forms, namely, either by Eq. (B.19) or by Eq. (B.1).

B.7 Illustrative Example

Let us consider an example of how an analytical representation of a surface in a Cartesian reference system can be converted into the natural representation of the same surface.

A Cartesian coordinate system $X_gY_gZ_g$ is associated with a gear tooth flank $\mathcal{G}$ as schematically shown in Fig. B.3.

The position vector of a point $\mathbf{r}_g$ of the gear tooth flank $\mathcal{G}$ can be represented as the sum of three vectors:

$$\mathbf{r}_g = \mathbf{A} + \mathbf{B} + \mathbf{C} \tag{B.61}$$

Each of the vectors **A**, **B**, and **C** can be expressed in terms of projections onto the axes of the reference system $X_gY_gZ_g$. Then, Eq. (B.61) is cast into the equation:

[10]Two surfaces with the identical first and second fundamental forms might also be symmetrical. Refer to the literature – Koenderink, J.J., *Solid Shape*, The MIT Press, Cambridge, MA, 1990, 699 pages – on the differential geometry of surfaces for details about this specific issue.

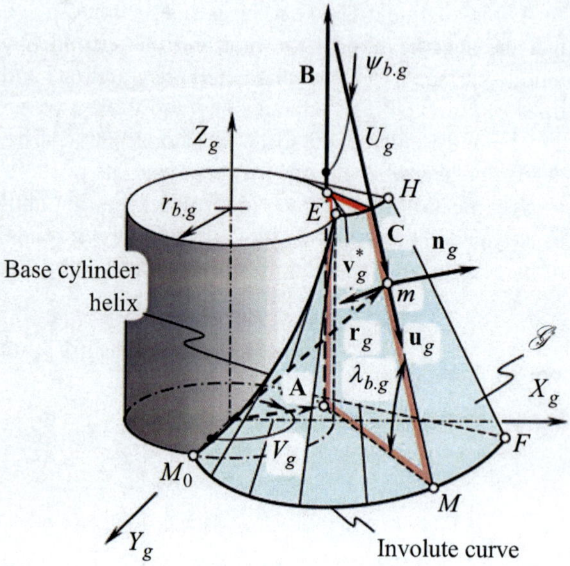

Fig. B.3 Derivation of the natural form of the representation of a gear tooth flank, *M*

$$\mathbf{r}_g(U_g, V_g) = \begin{bmatrix} r_{b.g} \cos V_g + U_g \cos \tau_{b.g} \sin V_g \\ r_{b.g} \sin V_g - U_g \sin \tau_{b.g} \sin V_g \\ r_{b.g} \tan \tau_{b.g} - U_g \sin \tau_{b.g} \\ 1 \end{bmatrix} \tag{B.62}$$

This yields the calculation of two tangent vectors $\mathbf{U}_g(U_g, V_g)$ and $\mathbf{V}_g(U_g, V_g)$, which are correspondingly equal:

$$\mathbf{U}_g(U_g, V_g) = \begin{bmatrix} \cos \tau_{b.g} \sin V_g \\ -\cos \tau_{b.g} \cos V_g \\ -\sin \tau_{b.g} \\ 0 \end{bmatrix} \tag{B.63}$$

$$\mathbf{V}_g(U_g, V_g) = \begin{bmatrix} -r_{b.g} \sin V_g + U_g \cos \tau_{b.g} \cos V_g \\ r_{b.g} \cos V_g + U_g \cos \tau_{b.g} \sin V_g \\ r_{b.g} \tan \tau_{b.g} \\ 0 \end{bmatrix} \tag{B.64}$$

By substituting the derived vectors $\mathbf{U}_g$ and $\mathbf{V}_g$ in Eq. (B.14), one can come up with the formulae for the calculation of the fundamental magnitudes of the first order:

$$E_g = 1 \tag{B.65}$$

$$F_g = -\frac{r_{b.g}}{\cos \tau_{b.g}} \tag{B.66}$$

$$G_g = \frac{U_g^2 \cos^4 \tau_{b.g} + r_{b.g}^2}{\cos^2 \tau_{b.g}} \tag{B.67}$$

These expressions can be directly substituted in Eq. (B.12) for the first fundamental form $\mathbf{\Phi}_{1.\ g}$ of the gear tooth flank $\mathcal{G}$:

$$\mathbf{\Phi}_{1.g} \quad \Rightarrow \quad dU_g^2 - 2\frac{r_{b.g}}{\cos \tau_{b.g}} dU_g dV_g + \frac{U_g^2 \cos^4 \tau_{b.g} + r_{b.g}^2}{\cos^2 \tau_{b.g}} dV_g^2 \tag{B.68}$$

The derived expressions for the fundamental magnitudes E_g, F_g, and G_g (see Eqs. (B.65) through (B.67)) can also be substituted in Eq. (B.13). In this manner, a corresponding matrix representation of the first fundamental form $\mathbf{\Phi}_{1.\ g}$ of the gear tooth flank $\mathcal{G}$ can be calculated. The interested reader may wish to complete these formulae on his or her own.

The discriminant H_g of the first fundamental form of the gear tooth flank $\mathcal{G}$ can be calculated from the expression:

$$H_g = U_g \cos \tau_{b.g} \tag{B.69}$$

In order to derive an equation for the second fundamental form $\mathbf{\Phi}_{2.\ g}$ of the gear tooth flank $\mathcal{G}$, the second derivatives of the position vector of the point $\mathbf{r}_g(U_g, V_g)$ with respect to the U_g- and V_g- parameters are necessary. The above-derived equations for the tangent vectors $\mathbf{U}_g$ and $\mathbf{V}_g$ (see Eqs. (B.63) and (B.64)) make the following expressions for the derivatives under consideration possible:

$$\frac{\partial \mathbf{U}_g}{\partial U_P} = \begin{bmatrix} 0 \\ 0 \\ 0 \\ 1 \end{bmatrix} \tag{B.70}$$

$$\frac{\partial \mathbf{U}_g}{\partial V_g} \equiv \frac{\partial \mathbf{V}_g}{\partial U_g} = \begin{bmatrix} \cos \tau_{b.g} \cos V_g \\ \cos \tau_{b.g} \sin V_g \\ 0 \\ 1 \end{bmatrix} \tag{B.71}$$

$$\frac{\partial \mathbf{V}_g}{\partial V_g} = \begin{bmatrix} -r_{b.g} \cos V_g - U_g \cos \tau_{b.g} \sin V_g \\ -r_{b.g} \sin V_g + U_g \cos \tau_{b.g} \cos V_g \\ 0 \\ 1 \end{bmatrix} \tag{B.72}$$

Furthermore, these expressions (see Eqs. (B.70) through (B.72)) are substituted in Eqs. (B.36) through (B.38). After the necessary formula transformations are complete, Eqs. (B.36) through (B.38) are cast into the set of formulae for the calculation of the fundamental magnitudes of the second order of the gear tooth flank $\mathcal{G}$. This set of formulae is as follows:

$$L_g = 0 \tag{B.73}$$

$$M_g = 0 \tag{B.74}$$

$$N_g = -U_g \sin\tau_{b.g} \cos\tau_{b.g} \tag{B.75}$$

Furthermore, after substituting Eqs. (B.73) through (B.75) in Eq. (B.31), an equation for the calculation of the second fundamental form of the gear tooth flank $\mathcal{G}$ can be represented in the form:

$$\mathbf{\Phi}_{2.g} \quad \Rightarrow \quad -d\mathbf{r}_g \cdot d\mathbf{N}_g = -U_g \sin\tau_{b.g} \cos\tau_{b.g} dV_g^2 \tag{B.76}$$

Similar to Eq. (B.68), the derived expressions for the fundamental magnitudes L_g, M_g, and N_g of the second order can be substituted in Eq. (B.32) for the second fundamental form $\mathbf{\Phi}_{2.g}$. In this manner, a corresponding matrix representation of the second fundamental form $\mathbf{\Phi}_{2.g}$ of the surface $\mathcal{G}$ can be derived. The interested reader may wish to complete this formula transformation on his or her own.

For the calculation of the discriminant T_g of the second fundamental form $\mathbf{\Phi}_{2.g}$ of the gear tooth flank $\mathcal{G}$, the following expression can be used:

$$T_g = U_g \sin\tau_{b.g} \cos\tau_{b.g} \tag{B.77}$$

The natural representation of the gear tooth flank $\mathcal{G}$ can be expressed in terms of the derived set of six equations for the calculation of the fundamental magnitudes of the first E_g, F_g, G_g and second L_g, M_g, and N_g orders (see Table B.1).

All major elements of the local geometry of the gear tooth flank $\mathcal{G}$ can be calculated based on the fundamental magnitudes E_g, F_g, G_g and L_g, M_g, N_g of the first $\mathbf{\Phi}_{1.P}$ and second $\mathbf{\Phi}_{2.g}$ fundamental forms, respectively. The location and orientation of the gear tooth flank $\mathcal{G}$ are the two parameters that remain indefinite.

Once a part surface is represented in its natural form – that is, it is expressed in terms of the six fundamental magnitudes of the first and second orders –further calculation of the parameters of a gear tooth flank, $\mathcal{G}$, becomes much easier. In order to demonstrate the significant simplification of the calculation of the parameters of a gear tooth flank, $\mathcal{G}$, several useful equations are presented below as examples.

Table B.1 The fundamental magnitudes of the first and second orders of the involute gear tooth flank $\mathcal{G}$

$E_g = 1$	$L_g = 0$
$F_g = -\frac{r_{b.g}}{\cos\tau_{b.g}}$	$M_g = 0$
$G_g = \frac{U_g^2 \cos^4\tau_{b.g} + r_{b.g}^2}{\cos^2\tau_{b.g}}$	$N_g = -U_g \sin\tau_{b.g} \cos\tau_{b.g}$

B.8 Few More Useful Equations

Many calculations of the parameters of geometry can be significantly simplified by use of the first and second fundamental forms of a smooth, regular part surface, $\mathcal{G}$.

1. For the calculation of the radius R_g of normal curvature within a normal plane section through a current point m on a gear tooth flank $\mathcal{G}$ and at a given direction, the following equation can be used:

$$R_g = \frac{\Phi_{1.g}}{\Phi_{2.g}} \tag{B.78}$$

2. Euler's formula for the calculation of the normal curvature $k_{\theta.\,g}$ at a point m in a direction that is specified by the angle θ can be represented as follows:

$$k_{\theta.g} = k_{1.g} \cos^2\theta + k_{2.g} \sin^2\theta \tag{B.79}$$

Here, in Eq. (B.79), θ is the angle that the normal plane section C_g makes with the first principal plane section $C_{1.\,g}$. In other words, $\theta = \angle(\mathbf{t}_g, \mathbf{t}_{1.\,g})$; here, $\mathbf{t}_g$ is the unit tangent vector within the normal plane section C_g.

Equation (B.79) is also a good illustration of the significant simplification of the calculations when the fundamental magnitudes E_g, F_g, G_g and L_g, M_g, N_g of the first and second orders, respectively, are used.

In order to gain a profound understanding of the differential geometry of surfaces, the interested reader may wish to refer to advanced monographs in the field. Systematic discussions of the topic are available from many sources. The author would like to turn the reader's attention to the books by doCarmo, Eisenhart, Stuik, and others.

Appendix C: Contact Geometry of the Tooth Flanks of a Gear and a Mating Pinion

In the theory of gearing, the kinematics of gearing is considered as the prime element of a gear pair. Other important elements of gearing, namely, (a) The shape and geometry of a gear tooth flank, $\mathcal{G}$ (b) The shape and geometry of a mating pinion tooth flank, $\mathcal{P}$ (as well as numerous others) are considered as the secondary elements of gearing. This does not mean that the importance of the secondary elements is lower than that of the primary element. No, this is incorrect. This just means that the most favorable parameters of the secondary elements can be expressed in terms of the parameters of the prime element. Ultimately, an entire gear pair can be synthesized on just the premise of the prime element – that is, on the premise of the desirable kinematics of the gear pair. In other words, having just the desirable kinematics of the gear pair to be designed, it is possible to synthesize the rest of the design parameters of the gear pair. Only the kinematics of gearing is used for the purposes of synthesizing the best possible gear pair for transmitting the input rotation and torque.

In order to solve the problem of synthesizing the most favorable gear pair, an appropriate analytical description of the contact geometry of the gear tooth flank $\mathcal{G}$ and the mating pinion tooth flank $\mathcal{P}$ is required. The problem of an analytical description of the contact geometry between two smooth regular surfaces in the first order of tangency is a sophisticated one.

Investigation of the contact geometry of curves and surfaces can be traced back to the eighteenth century. In considerable detail, the study of the contact of curves and surfaces was undertaken by J.L. Lagrange[11] in his *Theorié des Fonctions Analytiques* (1797) and by A.L. Cauchy[12] in his *Leçons sur les Applications du Calcul Infinitésimal á la Geometrie* (1826). Later on, in the twentieth century, an investigation into the realm of the contact geometry of curves and surfaces was undertaken by J. Favard[13] in his *Course de Gèomètrie Differentialle Locale* (1957). A few more names of the researchers in the field are to be mentioned.

The results of the research obtained from the field of contact geometry of two smooth regular surfaces are widely used in the theory of gearing. The problem of synthesizing the most favorable gear pair can be solved on the premise of the analysis of the topology of the contacting surfaces in the differential vicinity of the point of their contact.

Various methods for the analytical description of the contact geometry between two smooth regular surfaces have been developed by now. The latest achievements in the field are discussed in numerous papers and monographs available in the public domain.

An in-detail analysis of the known methods of the analytical description of the geometry of contact between two smooth regular surfaces has uncovered the poor capability of the known methods to solve problems in the field of designing efficient gear pairs. Therefore, an accurate method for the analytical description of the contact geometry between two smooth regular surfaces, $\mathcal{G}$ and $\mathcal{P}$, in the first order of tangency, which fits the needs of the theory of gearing, is necessary. Such a method is worked out in this chapter.

It is convenient to begin the discussion starting from an analytical description of the local relative orientation of the gear tooth flank $\mathcal{G}$ and that of the mating pinion tooth flank $\mathcal{P}$. The proposed analytical description is relevant to the differential vicinity of the point of contact K of the tooth flanks $\mathcal{G}$ and $\mathcal{P}$.

---

[11] Joseph-Louis Lagrange (January 25, 1736–April 10, 1813), an Italian-born (born Giuseppe Lodovico (Luigi) Lagrangia) famous French mathematician and mechanician.

[12] Augustin-Louis Cauchy (August 21, 1789–May 23, 1857), a famous French mathematician.

[13] Jean Favard (August 28, 1902–January 21, 1965), a French mathematician.

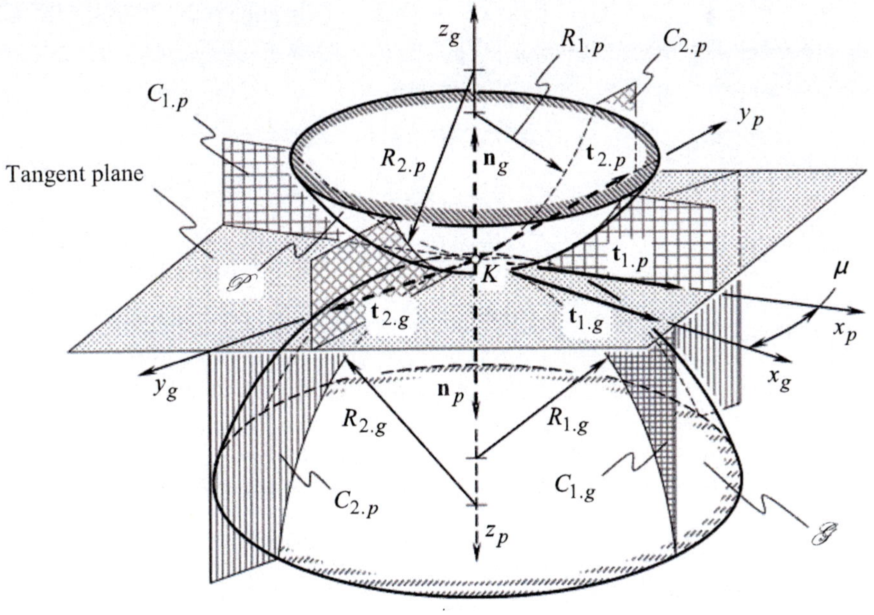

Fig. C.1 Local configuration of two quadrics that are tangent to a gear tooth flank $\mathcal{G}$ and to a mating pinion tooth flank $\mathcal{P}$ at a contact point K. (After S.P. Radzevich, ~2008)

C.1 Local Relative Orientation at a Point of Contact of the Tooth Flanks of a Gear and a Mating Pinion

When the gears rotate, the gear tooth flank $\mathcal{G}$ and the mating pinion tooth flank $\mathcal{P}$ are in permanent tangency to one another. The contacting surfaces $\mathcal{G}$ and $\mathcal{P}$ can be locally approximated by the corresponding quadrics as schematically illustrated in Fig. C.1. The requirement to be permanently in tangency to each other imposes a kind of restriction on the relative configuration (location and orientation) of the tooth flanks $\mathcal{G}$ and $\mathcal{P}$ on their instantaneous relative motion.

In the theory of gearing, a quantitative measure of the relative orientation of a gear tooth flank, $\mathcal{G}$, and of the mating pinion tooth flank, $\mathcal{P}$, is established.

The relative orientation of a point of contact of a gear tooth flank, $\mathcal{G}$, and of a mating pinion tooth flank, $\mathcal{P}$, is specified by the angle μ of the local[14] relative orientation of the surfaces. By definition, angle μ is equal to the angle that the unit tangent vector $\mathbf{t}_{1.g}$ of the first principal direction of the gear tooth flank $\mathcal{G}$ forms with the unit tangent vector $\mathbf{t}_{1.p}$ of the first principal direction of the mating pinion tooth flank $\mathcal{P}$. This same angle μ can also be determined as the angle that makes the unit

[14]The surface orientation is "local" by nature, as it relates only to the differential vicinity of point K of contact of the tooth flanks $\mathcal{G}$ and $\mathcal{P}$.

tangent vectors $\mathbf{t}_{2.\,g}$ and $\mathbf{t}_{2.\,p}$ of the second principal directions of the surfaces $\mathcal{G}$ and $\mathcal{P}$ at contact point K. This immediately yields equations for the calculation of the angle μ:

$$\sin\mu = |\mathbf{t}_{1.g} \times \mathbf{t}_{1.p}| = |\mathbf{t}_{2.g} \times \mathbf{t}_{2.p}|, \tag{C.1}$$

$$\cos\mu = \mathbf{t}_{1.g} \cdot \mathbf{t}_{1.p} = \mathbf{t}_{2.g} \cdot \mathbf{t}_{2.p}, \tag{C.2}$$

$$\tan\mu = \frac{|\mathbf{t}_{1.g} \times \mathbf{t}_{1.p}|}{\mathbf{t}_{1.g} \cdot \mathbf{t}_{1.p}} \equiv \frac{|\mathbf{t}_{2.g} \times \mathbf{t}_{2.p}|}{\mathbf{t}_{2.g} \cdot \mathbf{t}_{2.p}} \tag{C.3}$$

where

$\mathbf{t}_{1.\,g}$, $\mathbf{t}_{2.\,g}$ are the unit vectors of the principal directions on the gear tooth flank $\mathcal{G}$ measured at contact point K.

$\mathbf{t}_{1.\,p}$, $\mathbf{t}_{2.\,p}$ are the unit vectors of the principal directions on the mating pinion tooth flank $\mathcal{P}$ at the same contact point K of the tooth flanks $\mathcal{G}$ and $\mathcal{P}$.

The directions of the unit tangent vectors $\mathbf{t}_{1.\,g}$ and $\mathbf{t}_{2.\,g}$ of the principal directions on the gear tooth flank $\mathcal{G}$ (as well as the directions of the unit tangent vectors $\mathbf{t}_{1.\,p}$ and $\mathbf{t}_{2.\,p}$ of the principal directions on the pinion tooth flank $\mathcal{P}$) can be specified in terms of the ratio dU_g/dV_g (or in terms of the ratio dU_p/dV_p in the case of the pinion tooth flank $\mathcal{P}$). The corresponding values of the ratio $dU_{g(p)}/dV_{g(p)}$ are calculated as roots of the quadratic equation:

$$\begin{vmatrix} E_{g(p)} \dfrac{dU_{g(p)}}{dV_{g(p)}} + F_{g(p)} & F_{g(p)} \dfrac{dU_{g(p)}}{dV_{g(p)}} + G_{g(p)} \\ L_{g(p)} \dfrac{dU_{g(p)}}{dV_{g(p)}} + M_{g(p)} & M_{g(p)} \dfrac{dU_{g(p)}}{dV_{g(p)}} + N_{g(p)} \end{vmatrix} = 0 \tag{C.4}$$

In the case of point contact of the surfaces $\mathcal{G}$ and $\mathcal{P}$, the actual value of the angle μ is calculated at the contact point K. If the tooth flanks $\mathcal{G}$ and $\mathcal{P}$ are in line contact, then the actual value of the angle μ can be calculated at every point within the line of contact.[15] The line of contact of the tooth flanks $\mathcal{G}$ and $\mathcal{P}$ is commonly referred to as the "characteristic line" $\mathcal{E}$ or just as the characteristic $\mathcal{E}$, for simplicity.

Determination of the angle μ of the local relative orientation of the tooth flanks $\mathcal{G}$ and $\mathcal{P}$ of a gear and a mating pinion, respectively, is illustrated in Fig. C.1.

In order to calculate the actual value of the angle μ of the local relative orientation of the tooth flanks $\mathcal{G}$ and $\mathcal{P}$, the unit vectors of the principal directions $\mathbf{t}_{1.\,g}$ and $\mathbf{t}_{1.\,p}$ are employed.

[15] It is worthy pointing out here that in the case of line contact, the relative orientation of the two surfaces $\mathcal{G}$ and $\mathcal{P}$ is predetermined in a "global" sense. However, the actual value of the angle μ of the surface "local" relative orientation at different points of the characteristic $\mathcal{E}$ is different.

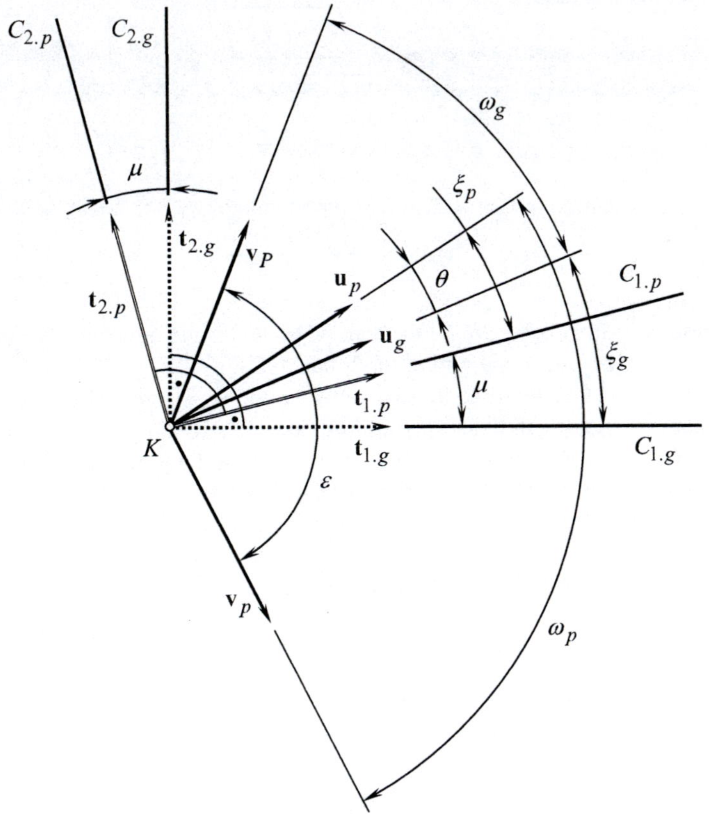

Fig. C.2 Local relative orientation at the contact point of the tooth flanks of a gear, $\mathcal{G}$, and a mating pinion, $\mathcal{P}$, considered in a common tangent plane

Let us consider the tooth flanks $\mathcal{G}$ and $\mathcal{P}$ in point contact, which are represented in a common reference system. The surfaces make contact at a point K. For further analysis, an equation of the common tangent plane to the tooth flanks $\mathcal{G}$ and $\mathcal{P}$ at the contact point K is necessary (see Figs. C.1 and C.2).

$$\left(\mathbf{r}_{\text{tp}} - \mathbf{r}_K\right) \cdot \mathbf{u}_g \cdot \mathbf{v}_g = 0 \tag{C.5}$$

Here,

$\mathbf{r}_{\text{tp}}$ is the position vector of a point of the common tangent plane.
$\mathbf{r}_K$ is the position vector of the contact point K.
$\mathbf{u}_g$ and $\mathbf{v}_g$ are the unit vectors that are tangent to the U_g- and V_g- coordinate lines, respectively, on the gear tooth flank $\mathcal{G}$ at the contact point K.

The angle ω_g is one that is formed by the unit vectors $\mathbf{u}_g$ and $\mathbf{v}_g$. The actual value of the angle ω_g can be calculated from one of the following equations:

$$\sin\omega_g = \frac{\sqrt{E_g G_g - F_g^2}}{\sqrt{E_g G_g}} \tag{C.6}$$

$$\cos\omega_g = \frac{F_g}{\sqrt{E_g G_g}} \tag{C.7}$$

$$\tan\omega_g = \frac{\sqrt{E_g G_g - F_g^2}}{F_g} \tag{C.8}$$

Equations similar to Eqs. (C.6) through (C.8) are also valid for the calculation of the angle ω_p at a point of the pinion tooth flank $\mathcal{P}$.

The respective tangent directions $\mathbf{u}_g$ and $\mathbf{v}_g$ to the U_g- and V_g- coordinate lines at a point of the gear tooth flank $\mathcal{G}$ as well as the respective tangent directions $\mathbf{u}_p$ and $\mathbf{v}_p$ to the U_p- and V_p- coordinate lines at a point on the pinion tooth flank $\mathcal{P}$ are specified in terms of the angles θ and ε, respectively. For the calculation of the actual values of the angles θ and ε, the following equations can be used:

$$\cos\theta = \mathbf{u}_g \cdot \mathbf{u}_p \tag{C.9}$$

$$\cos\varepsilon = \mathbf{v}_g \cdot \mathbf{v}_p \tag{C.10}$$

The angle ξ_g is one that the first principal direction $\mathbf{t}_{1.\,g}$ on the gear tooth surface $\mathcal{G}$ forms with the unit tangent vector $\mathbf{u}_g$ (see Fig. C.2). The equation for the calculation of the actual value of the angle ξ_g was derived by Prof. S.P. Radzevich:

$$\sin\xi_g = \frac{\eta_g}{\sqrt{{\eta_g}^2 - 2\eta_g \cos\omega_g + 1}} \sin\omega_g \tag{C.11}$$

where η_g is the ratio $\eta_g = \frac{\partial U_g}{\partial V_g}$.

In the event $F_g = 0$, the equality $\tan\xi_g = \eta_g$ is observed. Here, the ratio η_g is equal to the root of the quadratic equation:

$$\left(F_g L_g - E_g M_g\right)\eta_g^2 + \left(G_g L_g - E_g N_g\right)\eta_g + \left(G_g M_g - F_g N_g\right) = 0 \tag{C.12}$$

which immediately follows from the equation:

$$\begin{vmatrix} E_g\, dU_g + F_g\, dV_g & F_g\, dU_g + G_g\, dV_g \\ L_g\, dU_g + M_g\, dV_g & M_g\, dU_g + N_g\, dV_g \end{vmatrix} = 0 \tag{C.13}$$

The equation for the calculation of the actual value of the angle ξ_g allows for another representation. Following the chain rule, $d\mathbf{r}_g$ can be represented in the form:

$$d\mathbf{r}_g = \mathbf{U}_g dU_g + \mathbf{V}_g dV_g \tag{C.14}$$

By definition, $\tan \xi_g = \frac{\sin \xi_g}{\cos \xi_g}$. The functions $\sin\xi_g$ and $\cos\xi_g$ yield representations such as:

$$\sin \xi_g = \frac{|\mathbf{U}_g \times d\mathbf{r}_g|}{|\mathbf{U}_g| \cdot |d\mathbf{r}_g|} \tag{C.15}$$

$$\cos \xi_P g = \frac{\mathbf{U}_g \cdot d\mathbf{r}_g}{|\mathbf{U}_g| \cdot |d\mathbf{r}_g|} \tag{C.16}$$

The last expressions yield:

$$\begin{aligned} \tan \xi_g &= \frac{\sin \xi_g}{\cos \xi_g} = \frac{|\mathbf{U}_g \times d\mathbf{r}_g|}{\mathbf{U}_g \cdot d\mathbf{r}_g} = \frac{|\mathbf{U}_g \times d\mathbf{r}_g|}{\mathbf{U}_g \cdot (\mathbf{U}_g dU_g + \mathbf{V}_g dV_g)} \\ &= \frac{|\mathbf{U}_g \times d\mathbf{r}_g| \cdot dV_g}{\mathbf{U}_g \cdot \mathbf{U}_g dU_g + \mathbf{U}_g \cdot \mathbf{V}_g dV_g} \end{aligned} \tag{C.17}$$

By definition:

$$\mathbf{U}_g \cdot \mathbf{U}_g = E_g \tag{C.18}$$

$$\mathbf{U}_g \cdot \mathbf{V}_g = F_g \tag{C.19}$$

$$|\mathbf{U}_g \times \mathbf{V}_g| = \sqrt{E_g G_g - F_g^2} \tag{C.20}$$

Equations (C.14) through (C.20) yield the formula:

$$\tan \xi_g = \frac{\sqrt{E_g G_g - F_g^2}}{\eta_g \cdot E_g + F_g} \tag{C.21}$$

for the calculation of the actual value of the angle ξ_g.

Equations similar to Eqs. (C.11) and (C.21) are also valid for the calculation of the actual value of the angle ξ_p that the first principal direction $\mathbf{t}_{1.\,p}$ at a point of the pinion tooth flank $\mathcal{P}$ forms with the unit tangent vector $\mathbf{u}_p$.

The performed analysis makes possible the following equations for the calculation of the unit vectors of the principal directions $\mathbf{t}_{1.\,g}$ and $\mathbf{t}_{2.\,g}$:

$$\mathbf{t}_{1.g} = \mathbf{Rt}\left(\xi_g, \mathbf{n}_g\right) \cdot \mathbf{u}_g \tag{C.22}$$

$$\mathbf{t}_{2.g} = \mathbf{Rt}\left[\left(\xi_g + \frac{\pi}{2}\right), \mathbf{n}_g\right] \cdot \mathbf{u}_g \tag{C.23}$$

for the gear tooth flank $\mathcal{G}$ and similar equations for the calculation of the unit vectors of the principal directions $\mathbf{t}_{1.\,p}$ and $\mathbf{t}_{2.\,p}$:

$$\mathbf{t}_{1.p} = \mathbf{Rt}\left(\xi_p, \mathbf{n}_g\right) \cdot \mathbf{u}_p \tag{C.24}$$

$$\mathbf{t}_{2.p} = \mathbf{Rt}\left[\left(\xi_p + \frac{\pi}{2}\right), \mathbf{n}_g\right] \cdot \mathbf{u}_p \tag{C.25}$$

for the pinion tooth flank $\mathcal{P}$.

The operator of rotation $\mathbf{Rt}\,(\varphi_A, A_0)$ through an angle φ_A about an axis A_0 is employed for the calculation of the operators of rotation in Eqs. (C.22) through (C.25).

C.2 The Second-Order Analysis: Planar Characteristic Images

For a more accurate analytical description of the contact geometry of the gear tooth flank $\mathcal{G}$ and the pinion tooth flank $\mathcal{P}$, consideration of the second-order parameters is necessary. A second-order analysis incorporates elements of both the first- and second-order analyses. For performing a second-order analysis, familiarity with the "Dupin indicatrix" is highly desirable.[16] This is a perfect starting point for consideration of the second-order analysis.

C.2.1 Preliminary Remarks: Dupin Indicatrix

At any point of a smooth regular gear tooth flank, $\mathcal{G}$ (as well as at any point of a smooth regular pinion tooth flank, $\mathcal{P}$), a corresponding Dupin indicatrix can be constructed. The Dupin indicatrix $\text{Dup}\,(\mathcal{G})$ at a point of the gear tooth flank $\mathcal{G}$ as well as the Dupin indicatrix $\text{Dup}\,(\mathcal{P})$ at a point of the pinion tooth flank $\mathcal{P}$ are planar characteristic curves of the second order. They are used for graphical interpretation of the distribution of the normal radii of curvature of a surface in the differential vicinity of a surface point.

A Dupin indicatrix at a point of the tooth flank $\mathcal{G}$ (as well as at a point of the tooth flank $\mathcal{P}$) is of critical importance in the theory of gearing. Generation of this planar characteristic curve is illustrated with a diagram shown in Fig. C.3.

A plane W through the unit normal vector $\mathbf{n}_g$ to the gear tooth flank $\mathcal{G}$ at a point m rotates about the unit normal vector $\mathbf{n}_g$. While rotating, the plane occupies consecutive positions W_1, W_2, W_3, and others. The radii of normal curvature of the line of intersection of the gear tooth flank $\mathcal{G}$ by the normal planes W_1, W_2, W_3 are equal to $R_{g,\,1}$, $R_{g,\,2}$, $R_{g,\,3}$, respectively, and so forth. The gear tooth flank $\mathcal{G}$ is intersected by a plane Q. The plane Q is orthogonal to the unit normal vector $\mathbf{n}_g$. This plane is at a certain small distance δ from the point m. When the distance δ approaches zero ($\delta \rightarrow 0$) and when the scale of the line of intersection of the gear tooth flank $\mathcal{G}$ by the plane Q approaches infinity, the line of intersection of the gear

[16] Fransua Pier Charles Dupin (October 6, 1784–January 18, 1873), a French mathematician.

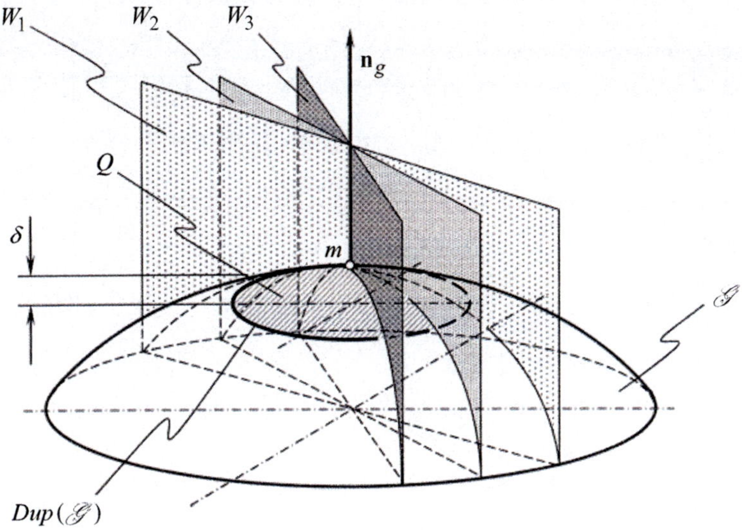

Fig. C.3 A Dupin indicatrix at a point of a smooth regular gear tooth flank, $\mathcal{G}$

surface $\mathcal{G}$ by the plane Q approaches the planar characteristic curve that is commonly referred to as the Dupin indicatrix Dup $(\mathcal{G})$.

In the differential geometry of surfaces, a surface is construed as a zero-thickness film. Because of this, the Dupin indicatrices of the following five different types are distinguished in the differential geometry of surfaces (see Fig. C.4):

- Elliptic (see Fig. C.4a)
- Umbilic (see Fig. C.4b)
- Parabolic (see Fig. C.4c)
- Hyperbolic (see Fig. C.4d)
- Minimal (see Fig. C.4e)

A Dupin indicatrix for a plane local surface patch does not exist. In the case of a plane, all points of the Dupin indicatrix are remote to infinity.

For local surface patches having a negative full curvature ($\mathcal{G}_g < 0$), the phantom branches (that is, the branches that are not intersected by a plane perpendicular to the unit normal vector $\mathbf{n}_g$ to the gear tooth flank $\mathcal{G}$ at a point, m) of the characteristic curve Dup $(\mathcal{G})$, in Fig. C.4d and e, are shown in dashed lines.

An easy way to derive an equation of the characteristic curve Dup $(\mathcal{G})$ is discussed immediately below.

Euler's formula:

$$k_{1.g} \cos^2\varphi + k_{2.g} \sin^2\varphi = k_g \tag{C.26}$$

yields a representation in the form:

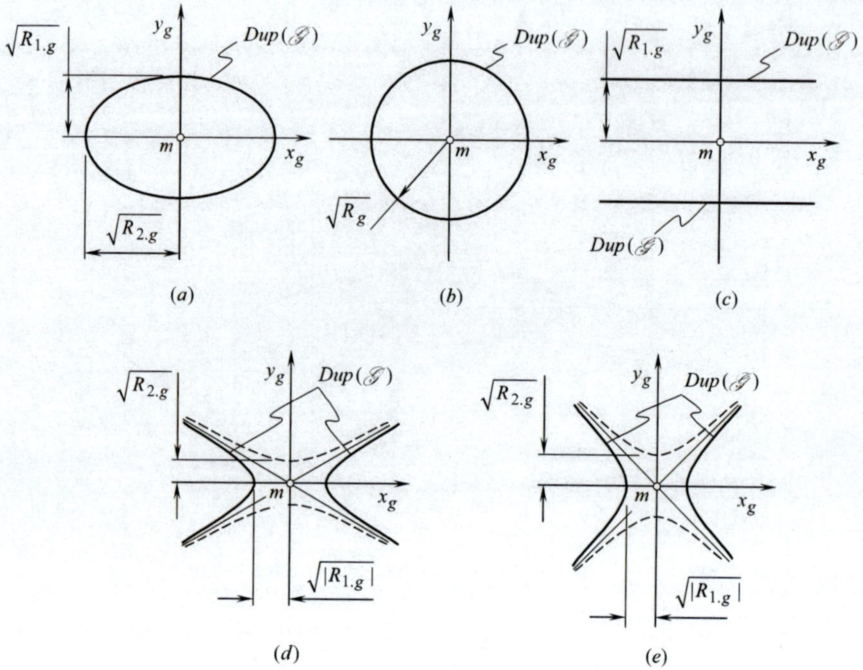

Fig. C.4 Five different kinds of Dupin indicatrices, Dup $(\mathcal{G})$, at a point m of a smooth regular gear tooth flank, $\mathcal{G}$

$$\frac{k_{1.g}}{k_g}\cos^2\varphi + \frac{k_{2.g}}{k_g}\sin^2\varphi = 1 \tag{C.27}$$

The transition from polar coordinates to Cartesian coordinates can be performed using well-known formulas:

$$x_g = \rho\cos\varphi \tag{C.28}$$

$$y_g = \rho\sin\varphi \tag{C.29}$$

These formulas make the following expressions for $\cos^2\varphi = x_g^2/\rho^2$ and $\sin^2\varphi = y_g^2/\rho^2$ possible. After substituting the last formulas in the above equation (C.27), one can come up with the equation:

$$\frac{k_{1.g}}{k_g}\cdot\frac{x_g^2}{\rho^2} + \frac{k_{2.g}}{k_g}\cdot\frac{y_g^2}{\rho^2} = 1. \tag{C.30}$$

It is convenient to designate $\rho = \sqrt{k_g^{-1}}$. The principal curvatures $k_{1.g}$ and $k_{2.g}$ are the roots of the quadratic equation:

$$\begin{vmatrix} L_g - E_g k_g & M_g - F_g k_g \\ M_g - F_g k_g & N_g - G_g k_g \end{vmatrix} = 0 \qquad \text{(C.31)}$$

Substituting the calculated values of the principal curvatures $k_{1.\ g}$ and $k_{2.\ g}$ in Eq. (C.30), and after performing the necessary formula transformations, an equation[17] for the Dupin indicatrix Dup $(\mathcal{G})$ can be represented in the form:

$$k_{1.g}\, x_g^2 + k_{2.g}\, y_g^2 = 1 \qquad \text{(C.32)}$$

Equation (C.32) describes a particular case of the Dupin indicatrix, which is represented in a Darboux frame.[18]

The general form of the equation of a Dupin indicatrix at a point m of a gear tooth flank, $\mathcal{G}$, is often represented as:

$$\text{Dup}\,(\mathcal{G}) \quad \Rightarrow \quad \frac{L_g}{E_g} x_g^2 + \frac{2M_g}{\sqrt{E_g\, G_g}} x_g\, y_g + \frac{N_g}{G_g} y_g^2 = 1 \qquad \text{(C.33)}$$

In Eq. (C.33), the characteristic curve Dup $(\mathcal{G})$ is expressed in terms of the fundamental magnitudes E_g, F_g, and G_g and L_g, M_g, and N_g of the first $\mathbf{\Phi}_{1.\ g}$ and second order $\mathbf{\Phi}_{2.\ g}$ fundamental magnitudes, respectively, at a point of the gear tooth flank $\mathcal{G}$.

C.2.2 Matrix Representation of the Equation of a Dupin Indicatrix at a Point of a Gear Tooth Flank

Like any other quadratic form, the equation of a Dupin indicatrix of the gear tooth flank $\mathcal{G}$ can be represented in matrix form:

[17] The same equation of the Dupin indicatrix could be derived in another way. Coxeter considers a pair of conics obtained by expanding an equation in Monge's form $z = z(x, y)$ in a McLaurin series:

$$z = z(0,\, 0) + z_1 x + z_2 y + \frac{1}{2}\left(z_{11} x_1^2 + 2 z_{12} xy + z_{22} y^2\right) + \cdots = \frac{1}{2}\left(b_{11} x^2 + 2 b_{12} xy + b_{22} y^2\right).$$

This gives the equation $(b_{11}x^2 + 2b_{12}xy + b_{22}y^2) = \pm 1$ of the Dupin indicatrix.

[18] Jean Gaston Darboux (August 13, 1842–February 23, 1917), a French mathematician.

$$
\mathrm{Dup}\,(\mathcal{G}) \quad \Rightarrow \quad \begin{bmatrix} x_g & y_g & 0 & 0 \end{bmatrix} \cdot \begin{bmatrix} \dfrac{L_g}{E_g} & \dfrac{2M_g}{\sqrt{E_g G_g}} & 0 & 0 \\ \dfrac{2M_g}{\sqrt{E_g G_g}} & \dfrac{N_g}{G_g} & 0 & 0 \\ 0 & 0 & \mp 1 & 0 \\ 0 & 0 & 0 & 1 \end{bmatrix} \cdot \begin{bmatrix} x_g \\ y_g \\ 0 \\ 0 \end{bmatrix} = \pm 1 \tag{C.34}
$$

In a Darboux frame, this equation reduces to:

$$
\mathrm{Dup}\,(\mathcal{G}) \quad \Rightarrow \quad \begin{bmatrix} x_g & y_g & 0 & 0 \end{bmatrix} \cdot \begin{bmatrix} L_g & M_g & 0 & 0 \\ M_g & N_g & 0 & 0 \\ 0 & 0 & \mp 1 & 0 \\ 0 & 0 & 0 & 1 \end{bmatrix} \cdot \begin{bmatrix} x_g \\ y_g \\ 0 \\ 0 \end{bmatrix} = \pm 1 \tag{C.35}
$$

It is convenient to implement a matrix representation of the equation of the Dupin indicatrix (see above), for instance, when investigating spatial gearings, that is, crossed-axis gearings, when multiple coordinate system transformations are required.

The equation of the Dupin indicatrix can be represented in the form:

$$
r_{\mathrm{Dup}}(\varphi) = \sqrt{|R_g(\varphi)|} \cdot \mathrm{sgn}\, \Phi_{2.g}^{-1} \tag{C.36}
$$

This equation reveals that the position vector of a point of the Dupin indicatrix $\mathrm{Dup}\,(\mathcal{G})$ in any direction is equal to the square root of the radius of curvature in the same direction.[19]

[19]Similar to the Dupin indicatrix $\mathrm{Dup}(\mathcal{G})$, a planar characteristic curve of another type can be introduced. The equation of this characteristic curve can be postulated in the form: $r_{\mathrm{Dup}.k}(\varphi) = \sqrt{|k_g(\varphi)|} \cdot \mathrm{sgn}\, \Phi_{2.g}^{-1}$. Application of the curvature indicatrix in the form $r_{\mathrm{Dup.}\ k}(\varphi)$ makes avoiding uncertainty in cases of the plane surface possible. For a plane surface, the characteristic curve $\mathrm{Dup}(\mathcal{G})$ does not exist, whereas $r_{\mathrm{Dup.}\ k}(\varphi)$ exists. It shrinks to the point m on the gear tooth flank $\mathcal{G}$.

C.3 Degree of Conformity at a Point of Contact of the Tooth Flanks of a Gear and a Mating Pinion (in the First Order of Tangency)

For an accurate analytical description of the contact geometry of the gear and mating pinion tooth flanks in the first order of tangency, it is necessary to perform a higher-order analysis.

The method discussed below of a higher-order analysis targets the development of an analytical description of the degree of conformity of the pinion tooth flank $\mathcal{P}$ to the gear tooth flank $\mathcal{G}$ at a current point K of their contact. The higher the degree of conformity of the tooth flanks $\mathcal{G}$ and $\mathcal{P}$, the closer are these surfaces to each other in the differential vicinity of the point K. This qualitative ("intuitive") definition of the degree of conformity of two smooth regular surfaces needs a corresponding quantitative measure.

C.3.1 Preliminary Remarks

Implementation of the resultant deviation $l_{\mathbf{cnf}}$ (see Fig. C.5) of two smooth regular surfaces in contact for an analytical description of their contact geometry is a type of straightforward solution to the problem under consideration. This approach is proven to be computationally ineffective. However, the approach provides an insight into how an effective method for solving the problem under consideration can be developed.

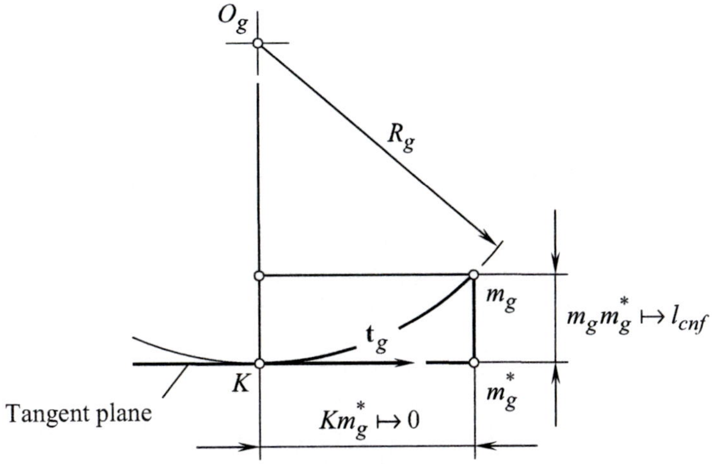

Fig. C.5 Transition from the resultant deviation $l_{\mathbf{cnf}}$ to the indicatrix of conformity $\mathrm{Cnf}(\mathcal{G}/\mathcal{P})$ at the contact point K between two smooth regular tooth flanks $\mathcal{G}$ and $\mathcal{P}$

As seen in Fig. C.5, three geometrical parameters are interrelated when a deviation of a surface from the tangent plane is considered in the differential vicinity of a surface point. They are as follows:

(a) The measure of the deviation $m_g m_g^*$ of a gear tooth flank, $\mathcal{G}$, from the tangent plane l_{cnf}
(b) The distance Km_g^* of a current point m_g from the contact point K
(c) The radius of normal curvature R_g of the gear tooth flank $\mathcal{G}$ at the contact point K

As a consequence of this relationship among the parameters $m_g m_g^*$, Km_g^*, and R_g, any one of them can be used for the purposes of quantitative evaluation of the degree of conformity of the contacting tooth flanks $\mathcal{G}$ and $\mathcal{P}$ of the gear and mating pinion, respectively. It follows from Fig. C.5:

$$m_g m_g^* = R_g - \sqrt{R_g^2 - \left(Km_g^*\right)^2}\Bigg|_{m_g \to K} \mapsto l_{\text{cnf}} \tag{C.37}$$

Inversely, for the radius of normal curvature R_g at a point of the gear tooth flank $\mathcal{G}$, the following expression is valid:

$$R_g = \frac{\left(m_g m_g^*\right)^2 + \left(Km_g^*\right)^2}{2 \cdot m_g m_g^*} \tag{C.38}$$

Ultimately, one may conclude that any legitimate analytical function of the normal radii of curvature R_g and R_p at a point of contact of the gear tooth flank $\mathcal{G}$ and the pinion tooth flank $\mathcal{P}$ can be used for this particular purpose.

Let us consider two smooth regular tooth flanks $\mathcal{G}$ and $\mathcal{P}$ in the first order of tangency that make contact at a point K. The degree of conformity of the tooth flanks $\mathcal{G}$ and $\mathcal{P}$ can be construed as a function of the radii of normal curvature R_g and R_p of the contacting surfaces. The radii of normal curvature R_g and R_p of the tooth flanks $\mathcal{G}$ and $\mathcal{P}$, respectively, are taken in a common normal plane section through the point K. For a specified radius of normal curvature R_g of the tooth flank $\mathcal{G}$, the degree of conformity of the tooth flanks depends upon the corresponding value of the radius of normal curvature R_p of the pinion tooth flank $\mathcal{P}$.

In most cases of gear meshing, the degree of conformity at a point of contact of the tooth flanks $\mathcal{G}$ and $\mathcal{P}$ is not constant and changes as the coordinates of the contact point change. The degree of conformity of the surfaces relative to one another depends on the orientation of the normal plane section through the contact point K and changes as the normal plane section turns about the common perpendicular $\mathbf{n}_g$. This statement immediately follows from the above-made conclusion that the degree of conformity at a point of contact of the tooth flanks $\mathcal{G}$ and $\mathcal{P}$ yields interpretation in terms of the radii of normal curvature R_g and R_p, respectively.

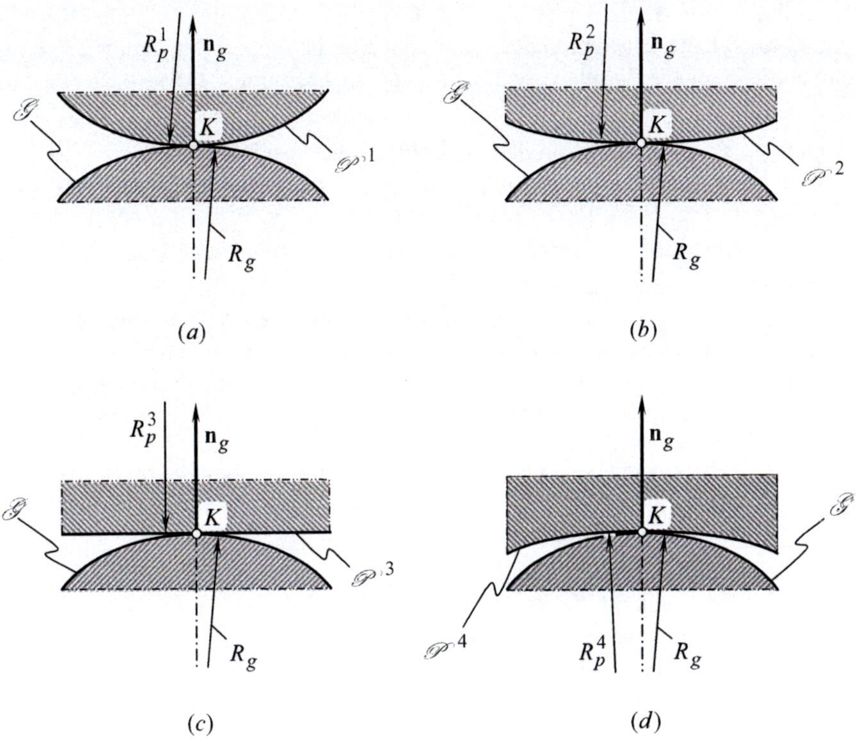

Fig. C.6 Sections of two smooth regular tooth flanks $\mathcal{G}$ and $\mathcal{P}$ in contact by a plane through the common perpendicular $\mathbf{n}_g$

The change of the degree of conformity of a gear tooth flank $\mathcal{G}$ and a mating pinion tooth flank $\mathcal{P}$ due to turning of the normal plane section about the common perpendicular $\mathbf{n}_g$ is illustrated in Fig. C.6. Here, in Fig. C.6, just two-dimensional examples are shown, for which the same normal plane section of the gear tooth flank $\mathcal{G}$ makes contact with different plane sections $\mathcal{P}^i$ of the pinion tooth flank $\mathcal{P}$.

In the example shown in Fig. C.6a, the radius of normal curvature R_p^1 of the convex plane section $\mathcal{P}^1$ of the pinion tooth flank $\mathcal{P}$ is positive ($R_p^1 > 0$). The convex normal plane section of the pinion tooth flank $\mathcal{P}$ makes contact with the convex normal plane section ($R_g > 0$) of the gear tooth flank $\mathcal{G}$. The degree of conformity of the pinion tooth flank $\mathcal{P}$ to the gear tooth flank $\mathcal{G}$ in Fig. C.6a is relatively low as both the contacting surfaces are convex.

Another example is shown in Fig. C.6b. The radius of normal curvature R_p^2 of the convex plane section $\mathcal{P}^2$ of the pinion tooth flank $\mathcal{P}$ is also positive ($R_p^2 > 0$). However, its value exceeds the value R_p^1 of the radius of normal curvature in the first example ($R_p^2 > R_p^1$). This indicates that the degree of conformity of the pinion tooth flank $\mathcal{P}$ to the gear tooth flank $\mathcal{G}$ (Fig. C.6b) is greater compared to that shown in Fig. C.6a.

In the next example depicted in Fig. C.6c, the normal plane section $\mathcal{P}^3$ of the pinion tooth flank $\mathcal{P}$ is represented with a locally flat section. The radius of normal curvature R_p^3 of the flat plane section $\mathcal{P}^3$ approaches infinity ($R_p^3 \to \infty$). Thus, the inequality $R_p^3 > R_p^2 > R_p^1$ is valid. Therefore, the degree of conformity of the pinion tooth flank $\mathcal{P}$ to the gear tooth flank $\mathcal{G}$ in Fig. C.6c, also increases.

Finally, for a concave normal plane section, $\mathcal{P}^4$, of the pinion tooth flank $\mathcal{P}$ that is illustrated in Fig. C.6d, the radius of normal curvature R_p^4 is of a negative value ($R_p^4 < 0$). In this case, the degree of conformity of the pinion tooth flank $\mathcal{P}$ to the gear tooth flank $\mathcal{G}$ is the greatest of four examples considered in Fig. C.6.

The examples shown in Fig. C.6 qualitatively illustrate what is known intuitively regarding the different degrees of conformity of two smooth regular surfaces in the first order of tangency. Intuitively, one can realize that in the examples shown in Fig. C.6a–d, the degree of conformity at a point of contact of the two tooth flanks $\mathcal{G}$ and $\mathcal{P}$ is gradually increased.

A similar observation is made for a given pair of the tooth flanks $\mathcal{G}$ and $\mathcal{P}$, when different sections of the surfaces by a plane surface through the common perpendicular $\mathbf{n}_g$ are considered (see Fig. C.7a). When rotating the plane section about the common perpendicular $\mathbf{n}_g$, it can be observed that the degree of conformity of the gear and the pinion tooth flanks, i.e., $\mathcal{G}$ and $\mathcal{P}$, respectively, is different in different configurations of the cross-sectional plane (see Fig. C.7b).

The above examples provide an intuitive understanding of what the degree of conformity at a point of contact of two smooth regular tooth flanks, $\mathcal{G}$ and $\mathcal{P}$, means. These examples cannot be directly employed to quantitatively evaluate the degree of conformity at a point of contact of two smooth regular tooth flanks, namely, $\mathcal{G}$ and $\mathcal{P}$. The next necessary step is to introduce an appropriate quantitative evaluation of the degree of conformity of two smooth regular surfaces in the first order of tangency. In other words, how can a certain degree of conformity of two smooth regular surfaces be described analytically?

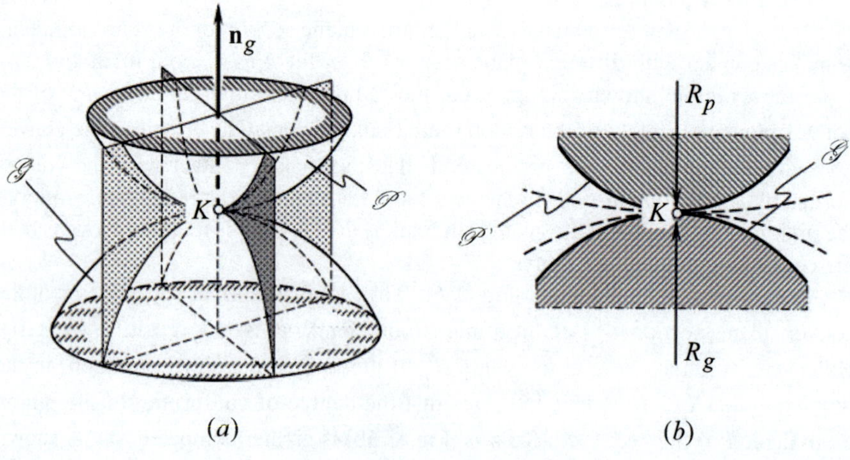

Fig. C.7 An analytical description of the contact geometry of two smooth regular tooth flanks, $\mathcal{G}$ and $\mathcal{P}$, of a gear and a mating pinion, respectively

C.3.2 The Indicatrix of Conformity at a Point of Contact of the Tooth Flanks of a Gear and a Mating Pinion

This section is aimed at introducing a quantitative measure of the degree of conformity at a point of contact between two smooth regular surfaces. The degree of conformity at a point of contact of two tooth flanks, $\mathcal{G}$ and $\mathcal{P}$, indicates how the pinion tooth flank $\mathcal{P}$ is close to the gear tooth flank $\mathcal{G}$ in the differential vicinity of a point K of their contact, i.e., how much the surface $\mathcal{P}$ is "congruent" to the surface $\mathcal{G}$ in the differential vicinity of the contact point K. This particular type of congruency between the contacting surfaces $\mathcal{G}$ and $\mathcal{P}$ can also be construed as the "local congruency" of the contacting surfaces.

Quantitatively, the degree of conformity at a point of contact of a smooth regular surface, $\mathcal{P}$, to another surface, $\mathcal{G}$, can be expressed in terms of the difference between the corresponding radii of normal curvature of the contacting surfaces. In order to develop a quantitative measure of the degree of conformity of the tooth flanks $\mathcal{G}$ and $\mathcal{P}$, it is convenient to implement Dupin indicatrices, $\text{Dup}(\mathcal{G})$ and $\text{Dup}(\mathcal{P})$, constructed at a point of contact of the gear tooth flank $\mathcal{G}$ and the pinion tooth flank $\mathcal{P}$, respectively.

It is natural to assume that the smaller difference between the normal curvatures of the surfaces $\mathcal{G}$ and $\mathcal{P}$, in a common cross section by a plane through the common normal vector $\mathbf{n}_g$, results in a greater degree of conformity at a point of contact of the tooth flanks $\mathcal{G}$ and $\mathcal{P}$.

The Dupin indicatrix $\text{Dup}(\mathcal{G})$ indicates the distribution of the radii of normal curvature at a point of the gear tooth flank $\mathcal{G}$, as it had been shown, for example, for a concave elliptic patch of the surface $\mathcal{G}$ (see Fig. C.8). For a gear tooth flank, $\mathcal{G}$, the equation of this characteristic curve in polar coordinates can be represented in the form:

$$\text{Dup}(\mathcal{G}) \quad \Rightarrow \quad r_g(\varphi_g) = \sqrt{|R_g(\varphi_g)|} \tag{C.39}$$

where

r_g is the position vector of a point of the Dupin indicatrix $\text{Dup}(\mathcal{G})$ at a point of the gear tooth flank $\mathcal{G}$.
φ_g is the polar angle of the indicatrix $\text{Dup}(\mathcal{G})$.

The similar is true with respect to the Dupin indicatrix $\text{Dup}(\mathcal{G})$ at a point of the pinion tooth flank $\mathcal{P}$, as was shown earlier, for instance, for a convex elliptical patch of the pinion tooth flank $\mathcal{P}$ (see Fig. C.8). The equation of this characteristic curve in polar coordinates can be represented in the form:

$$\text{Dup}(\mathcal{P}) \quad \Rightarrow \quad r_p(\varphi_p) = \sqrt{|R_p(\varphi_p)|} \tag{C.40}$$

where

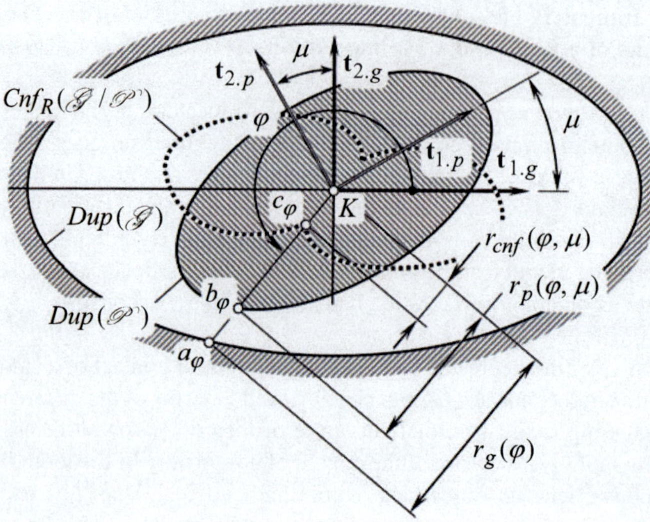

Fig. C.8 The indicatrix of conformity $\text{Cnf}_R(\mathcal{P} \mapsto \mathcal{G})$ at a point of contact of the tooth flanks $\mathcal{G}$ and $\mathcal{P}$. (After Prof. S.P. Radzevich: Radzevich, S.P., *Differential-Geometric Method of Surface Generation*, Dr. Sci. Thesis, Tula, Tula Polytechnic Institute, 1991, 300 pages)

r_p is the position vector of a point of the Dupin indicatrix Dup($\mathcal{G}$) at a point of the pinion tooth flank $\mathcal{P}$.

φ_p is the polar angle of the indicatrix Dup($\mathcal{P}$).

In the coordinate plane $x_g y_g$ of the local reference system $x_g y_g z_g$, the equalities $\varphi_g = \varphi$ and $\varphi_p = \varphi + \mu$ are valid. Therefore, in the coordinate plane $x_g y_g$, Eqs. (C.39) and (C.40) are cast into:

$$\text{Dup}(\mathcal{G}) \quad \Rightarrow \quad r_g(\varphi) = \sqrt{|R_g(\varphi)|} \tag{C.41}$$

$$\text{Dup}(\mathcal{P}) \quad \Rightarrow \quad r_p(\varphi, \mu) = \sqrt{|R_p(\varphi, \mu)|} \tag{C.42}$$

When the degree of conformity at a point of contact of the gear tooth flank $\mathcal{G}$ is greater, the difference between the functions $r_g(\varphi)$ and $r_p(\varphi, \mu)$ becomes smaller, and vice versa. The last part makes the following conclusion valid:

Conclusion C.1 The distance between the corresponding[20] points of the Dupin indicatrices Dup($\mathcal{G}$) and Dup($\mathcal{P}$) constructed at a point of contact of a gear tooth

[20]The corresponding points of the Dupin indicatrices Dup(P) and Dup(T) share the same straight line through the contact point K of the tooth flanks $\mathcal{G}$ and $\mathcal{P}$ and are located at the same side of the point K.

flank, $\mathcal{G}$, and a mating pinion tooth flank, $\mathcal{P}$, can be employed for the purpose of indication of the degree of conformity at a point of contact of the gear tooth flank $\mathcal{G}$ and of the pinion tooth flank $\mathcal{P}$ at the contact point K.

The equation of the indicatrix of conformity $\text{Cnf}_R(\mathcal{G}/\mathcal{P})$ at a point of contact of a gear tooth flank, $\mathcal{G}$, and a mating pinion tooth flank, $\mathcal{P}$, is defined to be of the following structure:

$$\begin{aligned}\text{Cnf}_R(\mathcal{G}/\mathcal{P}) \quad \Rightarrow \quad r_{\text{cnf}}(\varphi,\mu) &= \sqrt{|R_g(\varphi)|}\,\text{sgn}\,R_g(\varphi) + \sqrt{|R_p(\varphi,\mu)|}\,\text{sgn}\,R_p(\varphi,\mu) \\ &= r_g(\varphi)\,\text{sgn}\,R_g(\varphi) + r_p(\varphi,\mu)\,\text{sgn}\,R_p(\varphi,\mu)\end{aligned} \tag{C.43}$$

Because the location of a point a_φ of the Dupin indicatrix $\text{Dup}(\mathcal{G})$ at a point of the gear tooth flank $\mathcal{G}$ is specified by the position vector $r_g(\varphi)$ and the location of a point b_φ of the Dupin indicatrix $\text{Dup}(\mathcal{P})$ at a point of the pinion tooth flank $\mathcal{P}$ is specified by the position vector $r_p(\varphi,\mu)$, the location of a point c_φ (see Fig. C.8) of the indicatrix of conformity $\text{Cnf}_R(\mathcal{G}/\mathcal{P})$ at a point of contact K of the tooth flanks $\mathcal{G}$ and $\mathcal{P}$ is specified by the position vector $r_{\mathbf{cnf}}(\varphi,\mu)$. Therefore, the equality $r_{\mathbf{cnf}}(\varphi,\mu) = Kc_\varphi$ is observed, and the length of the straight-line segment Kc_φ is equal to the distance $a_\varphi b_\varphi$.

Here, in Eq. (C.43),

$r_g = \sqrt{|R_g|}$ is the position vector of a point of the Dupin indicatrix of the gear tooth flank $\mathcal{G}$ at a point K of contact with the pinion tooth flank $\mathcal{P}$ and

$r_p = \sqrt{|R_p|}$ is the position vector of a corresponding point of the Dupin indicatrix of the pinion tooth flank $\mathcal{P}$.

Here, in Eq. (C.43), the multipliers $\text{sgn}R_g(\varphi)$ and $\text{sgn}R_p(\varphi,\mu)$ are assigned to each of the functions $r_g(\varphi) = \sqrt{|R_g(\varphi)|}$ and $r_p(\varphi,\mu) = \sqrt{|R_p(\varphi,\mu)|}$, respectively, for just retaining the corresponding sign of the functions, that is, for retaining the same sign that the radii of normal curvature $R_g(\varphi)$ and $R_p(\varphi,\mu)$ have.

Ultimately, one can conclude that the position vector $r_{\mathbf{cnf}}$ of a point of the indicatrix of conformity $\text{Cnf}_R(\mathcal{G}/\mathcal{P})$ can be expressed in terms of the position vectors r_g and r_p of the Dupin indicatrices $\text{Dup}(\mathcal{G})$ and $\text{Dup}(\mathcal{P})$, respectively.

For the calculation of the current value of the radius of normal curvature $R_g(\varphi)$ at a point of the gear tooth flank $\mathcal{G}$, the equality:

$$R_g(\varphi) = \frac{\Phi_{1.g}}{\Phi_{2.g}} \tag{C.44}$$

can be used.

Similarly, for the calculation of the current value of the radius of normal curvature $R_p(\varphi,\mu)$ at a point of the pinion tooth flank $\mathcal{P}$, the equality:

$$R_p(\varphi, \mu) = \frac{\Phi_{1.p}}{\Phi_{2.p}} \tag{C.45}$$

can be employed.

Use of the angle μ of the local relative orientation of the tooth flanks $\mathcal{G}$ and $\mathcal{P}$ indicates that the radii of normal curvature $R_g(\varphi)$ and $R_p(\varphi, \mu)$ are taken in a common normal plane section through the contact point K.

Furthermore, it is well-known that the inequalities $\Phi_{1.\,g} \geq 0$ and $\Phi_{1.\,p} \geq 0$ are always valid. Therefore, Eq. (C.43) can be rewritten in the following form:

$$r_{\text{cnf}} = r_g(\varphi)\,\text{sgn}\,\Phi_{2.g}^{-1} + r_p(\varphi, \mu)\,\text{sgn}\,\Phi_{2.p}^{-1} \tag{C.46}$$

For the derivation of an equation of the indicatrix of conformity $\text{Cnf}_R(\mathcal{G}/\mathcal{P})$, it is convenient to use the Euler equation for the normal radius of curvature $R_g(\varphi)$ at a point of the gear tooth flank $\mathcal{G}$:

$$R_g(\varphi) = \frac{R_{1.g} \cdot R_{2.g}}{R_{1.g} \cdot \sin^2\varphi + R_{2.g} \cdot \cos^2\varphi} \tag{C.47}$$

Here, the radii of principal curvature $R_{1.\,g}$ and $R_{2.\,g}$ are the roots of the quadratic equation:

$$\begin{vmatrix} L_g \cdot R_g - E_g & M_g \cdot R_g - F_g \\ M_g \cdot R_g - F_g & N_g \cdot R_g - G_g \end{vmatrix} = 0 \tag{C.48}$$

One should recall that the inequality $R_{1.\,g} < R_{2.\,g}$ is always observed.

Equations (C.47) and (C.48) allow for expression of the radius of normal curvature $R_g(\varphi)$ at a point of the gear tooth flank $\mathcal{G}$ in terms of the fundamental magnitudes of the first order E_g, F_g, and G_g and those of the second order L_g, M_g, and N_g.

A similar consideration is applicable to the pinion tooth flank $\mathcal{P}$. Omitting routine analysis, one can conclude that the radius of normal curvature $R_p(\varphi, \mu)$ of a point of the pinion tooth flank $\mathcal{P}$ can be expressed in terms of the fundamental magnitudes of the first order E_p, F_p, and G_p and those of the second order L_p, M_p, and N_p.

Finally, on the premise of the above-performed analysis, the following equation for the indicatrix of conformity $\text{Cnf}_R(\mathcal{G}/\mathcal{P})$ at a point of contact of the tooth flanks $\mathcal{G}$ and $\mathcal{P}$ can be derived:

$$\begin{aligned} r_{\text{cnf}}(\varphi, \mu) = & \sqrt{\left| \frac{E_g G_g}{L_g G_g \cos^2\varphi - M_g \sqrt{E_g G_g} \sin 2\varphi + N_g E_g \sin^2\varphi} \right|}\,\text{sgn}\,\Phi_{2.g}^{-1} \\ & + \sqrt{\left| \frac{E_p G_p}{L_p G_p \cos^2(\varphi+\mu) - M_p \sqrt{E_p G_p} \sin 2(\varphi+\mu) + N_p E_p \sin^2(\varphi+\mu)} \right|}\,\text{sgn}\,\Phi_{2.p}^{-1} \end{aligned} \tag{C.49}$$

Equation (C.49) of the characteristic curve[21] $\mathrm{Cnf}_{R}(P/T)$ is known from the late 1970s.

Analysis of Eq. (C.49) reveals that the indicatrix of conformity $\mathrm{Cnf}_{R}(\mathcal{G}/\mathcal{P})$ at a point of contact of a gear tooth flank $\mathcal{G}$ and the mating pinion tooth flank $\mathcal{P}$ is represented by a planar centro-symmetrical curve of the fourth order. In particular cases, this characteristic curve also possesses a property of mirror symmetry. Mirror symmetry of the indicatrix of conformity is observed, for example, when the angle μ of the local relative orientation of the tooth flanks $\mathcal{G}$ and $\mathcal{P}$ is equal to $\mu = \pm\pi \cdot n/2$, where n is an integer.

It is important to observe here that even for the most general case of gearing, the position vector of a point, $r_{\mathbf{cnf}}(\varphi, \mu)$, of the indicatrix of conformity $\mathrm{Cnf}_{R}(\mathcal{G}/\mathcal{P})$ is not dependent on the fundamental magnitudes F_g and F_p. The independence of the indicatrix of conformity $\mathrm{Cnf}_{R}(\mathcal{G}/\mathcal{P})$ of the fundamental magnitudes F_g and F_p is because of the following:

The coordinate angle ω_g at a point of the gear tooth flank $\mathcal{G}$ can be calculated from the formula:

$$\omega_g = \arccos \frac{F_g}{\sqrt{E_g G_g}} \tag{C.50}$$

It is natural that the position vector $r_{\mathbf{cnf}}(\varphi, \mu)$ of a point of the indicatrix of conformity $\mathrm{Cnf}_{R}(\mathcal{G}/\mathcal{P})$ is not a function of the coordinate angle ω_g. Although the position vector $r_{\mathbf{cnf}}(\varphi, \mu)$ depends on the fundamental magnitudes E_g, G_g and E_p, G_p, the above analysis makes it clear why it does not depend upon the fundamental magnitudes F_g and F_p.

Two illustrative examples of the indicatrix of conformity $\mathrm{Cnf}_{R}(\mathcal{G}/\mathcal{P})$ at a point of contact of a gear tooth flank, $\mathcal{G}$, and a mating pinion tooth flank, $\mathcal{P}$, are shown in Fig. C.9. The first example (see Fig. C.9a) relates to the cases of contact of a saddle-like local patch of the tooth surface $\mathcal{G}$ and of a convex elliptic-like local patch of the tooth surface $\mathcal{P}$. The second one (see Fig. C.9b) is for the case of contact of a convex parabolic-like local patch of the tooth surface $\mathcal{G}$ and of a convex elliptic-like local patch of the tooth surface $\mathcal{P}$. For both cases (see Fig. C.9), the corresponding curvature indicatrices $\mathrm{Crv}(\mathcal{G})$ and $\mathrm{Crv}(\mathcal{P})$ at the point of contact of the tooth flanks $\mathcal{G}$ and $\mathcal{P}$, respectively, are also depicted in Fig. C.9. The imaginary (phantom) branches of the Dupin indicatrix $\mathrm{Dup}(\mathcal{G})$ (not labeled in Fig. C.9a) for the saddle-like local patch of the gear tooth flank $\mathcal{G}$ are shown as a dashed line (see Fig. C.9a).

The gear tooth flank $\mathcal{G}$ and the pinion tooth flank $\mathcal{P}$ can make contact geometrically; however, the physical conditions of their contact could be violated. Violation

[21] The equation of this characteristic curve is known from:

(a) Pat. No.1249787, USSR, *A Method of Sculptured Part Surface Machining on a Multi-Axis NC Machine, S.P. Radzevich*, B23C 3/16, Filed: December 27, 1984, and (in hidden form) from:

(b) Pat. No.1185749, USSR, *A Method of Sculptured Part Surface Machining on a Multi-Axis NC Machine, S.P. Radzevich*, B23C 3/16, Filed: October 24, 1983.

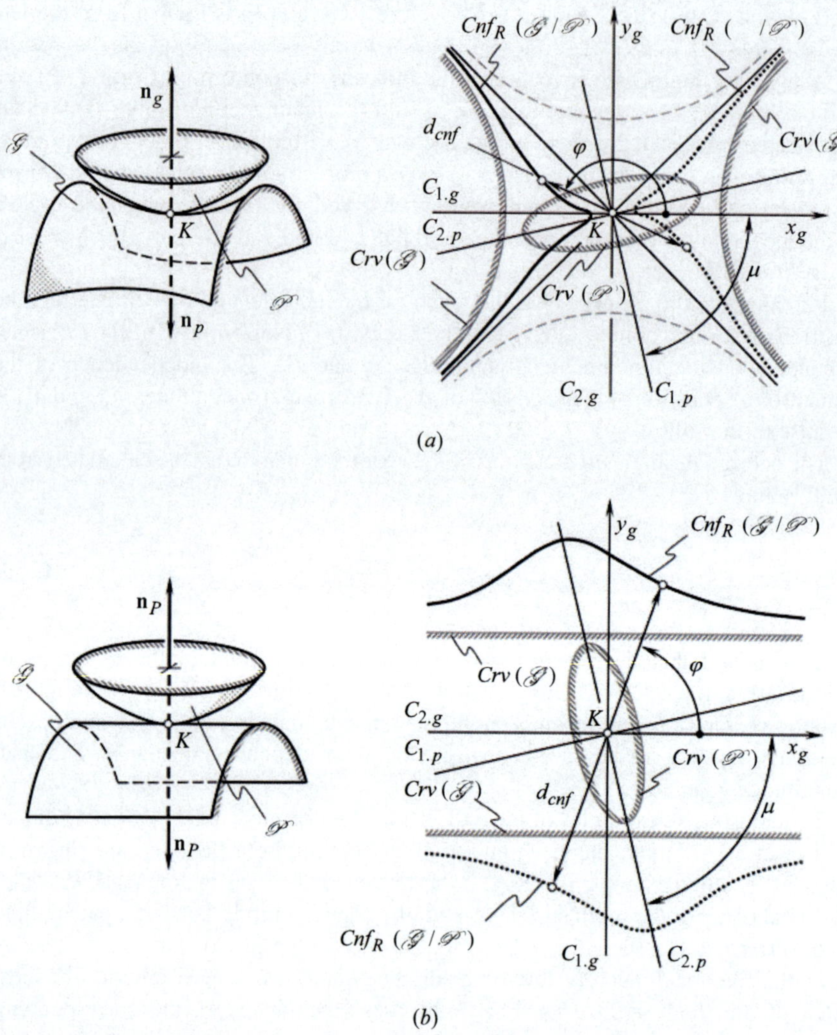

Fig. C.9 Examples of the indicatrix of conformity $\mathrm{Cnf}_R(\mathcal{G}/\mathcal{P})$ at a point of contact K of a smooth regular gear tooth flank, $\mathcal{G}$, and a mating pinion tooth flank, $\mathcal{P}$

of the physical conditions of the contact indicates that the bodies bounded by the contacting surfaces $\mathcal{G}$ and $\mathcal{P}$ interfere with one another. Implementation of the indicatrix of conformity $\mathrm{Cnf}_R(\mathcal{G}/\mathcal{P})$ immediately uncovers the surface interference if there is any. Three illustrative examples of the violation of the physical conditions of contact are illustrated in Fig. C.10. When the correspondence between the radii of normal curvature of the contacting tooth flanks $\mathcal{G}$ and $\mathcal{P}$ is inappropriate, the indicatrix of conformity $\mathrm{Cnf}_R(\mathcal{G}/\mathcal{P})$ either intersects itself (see Fig. C.10a) or all of it diameters become negative (see Fig. C.10b and c).

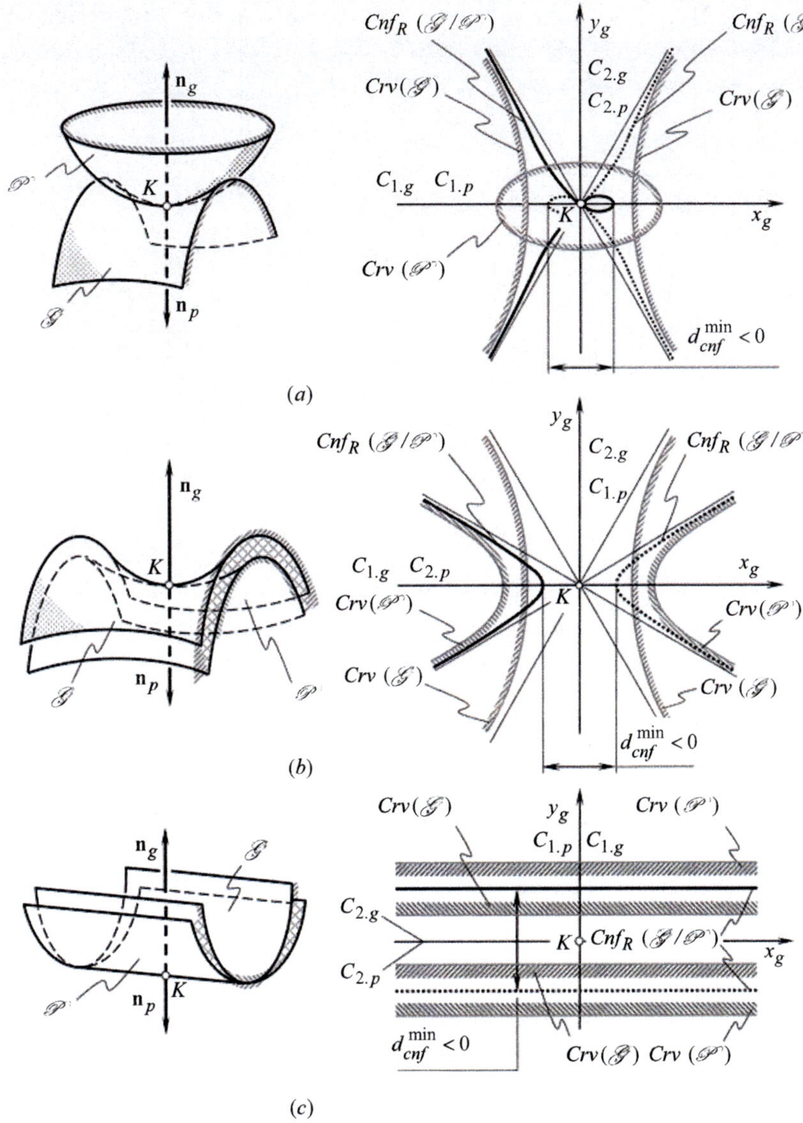

Fig. C.10 Examples of violation of the physical conditions of contact of a smooth regular gear tooth flank, $\mathcal{G}$, and a mating pinion tooth flank, $\mathcal{P}$

Another interpretation of satisfaction and violation of the physical conditions of contact of two smooth regular tooth flanks, $\mathcal{G}$ and $\mathcal{P}$, is illustrated in Fig. C.11. The physical condition of contact is fulfilled when all diameters of the indicatrix of conformity $\text{Cnf}_R(\mathcal{G}/\mathcal{P})$ are positive. In this case, the gear tooth flank $\mathcal{G}$ and the mating pinion tooth flank $\mathcal{P}$ may contact one another like two rigid bodies do. An example of the indicatrix of conformity $\text{Cnf}_R(\mathcal{G}/\mathcal{P})$ for such a case is depicted in Fig. C.11a. In cases in which this planar characteristic curve has negative diameters,

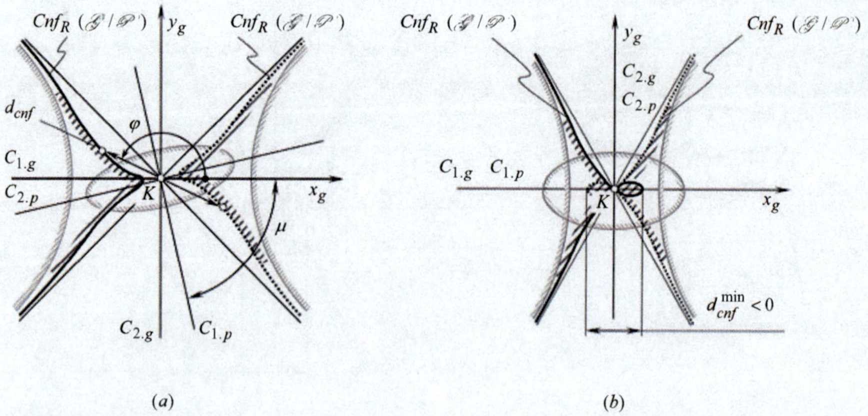

Fig. C.11 Another interpretation of satisfaction (**a**) and violation (**b**) of the physical conditions of contact of a smooth regular gear tooth flank, $\mathcal{G}$, and a mating pinion tooth flank, $\mathcal{P}$

as schematically shown in Fig. C.11b, the physical contact between the tooth flanks $\mathcal{G}$ and $\mathcal{P}$ is infeasible.

The value of the current diameter[22] $d_{\mathbf{cnf}}$ of the indicatrix of conformity $\text{Cnf}_R(\mathcal{G}/\mathcal{P})$ indicates that the degree of conformity to each of the gear tooth flank $\mathcal{G}$ and the mating pinion tooth flank $\mathcal{P}$ is in a corresponding cross section of the surfaces by a normal plane through the common perpendicular. The orientation of the normal plane section with respect to the tooth flanks $\mathcal{G}$ and $\mathcal{P}$ is specified by the corresponding central angle $\boldsymbol{\phi}$.

For the orthogonally parameterized gear tooth flank $\mathcal{G}$ and the mating pinion tooth flank $\mathcal{P}$, the equation of the Dupin indicatrices $\text{Dup}(\mathcal{G})$ and $\text{Dup}(\mathcal{P})$ is simplified to:

$$L_g x_g^2 + 2M_g x_g y_g + N_g y_g^2 = \pm 1 \tag{C.51}$$

$$L_p x_p^2 + 2M_p x_p y_p + N_p y_p^2 = \pm 1 \tag{C.52}$$

After being represented in a common reference system, use of Eqs. (C.51) and (C.52) makes a simplified equation of the indicatrix of conformity $\text{Cnf}_R(\mathcal{G}/\mathcal{P})$ at a point of contact of the tooth flanks $\mathcal{G}$ and $\mathcal{P}$ possible:

[22] The diameter of a symmetrical curve that possesses a property of central symmetry can be defined as a distance between two points of the curve, measured along the corresponding straight line through the center of symmetry of the curve.

$$r_{\mathrm{cnf}}(\varphi, \mu) = \left(L_g \cos^2\varphi - M_g \sin 2\varphi + N_g \sin^2\varphi\right)^{-0.5} \operatorname{sgn} \Phi_{2.g}^{-1} + \left[L_p \cos^2(\varphi+\mu) - M_p \sin 2(\varphi+\mu) + N_p \sin^2(\varphi+\mu)\right]^{-0.5} \operatorname{sgn} \Phi_{p.T}^{-1} \tag{C.53}$$

Equation (C.53) is valid for the orthogonally parameterized tooth flanks $\mathcal{G}$ and $\mathcal{P}$.

C.3.3 Directions of the Extremum Degree of Conformity at a Point of Contact of the Tooth Flanks of a Gear and a Mating Pinion

The directions along which the degree of conformity at a point of contact of a gear tooth flank, $\mathcal{G}$, and a mating pinion tooth flank, $\mathcal{P}$, is extremum – that is, the degree of conformity reaches either its maximal or minimal value – is of prime importance for engineering applications. This issue is especially important for synthesizing a favorable gear pair.

The directions of the extremal degree of conformity of the contacting smooth regular tooth flanks $\mathcal{G}$ and $\mathcal{P}$, that is, the directions pointed along the extremal diameters $d_{\mathrm{cnf}}^{\min}$ and $d_{\mathrm{cnf}}^{\max}$ of the indicatrix of conformity $\mathrm{Cnf}_R(\mathcal{G}/\mathcal{P})$, can be found from the equation of the indicatrix of conformity $\mathrm{Cnf}_R(\mathcal{G}/\mathcal{P})$. For the reader's convenience, the equation of this characteristic curve is transformed and is represented in the form:

$$r_{\mathrm{cnf}}(\varphi, \mu) = \sqrt{|r_{1.g} \cos^2\varphi + r_{2.g} \sin^2\varphi|} \operatorname{sgn} \Phi_{2.g}^{-1} + \sqrt{|r_{1.p} \cos^2(\varphi+\mu) + r_{2.p} \sin^2(\varphi+\mu)|} \operatorname{sgn} \Phi_{2.p}^{-1} \tag{C.54}$$

Two directions within the common tangent plane are specified by the angles $\varphi_{\mathbf{min}}$ and $\varphi_{\mathbf{max}}$. These directions feature an extremum degree of conformity of the pinion tooth flank $\mathcal{P}$ to the gear tooth flank $\mathcal{G}$. Actually, the angles are the roots of the equation:

$$\frac{\partial}{\partial \varphi} r_{\mathrm{cnf}}(\varphi, \mu) = 0. \tag{C.55}$$

It can be easily proved that in the general case of contact of two smooth regular tooth flanks, $\mathcal{G}$ and $\mathcal{P}$, the difference between the angles $\varphi_{\mathbf{min}}$ and $\varphi_{\mathbf{max}}$ is not equal to 0.5π. This means that the equality

$$\varphi_{\min} - \varphi_{\max} = \pm 0.5\pi n \tag{C.56}$$

is not always observed, and, in most cases, the relationship:

$$\varphi_{\min} - \varphi_{\max} \neq \pm 0.5\pi n \tag{C.57}$$

is valid (here, n is an integer). The condition (see Eq. (C.56)): $\varphi_{\mathbf{min}} = \varphi_{\mathbf{max}} \pm 0.5\pi n$ is fulfilled only in cases in which the angle μ of the local relative orientation of the contacting surfaces $\mathcal{G}$ and $\mathcal{P}$ is $\mu = \pm 0.5\pi n$, and, thus, the principal directions $\mathbf{t}_{1.\,g}$ and $\mathbf{t}_{2.\,g}$ of the gear tooth flank $\mathcal{G}$ and the principal directions $\mathbf{t}_{1.\,p}$ and $\mathbf{t}_{2.\,p}$ of the mating pinion tooth flank $\mathcal{P}$ are either aligned with each other or are oppositely directed.

This enables one to making the following statement:

Statement C.1 In the general case of contact of two smooth regular tooth flanks, the directions along which the degree of conformity of the tooth flanks $\mathcal{G}$ and $\mathcal{P}$ is extremal are not orthogonal to one another.

This statement is important for engineering applications.

The solution to equation $\partial r_{\text{rel}}(\varphi)/\partial\varphi = 0$ returns two extremal angles $\varphi_{\mathbf{min}}$ and $\varphi_{\mathbf{max}} = \varphi_{\mathbf{min}} + 90°$ (here, $r_{\mathbf{rel}}(\varphi)$ denotes the position vector of a point of the Dupin indicatrix at a point on the surface of relative curvature). Equation (C.55) allows for two solutions, $\varphi^*_{\min}$ and $\varphi^*_{\max}$. Therefore, the extremal difference:

$$\Delta\varphi_{\min} = \varphi_{\min} - \varphi^*_{\min} \tag{C.58}$$

as well as the extremal difference:

$$\Delta\varphi_{\max} = \varphi_{\max} - \varphi^*_{\max} \tag{C.59}$$

can be easily calculated.

Generally speaking, neither the extremal difference $\Delta\varphi_{\mathbf{min}}$ nor the extremal difference $\Delta\varphi_{\mathbf{max}}$ is equal to zero. They are equal to zero only in particular cases, e.g., when the angle μ of the local relative orientation of the surfaces $\mathcal{G}$ and $\mathcal{P}$ fulfills the relationship $\mu = \pm 0.5\pi n$.

Let us consider an example that illustrates Statement C.1.

Example As an illustrative example, let us analytically describe the contact geometry of two convex parabolic patches of the contacting tooth flanks $\mathcal{G}$ and $\mathcal{P}$ (see Fig. C.12). The example pertains to finish cutting a helical involute gear by a disk-type shaving cutter. In the example under consideration, the design parameters of the gear and of the shaving cutter, along with the specified corresponding configurations, yield the following numerical data for the calculation: At the point K of the tooth flanks' contact, the principal curvatures of the gear tooth flank $\mathcal{G}$ are $k_{1.\,g} = 4\,\text{mm}^{-1}$ and $k_{2.\,g} = 0$. The principal curvatures of the mating pinion tooth flank $\mathcal{P}$ are $k_{1.\,p} = 1\,\text{mm}^{-1}$ and $k_{2.\,p} = 0$. The angle μ of the local relative orientation of the tooth flanks $\mathcal{G}$ and $\mathcal{P}$ is $\mu = 45°$.

Two approaches can be implemented for the analytical description of the contact geometry of the tooth flanks $\mathcal{G}$ and $\mathcal{P}$. The first one is based on the implementation of the Dupin indicatrix of the surface of relative curvature. The second one is based on

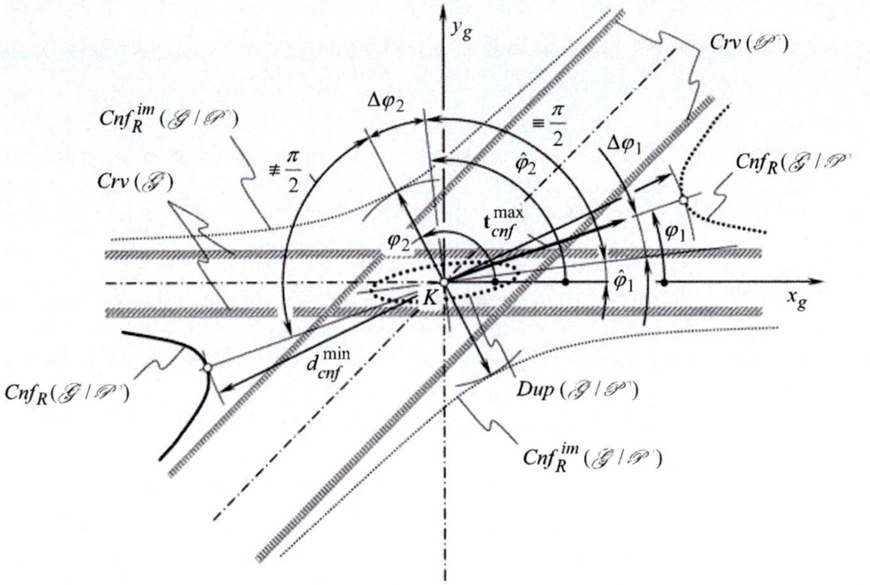

Fig. C.12 Example: determination of the optimum instantaneous kinematics in a gear shaving operation

the application of the indicatrix of conformity $Cnf_R(\mathcal{G}/\mathcal{P})$ constructed at the contact point K of the interacting tooth flanks $\mathcal{G}$ and $\mathcal{P}$.

The first approach In the case under consideration, the normal curvature $k_{\boldsymbol{R}}$ of the surface of relative curvature $\boldsymbol{R}$ can be analytically expressed as:

$$k_{\boldsymbol{R}} = k_{1.g} \cos^2 \varphi - k_{1.p} \cos^2 (\varphi + \mu) \tag{C.60}$$

Therefore, the following equality:

$$\frac{\partial k_{\boldsymbol{R}}}{\partial \varphi} = -2k_{1.g} \sin \varphi \cos \varphi + 2k_{1.p} \sin (\varphi + \mu) \cos (\varphi + \mu) = 0 \tag{C.61}$$

is valid for the directions of the extremum degree of conformity of the tooth flanks $\mathcal{G}$ and $\mathcal{P}$ at every point of their contact.

For the directions of the extremal degree of conformity at the point of contact of the gear tooth flank $\mathcal{G}$ and the mating pinion tooth flank $\mathcal{P}$, Eq. (C.61) yields extremal values $\varphi_{\mathbf{min}} = 7°$ and $\varphi_{\mathbf{max}} = \varphi_{\mathbf{min}} + 90°= 97°$ of the angles $\varphi_{\mathbf{min}}$ and $\varphi_{\mathbf{max}}$, respectively.

The direction that is specified by the angle $\varphi_{\mathbf{min}} = 7°$ indicates the direction of the minimal diameter of the Dupin indicatrix of the surface of relative curvature. This same direction corresponds to the maximal degree of conformity at the point of contact of the tooth flanks $\mathcal{G}$ and $\mathcal{P}$. Another direction that is specified by the angle $\varphi_{\mathbf{max}} = 97°$ indicates the direction of the minimum degree of conformity of the contacting tooth flanks $\mathcal{G}$ and $\mathcal{P}$ at the same contact point.

The second approach For the case under consideration, use of Eq. (C.49) of the indicatrix of conformity $\mathrm{Cnf}_R(\mathcal{G}/\mathcal{P})$ at a point of contact of the gear tooth flank $\mathcal{G}$ and the mating pinion tooth flank $\mathcal{P}$ makes calculation of the extremal angles $\varphi^*_{\mathrm{min}} = 19\,°$ and $\varphi^*_{\mathrm{max}} = 118\,°$ possible.

Imaginary branches of the indicatrix of conformity $\mathrm{Cnf}_R(\mathcal{G}/\mathcal{P})$ at the point of contact of the tooth flanks $\mathcal{G}$ and $\mathcal{P}$ in Fig. C.12 are depicted by a dashed line.

It is important to focus the readers' attention here on two issues.

First, the extremal angles $\varphi_{\mathbf{min}}$ and $\varphi_{\mathbf{max}}$, which are calculated using the first approach, are not equal to the corresponding extremal angles φ^*_{min} and φ^*_{max}, which are calculated using the second approach. The relationships $\varphi_{\mathrm{min}} \neq \varphi^*_{\mathrm{min}}$ and $\varphi_{\mathrm{max}} \neq \varphi^*_{\mathrm{max}}$ are generally observed.

Second, the difference $\Delta\varphi^*$ between the extremal angles φ^*_{min} and φ^*_{max} is not equal to half of π. Therefore, the relationship $\varphi^*_{\mathrm{max}} - \varphi^*_{\mathrm{min}} \not\equiv 90°$ between the extremal angles φ^*_{min} and φ^*_{max} is observed. In the general case of contact of two sculptured surfaces, the directions of the extremal degree of conformity of the gear tooth flank $\mathcal{G}$ and the mating pinion tooth flank $\mathcal{P}$ are not orthogonal to one another.

The discussed example reveals that in general cases of contact of two smooth regular tooth flanks, the indicatrix of conformity $\mathrm{Cnf}_R(\mathcal{G}/\mathcal{P})$ can be implemented for the purpose of an accurate analytical description of the contact geometry of the surfaces. The Dupin indicatrix of the surface of relative curvature can be implemented for this purpose only in the relative orientation of particular cases of the surfaces $\mathcal{G}$ and $\mathcal{P}$. Application of the Dupin indicatrix of the surface of relative curvature enables only an approximate analytical description of the geometry of contact of the surfaces. The Dupin indicatrix of the surface of relative curvature could be equivalent to the indicatrix of conformity only in degenerate cases of contact of the surfaces. Advantages of the indicatrix of conformity over the Dupin indicatrix of the surface of relative curvature are because this characteristic curve $\mathrm{Cnf}_R(\mathcal{G}/\mathcal{P})$ is a curve of the fourth order.

C.3.4 Important Properties of the Indicatrix of Conformity $\mathrm{Cnf}_R(\mathcal{G}/\mathcal{P})$ at a Point of Contact of the Tooth Flanks of a Gear and a Mating Pinion

The performed analysis of Eq. (C.49) of the indicatrix of conformity $\mathrm{Cnf}_R(\mathcal{G}/\mathcal{P})$ at a point of contact of a gear tooth flank and a mating pinion tooth flank reveals that this characteristic curve possesses the following important properties.

1. The indicatrix of conformity $\mathrm{Cnf}_R(\mathcal{G}/\mathcal{P})$ at a point of contact of the tooth flanks $\mathcal{G}$ and $\mathcal{P}$ is a planar characteristic curve of the fourth order. It possesses the property of central symmetry and, in particular cases, also the property of mirror symmetry.

2. The indicatrix of conformity $\mathrm{Cnf}_R(\mathcal{G}/\mathcal{P})$ is closely related to the second fundamental forms $\Phi_{2.\,g}$ and $\Phi_{2.\,p}$ of the surfaces $\mathcal{G}$ and $\mathcal{P}$, respectively. This characteristic curve is invariant with respect to the kind of parameterization of the tooth flanks $\mathcal{G}$ and $\mathcal{P}$ but not its equation. A change in the parameterization of the surfaces $\mathcal{G}$ and $\mathcal{P}$ leads to a change in the equation of the indicatrix of conformity $\mathrm{Cnf}_R(\mathcal{G}/\mathcal{P})$, whereas the shape and parameters of this characteristic curve remain unchanged.
3. The characteristic curve $\mathrm{Cnf}_R(\mathcal{G}/\mathcal{P})$ is independent of the actual value of the coordinate angle ω_g that forms the coordinate lines U_g and V_g on the gear tooth flank $\mathcal{G}$. It is also independent of the actual value of the coordinate angle ω_p that forms the coordinate lines U_p and V_p on the mating pinion tooth flank $\mathcal{P}$. However, the parameters of the indicatrix of conformity $\mathrm{Cnf}_R(\mathcal{G}/\mathcal{P})$ depend upon the angle μ of the local relative orientation of the tooth flanks $\mathcal{G}$ and $\mathcal{P}$. Therefore, for a given pair of the tooth flanks $\mathcal{G}$ and $\mathcal{P}$, the degree of conformity of the surface varies correspondingly to the variation of the angle μ, whereas the pinion tooth flank $\mathcal{P}$ spins around the unit vector of the common perpendicular.

More properties of the indicatrix of conformity $\mathrm{Cnf}_R(\mathcal{G}/\mathcal{P})$ at a point of contact of a gear tooth flank, $\mathcal{G}$, and a mating pinion tooth flank, $\mathcal{P}$, can be outlined.

C.3.5 Converse Indicatrix of Conformity at a Point of Contact of the Tooth Flanks of a Gear and a Mating Pinion

For the Dupin indicatrix $\mathrm{Dup}(\mathcal{G}/\mathcal{P})$ at a point on the surface of relative curvature R there exists a corresponding inverse Dupin indicatrix, $\mathrm{Dup}_k(\mathcal{G}/\mathcal{P})$. Similarly, for the indicatrix of conformity $\mathrm{Cnf}_R(\mathcal{G}/\mathcal{P})$, at a point of contact of the tooth flanks $\mathcal{G}$ and $\mathcal{P}$, there exists a corresponding converse indicatrix of conformity $\mathrm{Cnf}_k(\mathcal{G}/\mathcal{P})$. This characteristic curve can be directly expressed in terms of the normal curvatures k_g and k_p of the surfaces $\mathcal{G}$ and $\mathcal{P}$, respectively:

$$\mathrm{Cnf}_k(\mathcal{G}/\mathcal{P}) \quad \Rightarrow \quad r_{\mathrm{cnf}}^{\mathrm{cnv}}(\varphi, \mu) = \sqrt{|k_g(\varphi)|} \cdot \mathrm{sgn}\,\Phi_{2.g}^{-1} - \sqrt{|k_p(\varphi, \mu)|} \cdot \mathrm{sgn}\,\Phi_{2.p}^{-1} \tag{C.62}$$

For derivation of an equation of the converse indicatrix of conformity $\mathrm{Cnf}_k(\mathcal{G}/\mathcal{P})$, Euler's formula for a surface normal curvature is used in the following representation:

$$k_g(\varphi) = k_{1.g} \cos^2\varphi + k_{2.g} \sin^2\varphi \tag{C.63}$$

$$k_p(\varphi, \mu) = k_{1.p} \cos^2(\varphi + \mu) + k_{2.p} \sin^2(\varphi + \mu) \tag{C.64}$$

Here, in Eqs. (C.63) and (C.64), the principal curvatures of the gear tooth flank $\mathcal{G}$ are designated as $k_{1.\,g}$ and $k_{2.\,g}$, whereas $k_{1.\,p}$ and $k_{2.\,p}$ are the principal curvatures of the mating pinion tooth flank $\mathcal{P}$.

After substituting Eqs. (C.63) and (C.64) in Eq. (C.62), one can come up with the equation:

$$
\begin{aligned}
r_{\text{cnf}}^{\text{cnv}}(\varphi,\mu) = & \sqrt{|k_{1.g}\cos^2\varphi + k_{2.g}\sin^2\varphi|\,\text{sgn}\,\Phi_{2.g}^{-1}} \\
& - \sqrt{|k_{1.p}\cos^2(\varphi+\mu) + k_{2.p}\sin^2(\varphi+\mu)|\,\text{sgn}\,\Phi_{2.p}^{-1}}
\end{aligned}
\tag{C.65}
$$

for the converse indicatrix of conformity $\text{Cnf}_k(\mathcal{G}/\mathcal{P})$ at a point of contact of the tooth flanks $\mathcal{G}$ and $\mathcal{P}$ in the first order of tangency.

Here, in Eq. (C.65), the principal curvatures $k_{1.\,g}$, $k_{2.\,g}$ and $k_{1.\,p}$, $k_{2.\,p}$ can be expressed in terms of the corresponding fundamental magnitudes E_g, F_g, and G_g of the first and L_g, M_g, and N_g of the second orders of the gear tooth flank $\mathcal{G}$ and in terms of the corresponding fundamental magnitudes E_p, F_p, and G_p of the first and L_p, M_p, and N_p of the second orders of the mating pinion tooth flank $\mathcal{P}$. In this manner, Eq. (C.65) of the converse indicatrix of conformity $\text{Cnf}_k(\mathcal{G}/\mathcal{P})$ can be cast into a form that is similar to Eq. (C.49) of the ordinary indicatrix of conformity $\text{Cnf}_R(\mathcal{G}/\mathcal{P})$ at a point of contact of the tooth flanks $\mathcal{G}$ and $\mathcal{P}$.

It can be shown that similar to the indicatrix of conformity $\text{Cnf}_R(\mathcal{G}/\mathcal{P})$, the characteristic curve $\text{Cnf}_k(\mathcal{G}/\mathcal{P})$ also possess the property of central symmetry. In particular cases of surface contact, it also possesses the property of mirror symmetry. The directions of the extremal degree of conformity of the gear tooth flank $\mathcal{G}$ and the mating pinion tooth flank $\mathcal{P}$ are orthogonal to one another only in degenerate cases of surface contact.

Equation (C.65) of the converse indicatrix of conformity $\text{Cnf}_k(\mathcal{G}/\mathcal{P})$ is convenient for implementation when:

(a) The gear tooth flank $\mathcal{G}$
(b) The mating pinion tooth flank $\mathcal{P}$
(c) Both of them

feature point(s) or line(s) of inflection. In the point(s) or (line(s)) of inflection, the radii of normal curvature $R_{g(p)}$ of the surfaces $\mathcal{G}$ and $\mathcal{P}$ are equal to infinity. Points/lines of inflection cause indefiniteness when calculating the position vector $r_{\text{cnf}}(\varphi,\mu)$ of a point of the characteristic curve $\text{Cnf}_R(\mathcal{G}/\mathcal{P})$. Equation (C.65) of the converse indicatrix of conformity $\text{Cnf}_k(\mathcal{G}/\mathcal{P})$ is free of the disadvantages of such kind and is therefore recommended for practical applications.

Appendix D: Applied Coordinate Systems and Linear Transformations

Consequent coordinate system transformations can be easily described analytically with implementation of matrices. The use of matrices for coordinate system transformation[23] can be traced back to the mid-1940s[24] when Dr. S.S. Mozhayev[25] began describing coordinate system transformations by means of matrices.

Below, coordinate system transformation is briefly discussed from the standpoint of its implementation in the theory of gearing.

D.1 Coordinate System Transformation

Homogeneous coordinates utilize a mathematical trick to embed three-dimensional coordinates and transformations into a four-dimensional matrix format. As a result, inversions or combinations of linear transformations are simplified to inversions or multiplication of the corresponding matrices.

D.1.1 Homogeneous Coordinate Vectors

Instead of representing each point $\mathbf{r}(x, y, z)$ in three-dimensional space with a single three-dimensional vector,

$$\mathbf{r} = \begin{bmatrix} x \\ y \\ z \end{bmatrix} \tag{D.1}$$

homogeneous coordinates allow each point $\mathbf{r}(x, y, z)$ to be represented by any of an infinite number of four-dimensional vectors:

---

[23] Matrices were introduced to mathematics by A. Cayley in 1857. They provide a compact and flexible notation particularly useful in dealing with linear transformations, and they presented an organized method for the solution of systems of linear differential equations.

[24] Application of matrices for the purposes of analytical representation of coordinate system transformation should be credited to Dr. S.S. Mozhayev (Mozhayev, S.S., *General Theory of Cutting Tools, Doctoral Thesis*, Leningrad, Leningrad Polytechnic Institute, 1951, 295 pages). Dr. S.S. Mozhayev began using matrices for this purpose in the mid-1940s. Later on, the matrix approach to coordinate system transformation was used by Denavit and Hartenberg as well as by many other researchers.

[25] S.S. Mozhayev is a soviet scientist mostly known for his accomplishments in the theory of cutting tool design.

$$\mathbf{r} = \begin{bmatrix} T \cdot x \\ T \cdot y \\ T \cdot z \\ T \end{bmatrix} \tag{D.2}$$

The three-dimensional vector corresponding to any four-dimensional vector can be calculated by dividing the first three elements by the fourth element, and a four-dimensional vector corresponding to any three-dimensional vector can be created by simply adding a fourth element and setting it equal to one.

D.1.2 Homogeneous Coordinate Transformation Matrices of the Dimension 4 × 4

Homogeneous coordinate transformation matrices operate on four-dimensional homogeneous vector representations of traditional three-dimensional coordinate locations. Any three-dimensional linear transformation (translation, rotation, and so forth) can be represented by a 4 × 4 homogeneous coordinate transformation matrix. In fact, because of the redundant representation of a three-dimensional space in a homogeneous coordinate system, an infinite number of different 4 × 4 homogeneous coordinate transformation matrices are available to perform any given linear transformation. This redundancy can be eliminated to provide a unique representation by dividing all elements of a 4 × 4 homogeneous transformation matrix by the last element (which will become equal to one). This means that a 4 × 4 homogeneous transformation matrix can incorporate as many as 15 independent parameters. The generic format representation of a homogeneous transformation equation for mapping the three-dimensional coordinate (x_1, y_1, z_1) to the three-dimensional coordinate (x_2, y_2, z_2) is:

$$\begin{bmatrix} T^* \cdot x_2 \\ T^* \cdot y_2 \\ T^* \cdot z_2 \\ T^* \end{bmatrix} = \begin{bmatrix} T^* \cdot a & T^* \cdot b & T^* \cdot c & T^* \cdot d \\ T^* \cdot e & T^* \cdot f & T^* \cdot g & T^* \cdot h \\ T^* \cdot i & T^* \cdot j & T^* \cdot k & T^* \cdot m \\ T^* \cdot n & T^* \cdot p & T^* \cdot q & T^* \end{bmatrix} \cdot \begin{bmatrix} T \cdot x_2 \\ T \cdot y_2 \\ T \cdot z_2 \\ T \end{bmatrix} \tag{D.3}$$

If any two matrices or vectors of this equation are known, then the third matrix (or vector) can be calculated, and, then, the redundant T element in the solution can be eliminated by dividing all elements of the matrix by the last element.

Various transformation models can be used to constrain the form of the matrix to transformations with fewer degrees of freedom.

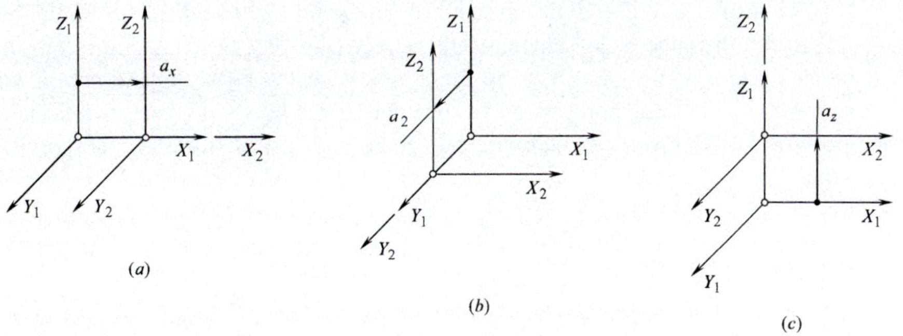

Fig. D.1 Analytical description of the operators of translations $\mathbf{Tr}(a_x, X)$, $\mathbf{Tr}(a_y, Y)$, and $\mathbf{Tr}(a_z, Z)$ along the coordinate axes of a Cartesian reference system XYZ

D.1.3 Translations

The translation of a coordinate system is one of the major linear transformations used in the theory of part surface generation. Translations of the coordinate system $X_2Y_2Z_2$ along the axes of the coordinate system $X_1Y_1Z_1$ are depicted in Fig. D.1. Translations can be analytically described by the homogeneous transformation matrix of dimension 4×4.

For an analytical description of translation along the coordinate axes, the operators of translation $\mathbf{Tr}(a_x, X)$, $\mathbf{Tr}(a_y, Y)$, and $\mathbf{Tr}(a_z, Z)$ are used. These operators yield a matrix representation in the form:

$$\mathbf{Tr}(a_x, X) = \begin{bmatrix} 1 & 0 & 0 & a_x \\ 0 & 1 & 0 & 0 \\ 0 & 0 & 1 & 0 \\ 0 & 0 & 0 & 1 \end{bmatrix} \tag{D.4}$$

$$\mathbf{Tr}\left(a_y, Y\right) = \begin{bmatrix} 1 & 0 & 0 & 0 \\ 0 & 1 & 0 & a_y \\ 0 & 0 & 1 & 0 \\ 0 & 0 & 0 & 1 \end{bmatrix} \tag{D.5}$$

$$\mathbf{Tr}(a_z, Z) = \begin{bmatrix} 1 & 0 & 0 & 0 \\ 0 & 1 & 0 & 0 \\ 0 & 0 & 1 & a_z \\ 0 & 0 & 0 & 1 \end{bmatrix} \tag{D.6}$$

Here, in Eqs. (D.4) through (D.6), the parameters a_x, a_y, and a_z are assigned values that denote the distance of the translation along the corresponding axis.

Let us consider two coordinate systems, $X_1Y_1Z_1$ and $X_2Y_2Z_2$, displaced along the X_1- axis at a distance a_x as schematically depicted in Fig. D.1a. A point m in the reference system $X_2Y_2Z_2$ is given by the position vector $\mathbf{r}_2(m)$. In the coordinate system $X_1Y_1Z_1$, the same point m can be specified by the position vector $\mathbf{r}_1(m)$. Then, the position vector $\mathbf{r}_1(m)$ can be expressed in terms of the position vector $\mathbf{r}_2(m)$ by the equation:

$$\mathbf{r}_1(m) = \mathbf{Tr}(a_x, X) \cdot \mathbf{r}_2(m) \tag{D.7}$$

Equations similar to Eq. (D.7) are valid for the operators $\mathbf{Tr}(a_y, Y)$ and $\mathbf{Tr}(a_z, Z)$ of the coordinate system transformation. The latter is schematically illustrated in Fig. D.1b and c.

Use of the operators of translation $\mathbf{Tr}(a_x, X)$, $\mathbf{Tr}(a_y, Y)$, and $\mathbf{Tr}(a_z, Z)$ makes an introduction of the operator $\mathbf{Tr}(a, \mathbf{A})$ of a combined transformation possible. Suppose that the point p on a rigid body goes through a translation describing a straight line from a point p_1 to a point p_2 with a change of coordinates (a_x, a_y, a_z). This motion of the point p can be analytically described with a resultant translation operator $\mathbf{Tr}(a, \mathbf{A})$:

$$\mathbf{Tr}(a, \mathbf{A}) = \begin{bmatrix} 1 & 0 & 0 & a_x \\ 0 & 1 & 0 & a_y \\ 0 & 0 & 1 & a_z \\ 0 & 0 & 0 & 1 \end{bmatrix} \tag{D.8}$$

The operator $\mathbf{Tr}(a, \mathbf{A})$ of the resultant coordinate system transformation can be interpreted as the operator of translation along an arbitrary axis having the vector $\mathbf{A}$ as the direct vector.

An analytical description of the translation of the coordinate system $X_1Y_1Z_1$ in the direction of an arbitrary vector $\mathbf{A}$ to the position of $X_2Y_2Z_2$ can be obtained from Fig. D.2. The operator of translation $\mathbf{Tr}(a, \mathbf{A})$ of that particular kind can be expressed in terms of the operators $\mathbf{Tr}(a_x, X)$, $\mathbf{Tr}(a_y, Y)$, and $\mathbf{Tr}(a_z, Z)$ of elementary translations:

$$\mathbf{Tr}(a, \mathbf{A}) = \mathbf{Tr}(a_z, Z) \cdot \mathbf{Tr}(a_y, Y) \cdot \mathbf{Tr}(a_x, X) \tag{D.9}$$

Evidently, the axis along the vector $\mathbf{A}$ is always the axis through the origins of both the reference systems $X_1Y_1Z_1$ and $X_2Y_2Z_2$.

Any and all coordinate system transformations that do not change the orientation of a geometrical object are referred to as "orientation-preserving transformation" or "direct transformation." Therefore, transformation of translation is an example of a direct transformation.

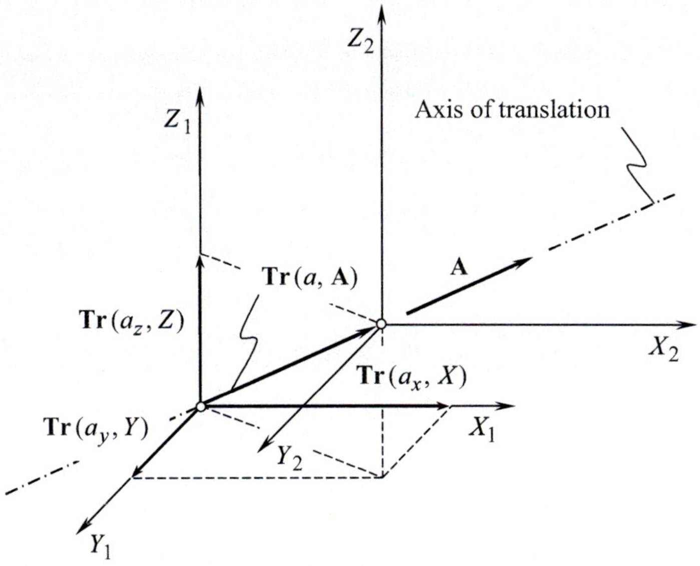

Fig. D.2 Analytical description of an operator, $\mathbf{Tr}(a, \mathbf{A})$, of translation along an arbitrary axis (vector **A** is the direct vector of the axis)

D.1.4 Rotation About a Coordinate Axis

Rotation of a coordinate system about a coordinate axis is another major linear transformation used in the theory of part surface generation. A rotation is specified by an axis of rotation and the angle of the rotation. It is a fairly simple trigonometric calculation to obtain a transformation matrix for a rotation about one of the coordinate axes.

Possible rotations of the coordinate system $X_2Y_2Z_2$ about the axis of the coordinate system $X_1Y_1Z_1$ are illustrated in Fig. D.3.

For an analytical description of rotation about a coordinate axis, the operators of rotation $\mathbf{Rt}(\varphi_x, X_1)$, $\mathbf{Rt}(\varphi_y, Y_1)$, and $\mathbf{Rt}(\varphi_z, Z_1)$ are used. These operators of linear transformations yield representation in the form of homogeneous matrices:

$$\mathbf{Rt}(\varphi_x, X_1) = \begin{bmatrix} 1 & 0 & 0 & 0 \\ 0 & \cos\varphi_x & \sin\varphi_x & 0 \\ 0 & -\sin\varphi_x & \cos\varphi_x & 0 \\ 0 & 0 & 0 & 1 \end{bmatrix} \tag{D.10}$$

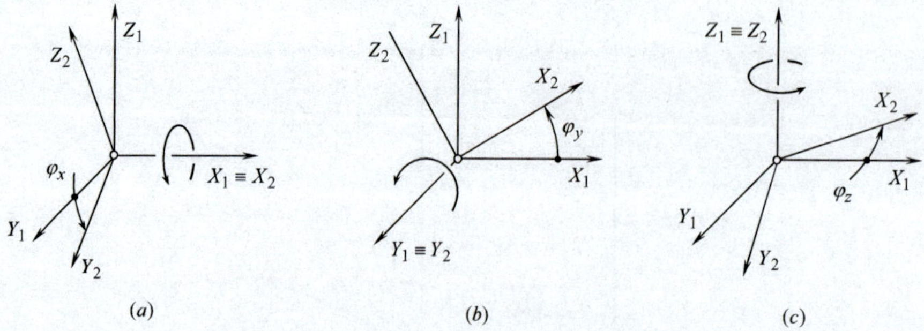

Fig. D.3 Analytical description of the operators of rotation $\mathbf{Rt}(\varphi_x, X)$, $\mathbf{Rt}(\varphi_y, Y)$, and $\mathbf{Rt}(\varphi_z, Z)$ about a coordinate axis of a reference system $X_1Y_1Z_1$

$$\mathbf{Rt}(\varphi_y, Y_1) = \begin{bmatrix} \cos\varphi_y & 0 & \sin\varphi_y & 0 \\ 0 & 1 & 0 & 0 \\ -\sin\varphi_y & 0 & \cos\varphi_y & 0 \\ 0 & 0 & 0 & 1 \end{bmatrix} \tag{D.11}$$

$$\mathbf{Rt}(\varphi_z, Z_1) = \begin{bmatrix} \cos\varphi_z & \sin\varphi_z & 0 & 0 \\ -\sin\varphi_z & \cos\varphi_z & 0 & 0 \\ 0 & 0 & 1 & 0 \\ 0 & 0 & 0 & 1 \end{bmatrix} \tag{D.12}$$

Here, φ_x, φ_y, and φ_z are assigned values that denote the corresponding angles of rotations about a corresponding coordinate axis: φ_x is the angle of rotation around the X_1- axis (pitch) of the Cartesian coordinate system $X_1Y_1Z_1$; φ_y is the angle of rotation around the Y_1- axis (roll); and φ_z is the angle of rotation around the Z_1- axis (yaw) of the same Cartesian reference system $X_1Y_1Z_1$.

The rotation about a coordinate axis is illustrated in Fig. D.3.

Let us consider two coordinate systems $X_1Y_1Z_1$ and $X_2Y_2Z_2$, which are turned about the X_1- axis through an angle φ_x, as shown in Fig. D.3a. In the reference system $X_2Y_2Z_2$, a point m is given by a position vector $\mathbf{r}_2(m)$. In the coordinate system $X_1Y_1Z_1$, the same point m can be specified by the position vector $\mathbf{r}_1(m)$. Then, the position vector $\mathbf{r}_1(m)$ can be expressed in terms of the position vector $\mathbf{r}_2(m)$ by the equation:

$$\mathbf{r}_1(m) = \mathbf{Rt}(\varphi_x, X) \cdot \mathbf{r}_2(m) \tag{D.13}$$

Equations similar to Eq. (D.13) are also valid for other operators $\mathbf{Rt}(\varphi_y, Y)$ and $\mathbf{Rt}(\varphi_z, Z)$ of the coordinate system transformation. These elementary coordinate system transformations are schematically illustrated in Fig. D.3b and c, respectively.

D.1.5 Rotation About an Arbitrary Axis Through the Origin

When a rotation is to be performed around an arbitrary vector based at the origin, the transformation matrix must be assembled from a combination of rotations about the Cartesian coordinates.

Two different approaches for an analytical description of a rotation about an arbitrary axis through the origin are discussed below.

Conventional Approach

An analytical description of rotation of the coordinate system $X_1Y_1Z_1$ about an arbitrary axis through the origin to the position of a reference system $X_2Y_2Z_2$ is illustrated in Fig. D.4. It is assumed here that the rotation is performed about the axis having a vector $\mathbf{A}_0$ as the direction vector. The operator $\mathbf{Rt}(\varphi_A, \mathbf{A}_0)$ of a rotation of this kind can be expressed in terms of the operators $\mathbf{Rt}(\varphi_x, X)$, $\mathbf{Rt}(\varphi_y, Y)$, and $\mathbf{Rt}(\varphi_z, Z)$ of elementary rotations:

$$\mathbf{Rt}(\varphi_A, \mathbf{A}_0) = \mathbf{Rt}(\varphi_z, Z) \cdot \mathbf{Rt}(\varphi_y, Y) \cdot \mathbf{Rt}(\varphi_x, X) \tag{D.14}$$

Evidently, the axis of rotation (a straight line along the vector $\mathbf{A}_0$) is always an axis through the origin.

The operators of translation and of rotation also yield linear transformations of other kinds as well.

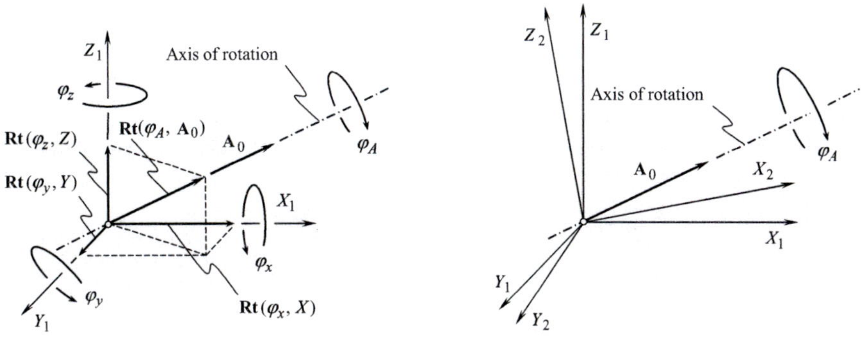

Fig. D.4 Analytical description of the operator $\mathbf{Rt}(\varphi_A, \mathbf{A})$ of rotation about an arbitrary axis through the origin of a Cartesian coordinate system $X_1Y_1Z_1$ (the vector $\mathbf{A}$ is the directing vector of the axis of rotation)

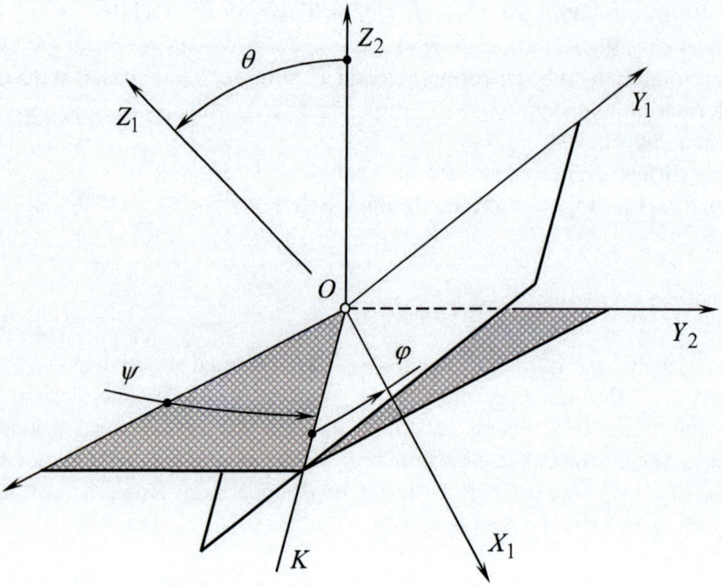

Fig. D.5 Euler angles

Eulerian Transformation

"Eulerian transformation" is a well-known kind of linear transformation used extensively in mechanical engineering. This kind of linear transformation is analytically described by the operator $\mathbf{Eu}(\psi, \theta, \varphi)$ of Eulerian[26] transformation.

The operator $\mathbf{Eu}(\psi, \theta, \varphi)$ is expressed in terms of three Euler angles (or Eulerian angles), namely, ψ, θ, and φ. The configuration of an orthogonal Cartesian coordinate system $X_1Y_1Z_1$ in relation to another orthogonal Cartesian coordinate system $X_2Y_2Z_2$ is defined by the Euler angles ψ, θ, and φ. These angles are shown in Fig. D.5.

The line of intersection of the coordinate plane X_1Y_1 of the first reference system by the coordinate plane X_2Y_2 of the second reference system is commonly referred to as the "line of nodes." In Fig. D.5, the line OK is the line of nodes. It is assumed here and below that the line of nodes OK and the axes Z_1 and Z_2 form a frame of the same orientation as do the reference systems $X_1Y_1Z_1$ and $X_2Y_2Z_2$.

The Euler angle φ is referred to as the "angle of pure rotation." This angle is measured between the X_1- axis and the line of nodes OK. The angle of pure rotation φ is measured within the coordinate plane X_1Y_1 in the direction of the shortest rotation from the axis X_1 to the axis Y_1.

[26]Leonhard Euler (April 15, 1707–September 18, 1783), a famous Swiss mathematician and physicist who spent most of his life in Russia and Germany.

The Euler angle θ is referred to as the "angle of nutation." The angle of nutation θ is measured between the axes Z_1 and Z_2. The actual value of this angle never exceeds 180°.

The Euler angle ψ is referred to as the "angle of precession." The angle of precession ψ is measured in the coordinate plane X_2Y_2. This angle lies between the line of nodes OK and the X_2- axis. The direction of the shortest rotation from the axis X_2 to the axis Y_2 is the direction in which the angle of precession is measured.

In case the angle of nutation is either $\theta = 0°$ or $\theta = 180°$, then the Euler angles are not defined.

The operator $\mathbf{Eu}(\psi, \theta, \varphi)$ of Eulerian transformation allows for the following matrix representation:

$$\mathbf{Eu}(\psi,\theta,\varphi) = \begin{bmatrix} -\sin\psi\cos\theta\sin\varphi + \cos\psi\cos\varphi & \cos\psi\cos\theta\sin\varphi + \sin\psi\cos\varphi & \sin\theta\sin\varphi & 0 \\ -\sin\psi\cos\theta\cos\varphi - \cos\psi\sin\varphi & \cos\psi\cos\theta\cos\varphi - \sin\psi\cos\varphi & \sin\theta\cos\varphi & 0 \\ \sin\theta\sin\varphi & -\cos\psi\cos\theta & \cos\theta & 0 \\ 0 & 0 & 0 & 1 \end{bmatrix} \tag{D.15}$$

Here, it is important to stress the difference between the operator $\mathbf{Eu}(\psi, \theta, \varphi)$ of Eulerian transformation the operator $\mathbf{Rt}(\psi_A, \mathbf{A}_0)$ of rotation about an arbitrary axis through the origin.

The operator $\mathbf{Rt}(\psi_A, \mathbf{A})$ of rotation about an arbitrary axis through the origin can result in the same final orientation of the coordinate system $X_2Y_2Z_2$ in relation to the coordinate system $X_1Y_1Z_1$ as does the operator $\mathbf{Eu}(\psi, \theta, \varphi)$ of Eulerian transformation. However, the operators of the linear transformations $\mathbf{Rt}(\psi_A, \mathbf{A}_0)$ and $\mathbf{Eu}(\psi, \theta, \varphi)$ are of a completely different nature. They can result in identical coordinate system transformation, but they are not equal to one another.

D.1.6 Rotation About an Arbitrary Axis Not Through the Origin

The transformation corresponding to the rotation of an angle φ around an arbitrary vector not through the origin cannot be readily written in a form similar to the rotation matrices about the coordinate axes.

The desired transformation matrix is obtained by combining a sequence of elementary translation and rotation matrices (once a single 4×4 matrix has been obtained, representing the composite transformations, it can be used in the same way as any other transformation matrix).

Rotation of the coordinate system $X_1Y_1Z_1$ to a configuration, which the coordinate system $X_2Y_2Z_2$ possesses, can be performed about a corresponding axis that features an arbitrary configuration in space (see Fig. D.6). The vector $\mathbf{A}$ is the direction vector of the axis of the rotation. The axis of the rotation is not a line through the origin.

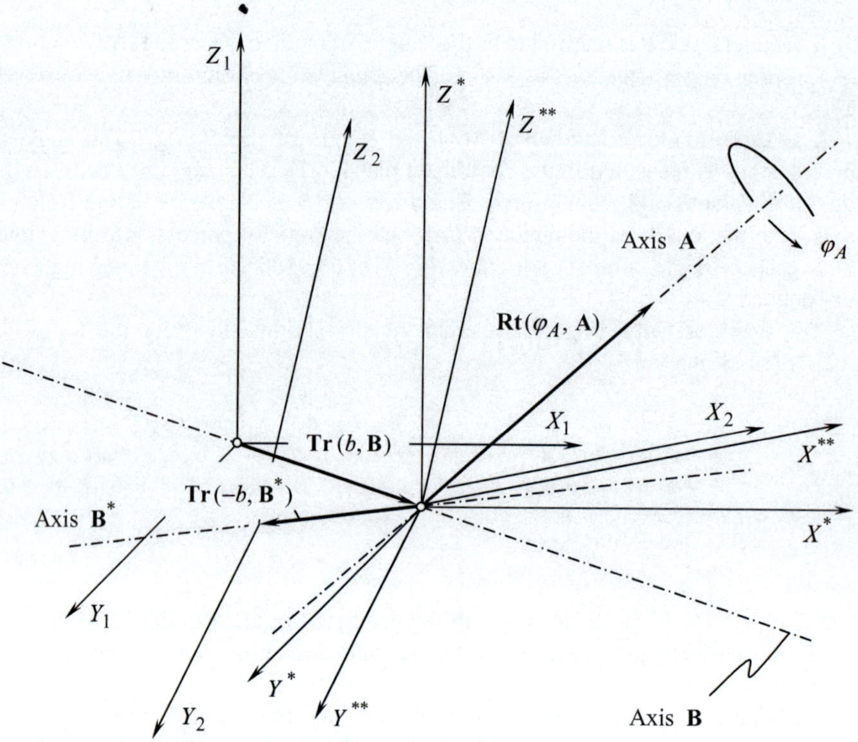

Fig. D.6 Analytical description of the operator $\mathbf{Rt}(\varphi_A, \mathbf{A})$ of rotation about an arbitrary axis not through the origin (vector **A** is the direct vector of the axis of the rotation)

The operator of linear transformation of this particular kind $\mathbf{Rt}(\psi_A, \mathbf{A})$ can be expressed in terms of the operator $\mathbf{Tr}(a, \mathbf{A})$ of translation along and of the operator $\mathbf{Rt}(\psi_A, \mathbf{A}_0)$ of rotation about an arbitrary axis through the origin:

$$\mathbf{Rt}(\varphi_A, \mathbf{A}) = \mathbf{Tr}(-b, \mathbf{B}^*) \cdot \mathbf{Rt}(\varphi_A, \mathbf{A}_0) \cdot \mathbf{Tr}(b, \mathbf{B}) \tag{D.16}$$

Here, in Eq. (D.16),

$\mathbf{Tr}(b, \mathbf{B})$ is the operator of translation along the shortest distance of approach of the axis of rotation and origin of the coordinate system.

$\mathbf{Tr}(-b, \mathbf{B}^*)$ is the operator of translation in the direction opposite to the translation $\mathbf{Tr}(b, \mathbf{B})$ after the rotation $\mathbf{Rt}(\psi_A, \mathbf{A})$ is completed.

In order to determine the shortest distance of approach B of the axis of rotation (that is, the axis along the directing vector **B**) and the origin of the coordinate system, let us consider the axis (**B**) through two given points $\mathbf{r}_{B.1}$ and $\mathbf{r}_{B.2}$.

The shortest distance between a certain point $\mathbf{r}_0$ and the straight line through the points $\mathbf{r}_{B.1}$ and $\mathbf{r}_{B.2}$ can be calculated from the following formula:

$$B = \frac{|(\mathbf{r}_2 - \mathbf{r}_1) \times (\mathbf{r}_1 - \mathbf{r}_0)|}{|\mathbf{r}_2 - \mathbf{r}_1|} \tag{D.17}$$

For the origin of the coordinate system, the equality $\mathbf{r}_0 = 0$ is observed. Then,

$$B = |\mathbf{r}_1| \cdot \sin \angle[\mathbf{r}_1, (\mathbf{r}_2 - \mathbf{r}_1)] \tag{D.18}$$

A matrix representation of the operators of translation $\mathbf{Tr}(a_x, X)$, $\mathbf{Tr}(a_y, Y)$, $\mathbf{Tr}(a_z, Z)$ along the coordinate axes, together with the operators of rotation $\mathbf{Rt}(\varphi_x, X)$, $\mathbf{Rt}(\varphi_y, Y)$, $\mathbf{Rt}(\varphi_z, Z)$ about the coordinate axes, is convenient for implementation of the theory of part surface generation. Moreover, use of the operators is the simplest possible way to analytically describe the linear transformations.

D.1.7 Resultant Coordinate System Transformation

The operators of translation $\mathbf{Tr}(a_x, X)$, $\mathbf{Tr}(a_y, Y)$, and $\mathbf{Tr}(a_z, Z)$ together with the operators of rotation $\mathbf{Rt}(\varphi_x, X)$, $\mathbf{Rt}(\varphi_y, Y)$, and $\mathbf{Rt}(\varphi_z, Z)$, respectively, are used for the purpose of composing the operator $\mathbf{Rs}(1 \mapsto 2)$ of the resultant coordinate system transformation. The transition from the initial Cartesian reference system $X_1Y_1Z_1$ to other Cartesian reference system $X_2Y_2Z_2$ is analytically described by the operator $\mathbf{Rs}(1 \mapsto 2)$ of the resultant coordinate system transformation.

For example, the expression:

$$\mathbf{Rs}(1 \mapsto 5) = \mathbf{Tr}(a_x, X) \cdot \mathbf{Rt}(\varphi_z, Z) \cdot \mathbf{Rt}(\varphi_x, X) \cdot \mathbf{Tr}(a_y, Y) \tag{D.19}$$

indicates that the transition from the coordinate system $X_1Y_1Z_1$ to the coordinate system $X_5Y_5Z_5$ is executed in the following four steps (see Fig. D.7):

- Translation $\mathbf{Tr}(a_y, Y)$
- Rotation $\mathbf{Rt}(\varphi_x, X)$
- Second rotation $\mathbf{Rt}(\varphi_z, Z)$
- Translation $\mathbf{Tr}(a_x, X)$

Ultimately, the equality:

$$\mathbf{r}_1(m) = \mathbf{Rs}(1 \mapsto 5) \cdot \mathbf{r}_5(m) \tag{D.20}$$

is valid.

When the operator $\mathbf{Rs}(1 \mapsto t)$ of the resultant coordinate system transformation is specified, the transition in the opposite direction can be performed by means of the operator $\mathbf{Rs}(t \mapsto 1)$ of the inverse coordinate system transformation. The following equality can be easily proven:

$$\mathbf{Rs}(t \mapsto 1) = \mathbf{Rs}^{-1}(1 \mapsto t) \tag{D.21}$$

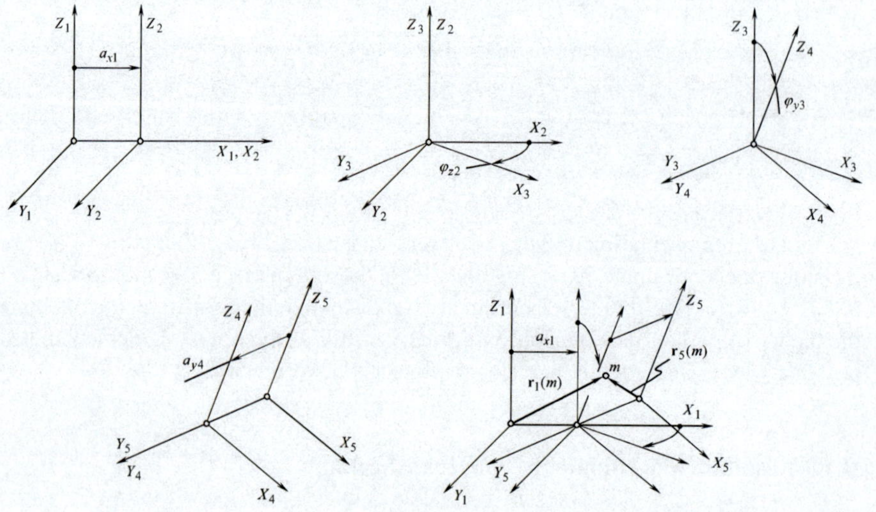

Fig. D.7 An example of the resultant coordinate system transformation analytically expressed by the operator $\mathbf{Rs}\,(1 \mapsto 5)$

In the above example illustrated in Fig. D.7, the operator $\mathbf{Rs}\,(5 \mapsto 1)$ of the inverse resultant coordinate system transformation can be expressed in terms of the operator $\mathbf{Rs}\,(1 \mapsto 5)$ of the direct resultant coordinate system transformation. Following Eq. (D.21), one can come up with the equation:

$$\mathbf{Rs}\,(5 \mapsto 1) = \mathbf{Rs}^{-1}(1 \mapsto 5) \tag{D.22}$$

It is easy to show that the operator $\mathbf{Rs}\,(1 \mapsto t)$ of the resultant coordinate system transformation allows for representation in the following form:

$$\mathbf{Rs}\,(1 \mapsto t) = \mathbf{Tr}\,(a, A) \cdot \mathbf{Eu}(\psi, \theta, \varphi) \tag{D.23}$$

The linear transformation $\mathbf{Rs}\,(1 \mapsto t)$ (see Eq. (D.23)) can also be expressed in terms of rotation about an axis $\mathbf{Rt}\,(\varphi_A, A)$ not through the origin (see Eq. (D.16)).

D.2 Complex Coordinate System Transformation

In particular cases of complex coordinate system transformations that are repeatedly used in practice, special purpose operators of coordinate system transformation can be composed of elementary operators of translation and operators of rotation.

D.2.1 Linear Transformation Describing a Screw Motion About a Coordinate Axis

Operators for an analytical description of screw motions about an axis of the Cartesian coordinate system are a particular case of the operators of the resultant coordinate system transformation.

By definition (see Fig. D.8), the operator $\mathbf{Sc}_x(\varphi_x, p_x)$ of a screw motion about the $X-$ axis of the Cartesian coordinate system XYZ is equal to:

$$\mathbf{Sc}_x(\varphi_x, p_x) = \mathbf{Rt}(\varphi_x, X) \cdot \mathbf{Tr}(a_x, X) \tag{D.24}$$

After substituting the operator of translation $\mathbf{Tr}(a_x, X)$ and the operator of rotation $\mathbf{Rt}(\varphi_x, X)$ (see Eq. (D.10)), Eq. (D.24) is cast into the expression:

$$\mathbf{Sc}_x(\varphi_x, p_x) = \begin{bmatrix} 1 & 0 & 0 & p_x \cdot \varphi_x \\ 0 & \cos\varphi_x & \sin\varphi_x & 0 \\ 0 & -\sin\varphi_x & \cos\varphi_x & 0 \\ 0 & 0 & 0 & 1 \end{bmatrix} \tag{D.25}$$

for the calculation of the operator of the screw motion $\mathbf{Sc}_x(\varphi_x, p_x)$ about the $X-$ axis.

The operators of the screw motions $\mathbf{Sc}_y(\varphi_y, p_y)$ and $\mathbf{Sc}_z(\varphi_z, p_z)$ about the $Y-$ and $Z-$ axes are correspondingly defined in a manner similar to that in which the operator of the screw motion $\mathbf{Sc}_x(\varphi_x, p_x)$ is defined:

$$\mathbf{Sc}_y(\varphi_y, p_y) = \mathbf{Rt}(\varphi_y, Y) \cdot \mathbf{Tr}(a_y, Y) \tag{D.26}$$

$$\mathbf{Sc}_z(\varphi_z, p_z) = \mathbf{Rt}(\varphi_z, Z) \cdot \mathbf{Tr}(a_z, Z) \tag{D.27}$$

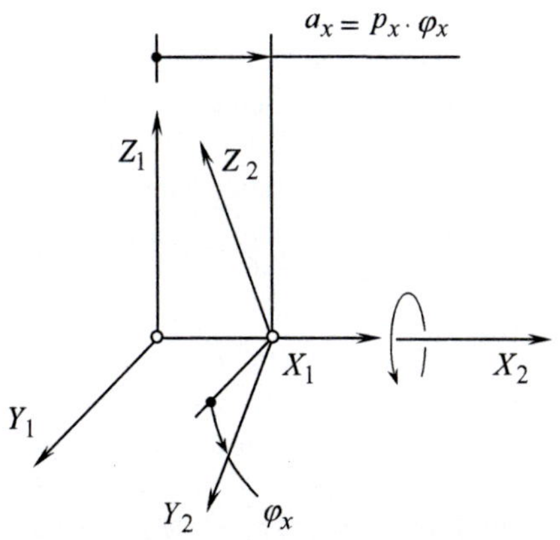

Fig. D.8 Analytical description of the operator of screw motion $\mathbf{Sc}_x(\varphi_x, p_x)$

Using Eqs. (D.5) and (D.6) together with Eqs. (D.11) and (D.12), one can come up with the expressions:

$$\mathbf{Sc}_y(\varphi_y, p_y) = \begin{bmatrix} \cos\varphi_y & 0 & -\sin\varphi_y & 0 \\ 0 & 1 & 0 & p_y \cdot \varphi_y \\ \sin\varphi_y & 0 & \cos\varphi_y & 0 \\ 0 & 0 & 0 & 1 \end{bmatrix} \quad \text{(D.28)}$$

$$\mathbf{Sc}_z(\varphi_z, p_z) = \begin{bmatrix} \cos\varphi_z & \sin\varphi_z & 0 & 0 \\ -\sin\varphi_z & \cos\varphi_z & 0 & 0 \\ 0 & 0 & 1 & p_z \cdot \varphi_z \\ 0 & 0 & 0 & 1 \end{bmatrix} \quad \text{(D.29)}$$

for the calculation of the operators of the screw motion $\mathbf{Sc}_y(\varphi_y, p_y)$ and $\mathbf{Sc}_z(\varphi_z, p_z)$ about the $Y-$ and $Z-$ axes, respectively.

Screw motions about a coordinate axis as well as screw surfaces are common in the theory of part surface generation. This makes it practical to use the operators of the screw motion $\mathbf{Sc}_x(\varphi_x, p_x)$, $\mathbf{Sc}_y(\varphi_y, p_y)$ and $\mathbf{Sc}_z(\varphi_z, p_z)$ in the theory of part surface generation.

In case of necessity, an operator of the screw motion about an arbitrary axis either through the origin of the coordinate system or not through the origin of the coordinate system can be derived following the method similar to that used for the derivation of the operators $\mathbf{Sc}_x(\varphi_x, p_x)$, $\mathbf{Sc}_y(\varphi_y, p_y)$, and $\mathbf{Sc}_z(\varphi_z, p_z)$.

D.2.2 Linear Transformation Describing the Rolling Motion of a Coordinate System

One more practical combination of a rotation and a translation is often used in the theory of part surface generation.

Let us consider a Cartesian coordinate system $X_1Y_1Z_1$ (see Fig. D.9). The coordinate system $X_1Y_1Z_1$ is traveling in the direction of the X_1- axis. The velocity of the translation is denoted by V. The coordinate system $X_1Y_1Z_1$ is rotating about its Y_1- axis simultaneously with the translation. The speed of the rotation is denoted by ω. Assume that the ratio V/ω is constant. Under such a scenario, the resultant motion of the reference system $X_1Y_1Z_1$ to its arbitrary position $X_2Y_2Z_2$ allows interpretation in the form of rolling with no sliding of a cylinder of radius R_w over the plane. The plane is parallel to the coordinate X_1Y_1- plane, and it is remote from it at a distance R_w. For the calculation of the radius of the rolling cylinder, the expression $R_w = V/\omega$ can be used.

Because rolling of the cylinder of a radius R_w over the plane is performed with no sliding, a certain correspondence between the translation and the rotation of the coordinate system is established. When the coordinate system turns through a certain

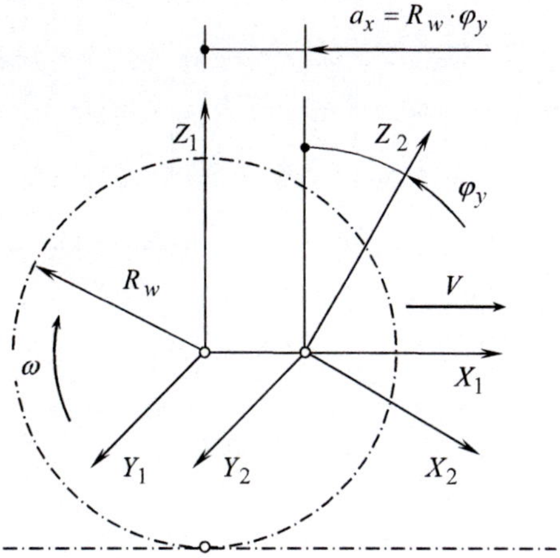

Fig. D.9 Illustration of the transformation of rolling $\mathbf{Rl}_x(\varphi_y, Y)$ of a coordinate system

angle φ_y, the translation of the origin of the coordinate system along the X_1- axis is $a_x = \varphi_r \cdot R_w$.

The transition from the coordinate system $X_1Y_1Z_1$ to the coordinate system $X_2Y_2Z_2$ can be analytically described by the operator of the resultant coordinate system transformation $\mathbf{Rs}\,(1 \mapsto 2)$, as follows:

$$\mathbf{Rs}\,(1 \mapsto 2) = \mathbf{Rt}\left(\varphi_y, Y_1\right) \cdot \mathbf{Tr}\,(a_x, X_1) \tag{D.30}$$

Here, $\mathbf{Tr}\,(a_x, X_1)$ is the operator of the translation along the X_1- axis and $\mathbf{Rt}\,(\varphi_y, Y_1)$ is the operator of the rotation about the Y_1- axis.

The operator of the resultant coordinate system transformation of the kind (see Eq. (D.30)) is referred to as the "operator of rolling motion over a plane."

When the translation is performed along the X_1- axis and the rotation is performed about the Y_1- axis, the operator of rolling is denoted by $\mathbf{Rl}_x(\varphi_y, Y)$. In this particular case, the equality $\mathbf{Rl}_x(\varphi_y, Y) = \mathbf{Rs}\,(1 \mapsto 2)$ (see Eq. (D.30) is valid. Based on this equality, the operator of rolling over a plane $\mathbf{Rl}_x(\varphi_y, Y)$ can be calculated from the equation:

$$\mathbf{Rl}_x\left(\varphi_y, Y\right) = \begin{bmatrix} \cos\varphi_y & 0 & -\sin\varphi_y & a_x \cdot \cos\varphi_y \\ 0 & 1 & 0 & 0 \\ \sin\varphi_y & 0 & \cos\varphi_y & a_x \cdot \sin\varphi_y \\ 0 & 0 & 0 & 1 \end{bmatrix} \tag{D.31}$$

While rotation remains about the Y_1- axis, the translation can be performed not along the X_1- axis but along the Z_1- axis instead. For rolling of this kind, the operator of rolling is:

$$\mathbf{Rl}_z(\varphi_y, Y) = \begin{bmatrix} \cos\varphi_y & 0 & -\sin\varphi_y & -a_z \cdot \sin\varphi_y \\ 0 & 1 & 0 & 0 \\ \sin\varphi_y & 0 & \cos\varphi_y & a_z \cdot \cos\varphi_y \\ 0 & 0 & 0 & 1 \end{bmatrix} \tag{D.32}$$

For the cases in which the rotation is performed about the X_1- axis, the corresponding operators of rolling are as follows:

$$\mathbf{Rl}_y(\varphi_x, X) = \begin{bmatrix} 1 & 0 & 0 & 0 \\ 0 & \cos\varphi_x & \sin\varphi_x & a_y \cdot \cos\varphi_x \\ 0 & -\sin\varphi_x & \cos\varphi_x & -a_y \cdot \sin\varphi_x \\ 0 & 0 & 0 & 1 \end{bmatrix} \tag{D.33}$$

for the case of rolling along the Y_1- axis and

$$\mathbf{Rl}_z(\varphi_x, X) = \begin{bmatrix} 1 & 0 & 0 & 0 \\ 0 & \cos\varphi_x & \sin\varphi_x & a_z \cdot \sin\varphi_x \\ 0 & -\sin\varphi_x & \cos\varphi_x & a_z \cdot \cos\varphi_x \\ 0 & 0 & 0 & 1 \end{bmatrix} \tag{D.34}$$

for the case of rolling along the Z_1- axis.

Similar expressions can be derived for the case of rotation about the Z_1- axis:

$$\mathbf{Rl}_x(\varphi_z, Z) = \begin{bmatrix} \cos\varphi_z & \sin\varphi_z & 0 & a_x \cdot \cos\varphi_z \\ -\sin\varphi_z & \cos\varphi_z & 0 & a_x \cdot \sin\varphi_z \\ 0 & 0 & 1 & 0 \\ 0 & 0 & 0 & 1 \end{bmatrix} \tag{D.35}$$

$$\mathbf{Rl}_y(\varphi_z, Z) = \begin{bmatrix} \cos\varphi_z & \sin\varphi_z & 0 & a_y \cdot \sin\varphi_z \\ -\sin\varphi_z & \cos\varphi_z & 0 & a_y \cdot \cos\varphi_z \\ 0 & 0 & 1 & 0 \\ 0 & 0 & 0 & 1 \end{bmatrix} \tag{D.36}$$

Use of the operators of rolling Eqs. (D.31) through (D.36) significantly simplifies the analytical description of the coordinate system transformations.

D.2.3 Linear Transformation Describing the Rolling of Two Coordinate Systems

In the theory of part surface generation, combinations of two rotations about parallel axes are of particular interest.

As an example, let us consider the two Cartesian coordinate systems $X_1Y_1Z_1$ and $X_2Y_2Z_2$ shown in Fig. D.10. The coordinate systems $X_1Y_1Z_1$ and $X_2Y_2Z_2$ are rotated about their axes Z_1 and Z_2, respectively. The axes of the rotations are parallel to each other ($Z_1 \parallel Z_2$). The rotations ω_1 and ω_2 of the coordinate systems can be interpreted so that a circle of a certain radius R_1 that is associated with the coordinates system $X_1Y_1Z_1$ is rolling with no sliding over a circle of the corresponding radius R_2 that is associated with the coordinate system $X_2Y_2Z_2$. When the center distance C is known, the radii R_1 and R_2 of the circles (that is, of the centrodes) can be expressed in terms of the center distance C and of the given rotations ω_1 and ω_2. For the calculations, the following formulae:

$$R_1 = C \cdot \frac{1}{1+u} \tag{D.37}$$

$$R_2 = C \cdot \frac{u}{1+u} \tag{D.38}$$

can be used. Here, the ratio ω_1/ω_2 is denoted by u.

In the initial configuration, the X_1- and X_2- axes align with each other. The Y_1- and Y_2- axes are parallel to each other. As shown in Fig. D.10, the initial configuration of the coordinate systems $X_1Y_1Z_1$ and $X_2Y_2Z_2$ is labeled as $X_1^*Y_1^*Z_1^*$ and $X_2^*Y_2^*Z_2^*$, respectively.

When the coordinate system $X_1Y_1Z_1$ turns through a certain angle φ_1, the coordinate system $X_2Y_2Z_2$ turns through the corresponding angle φ_2. When the angle φ_1 is known, the corresponding angle φ_2 is $\varphi_2 = \varphi_1/u$.

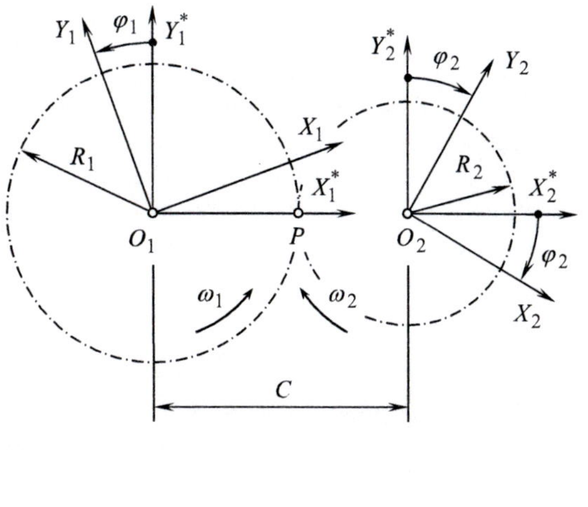

Fig. D.10 Derivation of the operator of rolling $\mathbf{Rr}_u(\varphi_1, Z_1)$ of two coordinate systems

The transition from the coordinate system $X_2Y_2Z_2$ to the coordinate system $X_1Y_1Z_1$ can be analytically described by the operator of the resultant coordinate system transformation **Rs** $(1 \mapsto 2)$. In the case under consideration, the operator **Rs** $(1 \mapsto 2)$ can be expressed in terms of the operators of the elementary coordinate system transformations:

$$\mathbf{Rs}\,(1 \mapsto 2) = \mathbf{Rt}\,(\varphi_1, Z_1) \cdot \mathbf{Rt}\,(\varphi_1/u, Z_1) \cdot \mathbf{Tr}\,(-C, X_1) \tag{D.39}$$

Other equivalent combinations of the operators of elementary coordinate system transformations can result in the same operator **Rs** $(1 \mapsto 2)$ of the resultant coordinate system transformation. The interested reader may wish to exercise his or her own way of deriving the equivalent expressions for the operator **Rs** $(1 \mapsto 2)$.

The operator of the resultant coordinate system transformations of the kind (see Eq. (D.39)) are referred to as the "operators of rolling motion over a cylinder."

When rotations are performed around the Z_1- and the Z_2- axes, the operator of the rolling motion over a cylinder is designated as $\mathbf{Rr}_u(\varphi_1, Z_1)$. In this particular case, the equality $\mathbf{Rr}_u(\varphi_1, Z_1) = \mathbf{Rs}\,(1 \mapsto 2)$ (see Eq. (D.39)) is valid. Based on this equality, the operator of rolling $\mathbf{Rr}_u(\varphi_1, Z_1)$ over a cylinder can be calculated from the equation:

$$\mathbf{Rr}_u(\varphi_1, Z_1) = \begin{bmatrix} \cos\left(\varphi_1 \cdot \dfrac{u+1}{u}\right) & \sin\left(\varphi_1 \cdot \dfrac{u+1}{u}\right) & 0 & -C \\ -\sin\left(\varphi_1 \cdot \dfrac{u+1}{u}\right) & \cos\left(\varphi_1 \cdot \dfrac{u+1}{u}\right) & 0 & 0 \\ 0 & 0 & 1 & 0 \\ 0 & 0 & 0 & 1 \end{bmatrix} \tag{D.40}$$

For the inverse transformation, the inverse operator of rolling of two coordinate systems $\mathbf{Rr}_u(\varphi_2, Z_2)$ can be used. It is $\mathbf{Rr}_u(\varphi_2, Z_2) = \mathbf{Rr}_u^{-1}(\varphi_1, Z_1)$. In terms of the operators of the elementary coordinate system transformations, the operator $\mathbf{Rr}_u(\varphi_2, Z_2)$ can be expressed as follows:

$$\mathbf{Rr}_u(\varphi_2, Z_2) = \mathbf{Rt}\,(\varphi_1/u, Z_2) \cdot \mathbf{Rt}\,(\varphi_1, Z_2) \cdot \mathbf{Tr}\,(C, X_1) \tag{D.41}$$

Other equivalent combinations of the operators of elementary coordinate system transformations can result in the same operator $\mathbf{Rr}_u(\varphi_2, Z_2)$ of the resultant coordinate system transformation. The interested reader may wish to exercise his or her own way of deriving the equivalent expressions for the operator $\mathbf{Rr}_u(\varphi_2, Z_2)$.

For the calculation of the operator of rolling of two coordinate systems $\mathbf{Rr}_u(\varphi_2, Z_2)$, the equation:

$$\mathbf{Rr}_u(\varphi_2, Z_2) = \begin{bmatrix} \cos\left(\varphi_1 \cdot \frac{u+1}{u}\right) & -\sin\left(\varphi_1 \cdot \frac{u+1}{u}\right) & 0 & C \\ \sin\left(\varphi_1 \cdot \frac{u+1}{u}\right) & \cos\left(\varphi_1 \cdot \frac{u+1}{u}\right) & 0 & 0 \\ 0 & 0 & 1 & 0 \\ 0 & 0 & 0 & 1 \end{bmatrix} \quad \text{(D.42)}$$

can be used.

Similar to that, the expression (see Eq. (D.40)) is derived for the calculation of the operator of rolling $\mathbf{Rr}_u(\varphi_1, Z_1)$ around the Z_1- and Z_2- axes, and the corresponding formulae can be derived for the calculation of the operators of rolling $\mathbf{Rr}_u(\varphi_1, X_1)$ and $\mathbf{Rr}_u(\varphi_1, Y_1)$ about the parallel axes X_1 and X_2 as well as about the parallel axes Y_1 and Y_2.

Use of the operators of rolling about two axes $\mathbf{Rr}_u(\varphi_1, X_1)$, $\mathbf{Rr}_u(\varphi_1, Y_1)$, and $\mathbf{Rr}_u(\varphi_1, Z_1)$ substantially simplifies the analytical description of the coordinate system transformations.

D.2.4 Coupled Linear Transformation

It can be observed here that a translation, $\mathbf{Tr}\,(a_x, X)$, along the $X-$ axis of a Cartesian reference system XYZ and a rotation $\mathbf{Rt}\,(\varphi_x, X)$ about the X-axis of the same coordinate system XYZ obey the commutative law, that is, these two coordinate system transformations can be performed in different order equally. It makes no difference whether the translation $\mathbf{Tr}\,(a_x, X)$ is initially performed, which is followed by the rotation $\mathbf{Rt}\,(\varphi_x, X)$, or whether the rotation $\mathbf{Rt}\,(\varphi_x, X)$ is initially performed, which is followed by the translation $\mathbf{Tr}\,(a_x, X)$. This is because the dot products $\mathbf{Rt}\,(\varphi_x, X) \cdot \mathbf{Tr}\,(a_x, X)$ and $\mathbf{Tr}\,(a_x, X) \cdot \mathbf{Rt}\,(\varphi_x, X)$ are identical to one another:

$$\mathbf{Rt}\,(\varphi_x, X) \cdot \mathbf{Tr}\,(a_x, X) \equiv \mathbf{Tr}\,(a_x, X) \cdot \mathbf{Rt}\,(\varphi_x, X) \quad \text{(D.43)}$$

This means that the translation from the coordinate system $X_1Y_1Z_1$ to the intermediate coordinate system $X^*Y^*Z^*$ followed by the rotation from the coordinate system $X^*Y^*Z^*$ to the final coordinate system $X_2Y_2Z_2$ produces the same reference $X_2Y_2Z_2$ as in a case in which the rotation from the coordinate system $X_1Y_1Z_1$ to the intermediate coordinate system $X^*Y^*Z^*$ followed by the translation from the coordinate system $X^*Y^*Z^*$ to the final coordinate system $X_2Y_2Z_2$.

The validity of Eq. (D.43) is illustrated in Fig. D.11. The translation $\mathbf{Tr}\,(a_x, X)$ that is followed by the rotation $\mathbf{Rt}\,(\varphi_x, X)$, as shown in Fig. D.11a, is equivalent to the rotation $\mathbf{Rt}\,(\varphi_x, X)$ that is followed by the translation $\mathbf{Tr}\,(a_x, X)$ as shown in Fig. D.11b.

Therefore, the two linear transformations $\mathbf{Tr}\,(a_x, X)$ and $\mathbf{Rt}\,(\varphi_x, X)$ can be coupled into a linear transformation:

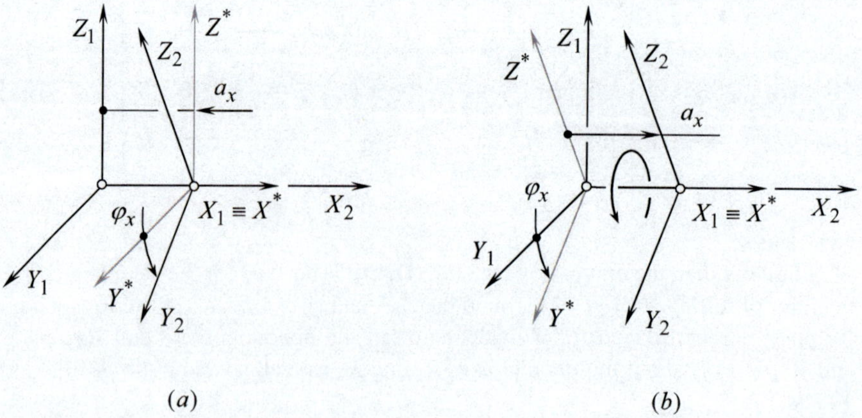

Fig. D.11 Equivalency of the linear transformations $\mathbf{Rt}(\varphi_x, X) \cdot \mathbf{Tr}(a_x, X)$ and $\mathbf{Tr}(a_x, X) \cdot \mathbf{Rt}(\varphi_x, X)$ that comprise the operator $\mathbf{Cp}_x(a_x, \varphi_x)$ of a coupled linear transformation of a Cartesian reference system *XYZ*

$$\mathbf{Cp}_x(a_x, \varphi_x) = \mathbf{Rt}(\varphi_x, X) \cdot \mathbf{Tr}(a_x, X) \equiv \mathbf{Tr}(a_x, X) \cdot \mathbf{Rt}(\varphi_x, X) \tag{D.44}$$

The operator of linear transformation $\mathbf{Cp}_x(a_x, \varphi_x)$ can be expressed in a matrix form (see Fig. D.12a) as follows:

$$\mathbf{Cp}_x(a_x, \varphi_x) = \begin{bmatrix} 1 & 0 & 0 & a_x \\ 0 & \cos\varphi_x & \sin\varphi_x & 0 \\ 0 & -\sin\varphi_x & \cos\varphi_x & 0 \\ 0 & 0 & 0 & 1 \end{bmatrix} \tag{D.45}$$

This expression is composed based on Eq. (D.4) for the linear transformation $\mathbf{Tr}(a_x, X)$ and on Eq. (D.10) for the linear transformation $\mathbf{Rt}(\varphi_x, X)$.

Two reduced cases of the operator of the linear transformation $\mathbf{Cp}_x(a_x, \varphi_x)$ are distinguished.

First, it could happen that in a particular case, the component a_x of the translation is zero, that is $a_x = 0$. Under such a scenario, the operator of linear transformation $\mathbf{Cp}_x(a_x, \varphi_x)$ reduces to the operator of rotation $\mathbf{Rt}(\varphi_x, X)$, and the equality $\mathbf{Cp}_x(a_x, \varphi_x) = \mathbf{Rt}(\varphi_x, X)$ is observed in the case under consideration.

Second, it could happen that in a particular case, the component φ_x of the rotation is zero, that is $\varphi_x = 0°$. Under such a scenario, the operator of linear transformation $\mathbf{Cp}_x(a_x, \varphi_x)$ reduces to the operator of translation $\mathbf{Tr}(a_x, X)$, and the equality $\mathbf{Cp}_x(a_x, \varphi_x) = \mathbf{Tr}(a_x, X)$ is observed in the case under consideration.

The is valid with respect to the translations and the rotations along and about the axes *Y* and *Z* of a Cartesian reference system *XYZ*. The corresponding coupled operators $\mathbf{Cp}_y(a_y, \varphi_y)$ and $\mathbf{Cp}_z(a_z, \varphi_z)$ for linear transformations of these kinds can also be composed (see Fig. D.12b and c):

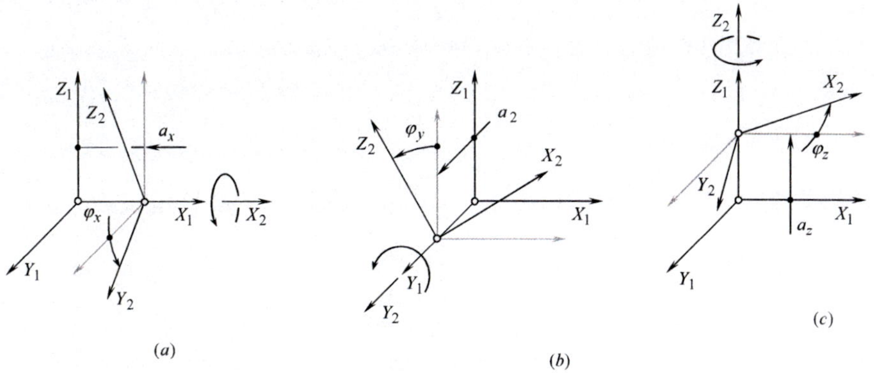

Fig. D.12 Equivalency of linear transformations $\mathbf{Rt}(\varphi_x, X) \cdot \mathbf{Tr}(a_x, X)$ and $\mathbf{Tr}(a_x, X) \cdot \mathbf{Rt}(\varphi_x, X)$, which comprise the operator $\mathbf{Cp}_x(a_x, \varphi_x)$ of the coupled linear transformation of a Cartesian reference system XYZ

$$\mathbf{Cp}_y(a_y, \varphi_y) = \begin{bmatrix} \cos\varphi_y & 0 & \sin\varphi_y & 0 \\ 0 & 1 & 0 & a_y \\ -\sin\varphi_y & 0 & \cos\varphi_y & 0 \\ 0 & 0 & 0 & 1 \end{bmatrix} \tag{D.46}$$

$$\mathbf{Cp}_z(a_z, \varphi_z) = \begin{bmatrix} \cos\varphi_z & \sin\varphi_z & 0 & 0 \\ -\sin\varphi_z & \cos\varphi_z & 0 & 0 \\ 0 & 0 & 1 & a_z \\ 0 & 0 & 0 & 1 \end{bmatrix} \tag{D.47}$$

In the operators of linear transformations $\mathbf{Cp}_x(a_x, \varphi_x)$, $\mathbf{Cp}_y(a_y, \varphi_y)$, and $\mathbf{Cp}_z(a_z, \varphi_z)$, the values of the translations a_x, a_y, and a_z, as well as the values of the rotations φ_x, φ_y, and φ_z, are finite values (and not continuous). The linear and angular displacements do not correlate with one another in time; thus, they do not comprise a screw. They are just a kind of translation along and a rotation about a coordinate axis of a Cartesian reference system.

Introduction of the operators of linear transformation $\mathbf{Cp}_x(a_x, \varphi_x)$, $\mathbf{Cp}_y(a_y, \varphi_y)$, and $\mathbf{Cp}_z(a_z, \varphi_z)$ makes the linear transformations easier as all the operators of the linear transformations become uniform.

The operators of linear transformation $\mathbf{Cp}_x(a_x, \varphi_x)$, $\mathbf{Cp}_y(a_y, \varphi_y)$, and $\mathbf{Cp}_z(a_z, \varphi_z)$ do not obey the commutative law. This is because rotation is not a vector by nature. Therefore, special care should be taken when treating rotations as vectors – when implementing coupled operators of linear transformations in particular.

The operators of coupled linear transformations $\mathbf{Cp}_x(a_x, \varphi_x)$, $\mathbf{Cp}_y(a_y, \varphi_y)$, and $\mathbf{Cp}_z(a_z, \varphi_z)$, (see Eqs. (D.45) through (D.47)) can be used for the purpose of an analytical description of a resultant coordinate system transformation. Under such the scenario, the operator $\mathbf{Rs}(1 \mapsto t)$ of a resultant coordinate system transformation can be expressed in terms of all the operators $\mathbf{Cp}_x(a_x, \varphi_x)$, $\mathbf{Cp}_y(a_y, \varphi_y)$, and $\mathbf{Cp}_z(a_z, \varphi_z)$ by the following expression:

$$\mathbf{Rs}(1 \mapsto t) = \prod_{\substack{i=1 \\ j=x,y,z}}^{t-1} \mathbf{Cp}_j^i\left(a_j^i, \varphi_j^i\right) \tag{D.48}$$

In Eq. (D.48), only operators of coupled linear transformations are used.

D.2.5 An Example of Non-orthogonal Linear Transformation

Let us consider a non-orthogonal reference system $X_1Y_1Z_1$ having a certain angle ω_1 between the axes X_1 and Y_1. The axis Z_1 is perpendicular to the coordinate plane X_1Y_1. Another reference system $X_2Y_2Z_2$ is identical to the first coordinate system $X_1Y_1Z_1$ and is turned about the Z_1- axis through a certain angle φ. The transition from the reference system $X_1Y_1Z_1$ to the reference system $X_2Y_2Z_2$ can be analytically described by the operator of linear transformation:

$$\mathbf{Rt}_\omega(1 \rightarrow 2) = \begin{bmatrix} \dfrac{\sin(\omega_1 + \varphi)}{\sin \omega_1} & \dfrac{\sin \varphi}{\sin \omega_1} & 0 & 0 \\ -\dfrac{\sin \varphi}{\sin \omega_1} & \dfrac{\sin(\omega_1 - \varphi)}{\sin \omega_1} & 0 & 0 \\ 0 & 0 & 1 & 0 \\ 0 & 0 & 0 & 1 \end{bmatrix} \tag{D.49}$$

In order to distinguish the operator of rotation in the orthogonal linear transformation $\mathbf{Rt}(1 \rightarrow 2)$ from the similar operator of rotation in a non-orthogonal linear transformation $\mathbf{Rt}_\omega(1 \rightarrow 2)$, the subscript "$\omega$" is assigned to the latter.

When $\omega = 90°$, Eq. (D.49) is cast into Eq. (D.12).

D.2.6 Conversion of a Coordinate System Hand

Application of the matrix method of coordinate system transformation presumes that both the reference systems "i" and "$(i \pm 1)$" are of the same hand. This means that it is assumed from the very beginning that both of them are either right-hand- or left-hand-oriented Cartesian coordinate systems. In the event the coordinate systems i and $(i \pm 1)$ are of an opposite hand, e.g., one of them is a right-hand-oriented coordinate system, whereas the other is a left-hand-oriented coordinate system, then one of the coordinate systems must be converted into the oppositely oriented Cartesian coordinate system.

For the conversion of a left-hand-oriented Cartesian coordinate system into a right-hand-oriented coordinate system, or vice versa, the operators of reflection are commonly used.

In order to change the direction of the X_i axis of the initial coordinate system i to the opposite direction (in this case, in the new coordinate system $(i \pm 1)$, the equalities $X_{i\pm 1} = -X_i$, $Y_{i\pm 1} \equiv Y_i$, and $Z_{i\pm 1} \equiv Z_i$ are observed), the operator of the reflection $\mathbf{Rf}_x(Y_i Z_i)$ can be applied. The operator of the reflection yields representation in a matrix form:

$$\mathbf{Rf}_x(Y_i Z_i) = \begin{bmatrix} -1 & 0 & 0 & 0 \\ 0 & 1 & 0 & 0 \\ 0 & 0 & 1 & 0 \\ 0 & 0 & 0 & 1 \end{bmatrix} \tag{D.50}$$

Similarly, implementation of the operators of reflections $\mathbf{Rf}_y(X_i Z_i)$ and $\mathbf{Rf}_z(X_i Y_i)$ changes the directions of the Y_i and Z_i axes onto opposite directions. The operators of reflections $\mathbf{Rf}_y(X_i Z_i)$ and $\mathbf{Rf}_z(X_i Y_i)$ can be expressed analytically in the form:

$$\mathbf{Rf}_y(X_i Z_i) = \begin{bmatrix} 1 & 0 & 0 & 0 \\ 0 & -1 & 0 & 0 \\ 0 & 0 & 1 & 0 \\ 0 & 0 & 0 & 1 \end{bmatrix} \tag{D.51}$$

$$\mathbf{Rf}_z(X_i Y_i) = \begin{bmatrix} 1 & 0 & 0 & 0 \\ 0 & 1 & 0 & 0 \\ 0 & 0 & -1 & 0 \\ 0 & 0 & 0 & 1 \end{bmatrix} \tag{D.52}$$

A linear transformation that reverses the direction of the coordinate axis is an "opposite transformation." Transformation of reflection is an example of an "orientation-reversing transformation."

D.3 Useful Equations

The sequence of the successive rotations can vary depending on the intention of the researcher. Several special types of successive rotations are known, including Eulerian transformation, Cardanian transformation, two kinds of Euler–Krylov transformations, and so forth. The sequence of the successive rotations can be chosen from a total of 12 different combinations. Even though the Cardanian transformation is different from the Eulerian transformation in terms of the combination of the rotations, they both use a similar approach to calculate the orientation angles.

D.3.1 *RPY* Transformation

A series of rotations can be performed in the order roll matrix (R), by the pitch matrix (P), and finally by the yaw matrix (Y). The linear transformation of this kind is commonly referred to as RPY− transformation. The resultant transformation of this kind can be represented by the homogeneous coordinate transformation matrix:

$$\mathbf{RPY}(\varphi_x, \varphi_y, \varphi_z) = \begin{bmatrix} \cos\varphi_y\cos\varphi_z + \sin\varphi_x\sin\varphi_y\sin\varphi_z & \cos\varphi_y\sin\varphi_z - \sin\varphi_x\sin\varphi_y\cos\varphi_z & \cos\varphi_x\sin\varphi_y & 0 \\ -\cos\varphi_x\sin\varphi_z & \cos\varphi_x\cos\varphi_z & \sin\varphi_x & 0 \\ \sin\varphi_x\cos\varphi_y\sin\varphi_z - \sin\varphi_y\cos\varphi_z & -\sin\varphi_x\cos\varphi_y\cos\varphi_z - \cos\varphi_y\sin\varphi_z & \cos\varphi_x\cos\varphi_y & 0 \\ 0 & 0 & 0 & 1 \end{bmatrix} \quad \text{(D.53)}$$

RPY− transformation can be used to solve problems in the field of part surface generation.

D.3.2 Operator of Rotation About an Axis in Space

A spatial rotation operator for the rotational transformation of a point about a unit axis $\mathbf{a}_0(\cos\alpha, \cos\beta, \cos\gamma)$ passing through the origin of the coordinate system can be described as follows, with $\mathbf{a}_0 = \mathbf{A}_0/|\mathbf{A}_0|$ designating the unit vector along the axis of rotation $\mathbf{A}_0$.

Suppose that the angle of rotation of the point about $\mathbf{a}_0$ is θ, then the "rotation operator" is expressed by:

$$\mathbf{Rt}(\theta, \mathbf{a}_0) = \begin{bmatrix} (1-\cos\theta)\cos^2\alpha + \cos\theta & (1-\cos\theta)\cos\alpha\cos\beta - \sin\theta\cos\gamma & (1-\cos\theta)\cos\alpha\cos\gamma + \sin\theta\cos\beta & 0 \\ (1-\cos\theta)\cos\alpha\cos\beta + \sin\theta\cos\gamma & (1-\cos\theta)\cos^2\beta + \cos\theta & (1-\cos\theta)\cos\beta\cos\gamma - \sin\theta\cos\alpha & 0 \\ (1-\cos\theta)\cos\alpha\cos\gamma - \sin\theta\cos\beta & (1-\cos\theta)\cos\beta\cos\gamma + \sin\theta\cos\alpha & (1-\cos\theta)\cos^2\gamma + \cos\theta & 0 \\ 0 & 0 & 0 & 1 \end{bmatrix} \quad \text{(D.54)}$$

Solutions to a problem in the field of part surface generation can be significantly simplified by implementation of the rotational operator $\mathbf{Rt}(\theta, \mathbf{a}_0)$ (see Eq. (D.54)).

D.3.3 Combined Linear Transformation

Suppose a point, p, on a rigid body rotates with an angular displacement, θ, about an axis along a unit vector, $\mathbf{a}_0$, passing through the origin of the coordinate system at first and then followed by a translation at a distance, B, in the direction of a unit vector, $\mathbf{b}$. The linear transformation of this kind can be analytically described by the homogeneous matrix

$$\mathbf{Rt}(\theta_{a0}, B_b) = \begin{bmatrix} (1-\cos\theta)\cos^2\alpha + \cos\theta & (1-\cos\theta)\cos\alpha\cos\beta - \sin\theta\cos\gamma & (1-\cos\theta)\cos\alpha\cos\gamma + \sin\theta\cos\beta & B\cos\alpha \\ (1-\cos\theta)\cos\alpha\cos\beta + \sin\theta\cos\gamma & (1-\cos\theta)\cos^2\beta + \cos\theta & (1-\cos\theta)\cos\beta\cos\gamma - \sin\theta\cos\alpha & B\cos\beta \\ (1-\cos\theta)\cos\alpha\cos\gamma - \sin\theta\cos\beta & (1-\cos\theta)\cos\beta\cos\gamma + \sin\theta\cos\alpha & (1-\cos\theta)\cos^2\gamma + \cos\theta & B\cos\gamma \\ 0 & 0 & 0 & 1 \end{bmatrix} \tag{D.55}$$

More operators of particular linear transformations can be found in the literature.

D.4 Chains of Consequent Linear Transformations and a Closed Loop of Consequent Coordinate System Transformations

Consequent coordinate system transformations form chains (circuits) of linear transformations. The elementary chain of coordinate system transformation is composed of two consequent transformations. Chains of linear transformations play an important role in the theory of part surface generation.

Two different kinds of chains of consequent coordinate system transformations are distinguished:

First is the transition from the coordinate system $X_g Y_g Z_g$ associated with the gear tooth flank $\mathcal{G}$ to the local Cartesian coordinate system $x_g y_g z_g$ having the origin at a point K of contact of the gear tooth flank $\mathcal{G}$ and of the pinion tooth flank $\mathcal{P}$. This linear transformation is also made up of numerous operators of intermediate coordinate system transformations ($X_{\text{in}} Y_{\text{in}} Z_{\text{in}}$). It forms a chain of direct consequent coordinate system transformations illustrated in Fig. D.13a.

The local coordinate system $x_g y_g z_g$ is associated with the gear tooth flank $\mathcal{G}$. The operator $\mathbf{Rs}\left(\mathcal{G} \rightarrow K_g\right)$ of the resultant coordinate system transformation for a direct chain of the linear transformations can be composed using a certain number of the operators of translations (see Eqs. (D.4) through (D.6)) and a corresponding number of the operators of rotations (see Eqs. (D.10) through (D.12)).

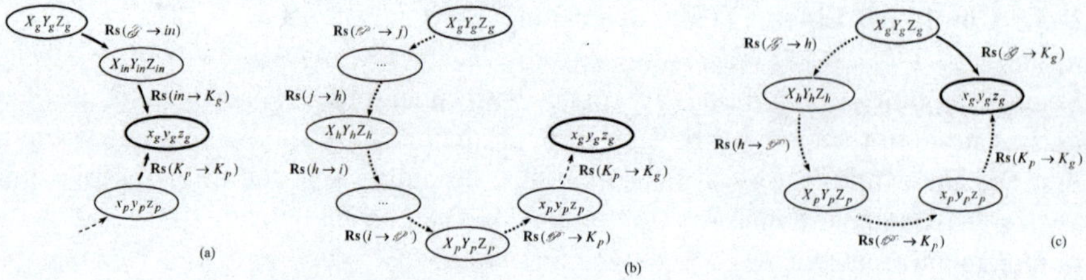

Fig. D.13 An example of a direct chain (**a**), a reverse chain (**b**), and a closed loop (**c**) of consequent coordinate system transformations

Second is transition from the coordinate system $X_gY_gZ_g$ to the local Cartesian coordinate system $x_py_pz_p$ with the origin at a point K of contact of the tooth flanks $\mathcal{G}$ and $\mathcal{P}$. The local coordinate system $x_py_pz_p$ is associated with the pinion tooth flank $\mathcal{P}$. This linear transformation is also made up of numerous intermediate coordinate system transformations ($X_jY_jZ_j$), for example, transitions from the coordinate system $X_hY_hZ_h$ associated with gear housing, to numerous intermediate coordinate system $X_iY_iZ_i$. The linear transformation of this kind forms a chain of inverse consequent coordinate system transformations shown in Fig. D.13b. The operator $\mathbf{Rs}\left(\mathcal{G} \rightarrow K_p\right)$ of the resultant coordinate system transformations for the inverse chain of transformations can be composed using a certain number of the operators of translations (see Eqs. (D.4), (D.5), and (D.6)) and a corresponding number of the operators of rotations (see Eqs. (D.10) through (D.12)).

Chains of the direct and reverse consequent coordinate system transformations together with the operator of transition from the local coordinate system $x_py_pz_p$ to the local coordinate system $x_gy_gz_g$ form a closed loop (a closed circuit) of the consequent coordinate system transformations depicted in Fig. D.13c.

If a closed loop of the consequent coordinate system transformations is complete, then implementation of a certain number of the operators of translations (see Eqs. (D.4), (D.5), and (D.6)) and a corresponding number of the operators of rotations (see Eqs. (D.10) through (D.12)) produces a result that is identical to the input data. This means that an analytical description of a meshing process specified in the original coordinate system remains the same after implementation of the operator of the resultant coordinate system transformations. This condition is the necessary and sufficient condition for the existence of a closed loop of consequent coordinate system transformations.

Implementation of the chains, as well as of the closed loops of consequent coordinate system transformations, makes it possible to consider the meshing process of the teeth flanks $\mathcal{G}$ and $\mathcal{P}$ in any and all of reference systems that make up the loop. Therefore, for consideration of a particular problem of part surface generation, the most convenient reference system can be chosen.

In order to complete the construction of a closed loop of consequent coordinate system transformations, an operator of transformation from the local coordinate system $x_py_pz_p$ to the local coordinate system $x_gy_gz_g$ must be composed. Usually, the local reference systems $x_gy_gz_g$ and $x_py_pz_p$ are a kind of semi-orthogonal

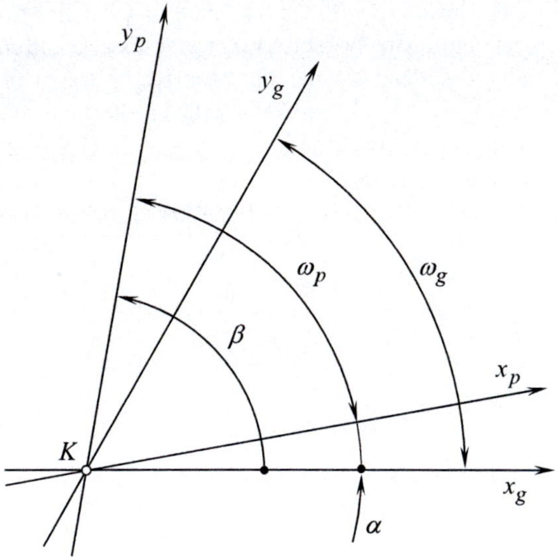

Fig. D.14 Local coordinate systems $x_g y_g z_g$ and $x_p y_p z_p$ with the origin at a contact point K

coordinate systems. This means that the axis z_P is always orthogonal to the coordinate plane $x_g y_g$, whereas the axes x_g and y_g can be either orthogonal to each other or not. The same is valid with respect to the local coordinate system $x_p y_p z_p$.

Two possible ways for performing the required transformation of the local reference systems $x_g y_g z_g$ and $x_p y_p z_p$ are considered below.

Following the first way, the operator $\mathbf{Rt}_\omega(p \to g)$ of the linear transformation of semi-orthogonal coordinate systems (see Fig. D.14) must be composed. The operator $\mathbf{Rt}_\omega(p \to g)$ can be represented in the form of the homogeneous matrix:

$$\mathbf{Rt}_\omega(p \to g) = \begin{bmatrix} \dfrac{\sin(\omega_p + \alpha)}{\sin \omega_p} & -\dfrac{\sin(\omega_g - \omega_p - \alpha)}{\sin \omega_p} & 0 & 0 \\ -\dfrac{\sin \alpha}{\sin \omega_p} & \dfrac{\sin(\omega_g - \alpha)}{\sin \omega_p} & 0 & 0 \\ 0 & 0 & -1 & 0 \\ 0 & 0 & 0 & 1 \end{bmatrix} \quad (D.56)$$

Here,

ω_g is the angle that makes the U_g and V_g coordinate lines on the gear tooth flank $\mathcal{G}$.
ω_p is the angle that makes the U_p and V_p coordinate lines on the pinion tooth flank, $\mathcal{P}$.
α is the angle that make the axes x_g and x_p of the local coordinate systems $x_g y_g z_g$ and $x_p y_p z_p$.

The auxiliary angle β in Fig. D.14 is $\beta = \omega_T + \alpha$.

The inverse coordinate system transformation, that is, the transformation from the local coordinate system $x_g y_g z_g$ to the local coordinate system $x_p y_p z_p$, can be analytically described by the operator $\mathbf{Rt}_\omega(g \to p)$ of the inverse coordinate system transformation. The operator $\mathbf{Rt}_\omega(g \to p)$ can be represented in the form of the homogeneous matrix:

$$\mathbf{Rt}_\omega(g \to p) = \begin{bmatrix} \dfrac{\sin(\omega_g - \alpha)}{\sin \omega_g} & \dfrac{\sin(\omega_g - \omega_p - \alpha)}{\sin \omega_g} & 0 & 0 \\ \dfrac{\sin \alpha}{\sin \omega_g} & \dfrac{\sin(\omega_p + \alpha)}{\sin \omega_g} & 0 & 0 \\ 0 & 0 & -1 & 0 \\ 0 & 0 & 0 & 1 \end{bmatrix} \tag{D.57}$$

Following the second way of transformation of the local coordinate systems, the auxiliary orthogonal local coordinate system must be constructed.

Let us consider an approach, according to which a closed loop (a closed circuit) of the consequent coordinate system transformations can be composed.

In order to construct an orthogonal normalized basis of the coordinate system $x_g y_g z_g$, an intermediate coordinate system $x_1 y_1 z_1$ is used. Axis x_1 of the coordinate system $x_1 y_1 z_1$ is pointed along the unit vector $\mathbf{u}_g$ that is tangent to the U_g- coordinate curve (see Fig. D.15). Axis y_1 is directed along the vector $\mathbf{v}_g$ that is tangent to the V_g- coordinate line on the gear tooth flank $\mathcal{G}$. Axis z_1 aligns with the unit normal vector $\mathbf{n}_g$ and is pointed outward of the gear tooth body.

For a gear tooth flank $\mathcal{G}$ having orthogonal parameterization (for which $F_g = 0$ and therefore $\omega_g = \pi/2$), an analytical description of coordinate system transformations is significantly simpler. Further simplification of the coordinate system transformation is possible when the coordinate U_g- and V_g- lines are congruent to the lines of curvature on the part surface $\mathcal{G}$. Under such a scenario, the local coordinate system is represented by a Darboux frame.

In order to construct a Darboux frame, the principal directions on the gear tooth flank $\mathcal{G}$ must be calculated. Determination of the unit tangent vectors $\mathbf{t}_{1.\,g}$ and $\mathbf{t}_{2.\,g}$ of the principal directions on the gear tooth flank $\mathcal{G}$ is considered in Appendix B.

In the common tangent plane, orientation of the unit vector $\mathbf{t}_{1.\,g}$ of the first principal direction on the gear tooth flank $\mathcal{G}$ can be uniquely specified by the included angle $\xi_{1.\,g}$ that the unit vector $\mathbf{t}_{1.\,g}$ forms with the U_g- coordinate curve. This angle depends on both the geometry of the gear tooth flank $\mathcal{G}$ and the parametrization of the gear tooth flank $\mathcal{G}$. Depicted in Fig. D.16 is the relationship between the tangent vectors $\mathbf{U}_g$ and $\mathbf{V}_g$ and the included angle $\xi_{1.\,g}$. From the law of sine's,

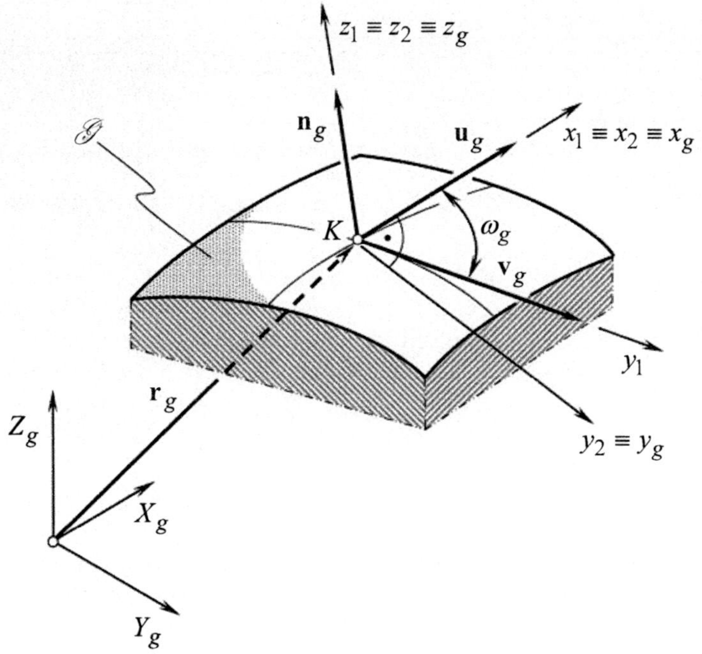

Fig. D.15 Local coordinate system $x_g y_g z_g$ associated with the gear tooth flank $\mathcal{G}$

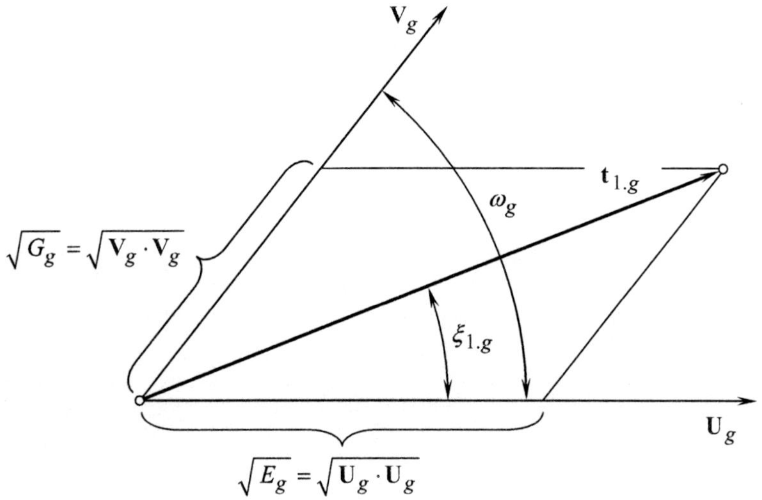

Fig. D.16 Differential relationships between the tangent vectors $\mathbf{U}_g$ and $\mathbf{V}_g$, the fundamental magnitudes of the first order, the included angle, $\xi_{1.g}$, and the direction of the unit tangent vector $\mathbf{t}_{1.g}$

$$\frac{\sqrt{G_g}}{\sin \xi_{1.g}} = \frac{\sqrt{E_g}}{\sin\left[\pi - \xi_{1.g} - \left(\pi - \omega_g\right)\right]} = \frac{\sqrt{F_g}}{\sin\left(\omega_g - \xi_{1.g}\right)} \tag{D.58}$$

Here, $\omega_g = \cos^{-1}\left(F_g/\sqrt{E_g G_g}\right)$.
Solving the above expression for the included angle $\xi_{1.\,g}$ results in:

$$\xi_{1.g} = \tan^{-1}\frac{\sqrt{E_g G_g - F_g^2}}{E_g + F_g} \tag{D.59}$$

Another possible way of constructing an orthogonal local basis of the local Cartesian coordinate system $x_P y_P z_P$ is based on the following consideration.

Let us consider an arbitrary non-orthogonal and not normalized basis, $\mathbf{x}_1\mathbf{x}_2\mathbf{x}_3$ (see Fig. D.17a). Let us construct an orthogonal and normalized basis based on the initial given basis, $\mathbf{x}_1\mathbf{x}_2\mathbf{x}_3$.

A cross product of any two of three vectors $\mathbf{x}_1$, $\mathbf{x}_2$, $\mathbf{x}_3$, for example, a cross product of the vectors $\mathbf{x}_1 \times \mathbf{x}_2$ determines a new vector $\mathbf{x}_4$ (see Fig. D.17b). Evidently, the vector $\mathbf{x}_4$ is orthogonal to the coordinate plane $\mathbf{x}_1\mathbf{x}_2$. Then, using the calculated vector $\mathbf{x}_4$ and one of two original vectors $\mathbf{x}_1$ or $\mathbf{x}_2$, for instance, vector, $\mathbf{x}_2$ yields the calculation of a new vector $\mathbf{x}_5 = \mathbf{x}_4 \times \mathbf{x}_2$ (see Fig. D.17c). The calculated basis $\mathbf{x}_1\mathbf{x}_4\mathbf{x}_5$ is orthogonal. In order to convert it into a normalized basis, each of the vectors $\mathbf{x}_1$, $\mathbf{x}_4$, and $\mathbf{x}_5$ must be divided by its magnitude:

$$\mathbf{e}_1 = \frac{\mathbf{x}_1}{|\mathbf{x}_1|} \tag{D.60}$$

$$\mathbf{e}_4 = \frac{\mathbf{x}_4}{|\mathbf{x}_4|} \tag{D.61}$$

$$\mathbf{e}_5 = \frac{\mathbf{x}_5}{|\mathbf{x}_5|} \tag{D.62}$$

The resultant basis $\mathbf{e}_1\mathbf{e}_4\mathbf{e}_5$ (see Fig. D.17d) is always orthogonal and normalized.

In order to complete the analytical description of a closed loop of consequent coordinate system transformations, it is necessary to compose the operator $\mathbf{Rs}\,(K_p \rightarrow K_g)$ of transformation from the local reference system $x_p y_p z_p$ to the local reference system $x_g y_g z_g$ (see Fig. D.13c).

In the case under consideration, the axes z_g and z_p align with the common unit normal vector $\mathbf{n}_g$. The axis z_g is pointed out from the bodily side to the void side of the gear tooth flank $\mathcal{G}$. The axis z_g is pointed oppositely. Due to that, the following equality is observed:

$$\mathbf{Rs}\left(K_p \rightarrow K_g\right) = \mathbf{Rt}\left(\varphi_z, z_p\right) \tag{D.63}$$

The inverse coordinate system transformation can be analytically described by the operator:

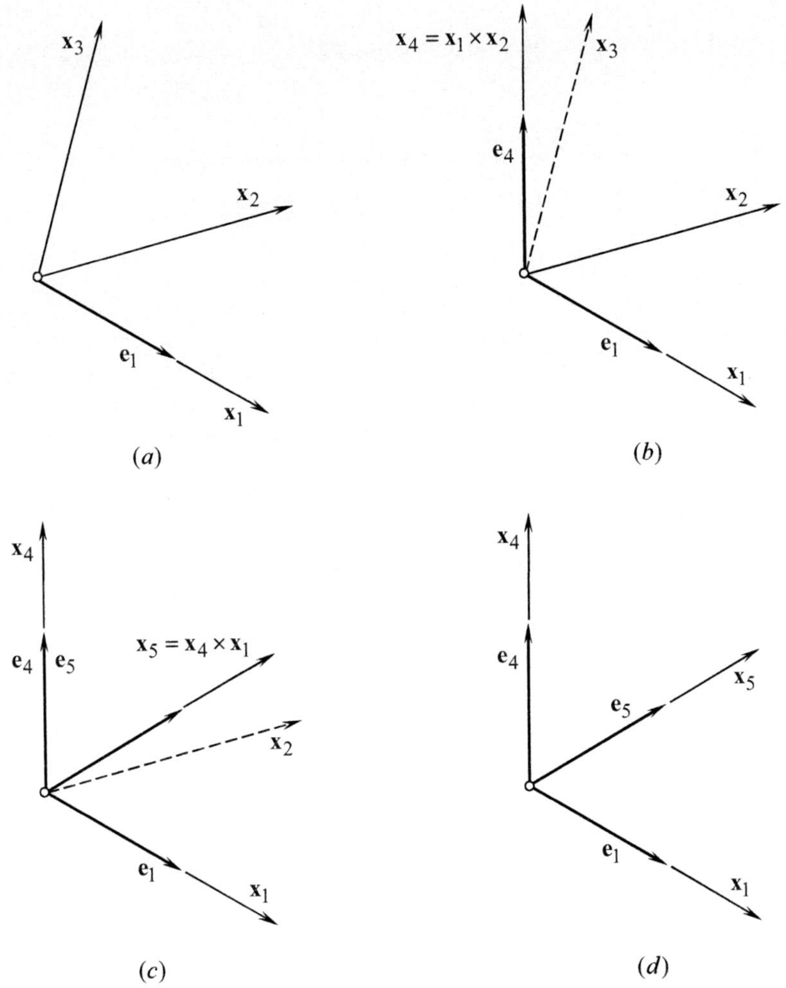

Fig. D.17 A normalized and orthogonally parameterized basis, $\mathbf{e}_1\mathbf{e}_4\mathbf{e}_5$, which is constructed from an arbitrary basis, $\mathbf{x}_1\mathbf{x}_2\mathbf{x}_3$

$$\mathbf{Rs}\left(K_g \rightarrow K_p\right) = \mathbf{Rs}^{-1}\left(K_p \rightarrow K_g\right) = \mathbf{Rt}\left(-\varphi_z, z_p\right) \tag{D.64}$$

The implementation of the discussed results allows for:

(a) Representation of the gear tooth flank $\mathcal{G}$ and the pinion tooth flank $\mathcal{P}$ of the form cutting tool as well as their relative motion in a common coordinate system
(b) Consideration of meshing of the gear tooth flank $\mathcal{G}$ in any desired coordinate system that is a component of the chain and/or the closed loop of consequent coordinate system transformations

The transition from one coordinate system to another coordinate system can be performed in both of two feasible directions, e.g., in direct and in inverse directions.

D.5 Impact of the Coordinate System Transformations on the Fundamental Forms of the Surface

Every coordinate system transformation entails a corresponding alteration to the equation of the gear tooth flank $\mathcal{G}$ and/or pinion tooth flank $\mathcal{P}$. Because of this, it is often necessary to recalculate the coefficients of the first $\Phi_{1.\ g}$ and of the second $\Phi_{2.\ g}$ fundamental forms of the gear tooth flank $\mathcal{G}$ (as many times as the coordinate system transformation is performed). This routine and time-consuming operation can be eliminated if the operators of coordinate system transformations are directly used to the fundamental forms $\Phi_{1.\ g}$ and $\Phi_{2.\ g}$.

After being calculated in an initial reference system, the fundamental magnitudes E_g, F_g, and G_g of the first $\Phi_{1.\ g}$ and the fundamental magnitudes L_g, M_g, and N_g of the second $\Phi_{2.\ g}$ fundamental forms can be determined in any new coordinate system using the operators of translation, of rotation, and of the resultant coordinate system transformation. A transformation of such kinds of the fundamental magnitudes $\Phi_{1.\ g}$ and $\Phi_{2.\ g}$ becomes possible by implementation of a formula, which can be found immediately below.

Let us consider a gear tooth flank, $\mathcal{G}$, which is given by equation $\mathbf{r}_g = \mathbf{r}_g(U_g, V_g)$, where $\left(U_g, V_g\right) \in \mathcal{G}$.

For the analysis below, it is convenient to use the equation of the first fundamental form $\Phi_{1.\ g}$ of the gear tooth flank $\mathcal{G}$, represented in the matrix form:

$$\left[\Phi_{1.g}\right] = \left[dU_g \quad dV_g \quad 0 \quad 0\right] \cdot \begin{bmatrix} E_g & F_g & 0 & 0 \\ F_g & G_g & 0 & 0 \\ 0 & 0 & 1 & 0 \\ 0 & 0 & 0 & 1 \end{bmatrix} \cdot \begin{bmatrix} dU_g \\ dV_g \\ 0 \\ 0 \end{bmatrix} \tag{D.65}$$

Similarly, the equation of the second fundamental form $\Phi_{2.\ g}$ of the gear tooth flank $\mathcal{G}$ can be given by:

$$\left[\Phi_{2.g}\right] = \left[dU_g \quad dV_g \quad 0 \quad 0\right] \cdot \begin{bmatrix} L_g & M_g & 0 & 0 \\ M_g & N_g & 0 & 0 \\ 0 & 0 & 1 & 0 \\ 0 & 0 & 0 & 1 \end{bmatrix} \cdot \begin{bmatrix} dU_g \\ dV_g \\ 0 \\ 0 \end{bmatrix} \tag{D.66}$$

The coordinate system transformation that is performed by the operator of linear transformation $\mathbf{Rs}\,(1 \rightarrow 2)$ transfers the equation $\mathbf{r}_g = \mathbf{r}_g(U_g, V_g)$ of the gear tooth flank $\mathcal{G}$, initially given in $X_1Y_1Z_1$, to the equation $\mathbf{r}_g^* = \mathbf{r}_g^*\left(U_g^*, V_g^*\right)$ of the same gear tooth flank $\mathcal{G}$ in a new coordinate system $X_2Y_2Z_2$. It is clear that $\mathbf{r}_g \neq \mathbf{r}_g^*$.

In the new coordinate system, the gear tooth flank $\mathcal{G}$ is analytically described by the following expression:

$$\mathbf{r}_g^*\left(U_g^*, V_g^*\right) = \mathbf{Rs}(1 \to 2) \cdot \mathbf{r}_g\left(U_g, V_g\right) \tag{D.67}$$

The operator of the resultant coordinate system transformation $\mathbf{Rs}(1 \to 2)$ casts the column matrices of variables in Eqs. (D.65) and (D.66) to the form:

$$\left[dU_g^* \quad dV_g^* \quad 0 \quad 0\right]^T = \mathbf{Rs}(1 \to 2) \cdot \left[dU_g \quad dV_g \quad 0 \quad 0\right]^T. \tag{D.68}$$

Substitution of Eq. (D.68) in Eqs. (D.65) and (D.66) makes the expressions for the fundamental forms $\mathbf{\Phi}_{1.g}^*$ and $\mathbf{\Phi}_{2.g}^*$ possible in the new coordinate system:

$$\left[\mathbf{\Phi}_{1.g}^*\right] = \left[\mathbf{Rs}(1 \to 2) \cdot \left[dU_g \quad dV_g \quad 0 \quad 0\right]^T\right]^T \cdot \left[\mathbf{\Phi}_{1.g}\right] \cdot \mathbf{Rs}(1 \to 2) \cdot \left[dU_g \quad dV_g \quad 0 \quad 0\right]^T \tag{D.69}$$

$$\left[\mathbf{\Phi}_{2.g}^*\right] = \left[\mathbf{Rs}(1 \to 2) \cdot \left[dU_g \quad dV_g \quad 0 \quad 0\right]^T\right]^T \cdot \left[\mathbf{\Phi}_{2.g}\right] \cdot \mathbf{Rs}(1 \to 2) \cdot \left[dU_g \quad dV_g \quad 0 \quad 0\right]^T \tag{D.70}$$

The following equation is valid for multiplication:

$$\left[\mathbf{Rs}(1 \to 2) \cdot \left[dU_g \quad dV_g \quad 0 \quad 0\right]^T\right]^T = \mathbf{Rs}^T(1 \to 2) \cdot \left[dU_g \quad dV_g \quad 0 \quad 0\right] \tag{D.71}$$

Therefore,

$$\left[\mathbf{\Phi}_{1.g}^*\right] = \left[dU_g \quad dV_g \quad 0 \quad 0\right]^T \cdot \left\{\mathbf{Rs}^T(1 \to 2) \cdot \left[\mathbf{\Phi}_{1.g}\right] \cdot \mathbf{Rs}(1 \to 2)\right\} \cdot \left[dU_g \quad dV_g \quad 0 \quad 0\right] \tag{D.72}$$

$$\left[\mathbf{\Phi}_{2.g}^*\right] = \left[dU_g \quad dV_g \quad 0 \quad 0\right]^T \cdot \left\{\mathbf{Rs}^T(1 \to 2) \cdot \left[\mathbf{\Phi}_{2.g}\right] \cdot \mathbf{Rs}(1 \to 2)\right\} \cdot \left[dU_g \quad dV_g \quad 0 \quad 0\right] \tag{D.73}$$

It can be easily shown that the matrices $\left[\mathbf{\Phi}_{1.g}^*\right]$ and $\left[\mathbf{\Phi}_{2.g}^*\right]$ in Eqs. (D.72) and (D.73) represent the quadratic forms with respect to dU_g and dV_g.

The operator of transformation $\mathbf{Rs}(1 \to 2)$ of the gear tooth flank $\mathcal{G}$ having the first $\mathbf{\Phi}_{1.\,g}$ and the second $\mathbf{\Phi}_{2.\,g}$ fundamental forms from the initial coordinate system

$X_1Y_1Z_1$ to the new coordinate system $X_2Y_2Z_2$, indicates that, in the new coordinate system, the corresponding fundamental forms are expressed in the form:

$$\left[\Phi^*_{1.g}\right] = \mathbf{Rs}^T(1 \to 2) \cdot \left[\Phi_{1.g}\right] \cdot \mathbf{Rs}(1 \to 2) \quad \text{(D.74)}$$

$$\left[\Phi^*_{2.g}\right] = \mathbf{Rs}^T(1 \to 2) \cdot \left[\Phi_{2.g}\right] \cdot \mathbf{Rs}(1 \to 2) \quad \text{(D.75)}$$

Equations (D.74) and (D.75) reveal that after the coordinate system transformation is completed, the first $\Phi^*_{1.g}$ and the second $\Phi^*_{2.g}$ fundamental forms of the gear tooth flank $\mathcal{G}$ in the coordinate system $X_2Y_2Z_2$ are expressed in terms of the first $\Phi_{1.\ g}$ and the second $\Phi_{2.\ g}$ fundamental forms initially represented in the coordinate system $X_1Y_1Z_1$. In order to do that, the corresponding fundamental form (either the form $\Phi_{1.\ g}$ or the form $\Phi_{2.\ g}$) must be pre-multiplied by $\mathbf{Rs}(1 \to 2)$ and after that it should be post-multiplied by $\mathbf{Rs}^T(1 \to 2)$.

Implementation of Eqs. (D.74) and (D.75) significantly simplifies formula transformations.

Equations similar to Eqs. (D.74) and (D.75) are valid with respect to the pinion tooth flank $\mathcal{P}$.

In case of use of the third $\Phi_{3.\ g}$ and fourth $\Phi_{4.\ g}$ fundamental forms, their coefficients can be expressed in terms of the fundamental magnitudes of the first and second orders.

Appendix E: Closest Distance of Approach Between the Tooth Flanks of a Gear and a Mating Pinion

Generally, the problem of the calculation of the closest distance of approach between two smooth regular surfaces is a sophisticated and challenging one. As per the author's knowledge, no general solution to the problem of calculation of the closest distance of approach between two smooth regular surfaces is available in the public domain. For the purpose of calculation of the deviation δ_g of the actual gear tooth flank $\mathcal{G}_{\text{ac}}$ with respect to the desired (nominal) gear tooth flank $\mathcal{G}_{\text{nom}}$, the problem under consideration can be reduced to the problem of computation of the closest distance of approach between two torus surfaces Tr_g and Tr_p.

Let us consider a gear tooth flank, $\mathcal{G}$, and a mating pinion tooth flank, $\mathcal{P}$, which are initially given in a common coordinate system $X_hY_hZ_h$, associated with the gear housing, as illustrated in Fig. E.1. The tooth flanks $\mathcal{G}$ and $\mathcal{P}$ are locally approximated by portions of the torus surfaces Tr_g and Tr_p, respectively. Again, not all points of the torus surfaces Tr_g and Tr_p can be used for the local approximation of the gear and the pinion tooth flanks $\mathcal{G}$ and $\mathcal{P}$, respectively. Only points that are located either within the biggest meridian or within the smallest meridian of the torus surface are employed for this purpose.

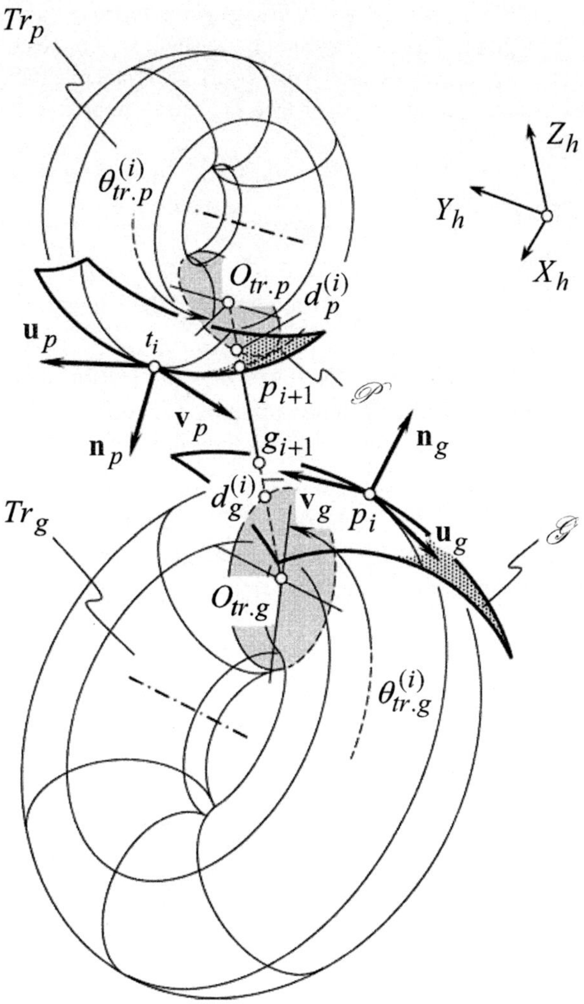

Fig. E.1 Calculation of the closest distance of approach of a gear tooth flank, $\mathcal{G}$, and a mating pinion tooth flank, $\mathcal{P}$

The points K_g and K_p^* are chosen as the first guess points on the torus surfaces Tr_g and Tr_p, respectively. For the analysis below, it is convenient to relabel the points K_g and K_p^* as g_i and p_i, respectively.

For a given configuration of the torus surfaces Tr_g and Tr_p, the closest distance of approach between these surfaces can be used as a first approximation to the closest distance of approach between the original gear and the pinion tooth flanks $\mathcal{G}$ and $\mathcal{P}$, respectively.

The closest distance of approach between the torus surfaces Tr_g and Tr_p is measured along the common perpendicular to these surfaces. The following equations can be composed on the premises of this property of the closest distance of approach.

The unit normal vector $\mathbf{n}_{\mathrm{Tr.}\ g}$ to the torus surface Tr_g is located within a plane through the axis of rotation of the surface Tr_g. In the coordinate system $X_{\mathrm{tr.}\ g}Y_{\mathrm{tr.}\ g}Z_{\mathrm{tr.}\ g}$ that is associated with the torus surface Tr_P, the equation of a plane through the axis of rotation of the torus surface Tr_g can be expressed in the form:

$$\left[\mathbf{r}_{\tau g} - \mathbf{r}_{\mathrm{tr}.g}^{(0)}\right] \times \mathbf{k}_{\mathrm{tr}.g} \times \mathbf{R}_{\mathrm{tr}.g} = 0 \tag{E.1}$$

where

$\mathbf{r}_{\tau g}$ is the position vector of a point of the plane through the axis of rotation of the torus Tr_g.

$\mathbf{r}_{\mathrm{tr}.g}^{(0)}$ is the position vector of a point within the plane $\mathbf{r}_{\tau g}$ (it is assumed below that this point coincides with the origin of the coordinate system $X_{\mathrm{tr.}\ g}Y_{\mathrm{tr.}\ g}Z_{\mathrm{tr.}\ g}$).

$\mathbf{k}_{\mathrm{tr.}\ g}$ is the unit vector of the $Z_{\mathrm{tr.}\ g}-$ axis.

Equation (E.1) is expressed in terms of the radius $\mathbf{R}_{\mathrm{tr.}\ g}$. This indicates that the set of all planes through the fixed $Z_{\mathrm{tr.}\ g}-$ axis forms a pencil of planes. The equation of the pencil of planes $\mathbf{r}_{\tau g}$ in the common coordinate system $X_hY_hZ_h$ can be represented in the form:

$$\mathbf{r}_{\tau g}\left(Z_{\mathrm{tr}.g}, V_{\tau r.g}, \theta_{\mathrm{tr}.g}\right) = \mathbf{Rs}\left(\mathrm{Tr}_g \mapsto h\right) \cdot \begin{bmatrix} V_{\mathrm{tr}.g} \cdot \cos\theta_{\mathrm{tr}.g} \\ V_{\mathrm{tr}.g} \cdot \sin\theta_{\mathrm{tr}.g} \\ Z_{\mathrm{tr}.g} \\ 1 \end{bmatrix} \tag{E.2}$$

The unit normal vector $\mathbf{n}_{\mathrm{Tr.}\ p}$ to the torus surface Tr_p is located within a plane through the axis of rotation of the surface Tr_p. In the coordinate system $X_{\mathrm{tr.}\ p}Y_{\mathrm{tr.}\ p}Z_{\mathrm{tr.}\ p}$, which is associated with the surface Tr_p, the equation of a plane through the axis of rotation of the torus surface Tr_p can be represented in the form:

$$\left[\mathbf{r}_{\tau p} - \mathbf{r}_{\mathrm{tr}.p}^{(0)}\right] \times \mathbf{k}_{\mathrm{tr}.p} \times \mathbf{R}_{\mathrm{tr}.p} = 0 \tag{E.3}$$

where

$\mathbf{r}_{\tau p}$ is the position vector of a point of the plane through the axis of rotation of the torus Tr_p.

$\mathbf{r}_{\mathrm{tr}.p}^{(0)}$ is the position vector of a point within the plane $\mathbf{r}_{\tau p}$ (it is assumed below that this point coincides with the origin of the coordinate system $X_{\mathrm{tr.}\ p}Y_{\mathrm{tr.}\ p}Z_{\mathrm{tr.}\ p}$).

$\mathbf{k}_{\mathrm{tr.}\ p}$ is the unit vector of the $Z_{\mathrm{tr.}\ p}-$ axis.

Equation (E.3) is expressed in terms of the radius $\mathbf{R}_{\mathrm{tr.}\ p}$. This indicates that the set of all planes through the fixed $Z_{\mathrm{tr.}\ p}-$ axis forms a pencil of planes. The equation of this pencil of planes $\mathbf{r}_{\tau p}$ in the common coordinate system $X_hY_hZ_h$ can be represented in the form:

$$\mathbf{r}_{\tau p}\left(Z_{\text{tr}.p}, V_{\text{tr}.p}, \theta_{\text{tr}.p}\right) = \mathbf{Rs}\left(\text{Tr}_p \mapsto h\right) \cdot \begin{bmatrix} V_{\text{tr}.p} \cdot \cos\theta_{\text{tr}.p} \\ V_{\text{tr}.p} \cdot \sin\theta_{\text{tr}.p} \\ Z_{\text{tr}.p} \\ 1 \end{bmatrix} \tag{E.4}$$

A straight line through the points $d_g^{(i)}$ and $d_p^{(i)}$, along which the shortest distance of approach $d_{gp}^{\min}$ of the torus surfaces Tr_g and Tr_p is measured, is the line of intersection of the planes $\mathbf{r}_{\tau g}$ and $\mathbf{r}_{\tau p}$. Therefore, this line $d_{gp}^{\min}$ must be aligned with both unit normal vectors $\mathbf{n}_{\text{tr}.\ g}$ and $\mathbf{n}_{\text{tr}.\ p}$.

In the coordinate system $X_hY_hZ_h$, the equation for the unit normal vector $\mathbf{n}_{\text{tr}.\ g}$ to the torus surface Tr_g yields representation in the matrix form:

$$\mathbf{n}_{\text{tr}.g} = \mathbf{Rs}\left(\text{Tr}_g \mapsto h\right) \cdot \begin{bmatrix} \left(C_{\text{tr}.g} + \cos\varphi_{\text{tr}.g}\right) \cdot \cos\varphi_{\text{tr}.g} \cdot \cos\theta_{\text{tr}.g} \\ \left(C_{\text{tr}.g} + \cos\varphi_{\text{tr}.g}\right) \cdot \cos\varphi_{\text{tr}.g} \cdot \sin\theta_{\text{tr}.g} \\ \left(C_{\text{tr}.g} + \cos\varphi_{\text{tr}.g}\right) \cdot \sin\varphi_{\text{tr}.g} \\ 1 \end{bmatrix} \tag{E.5}$$

where $C_{\text{tr}.\ g}$ is the parameter $C_{\text{tr}.g} = 1 - \frac{R_{2.g}}{R_{1.g}}$.

Similarly, in the coordinate system $X_hY_hZ_h$, the equation for the unit normal vector $\mathbf{n}_{\text{tr}.\ p}$ to the torus surface Tr_p yields the matrix representation in the form:

$$\mathbf{n}_{\text{tr}.p} = \mathbf{Rs}\left(\text{Tr}_p \mapsto h\right) \cdot \begin{bmatrix} \left(C_{\text{tr}.p} + \cos\varphi_{\text{tr}.p}\right) \cdot \cos\varphi_{\text{tr}.p} \cdot \cos\theta_{\text{tr}.p} \\ \left(C_{\text{tr}.p} + \cos\varphi_{\text{tr}.p}\right) \cdot \cos\varphi_{\text{tr}.p} \cdot \sin\theta_{\text{tr}.p} \\ \left(C_{\text{tr}.p} + \cos\varphi_{\text{tr}.p}\right) \cdot \sin\varphi_{\text{tr}.p} \\ 1 \end{bmatrix} \tag{E.6}$$

where $C_{\text{tr}.\ p}$ is the parameter $C_{\text{tr}.p} = 1 - \frac{R_{2.p}}{R_{1.p}}$.

Evidently, the points $O_{\text{tr}.\ g}$, $O_{\text{tr}.\ p}$, $d_g^{(i)}$, and $d_p^{(i)}$ (see Fig. E.1) are located within the straight line through the centers $O_{\text{tr}.\ g}$ and $O_{\text{tr}.\ p}$. The position vector $\mathbf{r}_{cd}$ of this straight line can be calculated from the equation:

$$\left(\mathbf{r}_{cd} - \mathbf{r}_{cg}\right) \times \left(\mathbf{r}_{cp} - \mathbf{r}_{cg}\right) = 0 \tag{E.7}$$

where

$\mathbf{r}_{cg}$ is the position vector of a point of the circle of a radius $R_{\text{tr}.\ g}$.
$\mathbf{r}_{cp}$ is the position vector of a point of the circle of a radius $R_{\text{tr}.\ p}$.

It is necessary that the straight line $\mathbf{r}_{cd}$ be along the unit normal vectors $\mathbf{n}_{tg}$ and $\mathbf{n}_{tp}$ to the torus surfaces Tr_g and Tr_p.

Considered together, Eqs. (E.2), (E.4), and (E.7) make the calculation of the closest distance of approach between the torus surfaces Tr_g and Tr_p possible. Then, the straight line $d_{gp}^{\min}$ intersects the part surface $\mathcal{G}$ and the generating surface $\mathcal{P}$ of the form cutting tool at the points g_{i+1} and p_{i+1}, respectively. The points g_{i+1} and p_{i+1} serve as the second guess to the closest distance of approach between the surfaces $\mathcal{G}$ and $\mathcal{P}$.

The cycle of the recursive calculations is repeated as many times as necessary for making the deviation of the calculation of the closest distance of approach between the surfaces $\mathcal{G}$ and $\mathcal{P}$ smaller than the maximal permissible value.

There is an alternative approach to the calculation of the closest distance of approach between two torus surfaces. The direction of the unit normal vector to an offset surface to Tr_g is identical to the direction of the unit normal vector $\mathbf{n}_{\text{tr.}\,g}$ to the torus surface Tr_g. This statement is also valid for the unit normal vector $\mathbf{n}_{\text{tr.}\,T}$ to the torus surface Tr_p. This property of the unit normal vectors $\mathbf{n}_{\text{tr.}\,g}$ and $\mathbf{n}_{\text{tr.}\,p}$ can be used for the modification of the method of calculation of the closest distance of approach between two torus surfaces.

The equation of the circle of radius $R_{\text{tr.}\,g}$ yields the matrix representation:

$$\mathbf{r}_{cg}(\theta_{\text{tr.}g}) = \mathbf{Rs}(\mathcal{G} \mapsto h) \cdot \begin{bmatrix} R_{\text{tr.}g} \cdot \cos\theta_{\text{tr.}g} \\ R_{\text{tr.}g} \cdot \sin\theta_{\text{tr.}g} \\ 0 \\ 1 \end{bmatrix} \tag{E.8}$$

The equation of the circle of radius $R_{\text{tr.}\,p}$ can be analytically described in a similar manner:

$$\mathbf{r}_{cp}(\theta_{\text{tr.}p}) = \mathbf{Rs}(\mathcal{P} \mapsto h) \cdot \begin{bmatrix} R_{\text{tr.}p} \cdot \cos\theta_{\text{tr.}p} \\ R_{\text{tr.}p} \cdot \sin\theta_{\text{tr.}p} \\ 0 \\ 1 \end{bmatrix} \tag{E.9}$$

The distance d_{gp} between two arbitrary points on the circles $\mathbf{r}_{cg}(\theta_{\text{tr.}\,g})$ and $\mathbf{r}_{cp}(\theta_{\text{tr.}\,p})$ is as follows:

$$d_{gp}(\theta_{\text{tr.}g}, \theta_{\text{tr.}p}) = |\mathbf{r}_{cg}(\theta_{\text{tr.}g}) - \mathbf{r}_{cp}(\theta_{\text{tr.}p})| \tag{E.10}$$

The distance d_{gp} is minimal for a specific (optimal) combination of the parameters $\theta_{\text{tr.}\,g}$ and $\theta_{\text{tr.}\,p}$. The favorable values of the parameters $\theta_{\text{tr.}\,g}$ and $\theta_{\text{tr.}\,p}$ can be calculated from the solution of the set of two equations:

$$\frac{\partial}{\partial \theta_{\text{tr.}g}} \mathbf{r}_{cg}(\theta_{\text{tr.}g}) = 0 \tag{E.11}$$

$$\frac{\partial}{\partial \theta_{\mathrm{tr}.p}} \mathbf{r}_{cp}\left(\theta_{\mathrm{tr}.p}\right) = 0 \tag{E.12}$$

From the solution of Eqs. (E.11) and (E.12), the optimal values $\theta_{\mathrm{tr}.g}^{(\mathrm{opt})}$ and $\theta_{\mathrm{tr}.p}^{(\mathrm{opt})}$, respectively, can be determined. These angles specify the direction of the closest distance of approach of the torus surfaces Tr_g and Tr_p, respectively.

Following this method, the three-dimensional problem of the calculation of the closest distance of approach of two torus surfaces is reduced to the problem of calculation of the closest distance of approach between two circles. Under a certain scenario, the latter approach could possess an advantage over the previous one.

The convergence of the disclosed algorithms for the computation of the closest distance of approach between two smooth regular surfaces is illustrated in Fig. E.2. The computation procedure is convergent regardless of the actual location of the first guess points on the surfaces $\mathcal{G}$ and $\mathcal{P}$.

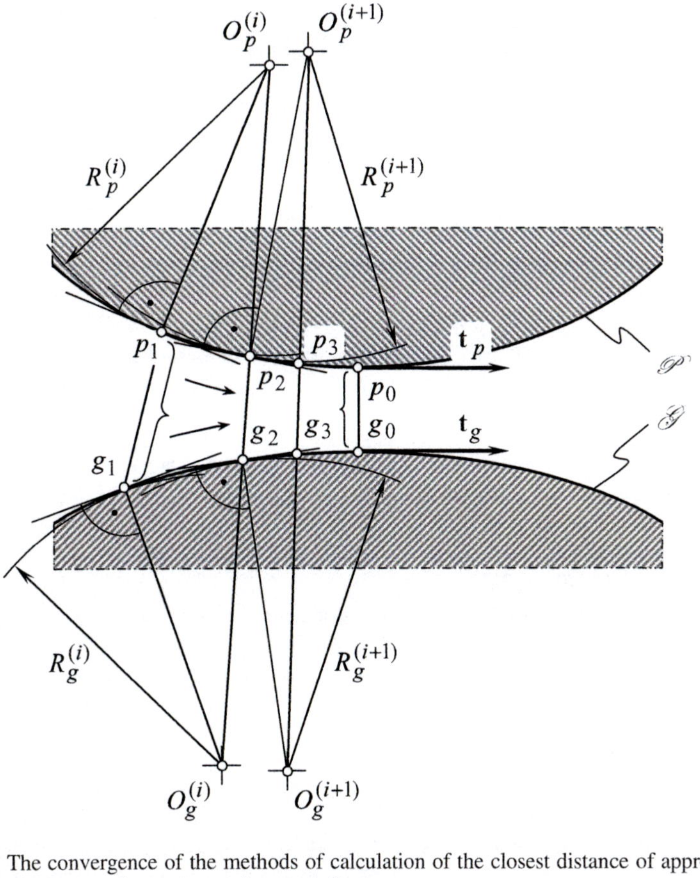

Fig. E.2 The convergence of the methods of calculation of the closest distance of approach of a gear tooth flank, $\mathcal{G}$, and a mating pinion tooth flank, $\mathcal{P}$

It is instructive to draw attention here to the similarities between the disclosed iterative method for the computation of the closest distance of approach between two smooth regular surfaces and Newton–Raphson's method, the iterative method of chords, and so forth. Many similarities can be found on this comparison.

Appendix F: Selected Bibliography on Novikov/Conformal Gearing

Immediately below in this section of the book, a list of publications on Novikov/conformal gearing is presented. The monographs, scientific papers, conference proceedings, patents on inventions, and so forth are covered by this list of publications. The list will be helpful, first of all, to those who are about to start new research in the field of Novikov/conformal gearing.

Abadjiev, V., Petrova, D., Abadjieva, E., "An Optimization of Type Wildhaber Gear Sets Based on Loading Capacity. A Software Estimate of the Hydrodynamic Loading Capacity", in: *Proceedings of 15th International Conference on Manufacturing Systems ICMaS*, 26–27 October 2006, Bucharest, Romania.

Akchurin, A., Bosman, R., Lugt, P.M., "Generation of Wear Particles and Running-in in Mixed Lubricated Sliding Contacts," *Tribology International*, Vol. 110, 2017, pp. 201–208.

Allan, T., "Some Aspects of the Design and Performance of Wildhaber-Novikov Gearing", *Proceedings of the Institution of Mechanical Engineers*, Vol. 179, Part I, No. 30, June 1, 1964, pp. 931–1935.

Amani, A., Rajabalinejad, M., Spitas, C., "Modeling of Elasto-Hydrodynamic Lubrication Problems in Gears", *Applied Mathematics and Computation*, Vol. 135, No. 1, 2003, pp. 3–79.

An'ishchenko, V.V., Koval'enko, G.D., "Investigation of Contact Stress in Novikov Gears Using Photo-Elastic Approach", in: *Novikov Gearing*, Vol. I, TsINTIAM, Moscow, 1964.

Ariga, Y., Nagata, S., "Load Capacity of a New W-N Gear with Basic Rack of Combined Circular and Involute Profile", *Journal of Mechanisms, Transmissions, and Automation in Design*, Vol. 107, No. 4, 1985, pp. 565–572.

Ariga, Y., Nagata, S., "W-N Type Gear Tooth Profile in View of the Radius Curvature", *Journal of the Japan Society for Precision Engineering*, Vol. 55, No. 11, 1989, pp. 2051–2056.

Astridge, D.G., *et al.*, "Tribology of High Conformity Gears", *Institution of Mechanical Engineers*, 1987, pp. 819–825.

Attia, A.Y., "Noise of Gears of Circular-Arc Tooth Profile", *Journal of Sound and Vibration*, Vol. 11, No. 4, April 1970, pp. 383–397.

Attia, A.Y., El-Bahloul, M.M., "Lubricatiopn Capacity of Gearas of Circular-Arc Tooth Profile", *ASME Journal of Tribology*, Vol. 110, No. 4, 1988, pp. 699–703.

Bair, B.-W., Sung, M.-H., Wang, J.-S., Chen, C.-F., "Tooth Profile Generation and Analysis of Oval Gears with Circular-Arc Teeth", *Mechanisms and Machine Theory*, Vol. 44, No. 6, 2009, pp. 1424–1442.

Baranov, G.G., "Some Problems of the Theory of Novikov Gearing for the Modified Center Distance", *Mechanics of Machines*, Vol. 31–32, 1972, pp. 93–107. (In Russian).

Batsch, M., "Comparative Fatigue Testing of Gears with Involute and Convexoconcave Teeth Profiles", *Advances in Manufacturing Science and Technology*, Vol. 40, No. 2, 2016, pp. 5–25.

Batsch, M., "The Method of Axes Distance Error Compensation in Convexo-Concave Novikov Gears", *Advances in Manufacturing Science and Technology*, Vol. 39, No. 4, 2015, 8 pages.

Batsch, M., Markowski, T., Legutko, S., Krolczyk, G.M., "Measurement and Mathematical Model of Convexo-Concave Novikov Gear Mesh", *Measurement*, Vol. 125, September 2018, pp. 516–525.

Bergmann, P., Grün, F., Summer, F., Gódor, I., Stadler, G., "Expansion of the Metrological Visualization Capability by the Implementation of Acoustic Emission Analysis," *Advances in Tribology*, Vol. 2017, Article ID 3718924, 2017, 17 pages.

Beskopylny, A., Onoshkov, N., Korotkin, V., "The Contact Mechanics of Novikov's Surface-Hardened Gearing during Running-in Process", *Advances in Tribology*, Vol. 2018, Article ID 9607092, 8 pages. https://doi.org/10.1155/2018/9607092.

Bobil'ov, D.K., "On a Motion of a Surface, that Retains in Tangency to another Surface, Motionless", Notes of Imperial Academy of Sciences, 1887.

Budika, Yu.N., "Theory of Meshing and Comparative Wear Resistance in Planar Meshing of General Type", in: *Proceedings of Seminar on Theory of Machines and Mechanisms*, Vol. X, No. 39, USSR Academy of Sciences, 1951, pp. 56–74.

Chen, C.-F., Tsay, C.-B., "Computerized Tooth Profile Generation and Analysis of Characteristics of Elliptical Gears with Circular-Arc Teeth", *Journal of Materials Processing Technology*, No. 148, 2004, pp. 226–234.

Chen, C.-K., Wang, C.-Y., "Compensating Analysis of a Double Circular-Arc Helical Gear by Computerized Simulation of Meshing", *Proceedings of the Institution of Mechanical Engineers, Part C: Journal of Mechanical Engineering Science*, Vol. 215, 2001, pp. 759–771.

Chen, D.-H., Zhang, G.-Q., Chen, X., "Research on Nonlinear Vibration and Bend Strength Character of WN Gear Pairs in Meshing", *Advanced Materials Research*, Vols. 706–708, 2013, pp. 1537–1540.

Chen, H., Duan, Z., *et al.*, "Study on the General Principle of Normal Circular-Arc Gear Transmission", *Mechanism and Machine Theory*, Vol. 41, No. 12, 2006, pp. 1424–1442.

Chen, H., Zhang, X., Cai, X., Ju, Z., Qu, C., Shi, D., "Computerized Design, Generation and Simulation of Meshing and Contact of Hyperboloidal-Type Normal Circular-Arc Gears", *Mechanism and Machine Theory*, Vol. 96, Part 1, 2016, pp. 127–145.

Chen, Z., Ding, H., Li, B., Luo, L., Zhang, L., Yang, J., "Geometry and Parameter Design of Novel Circular Arc Helical Gears for Parallel-Axis Transmission", *Advances in Mechanical Engineering*, Vol. 9, No. 2, 2017, pp. 1–11.

Cheng, D.X., *Chinese Standard*, China Machine Press, 1991, pp. 158–187.

Chironis, N., "Design of Novikov Gears", pp. 124–135, in: Nicholas P. Chironis, Editor, *Gear Design and Application*, McGraw-Hill Company, New York, 1967, 374 pages.

Chironis, N., "Design of Novikov Gears", *Product Engineering*, Vol. 33, No. 19, 17 September 1962, pp. 91–102.

Chironis, N.P., "New Tooth Shape Taking Over? Design of Novikov Gears", pp. 124–135, in: Nicholas P. Chironis, Editor, *Gear Design and Application*, Mc-Graw Hill Book Company, New York, 1967, 374 pages.

Colbourne, J.R., "The Contact Stress in Novikov Gears. La Pression Superficielle des Engrenages a Denture Novikov", *Mechanism and Machine Theory*, Vol. 24, No. 3, 1989, pp. 223–229.

Coy, J.J., Townsend, D.P., Zaretsky, E.V., *Gearing*, NASA Reference Publication 1152, AVSCOM, Technical Report 84-C-15, 1985, 69 pages.

Crook, A.W., "Developments in Elasto-Hydrodynamic Lubrication", *Journal of Institute of Petroleum*, Vol. 49, 1963, p. 295.

Davidov, Ya.S., "Memories . . .". http://referat.znate.ru/text/index-8600.html.

Davies, W.J., "Novikov Gearing", *Machinery*, Vol. 96, No. 4, 13 January 1960, pp. 64–73.

"Discussion on Novikov Gears", *Power Transmission*, London, 1960, pp. 29, 722, and 1961, pp. 30, 32.

Dobrovol'skii, A.V., "Some Potential Capabilities of the Novikov Engagement System", *Journal of Machinery Manufacture and Reliability*, Vol. 36, No. 6, December 2007, pp. 515–520.

Dobrovolsky, V., *et al.*, "Fundamentals of the Theory and Operation of Novikov Gears", in: *Machine Elements*, MIR Publishers, Moscow, 1968.

Duan, Z., Chen, H., Ju, Z., Liu, J., "Mathematical Model and Manufacture Programming of Loxodromic-Type Normal Circular-Arc Spiral Bevel Gear", *Frontiers of Mechanical Engineering*, Vol. 7, No. 3, September 2012, pp. 312–321.

Dyson, A., Evans, H.P., Snidle, W., "Wildhaber-Novikov Circular Arc Gears: Geometry and Kinematics", *Proceedings of The Royal Society, A, Mathematical Physical and Engineering Sciences*, Vol. 403, No. 1825, 1986, pp. 313–340.

Dyson, A., Evans, H.P., Snidle, R.W., "Wildhaber-Novikov Circular Arc Gears: Geometry and Kinematics", *Proceedings of the Royal Society of London. Series A*, Vol. 403, 1986, pp. 313–340.

Dyson, A., Evans, H.P., Snidle, W., "Wildhaber-Novikov Circular Arc Gears: Some Properties of Relevance to their Design", *Proceedings of The Royal Society, A, Mathematical Physical and Engineering Sciences*, Vol. 425, 1989, pp. 341–363.

Ellis, D.V., "The Westland Lynx", *The RUSI Journal*, Vol. 125, No. 4, 1980, pp. 70–73.

Evans, H.P., Snidle, R.W., "Wildhaber-Novikov Circular Arc Gears: Elastohydrodynamics", *Journal of Tribology*, Vol. 115, No. 3, July 1, 1993, pp. 487–492.

Fedyakin, R.V., *Investigation of Strength of Circular-Arc Gear Teeth*, Ph.D. Thesis, Moscow, Zhukovskiy Aviation Engineering Academy, 1955.

Fedyakin, R.V., *On Designing of Novikov Gearing*, Zhukovsky Air Force Engineering Academy Publishers, Moscow, 1957.

Fedyakin, R.V., Chesnokov, V.A., "Design of Tooth Gears of M.L. Novikov", *Vestnik Mashinostroyenia*, Vol. 38, May 1958, pp. 11–19.

Fedyakin, R.V., Chesnokov, V., "The Novikov Gear Tooth System," *Vestnik Mashinostroyenia*, Vol. 38, April 1958, pp. 3–11.

Feng, D.C., "A Technical Method of Processing End Relief Double W-N Gears on Hobbers", *Journal of Dalian Railway Institute*, Vol. 27, No. 2, 2006, pp. 29–33.

Fletcher, H.A.G., Berry, G., *Efficiency and Noise of a Novikov Gear Pair*, NEL Report No. 46.

FLIGHT International, 15 March 1975, pp. 398–399.

French, M.J., "Conformity of Circular-Arc Gears", *Journal of Mechanical Engineering Science*, Vol. 7, No. 2, 1965, pp. 220–223.

Gao, Y., Chen, B., Tan, R., Zhang, Y., "Design and Finite Element Analysis for Helical Gears with Pinion Circular Arc Teeth and Gear Parabolic Curve Teeth", *Bulletin of the JSME, Journal of Advanced Mechanical Design, Systems, and Manufacturing*, Vol. 10, No. 1, 2016, 16 pages.

GOST 15023-76, *Novikov Spur Gears with Double Line of Action, Basic Rack.*

Gribanov, V.M., Medintseva, Y.V., Ratov, D.V., *Novikov Gears: Accuracy Problems*, Lugansk, "Knowledge", 2010, 252 pages.

Grishel, I.N., "The Influence of Center Distance Errors on the Loading Capacity of Novikov Gears", *Vestnik Mashinostroyenia*, Vol. 39, No. 5, May 1959, pp. 16–19.

Gunner, A.M., "Comment on Critical Look at Novikov Gearing", *The Engineer*, 13 May 1960.

Gunner, A.M., "Comment on Critical Look at Novikov Gearing", *The Engineer*, 3 November 1961.

Gunner, A.M., "First AEI 'Circarc' Gear for the Science Museum, South Kensington", *Machinery*, 8 May 1963.

Gunner, A.M., Walker, H., "A Critical Look at the Novikov Gear", *Engineer*, London, 1960, pp. 209, 818, and 1961, pp. 212, 737.

Hageniers, O.L., *A Study of the Bending Stresses in a Wildhaber-Novikov Gear Set*, MS Thesis, University of Windsor, 1969, 97 pages.

Harrison, W.H., "Load Tests on Gears with Teeth to the Wildhaber-Novikov Design Basis," *VDI-Berichte*, No. 47, 1961, pp. 13–18.

Hlebanja, G., Hlebanja, J., "Uniform Power Transmission Gears", *Strojniški vestnik – Journal of Mechanical Engineering*, Vol. 55, No. 7–8, 2009, pp. 472–483.

http://en.cnki.com.cn/Article_en/CJFDTOTAL-CQSG200806022.htm

https://scholar.google.com/citations?user=5qZ8EBcAAAAJ&hl=en

http://www.zakgear.com/WN.html

https://www.amazon.com/State-research-design-Wildhaber-Novikov-circular/dp/B00072DPQW

Huston, R.L., "A Basis for Solid Modeling of Gear Teeth with Application in Design and Manufacture", *Mechanisms and Machine Theory*, Vol. 29, No. 5, 1994, pp. 713–723.

Investigation of the Conformal Gear for Helicopter Power Transmission, TRECOM Technical Report 64-28, Task 1D121401A14414, Contract DA. 44-177-AMC-101(T), prepared by: THE Boeing Company, Vertol Division, Morton, Pennsylvania, for U.S. Army Transportation Research Command, Fort Eustis, Virginia, June 1964, 128 pages.

Ishibashi, A., Yoshino, H., "Design, Manufacture and Load Carrying Capacity of Novikov Gear with 3–5 Pinion Teeth for High Gear Ratio", *The Japan Society of Mechanical Engineers*, Vol. 27, No. 229, 1984, pp. 1521–1528.

Itkis, M., Ya., *Geometrical Calculation of Novikov Cylindrical Gears*, Nizhnevolzhskoye knizhnoye izdatelstvo, Volgograd, 1973, 312 pages.

Johnson, D.C., "Novikov Gears: Gear Teeth with Circular Arc Profiles", *Engineering* (London, England), Vol. 188, October 2, 1959, p. 294.

Jonas, K., Holemar, A., "Novikov Gear Teeth", *Technical Digest*, April 1962, pp. 10–13.

Kaluzhnikov, A.N., "A New Point-Contact System for Gear Teeth", *Engineers' Digest*, 1958, pp. 19, 57.

Kang, H., Weidong, C., Qi, C., Yingming, L., "Research of Double Circular-Arc Gear Fast Modeling Based on CATIA Secondary Development", in: *2012 Third International Conference on Digital Manufacturing & Automation*, IEEE Conference Publication, 2012. https://ieeexplore.ieee.org/document/6298603.

Kasuya, K., Nogami, M., Matsunaga, T., Watanabe, M., "Tooth Bearing Analysis and Surface Durability of Symmetrical Conformal Gears", *Journal of Mechanical Design*, Vol. 103, No. 1, January 1, 1981, pp. 141–150 (10 pages).

Kiktyar, F.S., "Hobs for Cutting Novikov Gears", *Vestnik Mashinostroyenia*, Vol. 39, September 1959, pp. 8–13.

Klein, G.J., "The Wildhaber-Novikov System of Gearing", in: *DME/NAE Quarterly Bulletin* (1965-1), National Research Council of Canada, Ottawa, April 1965.

Kolchin, N.I., Oparin, G.V., "Novikov Gearing with Two Lines of Meshing and Helical Canal Surfaces of Teeth", *Izvestiya VUZov*, Vol. 8, 1968, pp. 29–32. (In Russian).

Kolodkin, M.S., "Comparative Experimental Investigation of Bearing Capacity of Involute Gearing and Novikov Gearing", in: *Proceedings of Zhukovsky Air Force Engineering Academy*, Vol. 313, Leningrad, 1960, pp. 9–23.

Korotkin, V.I., Gazzaev, D.A., "Modeling of the Contact Interaction of the Teeth of the Gear Wheels of Novikov," *Journal of Machinery Manufacture and Reliability*, Vol. 43, 2014, pp. 104–111.

Korotkin, V.I., Kharitonov, Yu.D., *Novikov Gearing*, RGU Publishers, Rostov-on-Don, 1991, 208 pages.

Korotkin, V.I., Onishkov, N.P., "To the Problem of Contact Fatigue Life of Surface Hardened Novikov Gears," *Vestnik Rostovskogo Gosudarstvennogo Universiteta Putey Soobshcheniya'* (Vestnik RGUPS), Vol. 3, 2007, p. 14.

Korotkin, V.I., Onishkov, N.P., Kharitonov, Yu.D., *Novikov Gearing: Achievements and Development*, Nova Science Publishers, Inc., New York, 2011, 249 pages.

Korotkin, V.I., Onishkov, N.P., Kharitonov, Yu.D., *Novikov Gearing: Achievements and Evolution*, Mashinostroyeniye-1, Moscow, 2007, 384 pages. [ISBN 978-5-94275-305-4/5-94275-305-9].

Korotkin, V.I., Onishkov, N.P., Kharitonov, Yu.D., "Stress State of Novikov Gearing Teeth under the Conditions of Real Multipair Engagement," *Spravochnik. Inzhenernyi zhurnal*, 2015, pp. 11–17.

Krasnoshchekov, N.S., "Inspection of Novikov Gears", *Vestnik Mashinostroyenia*, Vol. 38, October 1958, pp. 3–9.

Krasnoshchekov, N,S., "Sensitivity of Novikov Gearing to Basic Geometric Errors", *Izvestiya Vysshikh Uchebnykh Zavedeniy: Mashinostroyeniye*, No. 3–4, March–April 1958, pp. 52–63.

Krasnoshchekov, N.N., Fedyakin, R.V., Chesnokov, V.A., *Theory of Novikov Gearing*, Nauka, Moscow, 1976, 173 pages.

Kulikov, G.V., "Losses in Novikov Gearing", *Vestnik Mashinostroyenia*, Vol. 38, June 1958, pp. 11–16.

Kul'ikov, G.V., Fed'yakin, R.V., Chesnokov, V.A., "An Investigation of Contact Strength of Novikov Gears", in: *Novikov Gearing*, Vol. II, Zhukovskiy Aviation Engineering Academy, Moscow, 1962.

Kudriavtsev, V.N., *Calculation and Design of Novikov Gearing*, Leningrad Mozhaiskii Military Aviation Academy, Leningrad, 1959, 78 pages.

Kudryavtsev, V.N., "On the Issue of Novikov Gearing", in: *Research and Development of Novikov Gearing*, 1960, pp. 33–44. (In Russian).

Kumar, A., Bremble, G.R., "An Experimental Investigation into the Oil Film Thickness of Wildhaber/Novikov Gears Using a Simulation Machine. Part II: Measurement of Oil Film Thickness", *Wear*, Vol. 39, 1976, pp. 345–360.

Kuo, H.M., *A Study on the Bevel Gear with Circular-Arc Tooth Profiles*, Mater's Thesis, National Sun Yat-sen University, Taiwan, 2001.

Lagutin, S.A., Barmina, N., "Prof. F.L. Litvin: Contribution to the Formation of the Russian School of the Theory of Gearing", in: V. Goldfarb and N. Barmina, Editors, *Theory and Practice of Gearing and Transmissions*, Mechanisms and Machine Science 34, Springer International Publishing Switzerland, 2016.

Lemanski, A., Alberti, J., Mack, I., *An Evaluation of Conformal Contact*, USAAVLABS Technical Report 67-83, U. S. Army Aviation Material Laboratories, Fort Eustis, Virginia, Contract DA 44-177 AMC-389(T), The Boeing Company, Vertol Division, Philadelphia, PA, January 1968, 72 pages.

Lewis, W., "Investigation of the Strength of Gear Teeth", in: *Proceedings of the Engineers Club*, Philadelphia, 1893, pp. 16–23.

Li, Y., Wu, B.-L., Zhu, L.-L., "Analysis and Calculation of Double Circular Arc Gear Meshing Impact Model", *The Open Mechanical Engineering Journal*, No. 9, 2015, pp. 160–167.

Liang, D., Chen, B., Tan, R., Liao, R, "Geometric Design and Analysis of Gear Transmission with Double Circular Arc-Involute Tooth Profile", *Proceedings of the Institution of Mechanical Engineers, Part C: Journal of Mechanical Engineering Science*, Vol. 231, No. 11, 2015, pp. 2100–2109.

Liang, X.C., *Study on Gear and Cutter Manufacture*, Chongqing University Press, pp. 81–97.

Lingaiah, K., *State of Art at Research Level in the Design of Wildhaber-Novikov Circular Arc Gears*, American Society of Mechanical Engineers, 1976, 9 pages. ASIN: B00072DPQW.

Lingaiah, K., Ramachandra, K., "Conformity Factor in Wildhaber-Novikov Circular-Arc Gears", *ASME Journal of Mechanical Design*, Vol. 103, No. 1, 1981, pp. 134–140.

Lingaiah, K., Ramachandra, K., "Technology Transfer in the Design and Development of Wildhaber-Novikov Gears", in: *Proceedings of the Design Technology Transfer Conference, ASME*, April 1974.

Lingaiah, K., Ramachandra, K., "Three-Dimensional Photoelastic Study of the Load-Carrying Capacity/Face Width Ratio of Wildhaber-Novikov Gears for Automotive Applications", *Experimental Mechanics*, October, 1977, pp. 392–397. https://slideheaven.com/face-width-ratio-of-wildhaber-novikov-gears-for-automotive-applications.html.

Lingaiah, K., Ramachandra, K., "Photoelastic Optimization of the Profiles of Wildhaber-Novikov Gears", *Experimental Mechanics*, Vol. 16, No. 3, March, 1976, pp. 116–120.

Litvin, F.L., *New Kinds of Worm Gear Drives*, Mashgiz, Moscow, 1962, 103 pages.

Litvin, F.L., "The Investigation of the Geometric Properties of a Variety of Novikov Gearing", *The Proceedings of Leningrad Mechanical Institute*, No. 24, 1962. (in Russian).

Litvin, F.L., Feng, P.-H., Lagutin, S.A., *Computerized generation and Simulation of Meshing and Contact of New Type of Novikov-Wildhaber Helical Gears*, NASA/CR-2000-209415, ARL-CR-428, October 2000, 49 pages.

Litvin, F.L., Fuentes, A., Gonzalez-Perez, I., Carnevali, L., Sep, T.M., "New Version of Novikov–Wildhaber Helical Gears: Computerized Design, Simulation of Meshing and Stress Analysis", *Computer Methods in Applied Mechanics and Engineering*, Vol. 191, No. 49–50, 6 December 2002, pp. 5707–5740.

Litvin, F.L., Fuentes, A., Gonzalez-Perez, I., "Modified Involute Helical Gears: Computerized Design, Simulation of Meshing, and Stress Analysis", *Computer Methods in Applied Mechanics and Engineering*, Vol. 192, No. 33–34, 2003, pp. 3619–3655.

Litvin, F.L., Pin-Hao, F., Lagutin, S.A., "Computerized Generation and Simulation of Meshing and Contact of New Type of Novikov-Wildhaber Helical Gears", R – 2000209415. http://gearexpert.free.fr/fichiers_pdf/engrenage_Novikov_Wildhaber_NASA_report.pdf. Accessed: 30 June 2012.

Litvin, F.L., Lu, J., "Computerized Design and Generation of Double Circular-Arc Helical Gears with Low Transmission Errors", *Computer Methods in Applied Mechanics and Engineering*, Vol. 127, No. 1–4, 1995, pp. 57–86.

Litvin, F.L., Lu, J., "Computerized Simulation of Generation, Meshing and Contact of Double Circular-Arc Helical Gears", *Mathematical and Computer Modelling*, Vol. 18, No. 5, June 1993, pp. 31–47.

Litvin, F.L., Tsay, C.-B., "Helical Gears with Circular Arc Teeth: Simulation of Conditions of Meshing and Bearing Contact", *ASME Journal of Mechanisms, Transmissions, and Automation in Design*, Vol. 107, 1985, pp. 556–564.

Litvin, F.L., Tsay, C.-B., "Helical Gears with Circular Arc Teeth: Simulation of Condition of Meshing and Bearing Contact", *Gear Technology*, July/August 1987, pp. 22–23, 26–28, 30–31, 33–36.

Litvin, F.L., Tsay, C.-B., *Spiral Bevel and Circular Arc Helical Gears: Tooth Contact Analysis and the Effect of Misalignment on Circular Arc Helical Gears*, NASA Technical Memorandum 87013, USAAVSCOM, Technical Report 85-C-6, 1985, 16 pages.

Litvin, F.L., Tsay, C.-B., *Helical Gears with Circular Arc Teeth: Generation, Geometry, Precision and Adjustment to Errors, Computer Aided Simulation of Conditions of Meshing, and Bearing Contact*, NASA Contractor Report 4089, AVSCOM Technical Report 87-C-18, 1987, 92 pages.

Litvin, F.L., Tsung, W.-T., Tsay, C.-B., Coy, J.J., *Handschuh, R.F., Spiral Bevel and Circular Arc Helical Gears: Tooth Contact Analysis and the Effect of Misalignment on Circular Arc Helical Gears*, NASA Technical Memorandum 87013, USAAVSCOM, Technical Report 85-C-6, 1985, 19 pages.

Liu, T., Li, G., Wei, H., Sun, D., "Experimental Observation of Cross Correlation Between Tangential Friction Vibration and Normal Friction Vibration in a Running-in Process," *Tribology International*, Vol. 97, 2016, pp. 77–88.

Lu, J., Litvin, F.L., Chen, J.S., "Load Share and Finite Element Stress Analysis for Double Circular-Arc Helical Gears", *Mathematical and Computer Modelling*, Vol. 21, No. 10, 1995, pp. 13–30.

Lu, X.Z., Shang, J.K., *Meshing Principle of Circular-Arc Gears*, China Machine Press, 2003, pp. 46–53.

Lunin, S.V., *New Discoveries in WN Gear Geometry*, 2001. http://www.zakgear.com/WN.html.

Lunin, S., *Interactive Visualization with Parallel Computing for Gear Modeling*, 2012. http://www.zakgear.com/Parallel.html.

Lustenkov, M.Ye., *Transmissions with Intermediate Rolling Elements: Determination and Minimization of Power Losses*, Monograph, Mogil'ov, Belorus-Russian University Publishers, 2010, 274 pages.

Mack, J., Regl, R., *Preliminary Technical Evaluation of Novikov Gears*, ECD-359, 4 August 1961. (Unpublished source).

Maki, M., Watanabe, M., "A Study on WN Spiral Bevel Gear Manufactured by Machining Center", *Journal of Technol. Research*, Vol. 48, 2005, pp. 91–97.

Mallipeddi, D., Norell, M., Sosa, M., Nyborg, L., "Influence of Running-in on Surface Characteristics of Efficiency Tested Ground Gears," *Tribology International*, Vol. 115, 2017, pp. 45–58.

Mansoor, A., Hatem, R.W., Mohammad, Q.A. "Generation and Stress Analysis in New Version of Novikov Helical Gear Combining Double Circular Arc and Crowned Involute Profiles", *Innovative Systems Design and Engineering*, Vol. 7, No. 8, 2016, pp. 54–67.

Markowski, T., Batsch, M., "Influence of Novikov Convexo-Concave Gear Parameters on Contact Pattern", *Scientific Journal of Silesian University of Technology. Series: Transport*, Vol. 89, 2015, pp. 89–99. ISSN: 0209-3324. https://doi.org/10.20858/sjsutst.2015.89.10.

Markowski, T., Batsch, M., "Tooth Contact Analysis of Novikov Convexo-Concave Gears", *Advances in Manufacturing Science and Technology*, Vol. 39, No. 1, 2015, 17 pages. https://doi.org/10.2478/amst-2015-0004.

Matsuyama, T., "Extension of Novikov's Gearing on Skew Gears," *VDI-Berichte*, No. 47, 1961, pp. 23–24.

Moder, J., Grün, F., Stoschka, M., Gódor, I., "A Novel Two-Disc Machine for High Precision Friction Assessment," *Advances in Tribology*, Vol. 2017, Article ID 8901907, 2017, 16 pages.

Mu, X.-K., Li, Y.-G., Chen, D.-H., Li, C.-C., "Failure Analysis and Transmission Parameters Optimization Design of WN Gears Drive", *Applied Mechanics and Materials*, Vols. 229–231, 2012, pp. 449–452.

Mu, X.-K., Li, Y.-G., Chen, D.-H., Li, C.-C., "Transmission Parameters Optimization Design and Load Factor Formula Fitting of WN Gears Drive", *Advanced Materials Research*, Vols. 712–715, 2013, pp. 1050–1053.

Nacy, S.N., Abdullah, M.Q., Mohammed, M.N., "Generation of Crowned Parabolic Novikov Gears", *Engineering Letters*, 15:1, EL_15_1_4, 15 August 2007, 5 pages.

Nagata, S., Ariga, Y., "Development of a New Wildhaber-Novikov Gear with a Basic Rack of Combined Circular and Involute Profile", in: *International Symposium on Gearing & Power Transmission*, Tokyo, 1981.

Napier, P.C., *Novikov Gearing Inquiries at AEI*, Rugby, England and Rolls-Royce Aero, Derby, England, NM-736, 19 July 1963. (Unpublished source).

Nenadic, N.G., Wodenscheck, J.A., *et al.*, "Seeding Cracks Using a Fatigue Tester for Accelerated Gear Tooth Breaking", 2011.

Niemann, C., "Novikov Gear System and Other Special Gear Systems for High load Carrying Capacity," *VDI-Berichte*, No. 47, 1961, pp. 5–12.

Niemann, G., Rettig, H., Lechner, G., "Some Possibilities to Increase the Load-Carrying-Capacity of Gears", *SAE Transactions*, Vol. 71, 1963, pp. 169–184, 201.

Novikov, M.L., *A Novel System of Gear Meshing*, MDNTP, Moscow, 1959.

Novikov, M.L., *Fundamentals of Geometric Theory of Gearing with Point Meshing for High Power Density Transmissions*, Doctoral Thesis, Moscow, by Zhukovskii Military Aviation Engineering Academy, 1955.

Novikov, M.L., *Gearing with a Novel Type of Teeth Meshing*, Zhukovskii Military Aviation Engineering Academy, Moscow, 1958, 186 pages.

Novikov Gearing, Proceedings of Rostov-on-Don Research Institute (NIITM), Issue X, Rostov-on-Don, 1964.

"Novikov Gearing", *The Guardian*, 18th November, 1958.

PaDuan, Z., Chen, H., Ju, Z., Liu, J., "Mathematical Model and Manufacture Programming of Loxodromic-Type Normal Circular-Arc Spiral Bevel Gear", *Frontiers of Mechanical Engineering*, Vol. 7, No. 3, 2012, 4 pages.

Pakhomov, S.N., "Kinematic Schemes of Reducers with Radius Engagement", in: *Modern Engineering and Technology*, No. 2, 2016, 10 pages. http://technology.snauka.ru/2016/02/9558.

Park, S.-H., *Fundamental Development of Hypocycloidal Gear Transmissions*, Ph.D. Thesis, The University of Texas at Austin, 2005, 428 pages.

Pat. Bulgaria No. 48296, *Gear Drive*, G.A. Zhuravlev, Int. Cl. F16h, 55/08, Filed: September 16, 1986, Published: January 15, 1991.

Pat. Canada, No. 2,482,988, *Gear Tooth Profile*, J.R. Colbourn, Int. Cl. F16h 21/00, Filed: 2002.04.22, Patented: 2003.10.30. [Colboune, J.R., *Gear Tooth Profile*, Patent application No. 2,482,988, Int. Cl. F16 h 55/08, Filed: 2003.4.21, Published: 2003.10.30].

Pat. CN_1318783C, F16h 55/08, PTC/US2003/012170 2003.04.21, WO2003/089174, 2003.10.30, Published: 2004.10-22.

Pat. EP, No. 0,211,309, *Wide-Angle Gearing*, W.S. Rouverol, Int. Cl. F16h 55/08, Filed: July 21, 1985, Patented: February 25, 1987.

Pat. EP, No. 0,316,194, *Seat Mechanism*, B.A. Shotter, Int. Cl. B60N 1/06, Filed: November 11, 1987, Patented: May 17, 1989.

Pat. EP, No. 0,362,385, *Novikov Gearing*, L.S. Boiko, V.I. Korotkin, V.Ya. Veretennikov, Ye.G. Roslivker, R.V. Fedyakin, V.A. Chesnokov, A.S. Yakovl'ev, Yu.D. Kharitonov, and E.N. Galichenko, Int. Cl. F16h 55/08, Filed: 05.10.1989, Patented: 11.04.1990.

Pat. GB, No. 186.436, *Improvements in and Relating to Gear Teeth*, F.J. Bostock and S. Bramley-Moore, October 2, 1922. Filed July 2, 1921, No. 18,014/21.

Pat. GB No. 208,332, *Improvements in or Relating to Worm Gearing*, F.J. Bostock and S. Bramley-Moore, Filed: November 9, 1922, Patented: December 20, 1923.

Pat. GB, No. 1,157,562, *Improvements in Wildhaber-Novikov Gears*, D.C.A. Leggatt and B.A. Shotter, Int. Cl. F16h 55/08, Filed: October 26, 1967, Patented: July 9, 1969.

Pat. GB, No. 1,432,164, *Two-Point Contact Wildhaber-Novikov Gears*, Kabushaki Kaisha Hasegava Hagurume, Int. Cl. F16h 55/08, Filed: July 26, 1972, Patented: April 14, 1976.

Pat. DE, No. 2,336,570, *W-N Zahnrad der Doppellinieneingriffsbauweise*, H. Hagumura, Int. Cl. F16 h 55/08, Filed: July 26, 1972, Patented: February 7, 1974.

Pat. Germany No. 36 89 479 T2, Int. Cl. F16h 55/08, Filed: 31.10.1986, Patented: 05.05.1988.

Pat. Germany No. 10 2012 213 403 A1, *Generation of Crowned Parabolic Novikov Gears*, S.M. Somer, M.Q. Abdullah, and M.N. Mohammed, Int. Cl. F16H 55/08, F16H 48/10, Filed: July 31, 2012, Patented: February 6, 2014.

Pat. Japan, No. 60026860 A, *Train of Gears with Tooth Form in Circular Arc*, Y. Setsuyoshi, H. Megumi, and M. Kunihiko, Int. Cl. F16h 55/08, Filed: July 25, 1983, Patented: February 9, 1985.

Pat. Japan No. 4,449,045, Int. Cl. F16h 1/08, 55/08, Filed: September 5, 2008, Patented: June 23, 2009.

Pat. RU No. 2,087,774, *Cylindrical Gear Drive*, G.A. Zhuravlev, G.P. Smirnov, and A.V. Nikolayev, Int. Cl. F16h 1/04, Filed: 24.02.1995, Patented: 20.08.1997.

Pat. RU 2,092,726, *Worm Gearing with Balls*, V.G. Osetrov, V.I.Goldfarb, V.N. Mokretsov, and A.S. Kuniver, Int. Cl. F16H 1/16, Filed: January 24, 1995, Patented: October 10, 1997.

Pat. RU No. 2,318,150, *A Gear for Mixed-Type Meshing or Involute Meshing*, G.A. Zhuravlev, Int. Cl. F16h 55/08, F16h 1/06, Filed: 24.11.2005, Patented: 27.02.2008.

Pat. RU, No. 2,412,206 C2, *A Method of Generation of Gears by Means of Two Stage Rolling*, A.N. Petrovskij, Int. Cl. B23F 5/14, Filed: October 12, 2009, Patented: February 20, 2011.

Pat. RU, No. 2,426,023, *Gear Transmission*, I.Ye. Filimonov and P.A. Prokudin, Int. Cl. F16H 55/08, Filed: August 17, 2009, Patented: February 27, 2011.

Pat. UA No. 10358, *Novikov Gearing with Two Lines of Action and Elliptical Teeth*, O.P. Popov and L.O. Popova, Int. Cl. F16H 1/00, Filed: April 18, 2005, Patented: November 15, 2005.

Pat. UA No. 13562, *Novikov Gearing with Two Lines of Action*, O.P. Popov, Int. Cl. F16H 1/00, Filed: August 15, 2005, Patented: April 17, 2006.

Pat. UA No. 17818, *Transmission of Rotary Motion with Relatively Movable Cylindrical Surface of Action*, V.G. Fedorchenko and V.S. Krivun', F16H 1/00, Filed: April 12, 2006, Patented: October 16, 2006.

Pat. UA No. 28010, *Novikov Gearing with Elliptical-Circular Teeth*, O.P. Popov, Int. Cl. F16H 1/00, Filed: June 20, 2007, Patented: November 26, 2007.

Pat. UA No. 47661.

Pat. UA No. 77941, *Novikov Gearing*, O.P. Popov and L.O. Popova, Int. Cl. F16H 1/08, Filed: December 21, 2011, Patented: March 11, 2013.

Pat. UA No. 79647 C2, *Novikov Gearing with One Line of Action and Line Contact Between the Tooth Flanks*, O.P. Popov, Int. Cl. F16H 1/00, Filed: May 23, 2005, Patented: July 10, 2007.

Pat. UA No. 79810, *Novikov Gearing with Elliptical Teeth*, O.P. Popov and L.O. Popova, Int. Cl. F16H 1/00, Filed: February 21, 2005, Patented: July 25, 2007.

Pat. UA No. 84383, *Novikov Gearing with Point Meshing*, O.P. Popov, O.M. Medvedovsky, L.O. Popova, and O.I. Savenko, Int. Cl. F16H 1/00, Filed: February 11, 2013, Patented: October 25, 2013.

Pat. UA No. 85431, *Novikov Gearing with Line Meshing of Gear Teeth*, O.P. Popov and L.O. Popova, Int. Cl. F16H 1/00, Filed: March 11, 2013, Patented: November 25, 2013.

Pat. USA No. 407.437. *Machine for Planing Gear Teeth*, G.B. Grant, Filed: January 14, 1887 (serial No. 224,382), Patent issued: July 23, 1889.

Pat. USA No. 1,425,144, *Toothed Gearing*, H.J. Schmick, Patented: August 8, 1922, Filed: June 30, 1921, Serial No. 481.561.

Pat. USA, No. 1,601,750, *Helical Gearing*, E. Wildhaber, Patented: October 5, 1926, Filed: November 2, 1923.

Pat. USA No. 1,833,159, *Rolling Hinge Gear*, Filed: March 10, 1930, Patented: November 24, 1931.

Pat. USA, No. 1,858,568, *Method of Grinding Gears*, E. Wildhaber, Filed: September 25, 1926, Patented: May 17, 1932.

Pat. USA No. 1,946,358, *Articulated Differential Gear, Particularly for Motor Vehicle*, F. Porsche and K. Rabe, Filed: April 5, 1933, Patented: February 6, 1934.

Pat. USA, No. 1,973,185, *Gear and Gear Tooth*, N. Trbojevich, Cl. 74-41, Filed: August 14, 1931, Patented: September 11, 1934.

Pat. USA No. 2,010,899, *Universal Joint*, A.H. Rzeppa, Filed: June 12, 1933, Patented: August 13, 1935.

Pat. USA No. 2,046,584, *Universal Joint*, A.H. Rzeppa, Filed: August 8, 1934, Patented: July 7, 1936.

Pat. USA, No. 2,128,815, *Gearing*, J.J. Guest, Filed: January 16, 1936, Patented: August 30, 1938.

Pat. USA, No. 2,538,536, *Universal Joint*, E. Wildhaber, Filed: October 25, 1947, Patented: January 16, 1951.

Pat. USA No. 2,664,760, *Antifriction Gear Assembly*, J.H. Booth, Filed: February 20, 1951, Patented: January 5, 1954.

Pat. USA, No. 2,808,732, *Gearing*, W.D. Champion, Filed: September 5, 1944, Patented: October 8, 1957.

Pat. USA No. 2,990,724, *Internal-External Gears*, J. Edward, C. Anderson, and F.C. Haberland, Filed: April 6, 1956, Patented: July 4, 1961.

Pat. USA No. 2.994.230, *Gear Design*, F.C. Haberland and E.A. Moore, Filed: February 4, 1959, Patented: August 1, 1961.

Pat. USA No. 3,037,400, *Differential Gear Reducer*, E.V. Sundt, Filed: January 14, 1959, Patented: June 5, 1962.

Pat. USA, No. 3.180.172, *Gear Wheels and Racks*, D.C.A. Leggatt, Filled: November 30, 1962, Patented: April 27, 1965.

Pat. USA, No. 3,269,205, *Helical Gear Tooth Structure*, G. Niemann, Filed: December 31, 1963, Patented: August 30, 1966.

Pat. USA No. 3,365,974, *Anti-Friction Worm Gearing Drive*, E.M. Lieberman, Filed: October 7, 1965, Patented: January 30, 1968.

Pat. USA, No. 3,371,552, *Rolling Contact Gear*, G.E. Soper, Filed: August 30, 1965, Patented: March 5, 1968.

Pat. USA No. 3,454,394, *Circular Arc Gear for Smoothly Transmitting Power*, M. Honobe, Filed: September 1, 1967, Patented: July 8, 1969.

Pat. USA, No. 3,533,300, *Helical Gearing*, R.M. Studer, Int. Cl. B24b 55/04; F16h 55/06, Filed: August 29, 1968, Patented: October 13, 1970.

Pat. USA, No. 3,709,055, *Gear Tooth Profile*, L.H. Grove, Int. Cl. F16h 55/06, F01c 1/18, Filed: January 4, 1971, Patented: January 9, 1973.

Pat. USA No. 3,748,920, *Gear-Tooth Arrangement and Transmission Incorporating Same*, Y.G. Lambev, Int. Cl. F16h 55/06, F16h 1/20, Filed: May 24, 1971, Patented: July 31, 1973.

Pat. USA, No. 3,855,874, *Double Mesh Type W-N Gear*, S. Honman and Y. Fujii, Int. Cl. F16h 55/06, Filed: July 12, 1973, Patented: December 24, 1974.

Pat. USA, No. 3,937,098, *High Torque Gearing*, W.S. Rouverol, Int. l. F16h 55/06, Filed: December 4, 1974, Patented: February 10, 1976.

Pat. USA, No. 3,982,445, *High Torque Gearing*, W.S. Rouverol, Int. Cl. F16h 55/06, Filed: September 2, 1975, Patented: September 28, 1976.

Pat. USA, No. 4,031,770, *Dual Tooth Contact Type W-N Gear*, S. Ishikawa, Int. Cl. F16h 55/06, Filed: April 5, 1976, Patented: June 28, 1977.

Pat. USA, No. 4,051,745, *Multiple-Contact Type W-N Gear*, S. Ishikawa, Int. Cl. F16h 55/06, Filed: April 15, 1976, Patented: October 4, 1977.

Pat. USA, No. 4,121,482, *W-N Gear Having a Maximum Surface Strength*, S. Ishikawa, K. Kotake, and H. Kawasaki, Int. Cl. F16h 55/06, Filed: August 10, 1977; Patented: October 24, 1978.

Pat. USA, No. 4,140,026, *Conformal Gearing*, W.S. Rouverol, Int. Cl. F16h 55/14; F16h 55/06; F16h 57/00; Filed: February 25, 1977, Patented: February 20, 1979.

Pat. USA No. 4,149,431. *Preloaded Conformal Gearing*, W.S. Rouverol, Int. Cl. F16H 55/06, U.S. Cl. 74/462, Filed: January 13, 1978, Patented: April 17, 1979.

Pat. US No. 4,640,149, *High Profile Contact Ratio, Non-Involute Gear Tooth Form and Method*, R. Drago, Int. Cl. F16H 55/06, Filed: March 4, 1983, Patented: February 3, 1987.

Pat. USA No. 4,656,884, *Power Transmission Mechanism*, I. Memoto, Int. Cl. F16H 1/18, Filed: March 7, 1985, Patented: April 14, 1987.

Pat. USA No. 4,679,459, *Concave-Convex Gear Pair Having Staggered Teeth*, E.F. Geppert, Int. Cl. F16K 55/17, Filed: April 22, 1986, Patented: July 14, 1987.

Pat. USA, No. 5,022,280, *Novikov Gearing*, L.S. Boiko, V.I. Korotkin, V.Y. Veretennikov, Ye.G. Roslivker, R.V. Fedyakin, V.A. Chesnokov, A.S. Yakovlev, Yu.D. Kharitonov, V.M Fey, and E.N. Galichenko, Int. Cl. F16h 55/08, Filed: March 29, 1988, Patented: June 11, 1991.

Pat. USA No. 5,505,668, *Gear System*, B.A. Koriakov-Savoysky, I.V. Aleksashin, and I.P. Vlasov, Int. Cl. F16H 1/32, Filed: August 2, 1994, Patented: April 9, 1996.

Pat. USA No. 6,092,434, *Ball Screw Device of Ball-Circulating-Part Embedded Type*, T. Matsumoto, S. Kobayashi, and K. Miyaguchi, Int. Cl. F16H 25/22, Filed: November 4, 1997, Patented: January 25, 2000.

Pat. USA, No. 6,837,123 B2, *Non-Involute Gears with Conformal Contact*, R.M. Hawkins, Int. Cl. F16h 55/08, Filed: January 31, 2002, Patented: January 4, 2005.

Pat. USA No. 8,061,229, *Gear Drive*, G.A. Zhuravlev, Int. Cl. F16h 55/00, F16h, 55/08, Filed: July 5, 2005, Patented: November 22, 2011.

Pat. USA No. 2015/0198228 A1, *Gear Combination with a Planetary Differential in the Form of a Wildhaber-Novikov Spur Gear Differential*, T. Biermann, A. Kuerzdoerfer, and S. Welker, Int. Cl. F16H 48/11, F16H 48/40, Filed: June 4, 2013, Patented: January 29, 2015.

Pat. USSR, No. 95978, *Gear System*, Yu.N. Budika, Cl. 47h, 6, Filed: September 4, 1951 (No. 770/445917).

Pat. USSR, No. 109,113, *Gear Pairs and Cam Mechanisms Having Point System of Meshing*, M.L. Novikov, National Classification 47h, 6; Filed: April 19, 1956, published in Bull. of Inventions No. 10, 1957.

Pat. USSR, No. 162,743, *A Method for Cutting Circular-Form Teeth in a Pair of Bevel Gears for Novikov Gearing*, A.M. Badayev, Int. Cl. B23f 9/00, Filed: September 14, 1962, Patented: August 15, 1985.

Pat. USSR No. 163,857, *A Helical Gearing*, B.V. Shitikov and N.A. Bayazitov, Int. Cl. F06h, Filed: November 25, 1963, Published: July 22, 1964.

Pat. USSR, No. 182,462, *Gearing with Point System of Meshing and Having Multiple Paths of Contact*, R.V. Fed'akin and V.A. Chesnokov, National Cl. 47 h, 6, Filed: November 20, 1963, published in B.I. No. 7, 1966.

Pat. USSR, No. 532714, *Bevel Gear Drive*, Ya.S. Davidov and G.A. Solomakha, Int. Cl. F16h 1/00, Filed: July 31, 1973, Patented: October 25, 1976.

Pat. USSR No. 578519, *Gearing with Novikov System of Meshing*, V.P. Ostapchuk, L.P. Smirnov, R.V. Fed'akin, and V.A. Chesnokov, Int. Cl. F16H 55/08, Filed: April 23, 1973, Patented: January 3, 1978.

Pat. USSR No. 580391, *Cylindrical Gearing with Novikov Meshing*, G.A. Zhuravl'ev and V.I. Korotkin, Int. Cl. F16H 55/08, Filed: January 18, 1973, Filed: November 15, 1977.

Pat. USSR No. 735852, *Gearing with Novikov Meshing*, V.I. Korotkin and V.Ya. Veretennikov, Int. Cl. F16H 1/00, Filed: June 15, 1978, Patented: May 25, 1980.

Pat. USSR, No. 735,855, *Gear Train*, R.V. Fed'akin, V.A. Tchesnokov, Int. Cl. F16h 1/18, Filed: June 28, 1967, Patented: May 25, 1980.

Pat. USSR No. 800472, *Novikov Gearing with Parallel Axes*, Yu.A. Vereschagin, V.S. Gemega, and Ye.L. Nazarov, Int. Cl. F16H 55/08, Filed: May 4, 1978, Patented: January 30, 1981.

Pat. USSR, No. 819447, *Basic Rack for a Novikov Gear Drive*, G.A. Zhuravlev, Int. Cl. F16h 1/08, Filled: 25.11.1975, Patented: 07.04.1981.

Pat. USSR, No. 863938, *Gear Drive with Mixed-Type Meshing*, G.A. Zhuravlev and R.B. Iofis, Int. Cl. F16h 1/02, Filed: 02.08.1979, Patented: 15.09.1981.

Pat. USSR, No. 875,132, *Novikov Gearing*, B.D. Ivanov, A.U. Dergausov, O.A. Poltavsky, and V.L. Shinkarev, Int. Cl.F 16 h 1/06, Filed: August 4,1972, Patented: October 23, 1981.

Pat. USSR, No. 929,914, *A Base Rack of Double-Mesh Novikov Gearing*, G.A. Zhuravlev, Int. Cl. F16h 1/00, Filed: 08.07.1974, Patented: 23.05.1982.

Pat. USSR, No. 1,048,197, Gear Drive, G.A. Zhuravlev, *et al.*, Int. Cl. F16h 1/06, F16h 55/22, Filed: 17.10.1979, Patented: 15.10.1983.

Pat. USSR No. 1,184,994, *Bevel Gear Drive*, Ye.D. Baskakov, Yu.B. Ladogin, and S.N. Pakhomov, Int. Cl. F16h 1/14, Filed: October 4, 1983, Patented: October 15, 1985.

Pat. USSR, No. 1.185.749, *A Method of Sculptured Surface Machining on a Multi-Axis NC Machine*, S.P. Radzevich, B23C 3/16, Filed: October 24, 1983.

Pat. USSR, No. 1.249.787, *A Method of Sculptured Surface Machining on a Multi-Axis NC Machine*, S.P. Radzevich, B23C 3/16, Filed: December 27, 1984.

Pat. USSR No. 1,597,473 A1, *Novikov Gearing*, S.V. Shevchenko, N.L. Ututov, and A.A. Stoyanov, Int. Cl. F16H 1/08, Filed: August 5, 1988, Patented: October 7, 1990.

Pat. USSR, No. 1,663,283, *Power Gear Train with a Mix-Type of Meshing*, G.A. Zhuravlev, Int. Cl. F16h 55/08, Filed: 22.12.1986, Patented: 15.07.1991.

Pat. USSR, No. 1,700.319, *Novikov Gearing*, L.S. Bojko, V.I. Korotkin, V.Ya. Veretennikov, Ye. G. Roslivker, R.V. Chesnokov, A.S. Yakovlev, Yu.D. Haritonov, V.M. Fey, and E.N. Galichenko, F 16 h 55/06, Filed: 08.08.1985, Patented: 23.12.1991.

Pat. USSR, No. 1,700,320, *A Gear Drive of Mixed-Type Meshing*, G.A. Zhuravlev, Int. Cl. F16h 55/08, Filed: 23.01.1987, Patented: 23.12.1992.

Pat. USSR No. 1,710,889, *A Gear Drive*, G.A. Zhuravlev, Int. Cl. F16h 1/06, Filed: 06.12.1989, Patented: 07.02.1992.

Pat. WO 1979/000503 A1, *Preloaded Conformal Gearing*, W.S. Rouverol, Int. Cl. F16H 55/14, Filed: January 8, 1979, Published: August 9, 1979.

Pat. WO, No. 98/45623, *Gear Form Construction*, B.E. Berlinger and J.R. Colbourn, Int. Cl. F16H 55/08, Filed: April 10, 1997, Patented: October 15, 1998.

Pat. WO, No 2005/060650 A2, *Gear Tooth Profile Curvature*, J.R. Colbourne, Int. Cl.: Not classified, Filed: December 16, 2004, Patented: July 7, 2005.

Pat. WO, No. 2010/033090 A2, *Helical Cylindrical Gear Pair for Uniform Power Transmission*, G. Hlebanja and J. Hlebanja, Int. Cl. F16h 55/08, Filed: September 16, 2009, Patented: March 25, 2010.

Pat. WO 2017/100517 A1, *Conjugate Gears with Continuous Tooth Flank Contact*, B.E. Berlinger, Jr. and J.R. Colbourn, Filed: December 9, 2016, Patented: June 15, 2017.

Paulins, K., Irbe, A., Torims, T., "Spiral Bevel Gears with Optimized Tooth-End Geometry", *Proceedings of the 24th DAAAM International Symposium on Intelligent Manufacturing and Automation, 2013, Procedia Engineering*, Vol. 69, 2014, pp. 383–392.

Pavlov, A.I., "Contact of Convex and Concave Surfaces in Gearing", *Herald of the National Technical University "KhPI"*, Vol. 10, 2002, pp. 99–102.

PCT CN 2015100182, *Convex-Concave Arc Gear Mechanism Used for Parallel Axes Transmission*, Y. Chen and L. Yao, F 16H 2055/0893, F16H 55/08, F16H 55/08, F16H 55/17, Filed: 31.12.2015, Patented: 05.07.2018. (US20180187750).

PCT/SU86/00110, *Gear Drive with Mixed-Type Meshing*, G.A. Zhuravlev, Int. Cl. F16h 55/00, F16h, 55/08, Filed: November 3, 1986, Published: May 19, 1988.

"Progress Report on Wildhaber-Novikov Gearing", *Product Engineering*, Vol. 33, No. 13, 1962, p. 88.

Pub. No.: US 2002/0134184A1, *Non-Involute Gears with Conformal Contact*, R.M. Hawkins, Int. Cl. F16h 55/08, Filed: January 31, 2002, Published: September 26, 2002.

Qu, W.T., Peng, X.Q., *et al.*, "Finite Element Generalized Tooth Contact Analysis of Double Circular Arc Helical Gears", *Structural Engineering and Mechanics*, Vol. 43, No. 4, 2012, pp. 439–448.

Radzevich, S.P., "A Brief Overview on the Evolution of Gear Art: Design and Production of Gears, Gear Science", Chapter 11, pp. 418–485, in: S.P. Radzevich, Editor, *Advances in Gear Science and Manufacture*, CRC Press, Boca Raton, 2019, 549 pages.

Radzevich, S.P., "A Brief Overview on the Evolution of the Scientific Theory of Gearing: A Preliminary Discussion", in: *Proceedings of International Conference on Gears 2015*, October 5–7, 2015, Technical University of Munich (TUM), Garching (Near Munich), Germany, 2015, pp. 1035–1946.

Radzevich, S.P., "An Examination of High-Conformal Gearing", *Gear Solutions*, February 2018, pp. 31–39.

Radzevich, S.P., Editor, *Advances in Gear Science and Manufacture*, CRC Press, Boca Raton, 2019, 549 pages.

Radzevich, S.P., *Differential-Geometric Method of Surface Generation*, Doctoral Thesis, Tula, Tula Polytechnic Institute, 1991, 300 pages.

Radzevich, S.P., *Dudley's Handbook of Practical Gear Design and Manufacture*, 2nd Edition, CRC Press, Boca Raton, 2012, 896 pages.

Radzevich, S.P., *Dudley's Handbook of Practical Gear Design and Manufacture*, 3rd Edition, CRC Press, Boca Raton, 2016, 629 pages.

Radzevich, S.P., *Dudley's Handbook of Practical Gear Design and Manufacture*, 4th Edition, CRC Press, Boca Raton, 2021, 1170 pages.

Radzevich, S.P., *Gear Cutting Tools: Fundamentals of Design and Computation*, CRC Press, Boca Raton, 2010, 786 pages.

Radzevich, S.P., *Gear Cutting Tools: Science and Engineering*, 2nd Edition, Boca Raton, 2017, 564 pages.

Radzevich, S.P., *Generation of Surfaces: Kinematics Geometry of Surface Machining*, CRC Press, Boca Raton, 2013, 747 pages.

Radzevich, S.P., *Geometry of Surfaces: A Practical Guide for Mechanical Engineers*, Wiley, Chichester, 2013, 264 pages.

Radzevich, S.P., "High-Conformal Gearing: A New Look at the Concept of *Novikov Gearing*", in: *Proceedings of International Conference on Gears*, October 5–7, 2015, Technical University of Munich (TUM), Garching (Near Munich), Germany, 2015, pp. 1303–1314.

Radzevich, S.P., *High-Conformal Gearing: Kinematics and Geometry*, CRC Press, Boca Raton, 2015, 332 pages.

Radzevich, S.P., *Kinematic Geometry of Surface Machining*, CRC Press, Boca Raton, 2007, 508 pages.

Radzevich, S.P., *Methods for Investigation of the Conditions of Contact of Surfaces, Monograph*, UkrNIINTI, Kiev, №759–Uk88, 1987, 103 pages.

Radzevich, S.P., "On Analytical Description of Contact Geometry of Part Surfaces in Highest Kinematic Pairs", *Theory for Mechanisms and Machines*, St. Petersburg Polytechnic Institute, Vol. 3, No. 5, 2005, pp. 3–14.

Radzevich, S.P., "On the Priority of Dr. M.L. Novikov in the Development of *Novikov Gearing*", *Theory of Mechanisms and Machines*. http://tmm.spbstu.ru (In press).

Radzevich, S.P., *Theory of Gearing: Kinematics, Geometry, and Synthesis*, CRC Press, Boca Raton, 2012, 743 pages.

Radzevich, S.P., *Theory of Gearing: Kinematics, Geometry, and Synthesis*, 2nd Edition, CRC Press, Boca Raton, 2018, 934 pages.

Radzevich, S.P., "Vector Representation of Gear Pairs. Part I", *Theory of Mechanisms and Machines*, Vol. 6, No. 2, 2008, pp. 74–81. http://www.tmm.spbstu.ru/journal.html.

Radzevich, S.P., "Vector Representation of Gear Pairs. Part II", *Theory of Mechanisms and Machines*, Vol. 7, No. 1, 2009, pp. 17–26. http://www.tmm.spbstu.ru/journal.html.

Ramachandra, K., Lingaiah, K., "Photoelastic Investigation of the Load-Carrying Capacity of Wildhaber-Novikov Circular-Arc Gears", *Journal of the Institution of Engineers*, (India), Vol. 53, Part ME6, July 1974, pp. 313–321.

Regl, R., Mack, J., *Trip Report – Visit to Falk Corporation*, Milwaukee, Wisconsin, July 11, 1961 – ECD-355, 7 August 1961. (Unpublished source).

"Remarkable Gear System", *Flight*, June 18, 1954, p. 804.

Roano, A., *Getribe mit auf parallelen Achsen angeordneten, schraubverzahnten Rädern*. Patentschrift Nr. 939240, klasse 47 h, Gruppe 6, Ausgegeben am 16 Februar 1956, Bundesrepublik Deutschland.

Roano, A., *Ozubenà soukoli pro vratny chod, se sroubovicovymi zuby, s osami rovnobeznymi nebo mirne naklonenymi, pri cemz sklon zubü kola, jest odlisny od sklonu zubü pastorku*. Patentni spis c. 83732, Trida 47 h, 6. Vydano 3 ledna 1955. Republika Cheskoslovenska.

Roano, A., *Zahnradpaare für umsteuerbare Zahnradgetriebe mit parallelen order leicht gegeneinander geneigten Achsen*. Patentschrift Nr. 940194, klasse 47 b, Gruppe 23, Ausgegeben am März 1956. Bundesrepublik Deutschland.

Roslivker, E.G., "Hyperboloid Gears with Novikov Gearing", *Gears with Novikov Gearing*, Vol. 3, 1964, pp. 87–112. (In Russin).

Roslivker, Ye.G., "Hyperboloid Novikov Gearing", pp. 87–112, in: *Novikov Gearings*, Vol. 3, Zhukovskii Military Aviation Academy Publishers, Moscow, 1964.

Roslivker, E.G., "Investigation of Hyperboloid, Helical and Bevel Geras with Novikov Gearing", *Gears with Novikov Gearing*, Vol. 3, 1964, pp. 5–85. (In Russin).

Roth, K., "Stirnradpaarungen mit 1 bis 5-Zähnigen Ritzeln im Madchinenbau", *Konstrution*, Vol. 26, 1974, pp. 425–429.

"Russia Adopts the Novikov System of Gearing", *The Guardian*, 11th November 1958.

Silich, A.A., *Cylindrical Novikov Gearing*, LAP Lambert Academic Publishing, 2013, 100 pages.

Silich, A.A., *Novikov Gearing: Geometrical Calculation of Cylindrical Gearings*, Monograph, Tiumen', Publisher of the Tiumen' Institute of Industry, 2016, 79 pages.

Silich, A.A., *The Development of a Geometrical Theory of Design of Novikov Gearing, and of the Tooth Flanks Generation Process*, Doctoral Thesis, Kurgan State University, Kurgan, 1999, 534 pages.

Silich, A.A., Ishkina, E.G., "Application of Novikov Gears, Gear Processing Techniques and Enhanced Possibilities of their Production", *Key Engineering Materials*, Vol. 736, 2017, pp. 138–142.

Silich, A.A., Nekrasov, R.Yu., Zinchenko, A., "On the Peculiarities of the Circular Cutter Geometry Applied for Cutting the Wheels of Novikov's Gearings", *Key Engineering Materials*, Vol. 736, 2017, pp. 153–157.

Shao, J., *Double Circular Arc Gear*, China Machine Press, Beijing, 1994.

Small, N.C., "Bending of a Cantilever Plate Supported from an Elastic Half Space", *Journal of Applied Mechanics*, Paper No. 61-APM-11.

Snidle, R.W., Evans, H.R., "Some Aspects of Gear Tribology", *Proceedings of the IMechE, Part C: Journal of Mechanical Engineering Science,* Vol. 223, 2009, pp. 103–113.

Somer, M.N., Abdullah, M.Q., Mohammed, M.N., "Generation of Crowned Parabolic Novikov Gears", *Engineering Letters*, Vol. 15, No. 1, August 15, 2007, 5 pages. http://www.engineeringletters.com/issues_v15/issue_1/EL_15_1_4.pdf.

Somov, I.I., *Rational Mechanics*: *Part I. Kinematics*, Saint Petersburg, 1872.

Stebletsov, Yu.N., *An Increase of Efficiency of Hobbing of Gears for Novikov Gearing*, PhD Thesis, Orel State University, Orel, 2013, 16 pages.

Sun, D., Li, G., Wei, H., Liao, H., "Experimental Study on the Chaotic Attractor Evolvement of the Friction Vibration in a Running-in Process," *Tribology International*, Vol. 88, 2015, pp. 290–297.

Suslin, A., Pilla, C., "Study of Loading in Point-Involute Gears", *Procedia Engineering*, Vol. 176, 2017, pp. 12–18.

Suslov, G.K., "On Rolling of a Surface Over another Surfaces", *University News*, Kiev, 1892.

Tan, R., Chen, B., Xiang, D., Liang, D., "A Study on the Design and Performance of Epicycloid Bevels of Pure-Rolling Contact", *ASME Journal of Mechanical Design*, Vol. 140, No. 4, 2017, 11 pages.

The Novikov Gearing System – Review of Soviet and Western Literature 1957-1960, AID Report 60-3, Science and Technology Section, Air Information Division.

"The Westland WG.13 Lynx: A Brief Description of the Only International Aerospace Project in Which Britain has Leadership in Design, Production and Sales", *Aircraft Engineering and Aerospace Technology*, Vol. 43, No. 5, 1971, pp. 32–36. https://doi.org/10.1108/eb034769.

Technical Discussion on Novikov Gearing, Proceedings of the meeting of the British Gear Manufacturers Association, May 10, 1960.

Thompson, G.B., "The Shape of Gear Teeth", *New Scientist*, 1960, pp. 7, 665.

Tsay, C.-B., Fong, Z.H., "The Mathematical Model of Wildhaber-Novikov Gears Applicable to Finite Element Stress Analysis", *Mathematical and Computer Modeling*, Vol. 12, No. 8, 1989, pp. 939–946.

Tsay, C.-F., Liang, T.-L., Yang, S.-C., "Using Double Envelope Method on a Planetary Gear Mechanism with Double Circular-Arc Tooth", *Transactions of the CSME/de la SCGM*, Vol. 32, No. 2, 2008, pp. 267–282.

Tuplin, W.A., "The Novikov System of Gearing", *Machinery*, London, Vol. 93, 1958, p. 1260.

Veena Kumari, N., Shihari, P.V., Ramachandra, K., "Bending Stress Analysis of Involute Helical Gear and Wildagaber-Novikov Gear", *Trends in Mechanical Engineering & Technology*, Vol. 4, No. 2, 2014, pp. 17–21.

Verbitskiy, L.L., "The Curvature of the Mating Surfaces in Novikov Gearing", *Gears with Novikov Gearing*, Vol. 3, 1964, pp. 36–61. (In Russian).

Walker, H., "A Critical Look at the Novikov Gear", *The Engineer*, London, Vol. 209, 29 April 1960, pp. 725–729.

Wang, C.Y., Chen, C.K., "Compensating Analysis of a Double Circular-Arc Helical Gear by Computerized Simulation of Meshing", *Proceedings of the Institution of Mechanical Engineers, Part C: Journal of Mechanical Engineering Science*, Vol. 215, No. 7, 2001, pp. 759–771.

Wellauer, E.J., Seireg, A., "Bending Strength of Gear Teeth by Cantilever-Plate Theory", ASME Paper 59-A-50.

Wells, C.F., Shotter, B.A., "The Development of 'CirCarC' Gearing", *AEI Engineering*, Vol. 2, No. 2, March–April 1962, pp. 83–88.

Wildhaber, E., "Gears with Circular Tooth Profile Similar to the Novikov System," *VDI-Berichte*, No. 47, 1961, pp. 19–21.

Willis, R.J., "New Equations and Charts Pick Lightest-Weight Gears", *Product Engineering*, 21 January 1963.

Winter, H., Looman, J., "Tools for Making Helical Circular Arc Spur Gears," *VDI-Berichte*, No. 47, 1961, pp. 25–28.

Wu, B., Meng, H., Shao, J., "Vibration Analysis of a Doublecircular-Arc Helical Gearing System", *Chinese Journal of Mechanical Engineering*, Vol. 35, No. 5, 1999, pp. 85–89.

Wu, Y.C., Chen, K.Y., Tsay, C.B., Ariga, Y., "Contact Characteristics of Circular-Arc Curvilinear Tooth Gear Drives", *ASME Journal of Mechanical Design*, Vol. 131, August 2009, 8 pages.

Xie, X.-S., Yang, H.-C., "Kinematic Errors on Helical Gear of Triple Circular-Arc Teeth", *Journal of Mechanical Science and Technology*, Vol. 28, No. 8, August 2014, pp. 3137–3146.

Yakovlev, S.A., "Basic Profile and the Scope of Application of Novikov Gearing", *Vestnik Mashinostroeniya*, Vol. 69, No. 4, 1989, pp. 22–26.

Yakovl'ev, A.S., "*Novikov Mikhail Leontiyevich in Memories of Candidate of Engineering Sciences Yakovlev A.S.*", Visnik NTU "Kharkivskii Polytechnic Institute", No. 34, 2015, pp. 4–9.

Yakovl'ev, A.S., Pecheniy, V.I., "An Experimental Investigation of the Load Distribution within the Patch of Contact in Novikov Gears", in: *Reliability and Quality of Gearing*, Vol. 18-67-36, NIIINFORMT'AZhMASh, Moscow, 1967.

Yang, J.-B., Chen, C.-W., "Analytical and Numerical Calculations for Thermal Elastohydrodynamic Lubrication Problems of Helical W-N Gears", *Wear*, Vol. 145, No. 2, 1991, pp. 201–250.

Yang, J.-B., Chen, C.-W., "The Stiffness and Modification of Helical W-N Gear Teeth under Thermal EHL", *Tribology International*, Vol. 23, No. 4, August 1990, pp. 259–263.

Yang, J.-B., Qi, Y.-L., Chen, C.-W., "Measurement of Oil Film Thickness Between W-N Helical Gear Tooth Profiles Using Laser Transmission Method", *ASME Journal of Tribology*, Vol. 112, No. 4, 1990, pp. 708–711.

Yang, S.-C., "Mathematical Model of a Stepped Triple Circular-Arc Gear", *Mechanism and Machine Theory*, Vol. 44, No. 5, May 2009, pp. 1019–1031.

Ye, G., Ye, X., "A New Method for Seeking the Optimum Gear Tooth Profiles – The Theoretical Basis of Wildhaber–Novikov Gearing", *Mechanism and Machine Theory*, Vol. 37, No. 10, 2002, pp. 1087–1103.

Yu, M.S., *Gear Handbook*, China Machine Press, 2000, pp. 209–213.

Zhuravlev, G.A., "The Fallacy of the Physical Basis of Engagement Novikov as a Reason for the Limited Application," *Reducers and Gears*, Vol. 1, 2006.

Zhang, H., Hua, L., *et al.*, "Computerized Design and Simulation of Meshing of Modified Double Circular-Arc Helical Gears by Tooth end Relief with Helix", *Mechanism and Machine Theory*, Vol. 45, No. 1, 2016, pp. 46–64.

Zhang, T.J., "The Calculation of Tooth End Relief for Double Circular-Arc Gear", *Mechanical Transmission*, Vol. 21, No. 4, 1997, pp. 32–35.

Zhang, X.P., Cai, X., Chen, H.J., Qu, Ch., Liu, J., "A Parametric Modeling Method for Hyperboloid-Type Normal Circular-Arc Gear", *Journal of Applied Science*, Vol. 14, No. 3, 2014, pp. 229–236.

Zhang, H., Hua, L., Han, X., "Computerized Design and Simulation of Meshing of Modified Double Circular-Arc Helical Gears by Tooth End Relief with Helix", *Mechanism and Machine Theory*, Vol. 45, No. 1, January 2010, pp. 46–64.

Zhou, Y., Zuo, X., Zhu, H., Wei, T., "Development of Prediction Models of Running-In Attractor," *Tribology International*, Vol. 117, 2018, pp. 98–106.

Zhou, Y., *et al.*, "The Tooth Profile Design and Examination of Circular-Arc Gear Pump", *Applied Mechanics and Materials*, 2014, Vols. 672–674, pp. 1604–1607.

Index

S. P. Radzevich, *Novikov/Conformal Gearing*,
https://doi.org/10.1007/978-3-031-10019-2